AF651358

# MATERIA MEDICA

## SECTION XVI

## OLEO-RESINS


| | PAGE | | PAGE |
|---|---|---|---|
| Canada Turpentine | 489 | Balsam of Tolu | 495 |
| Copaiba | 490 | Balsam of Peru | 497 |
| Elemi | 494 | Storax | 499 |


## SECTION XVII

## TARS AND EMPYREUMATIC OILS


| | | | |
|---|---|---|---|
| Tar | 502 | Juniper-tar Oil | 504 |
| Coal-tar | 503 | | |


## SECTION XVIII

## FIXED OILS, FATS AND WAXES


| | | | |
|---|---|---|---|
| Poppy Oil | 506 | Sesamé Oil | 511 |
| Expressed Oil of Mustard | 506 | Croton Oil | 512 |
| Rape Oil | 507 | Castor Oil | 512 |
| Chaulmoogra Oil | 507 | Coco Nut Oil | 513 |
| Cotton-seed Oil | 507 | Cod-liver Oil | 513 |
| Oil of Theobroma | 508 | Suet | 515 |
| Linseed Oil | 508 | Lard | 515 |
| Arachis Oil | 509 | Beeswax | 516 |
| Almond Oil | 509 | Spermaceti | 518 |
| Olive Oil | 510 | Wool Fat | 519 |


## SECTION XIX

## SACCHARINE SUBSTANCES


| | | | |
|---|---|---|---|
| Manna | 521 | Honey | 522 |


## SECTION XX


Camphor . . . . . . . . 524


## SECTION XXI

## ANIMALS AND ANIMAL GLANDS AND SECRETIONS


| | | | |
|---|---|---|---|
| Leech | 529 | Ox Gall | 536 |
| Cantharides | 530 | Pepsinum | 537 |
| Cochineal | 532 | Gelatin | 538 |
| Suprarenal Gland | 533 | Musk | 538 |
| Thyroid Gland | 543 | Castor | 542 |
| Pancreas | 535 | Cuttle-fish Bone | 544 |

INDEX . . . . . . . . . . . . . 545

## SECTION X

## DRIED LATEX

| | PAGE | | PAGE |
|---|---|---|---|
| Opium | 415 | Caoutchouc | 424 |
| Lactucarium | 421 | Euphorbium | 427 |
| Guttapercha | 422 | | |

## SECTION XI

## DRIED JUICES

| | | | |
|---|---|---|---|
| Aloes | 431 | Butea Gum | 440 |
| Kino | 438 | Elaterium | 441 |
| Red Gum | 439 | | |

## SECTION XII

## EXTRACTS

| | | | |
|---|---|---|---|
| Gambier | 442 | Curare | 445 |
| Cutch | 444 | | |

## SECTION XIII

## GUMS

| | | | |
|---|---|---|---|
| Acacia Gum | 448 | Tragacanth Gum | 451 |
| Ghatti Gum | 450 | | |

## SECTION XIV

## RESINS

| | | | |
|---|---|---|---|
| Colophony | 454 | Benzoin | 464 |
| Burgundy Pitch | 458 | Dragon's Blood | 467 |
| Amber | 459 | Mastich | 468 |
| Guaiacum Resin | 460 | Sandarac | 470 |
| Araroba | 462 | Shellac | 471 |

## SECTION XV

## GUM-RESINS

| | | | |
|---|---|---|---|
| Gamboge | 474 | Ammoniacum | 482 |
| Scammony | 476 | Galbanum | 484 |
| Myrrh | 478 | Asafetida | 487 |
| Olibanum | 481 | | |

## SECTION VIII

## SUBTERRANEAN ORGANS

### Rhizomes, Corms, Bulbs and Roots

| | PAGE |
|---|---|
| Hydrastis | 290 |
| Black Hellebore | 292 |
| Aconite | 294 |
| Cimicifuga | 298 |
| Pareira Brava | 300 |
| Calumba | 302 |
| Podophyllum | 304 |
| Indian Podophyllum | 306 |
| Bloodroot | 307 |
| Horseradish | 309 |
| Senega | 309 |
| Rhatany | 312 |
| Marshmallow | 314 |
| Liquorice | 315 |
| Bryony | 320 |
| Sumbul | 321 |
| Ipecacuanha | 323 |
| Valerian | 329 |
| Indian Valerian | 332 |
| Elecampane | 332 |
| Pellitory | 334 |
| Arnica | 336 |
| Dandelion | 338 |
| Hemidesmus | 340 |
| Gelsemium | 342 |
| Picrorhiza | 343 |
| Gentian | 344 |
| Alkanet | 347 |
| Comfrey | 349 |
| Scammony | 349 |
| Jalap | 351 |
| Orizaba Jalap | 355 |
| Turpeth | 356 |
| Belladonna | 357 |
| Rhubarb | 360 |
| Kava | 369 |
| Serpentary | 370 |
| Sassafras | 372 |
| Turmeric | 373 |
| Ginger | 375 |
| Galangal | 379 |
| Orris | 381 |
| Sarsaparilla | 383 |
| Squill | 386 |
| Urginea | 388 |
| Colchicum | 389 |
| White Hellebore | 390 |
| American Veratrum | 392 |
| Sweet Flag | 392 |
| Couch Grass | 395 |
| Male Fern | 396 |
| Blue Cohosh | 399 |
| Leptandra | 399 |

## SECTION IX

## STARCHES, GLANDS, HAIRS, &c.

| | PAGE |
|---|---|
| Starch | 400 |
| Lycopodium | 405 |
| Lupulin | 407 |
| Kamala | 408 |
| Cotton | 410 |
| Galls | 412 |
| Cowhage | 414 |

## SECTION V

## HERBS AND ENTIRE PLANTS


| | PAGE | | PAGE |
|---|---|---|---|
| Aconite Herb | 192 | Savin Tops | 212 |
| Broom Tops | 194 | Carrageen | 213 |
| Hemlock Herb | 195 | Bladderwrack | 215 |
| Coltsfoot | 197 | Agar-agar | 216 |
| Grindelia | 198 | Cetraria | 217 |
| Lobelia | 200 | Ergot | 219 |
| Indian Pink | 202 | Pulsatilla | 222 |
| Chiretta | 203 | Pilewort | 222 |
| Bitter-sweet | 205 | Purging Flax | 222 |
| White Horehound | 206 | Wormwood | 222 |
| Indian Hemp | 207 | Lettuce | 222 |
| Euphorbia pilulifera | 210 | Centaury | 222 |


## SECTION VI

## WOODS


| | | | |
|---|---|---|---|
| Berberis | 224 | Sappan | 229 |
| Guaiacum | 224 | Red Sanders | 229 |
| Quassia | 226 | Sandal | 230 |
| Logwood | 227 | Tinospora | 232 |


## SECTION VII

## BARKS


| | | | |
|---|---|---|---|
| Barberry | 235 | Quebracho | 269 |
| Canella | 236 | Cinnamon | 269 |
| Winter's Bark | 237 | Cassia | 271 |
| Cotton Root Bark | 238 | Oliver | 273 |
| Cusparia | 238 | Bebeeru | 273 |
| Simaruba | 241 | Coto | 275 |
| Euonymus | 242 | Mezereon | 277 |
| Alder Buckthorn | 243 | Cascarilla | 279 |
| Cascara Sagrada | 246 | Slippery Elm | 281 |
| Acacia | 249 | Oak | 282 |
| Sassy | 249 | Willow | 285 |
| Wild Cherry | 251 | Azadirach | 287 |
| Quillaja | 253 | Piscidia | 288 |
| Witch-hazel | 255 | Calotropis | 238 |
| Pomegranate | 256 | Condurango | 288 |
| Black Haw | 258 | Elm | 288 |
| Cinchona | 259 | Larch | 288 |
| Alstonia | 268 | | |

## SECTION III

## FRUITS


| | PAGE | | PAGE |
|---|---|---|---|
| Star Anise | 85 | Fennel Fruits | 116 |
| Cocculus Indicus | 87 | Dill Fruits | 118 |
| Poppy Capsules | 89 | Anise Fruits | 119 |
| Bael Fruit | 91 | Embelia | 121 |
| Bitter Orange | 92 | Capsicum Fruits | 122 |
| Lemon | 94 | Laurel Berries | 125 |
| Raisins | 97 | Figs | 126 |
| Cassia Pods | 98 | Hops | 127 |
| Tamarinds | 99 | Cubebs | 128 |
| Senna Pods | 100 | Black Pepper | 132 |
| Prunes | 101 | Juniper Berries | 135 |
| Hips | 102 | Vanilla | 136 |
| Myrobalans | 102 | Cardamoms | 138 |
| Pimento | 103 | Annatto | 141 |
| Pomegranate Rind | 105 | Buckthorn Berries | 141 |
| Bitter Apple | 106 | Ajowan Fruits | 142 |
| Caraway Fruits | 108 | Chenopodium | 142 |
| Hemlock Fruits | 111 | Parsley Fruits | 142 |
| Coriander Fruits | 113 | Cashew Nuts | 142 |
| Cummin Fruits | 115 | Pearl Barley | 142 |


## SECTION IV

## SEEDS


| | | | |
|---|---|---|---|
| Stavesacre Seeds | 143 | Nux Vomica | 170 |
| Black Mustard Seeds | 145 | Ignatius Beans | 172 |
| White Mustard Seeds | 148 | Stramonium Seeds | 174 |
| Linseed | 149 | Henbane Seeds | 175 |
| Cola Seeds | 151 | Datura Seeds | 176 |
| Cocoa Seeds | 153 | Kaladana | 176 |
| Guarana | 154 | Ispaghula | 177 |
| Tonco Beans | 156 | Nutmegs | 178 |
| Calabar Beans | 157 | Mace | 181 |
| Butea Seeds | 159 | Castor Seeds | 182 |
| Fenugreek Seeds | 159 | Croton Seeds | 184 |
| Sweet Almonds | 161 | Grains of Paradise | 185 |
| Bitter Almonds | 163 | Colchicum Seeds | 186 |
| Melon Pumpkin Seeds | 164 | Cevadilla Seeds | 189 |
| Quince Seeds | 165 | Areca Nuts | 190 |
| Strophanthus Seeds | 167 | Jequirity Seeds | 191 |

# CONTENTS


| | PAGE | | PAGE |
|---|---|---|---|
| Definition | 1 | Drug Packages | 9 |
| History | 2 | Sale of Drugs | 12 |
| Commerce | 4 | Drying and Preservation | 17 |
| Drug Routes | 8 | | |


## SECTION I

## LEAVES


| | | | |
|---|---|---|---|
| Tea | 21 | Foxglove | 53 |
| Coca | 23 | Laurel | 56 |
| Buchu | 27 | Matico | 57 |
| Jaborandi | 30 | Betel | 58 |
| Senna, Alexandrian | 32 | Aconite | 59 |
| Senna, Indian | 35 | Maté | 59 |
| Cherry-laurel | 36 | Damiana | 59 |
| Witch-hazel | 38 | Wintergreen | 59 |
| Eucalyptus | 40 | Tylophora | 60 |
| Bearberry | 41 | Eriodictyon | 60 |
| Belladonna | 43 | Duboisia | 60 |
| Stramonium | 46 | Adhatoda | 60 |
| Datura | 49 | Boldo | 60 |
| Henbane | 49 | | |


## SECTION II

## FLOWERS


| | | | |
|---|---|---|---|
| Red Poppy Petals | 61 | Wormseed | 75 |
| Kousso | 62 | Calendula | 77 |
| Red-rose Petals | 65 | Insect Flowers | 78 |
| Cloves | 66 | Lavender Flowers | 79 |
| Elder Flowers | 69 | Saffron | 80 |
| Chamomile Flowers | 70 | Lily of the Valley Flowers | 83 |
| Arnica Flowers | 73 | | |

the chemistry of a number of the more important constituents. Several drugs not previously dealt with have also been included.

The general arrangement of the subject-matter has been retained, but in several cases sections have been united or divided. The correct pronunciation of a number of Latin plant-names has been indicated by marking the quantities by the syllables.

Many of the seventy-seven new figures have been reproduced from original photographs. For those illustrating the various forms of packages I am indebted to Messrs. Wright, Layman & Umney (Figs. 6, 7, 95, 117, 177, 179, 195, 204, 205, 255, 267), and to Dr. Weigel of Hamburg (Figs. 1 to 5). To the Editor of the *Pharmaceutical Journal*, the Editor of the *Chemist and Druggist*, and to Messrs. Chas. Smith, Gowland & Son, who have given permission for the reproduction of illustrations that have appeared in their respective publications, and to the London and India Docks Company, and the proprietors of Bull Wharf who have kindly allowed me to reproduce photographs taken on their premises, my thanks are also due.

Much information respecting the commerce in drugs has been furnished by Messrs. Cæsar & Loretz (Halle a. S.), Messrs. Davy, Hill & Hodgkinson, Messrs. Hearon, Squire & Francis, Messrs. Potter & Clark, and Messrs. Wright, Layman & Umney. The current literature has also been freely laid under contribution. In the work of proof-reading I have been most efficiently aided by Mr. Reginald R. Bennett, B.Sc.

The student is again strongly advised to make the study of crude drugs as far as possible a practical one. Before commencing any one of the first ten sections the morphology and anatomy of the respective organs should be studied in a text-book of botany. Each drug should be carefully compared with the description given; differences should be noted, and an explanation sought either from a teacher or from other works. Sketches should be frequently made. In the case of seeds, fruits, barks, roots and rhizomes, the prominent features of the transverse section should also be delineated, if necessary on an enlarged scale. Flowers and leaves may be expanded by soaking them in hot water. The meaning of botanical technical terms, if not precisely known, should be ascertained at the time from a suitable glossary, and a geographical atlas should also be kept at hand for reference.

H. G. G.

*May* 1909.

# PREFACE

## TO THE THIRD EDITION

The conditions under which this, the third edition, has been prepared, have been somewhat adverse. While it was considered desirable to add brief accounts of a number of drugs in less frequent use, due regard has had to be given to the high cost of production. A reduction in the number of pages has been made possible by the use of a slightly smaller type and by the omission of the classification according to natural orders and of a few of the illustrations. Notwithstanding this reduction in size, the number of drugs dealt with has been very considerably increased, and numerous notes have been added in order to increase the general utility of the work.

All temperatures are now stated in degrees Centigrade.

H. G. G.

# PREFACE

## TO THE SECOND EDITION

The present volume, published under the title of a 'Text-Book of Materia Medica,' constitutes a second edition of the author's 'Introduction to the Study of Materia Medica' which appeared in 1899. This change in title has been rendered desirable by the numerous additions that have been made to the work with the view of broadening its scope. To this end a chapter on the history and commerce of drugs has been added. Each section has been prefaced by a short introduction, including, in the section dealing with 'Animals and Animal Glands and Secretions,' a brief outline of modern zoological classification. For the convenience of advanced students the microscopical characters of, and assay processes for, a number of important drugs have been added. The commercial varieties have been more fully dealt with, and considerably more attention has been paid to

Reprint in India, 1999

*Published by :*
Scientific Publishers (India)
5-A, New Pali Road, P.B. No. 91
JODHPUR - 342 001
FAX : +91 291 613480

ISBN : 81-7233-213-0

First Indian Print : 1999

Originally Published in 1920 by J & A Churchill, London
Printed at Jagriti Offset, Delhi.

# MATERIA MEDICA

BEING AN ACCOUNT OF THE MORE IMPORTANT CRUDE DRUGS OF VEGETABLE AND ANIMAL ORIGIN

*DESIGNED for STUDENTS of PHARMACY and MEDICINE*

BY

HENRY G. GREENISH, F.I.C., F.L.S.

PROFESSOR OF PHARMACEUTICS TO THE PHARMACEUTICAL SOCIETY OF GREAT RRITAIN
AND IN THE UNIVERSITY OF LONDON
DIRECTOR OF THE PHARMACY RESEARCH LABORATORY
DOCTEUR (H.C.) DE L'UNIVERSITÉ DE PARIS
HANBURY GOLD MEDALLIST, 1917
JOINT EDITOR OF THE BRITISH PHARMACOPŒIA, 1914
MEMBER OF THE BOARD OF EXAMINERS OF THE PHARMACEUTICAL SOCIETY OF GREAT BRITAIN
LATE EXTERNAL EXAMINER IN MATERIA MEDICA AND PHARMACY
TO THE UNIVERSITY OF BIRMINGHAM

WITH 250 ILLUSTRATIONS

*THIRD EDITION*

SCIENTIFIC PUBLISHERS (INDIA)
P.O. BOX 91 JODHPUR

# MATERIA MEDICA

**Definition.**—Materia Medica or Pharmacognosy, for the terms as used by pharmacists are synonymous, may be defined as that science which aims at a complete and systematic knowledge of crude drugs of animal or vegetable origin. The term drugs may be understood to include all parts of animals or vegetables used in medicine, or such of their products as have not undergone any material elaboration or separation into their constituents; it is, however, not susceptible of exact limitation, and hence substances may be included which are of dietetic or technical rather than medicinal use.

A complete knowledge of a drug must include a knowledge of the morphological nature and the structure, both macroscopic and microscopic, of such as are organised, of the botanical or zoological source, of the constituents both active and inactive, their nature, the amounts in which they are present, and the relation they bear to the constituents of other drugs. The geographical source, the cultivation, mode of collection, preparation for the market, the trade route, commercial varieties, history, uses, &c., must also be known. To attain this knowledge the aid of chemistry and of chemical methods of investigation, of botany and botanical methods, of physiology, geography, commerce, history, &c., must be sought, or at least the results of other investigators in these respects utilised, and the information thus obtained combined to form a complete knowledge of each drug, and these systematised to build up the science of pharmacognosy.

Pharmacognosy may be subdivided into Scientific Pharmacognosy, as above described, and Applied Pharmacognosy. Applied Pharmacognosy is the application of scientific pharmacognosy to the solution of the pharmacognostical problems that are constantly arising in the daily work of the pharmacist. Thus Scientific Pharmacognosy includes an intimate knowledge of the structure of vegetable drugs; Applied Pharmacognosy utilises this knowledge to identify the drug entire or in powder and to determine its purity. Scientific Pharmacognosy treats of the constituents of drugs. Applied pharmacognosy ascertains the relative quality of a drug by determining the proportion in which one or more of these constituents is present.

**History.**—Although it is only within very recent years that pharmacognosy has substantiated its claim to rank as a science, the collection of herbs for medicinal use is of the remotest antiquity. Ancient Egyptian papyri, of which that of Ebers written about 1600 B.C. is the most important and most complete, prove that even at this early period the physicians were in possession of a very considerable number of drugs, including aloes, acacia gum, myrrh, Indian hemp, hemlock, opium, frankincense, cassia, cummin, coriander, castor seed, and many others. But as far as Western nations are concerned, Greece may be regarded as the cradle of medicine. About the middle of the twelfth century before the Christian era temples were built which served as centres for the exercise of the healing art. Most famous among these were the temples of the deified Æsculapius at Epidaurus, Cos, Trika, and Pergamos. At these temples, which were usually situated on high ground in pure air, cures were effected chiefly by hygienic means such as massage, baths, &c. Pythagoras (580 B.C.) added to this hygienic treatment the use of such drugs as squill and mustard. Hippocrates (B.C. 466), who placed the treatment of disease on a more rational basis, was familiar with numerous indigenous as well as exotic drugs, with many of which he doubtless became acquainted during his travels. Wormwood, acacia gum, chamomile, cinnamon, hemlock, gentian, henbane, myrrh, rhubarb, and many others were well known to him.

Theophrastus (about 370 B.C.) was a pupil of Aristotle, after whose death he became chief of the Aristotelian school. He enumerates some 500 plants which were known to him, distinguished cinnamon from cassia, and was acquainted with ergot. It was during this period (335–325 B.C.) that Alexander the Great undertook his expeditions to Persia, India, and Africa, as the result of which numerous treasures, and among them drugs, were brought back to Greece.

Alexander founded Alexandria, which he intended to become the seat of the learning and commerce of the world. Although science in Egypt had fallen so low that the Greeks learnt but little from the Egyptians, the Alexandrian school was invaluable in preserving and transmitting to posterity the writings of Greek and other authors.

After the death of Alexander the Roman Empire rose. During this period Dioscorides, who was a Greek by birth, travelled in Egypt, Africa, Spain, Italy, and Syria, where he became acquainted with a great variety of plants and drugs. His observations and knowledge he placed on record in a compendious work on pharmacognosy which served for many centuries as an inexhaustible source of information. This remarkable work contains classified descriptions of a great number of vegetable, animal, and mineral drugs, no fewer than 500 medicinal plants, including many exotic ones, being discussed. To Dioscorides belongs the credit of being the first to separate pharmacognosy from medicine.

Pliny, who lived about the same time as Dioscorides and evidently made use to some extent of the same sources of information, studied natural history; his work on this subject consisted of forty-seven books, many of which have been unfortunately lost, Pliny himself perishing in the eruption of Vesuvius (A.D. 79). His 'Natural History,' in addition to books dealing with zoology, meteorology, astrology, &c., treats of as many as 1,000 plants.

Galen (A.D. 131 to 200) was a physician and studied chiefly the production of various 'galenical' preparations, dealing only incidentally with drugs. He recognised the adulteration of cinnamon, pepper, saffron, and myrrh, and points out that exhausted rhubarb was sold in place of the genuine. He wrote twenty books containing a large number of receipts for various preparations.

With the subsequent decline of the Roman Empire, a long period of scientific inactivity, or perhaps rather retrogression, took place. The Arabian Empire then rose, and seats of learning were established and encouraged at Bagdad, Cordova, Toledo, &c. The Arabians acted chiefly as the transmitters of Greek and Roman knowledge, adding comparatively little of their own. The most important of their writers were Geber, Rhazes, Avicenna, and Ibn Baitar.

Geber (A.D. 850 ?), who studied at Cordova, wrote chiefly concerning salts of which a considerable number were known to him.

Rhazes (A.D. 875 ?) studied medicine in Bagdad, and with him a period of independent investigation began. He was acquainted with asafetida, aloes, dandelion, henbane, stavesacre, and numerous other drugs; he prepared extracts by evaporating infusions and made them into pills, which he coated with psyllium mucilage in order to render them less unpleasant to take.

Avicenna (A.D. 980) was perhaps the most gifted of the Arabian physicians and a very prolific writer. His 'Canon Medicinae' served as the chief source of medical knowledge down to the fifteenth century.

Ibn Baitar (A.D. 1197–1248) devoted himself chiefly to botanical work in which he excelled, his 'Corpus Simplicium Medicamentorum,' containing descriptions of about 2,000 drugs, of which 1,700 were of vegetable origin.

The Arabian school was followed by that of Salerno, near Naples, where a school of medicine, established in the twelfth century or possibly earlier, was instrumental in continuing scientific work until the dawn of science in Europe in the fifteenth century. During this period the knowledge at least of indigenous medicinal plants was kept alive and added to in the gardens attached to the convents and monasteries of Central Europe.

The beginning of the sixteenth century witnessed a revival in the study of drugs, which was stimulated by the discovery of America, of the West Indies, and of the sea route to India. Matthiolus (1501–1577), who studied in Italy, published a commentary on Dioscorides, which enjoyed a great reputation; he also collected and dried plants for pharmacognostical purposes and prepared careful illustrations. Monardes (1493–1578) collected drugs and other products brought from America, forming thus what was probably the first museum of materia medica. Clusius (1526–1609) also collected drugs from drug merchants and examined them; his 'Rarium Plantarum Historia' (1601) contains no fewer than 1,146 illustrations, many of which are of great excellence.

After the lapse of a considerable time, during which little of note was published, Pomet wrote his excellent and well-illustrated 'Histoire Générale des Drogues' (Paris, 1694), which was closely followed by Lémery's 'Traité Universel des Drogues Simples' (Paris, 1697) and 'Dictionnaire des Drogues Simples' (4th ed., Paris, 1727), and later by Geoffroy's 'Tractatus de Materia Medica' (Paris, 1741).

The period of modern pharmacognosy may be said to open with Guibourt's admirable 'Histoire Abrégée des Drogues Simples' (Paris, 1820), of which several editions were published. Guibourt relied largely on his own observations, and treated the subject from a purely pharmacognostical point of view. This was not the case with Pereira, whose classical 'Elements of Materia Medica and Therapeutics' (London, 1839) embraced, as its title indicates, both pharmacognosy and therapeutics. Schleiden was the first to employ the microscope to distinguish drugs by differences in their structure, a departure which was rapidly advanced by Berg and Schmidt ('Anatomischer Atlas,' Berlin, 1865), Vogl, and Moeller. Hanbury studied chiefly the botany aud commerce of drugs, while Flückiger devoted himself chiefly to the chemistry and history, the collaboration of these two celebrated pharmacognosists resulting in the publication of the classical 'Pharmacographia' (London, 1874). Meyer ('Wissenschaftliche Droguen-Kunde,' Berlin, 1891) followed the morphological development of drugs, Tschirch ('Anatomischer Atlas,' Leipzig, 1900) their structural development, and E. M. Holmes their botanical origin.

By these and other workers a mass of material has been accumulated, the systematical arrangement of which has been attempted by Tschirch in his 'Handbuch der Pharmakognosie.'

**Commerce.**—The oldest records of commercial intercourse between different peoples indicate that several thousand years before the

Europe during the fifteenth century. Towards the end of this period the carrying and distributing trade passed into the hands of the Italians. The Pisanese wrested Sardinia, Corsica, and the Balearic Islands from the Arabs, and, assisted by the Crusades, secured for Pisa a lucrative trade, which passed to Genoa and finally to Venice, the latter becoming the chief carrier of the world. She reopened the sea-trade of the Red Sea and Alexandria, which was safer and cheaper than the caravan routes; she received and distributed all the drugs and spices of the East, and thus acquired great power and great wealth, the zenith of her prosperity being reached towards the end of the fifteenth century.

About this time, however, Henry the Navigator, son of King John I. of Portugal, eager to annex the Indian trade, and convinced by information obtained from the Arabs of the possibility of reaching India by sea, explored the west coast of Africa, discovering Madeira and establishing colonies on the mainland. The way was thus paved for Vasco da Gama, who in 1498 discovered the sea route to India, and landed on the Malabar Coast. The Portuguese established ports on the coasts of India, Goa being the chief, colonised Mozambique, the Spice Islands (1511), Ceylon (1518), Java, Sumatra, and Celebes. About the same time Spain became a colonial power. In 1492 Columbus discovered San Salvador, and on his subsequent voyage the West Indies. In 1519–1521 Mexico was conquered by Cortes, and immediately afterwards (1529–1535) Peru and Chili by Pizarro. The cities of Vera Cruz, Cartagena, and Caracas on the east, and of Acapulco, Panama, and Lima on the west, were founded.

The rich commerce of India was now soon diverted to Portugal, and Lisbon succeeded Venice as the emporium for Europe, a position which it occupied for nearly a century. To Lisbon or to Portuguese merchants in Antwerp the Dutch and English repaired to obtain their supplies of spices. Antwerp became the emporium of the north. When, however, in 1580 Philip II. of Spain united Portugal with Spain, and soon afterwards declared war against England, he closed the ports of his empire to English merchants. This act was followed by the revolt of the Netherlands from Spanish dominion, and both the Dutch and the English found themselves compelled to engage in a direct trade with India.

They thus became rivals of the Portuguese and of one another. The Dutch were at first the more successful. They established themselves in India, in the East Indian Islands, the Moluccas and the Sunda Islands; they drove the Portuguese out, and took possession of the entire trade with India, the East Indies, Japan, and China. They also established colonies in the Cape, and in 1658 took Ceylon from the Portuguese. Several of the West Indian Islands, and part of the north-east of South America, fell into their hands. The commerce of all these colonies and countries flowed into Amsterdam, which thus supplanted Antwerp and Lisbon as the European emporium of Indian trade.

Meanwhile the English, who had met with less success, had nevertheless founded colonies in Calicut, Madras, Masulipatam, Calcutta, and Bencoolen, and had driven the Spaniards from Jamaica, Barbados, and other of the West Indian Islands. In 1600 the first English East Indian Company was formed which built forts in Java, Amboyna, and Banda. At first this company shared the spice trade, but eventually it was driven from the islands by the Dutch, who destroyed all spice trees but those on the island of Amboyna. The English were thus restricted to the coasts of India, where they competed successfully with the Portuguese. After the cessation of the Company's charter in 1833 trade rapidly increased. Cotton, sugar, tobacco, tea, coffee, and cinchona were introduced. Singapore, which had become a British possession in 1819, became, and still remains, the centre of East Indian trade. North America had already been colonised; here and in the West Indian Islands tobacco, allspice, cocoa, indigo, &c., had been introduced, and their cultivation had rapidly extended.

**Drug Routes.**—Foreign drugs reach the London market chiefly by the following routes :

I. *Central European.*—From Reval and Riga (Russian); from Hamburg and Bremen (German); from Amsterdam and Rotterdam (Dutch and German); from Antwerp (Belgian and Dutch); from Dieppe, Boulogne, Havre, and Bordeaux (French); in all cases direct to London.

II. *Mediterranean.*—From Lisbon (Portuguese); from Vigo, Seville, Malaga, Valencia, and Barcelona (Spanish); from Marseilles (French); from Genoa, Leghorn, and Bari (Italian); from Messina, Palermo, and Catania (Sicilian); from Malta; from Salonica, Constantinople, Smyrna, Alexandretta, Jaffa, Beyrout (Turkish); from Suez, Tripoli, Tangiers, Algiers (African); either direct to London or often (from the Eastern Mediterranean) viâ Trieste or viâ Marseilles.

III. *North-West African.*—From Rabat, Mazagan, Saffi, Mogadore, Bathurst to London, Liverpool, or Southampton.

IV. *West-Central African.*—From Sierra Leone, Bonny, Lagos, San Thomé, and Benguela to Liverpool.

V. *South African.*—From Cape Town, Mossel Bay, Port Elizabeth, East London, Natal, and Delagoa Bay to Southampton and Liverpool.

VI. *East-Central African.*—From Beira, Chinde, Quilimane, Mozambique, Zanzibar, Mombasa to London (often viâ Hamburg); from Mauritius, Réunion, Madagascar, and Seychelles to Liverpool.

VII. *Red Sea, Arabian, and Persian.*—From Suakin, Massowah, Aden, Bunder-Abbas, Bushire, Bussorah, and Bagdad to London.

VIII. *Indian.*—From Karachi, Bombay, Mangalore, Calicut,

Cochin, Aleppy, Colombo, Tuticorin, Madras, Calcutta, and Rangoon to London.

IX. *Malayan.*—From the various islands viâ Singapore and Penang to London; from Batavia viâ Amsterdam to London; from Saigon viâ Marseilles to London.

X. *Chinese.*—From Shanghai and Hong Kong, often viâ Havre or Hamburg, to London.

XI. *Japanese.*—From Nagasaki, Kobé, and Yokohama to London.

XII. *North American.*—From Quebec, Montreal, Boston, New York, Baltimore, Newport, New Orleans, and San Francisco to Liverpool, London, and Southampton, sometimes viâ Havre and Cherbourg.

XIII. *Central American.*—From Greytown and Belize to Liverpool and London, sometimes viâ Hamburg.

XIV. *West Indian.*—From the British islands to Liverpool Bristol, and Southampton; from the Dutch islands to London viâ Amsterdam.

XV. *South American.*—From La Guayra, Savanilla, Cartagena, and Maranham to Liverpool; from Pernambuco, Bahia, Rio Janeiro, and Buenos Ayres to London and Liverpool; from Montevideo to Southampton; from Valparaiso, Santiago, Mollendo, Truxillo, and Guayaquil to Liverpool.

XVI. *Australian.*—From Brisbane, Sydney, Melbourne, Adelaide, and Perth to London; from Tasmania viâ Melbourne.

XVII. *New Zealand.*—From Auckland and Wellington to London.

**Drug Packages.**—The various ways of packing drugs for exportation form an extremely interesting study, as may be gathered from the following examples:—

Aloes is exported from the West Indies, from East Africa, and from Cape Colony. That from the Dutch West Indian Islands usually arrives in wooden cases in which bottled spirits have been exported (compare 'Aloes'), but formerly West Indian aloes was frequently poured while still soft into empty gourds (fig. 1, A), which were then closed with a piece of cloth and packed in barrels; such gourd aloes is still occasionally seen. Socotrine aloes is commonly sent from East Africa in a pasty or semi-solid condition in kegs or barrels. Zanzibar aloes is filled while soft into goat-skins, which are afterwards packed in cases.

Cassia fistula is exported from Java in baskets made of plaited split cane (fig. 1, B). China root, much used in China, but now almost obsolete in Europe, and galangal root (fig. 2, E) arrive in mats made of plaited bast (fig. 1, C); a similar covering protects the cases of cassia bark and cassia buds (fig. 3, A, B, C), all of which are exported from China. Cloves are usually sent from Zanzibar in mats made from interlaced strips of coconut leaves (fig. 3, D).

'Seron' is the name generally given to packages made of raw hide, such as those in which Calisaya cinchona bark is still occasionally

A B C

FIG. 1.—Original packages of drugs (Weigel). A, gourd aloes; B, cassia fistula; C, China root.

G

C D E

FIG. 2.—Original packages of drugs (Weigel). A and F, seron of Calisaya bark; B, Honduras sarsaparilla; C, dragon's blood; D, compressed lobelia; E, galangal root; G, euphorbium.

imported (fig. 2, A, F); a similar protection is afforded to bales of Honduras sarsaparilla (fig. 2, B), the ends of which are covered with raw hide laced with strips of the same material. Jamaica and Lima sarsaparilla are made into large disc-shaped bales bound round with iron wire, while the native Jamaica drug is packed loose in bales.

FIG. 3.—Original packages of drugs (Weigel). A, small cassia bark; B, cassia bark in bundles; C, cassia buds; D, Zanzibar cloves; E, cinnamon bark.

FIG. 4.—Original packages of drugs (Weigel). A, Turkey opium; B, balsam of Peru; C, stick liquorice; D, musk; E, isinglass; F, lycopodium; G, vanilla.

Euphorbium (fig. 2, G) is exported in bales made of plaited reeds and provided with curious handles. Mogadore orris root arrives in somewhat similar packages. Dragon's blood is exported from

Sumatra in lumps (fig. 2, C), which usually bear the impress of matting, or in sticks, each of which is wrapped in palm leaf and secured by bands of flexible grass. Lobelia herb is sent either loose in bales (compare 'Lobelia') or in compressed packets (fig. 2, D). Cinnamon bark comes in very carefully made cylindrical bales consisting of sticks of the same length bound together and covered with sacking (fig. 3, E; fig. 11, A). Opium from Turkey (fig. 4, A) is packed in 'chests' with the fruits of a *Rumex* shaken between the cakes to prevent them adhering, while stick liquorice (fig. 4, C) is packed with bay leaves. Vanilla (fig. 4, G) comes in small bundles, several of

FIG. 5.—Original packages of drugs (Weigel). A, cassia oil; B, star-anise oil; C, lavender oil; D, lemon oil; E, lemongrass oil; F, Japanese peppermint oil; G, American peppermint oil; H, Japanese menthol.

which are enclosed in a tin, balsam of Peru and copaiba in old oil-canisters. Musk pods (fig. 4, D; compare also 'Musk') are wrapped singly in thin paper and packed in tins, which are sometimes enclosed in silk-covered 'caddies.'

Menthol is exported in tins, oils of cassia and star-anise (fig. 7) in leaden canisters, oil of lavender in tins, oils of lemon-grass and cajeput in wine-bottles, oils of lemon and bergamot in coppers of varying size (fig. 5, D), while otto of rose arrives in metal 'vases' covered with felt (fig. 6).

**Sale of Drugs.**—The bulk of the drugs sold in London arrive by sea. From the steamers bringing them they are conveyed either by lighters or by railway to the wharves that line the river bank or to the extensive warehouses of the Port of London Authority.

Most foreign drug growers or drug merchants have no offices in London in which to sell their goods; they therefore effect this through a drug broker as intermediary. In and around Mincing Lane a number of firms conduct the business of selling on commission not only drugs but also a great variety of other produce. To one of these the foreign drug merchant consigns his goods, advising the firm of their dispatch. On heir arrival in London the broker procures samples from the various packages and exhibits these at intervals on counters in his office. This exhibition constitutes what is known as a 'drug show' and is held normally on alternate Wednesdays. The show is attended by representatives of such drug merchants as desire to purchase, who note the parcels they wish to buy and the price they are prepared to give. On the day following the show the goods are sold by auction and this constitutes a 'drug sale.'

FIG. 6.—Vase of otto of rose.

Sometimes the broker sells direct to a customer at a price previously agreed upon; this is sale by private treaty. Not unfrequently the drug producer sells his produce to an agent in the same district or country, who, in his turn, deals direct with the drug merchant in London; such purchases are then shipped direct to the purchaser.

Although most drugs are sold by sample, in a number of instances the original packages are exhibited so as to allow of the inspection of the contents. These are shown in a well-lighted, lofty top-floor of the Cutler Street warehouse of the Port of London Authority, and also at the warehouses of the London and other docks, &c. Original packages of the following drugs are usually shown: aloes (excepting Cape aloes), benzoin, dragon's blood, elemi, gamboge, guaiacum resin, ipecacuanha, myrrh, sarsaparilla, scammony gum-resin, jalap, rhubarb, musk, castor, balsam of tolu, vanilla, &c. Other floors in the same warehouse serve for the storage of a variety of drugs pending their sale or removal; among these the following may be noted; aloes, aniseed, areca nuts, alkanet root, buchu leaves, calumba

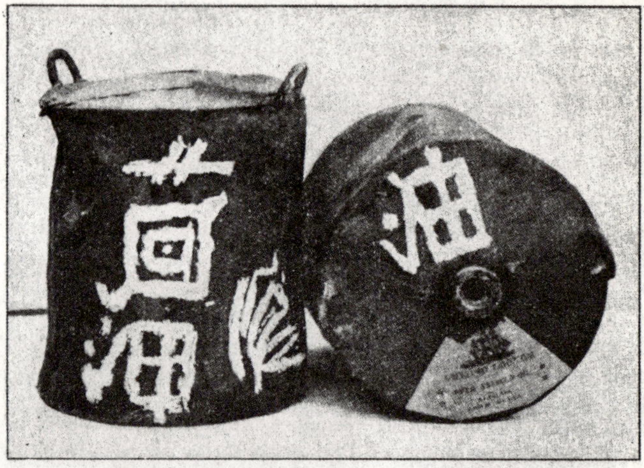


FIG. 7.—Leaden canisters of star-anise oil.

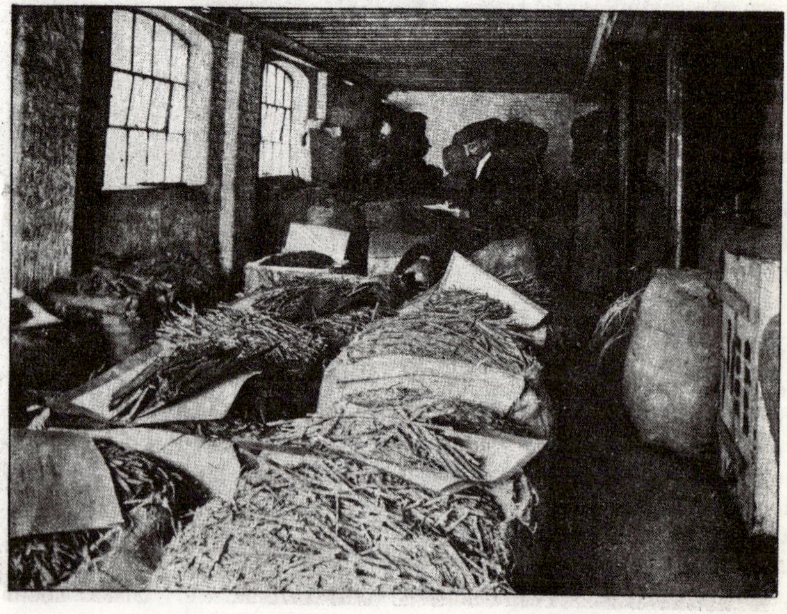

FIG. 8.—Sampling a few bales of bark. (Heap.)

FIG. 9.—Sale room of Messrs. Hale & Son. On the centre and window-tables various drugs; on the table to the left chiefly bristles; on the floor sample-bags of drugs. (Heap.)

FIG. 10.—A drug and gum show at London Docks. (Heap.)

fistula, castor seeds, chamomile flowers, chiretta, cocculus indicus root, canella bark, cantharides, cardamoms, cascara sagrada, cassia,

FIG. 11.—Show of drugs at London docks; A, cinnamon bark; B, cinchona bark; C, Zanzibar cloves; D, various bales of drugs. (Heap.)

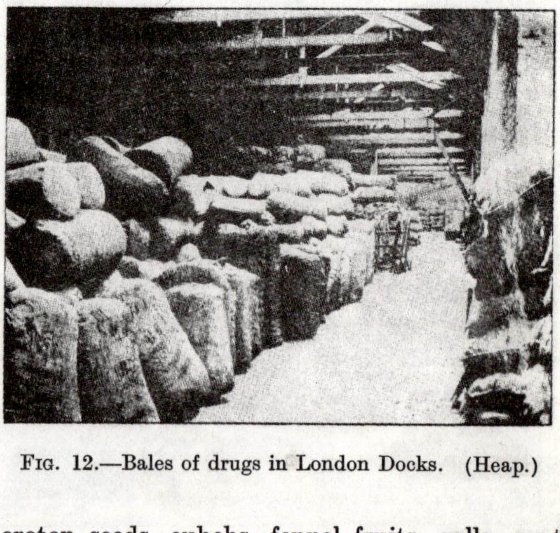

FIG. 12.—Bales of drugs in London Docks. (Heap.)

colocynth, croton seeds, cubebs, fennel fruits, galls, gentian, grains of paradise, Indian hemp, kola seeds, nux vomica, opium, orris root, quassia, saffron, sandal-wood, senna, squill, turmeric, valerian, &c.

Sometimes drugs are 'worked' before they are sold; they are then usually emptied on to the floor of the warehouse, mixed or sorted as required, and repacked in the cases. Spices such as pepper, cloves, ginger, nutmeg, cassia, pimento, and chillies, together with tapioca and arrowroot, are included in the spice sales. Cinnamon auctions are held once a month.

After arrival in the drug merchants' warehouse the drugs have to undergo further treatment before they are sold to the pharmacist.

FIG. 13.—Working tragacanth. (Heap.)

They are examined, freed from dust and dirt by sifting, and graded either by sifting or hand-picking; they have frequently to be further tested by assay. Few drugs, therefore, reach the pharmacist in the condition in which they are imported.

**Drying and Preservation of Drugs.**—The various parts of plants intended for medicinal use contain, when freshly collected, a large proportion of water from which they must be freed unless they are destined to be used in the fresh state. The following table,[1] which gives the approximate amounts of air-dry drug yielded by 100 parts of the fresh plant, will suffice to indicate the limits within which the percentage of moisture usually varies :—

[1] Tschirch, *Handbuch d. Pharmakognosie.*

| | | | |
|---|---|---|---|
| Red Poppy Petals | 18 | Hemlock Herb | 25 |
| Arnica Flowers | 20 | Dandelion Root | 22 |
| Chamomile Flowers | 25 | Liquorice Root | 33 |
| Belladonna Leaves | 18 | Belladonna Root | 38 |
| Foxglove Leaves | 20 | Sweet Flag Rhizome | 25 |
| Henbane Leaves | 20 | Colchicum Corm | 34 |
| Thornapple Leaves | 45 | Couch Grass | 40 |

During the process of drying certain changes occur which are produced, in part at least, by the action of enzymes contained in the cells. These changes may affect the colour of the drug and so become evident to the eye; thus the scarlet of the red poppy petal becomes dull violet, the green of leaves may turn to black, and gentian root, which is normally white, may acquire a reddish brown colour. They may also affect the odour of the drug; thus vanilla pods, valerian roots, orris rhizomes and other drugs acquire their characteristic odour during the process of drying. They may also affect certain constituents of the drug without producing any readily perceptible change; thus the gentiopicrin present in fresh gentian root may be completely decomposed. When the drug is quite free from moisture (not simply air-dry) these changes cease and so long as the drug is kept quite dry they cannot be resumed. The procedure adopted in drying drugs depends therefore upon whether the action of the enzymes is necessary and should be encouraged or whether it is prejudicial and should be avoided. In the former case slow drying and the maintenance of a temperature at which the enzymes are most active are indicated (compare vanilla, tea, cocoa, &c.). In the latter case the drying should be conducted as rapidly as possible in order to render the enzymes inactive before they have appreciably affected the constituents of the drug. Rapid drying is accomplished by exposing the drugs in thin layers to a free current of air at a temperature of about 35° to 40°. They may be scattered upon wire-netting trays (leaves not overlapping one another and roots sliced longitudinally or transversely) and placed in a well-ventilated loft; or vertical columns of such trays, 6 or 8 inches apart, may be formed in the open air and covered with roofing; or the trays may be assembled in a long narrow shed through which warm, dry air is driven. Open-air drying is often resorted to in hot climates.

Drugs that have been completely dried usually re-absorb about 10 per cent. of moisture when exposed to the atmosphere; they are then termed 'air-dry.' Even in air-dry drugs changes may occur; such changes are often prejudicial as they are for instance in Indian hemp and foxglove leaves. In the case of Indian hemp the change is probably simple oxidation of the cannabinol contained in it, but in the case of foxglove leaves and many other drugs enzyme action is certainly involved.

Two methods of obviating such enzyme action may be adopted. By the first the drugs are completely dried and maintained in that state in hermetically sealed containers or in well-closed vessels over quicklime. Flowers, kept absolutely dry over quicklime, will retain their natural colours for years and foxglove leaves their constituents unimpaired.

The second method consists in rendering the enzymes inactive before the drugs are dried. This may be effected by exposing them for a short time to the vapour of alcohol under moderate pressure at a temperature of about 95°. Life in the plants is arrested and all the enzymes killed without raising the temperature of the plants above about 80°. Fresh kola nuts may be sterilised by exposing them to steam under pressure at a temperature of about 110°. By such means the drugs are rendered stable and the process is termed 'stabilisation.'

Air-dry drugs kept in parcels, sacks, barrels, &c., are liable to be attacked by certain insect pests. Sayre[1] enumerates about 25.

The commonest of these insect pests belong either to the Coleoptera (beetles) or Tyroglyphidæ (mites).

Among the Coleoptera the most frequent and most destructive is the drug room bettle (*Sitodrepa panicea*). This beetle lays its eggs in the summer in crevices in the cork of roots, &c. The young larva drills a tunnel into the drug, hibernates through the winter, passes into the pupa stage in the early summer and finally emerges as a perfect insect about June or July. Dandelion root, burdock root, aconite root, belladonna root, ginger, nutmegs are particularly liable to attack.

The cigarette beetle (*Lasioderma serricorne*), the granary beetle (*Niptus hololeucus*), and *Ptinus brunneus* also do great damage.

Among the Tyroglyphidæ various species of *Tyroglyphus* appear to be the commonest. In this case the larva resembles the perfect insect but is smaller and has only six legs instead of eight. It possesses two pairs of strong mandibles with which it tears the drug into small pieces before ingestion. The mites are very minute but they often occur in prodigious numbers. Quince seed, ergot, cantharides, ground linseed and many powdered drugs are liable to be attacked.

Drugs may be protected from insect pests by keeping them quite dry, or by dusting them with lime which blocks up the breathing apparatus of the mature insect and of the larva. Drugs already infested may be freed from them by exposing them to the vapour of carbon tetrachloride, carbon disulphide, chloroform, &c., by which larvæ and mature insects are destroyed.

Vegetable drugs kept in too damp a place are liable to develop moulds, bacteria, yeasts, &c. These organisms are compelled to obtain part at least of their nourishment from organic substances and in doing so frequently produce ammonia, sulphuretted compounds, &c. The attacks of these organisms may also be warded off by keeping the drugs quite dry.

[1] *American Journ. Pharmacy*, 1893, p. 321.

## SECTION I

## LEAVES

Leaves may be defined as appendages borne upon an axis but differing from it in structure and organisation. They usually exhibit three regions, viz. a leaf-base, a stalk or petiole, and a lamina or blade. Each of these regions may offer characters that are useful in identifying the leaf. The following schedule may assist the student in examining leaves.

i. *Size and Shape.*—These may vary within certain, sometimes rather wide, limits. Should the leaf have shrivelled so much as to render them undeterminable it should be soaked in warm water and gently spread out on a white tile.

ii. *Venation.*—The mode of branching of the midrib, the angle at which the branches (lateral veins) leave it, the course they pursue, and their elevation above or depression below the surface of the leaf should be noted.

iii. *Margin.*—This margin may be entire, serrate, crenate, dentate, sinuate, &c.

iv. *Apex.*—This may be rounded, pointed, notched, &c.

v. *Base.*—The appearance of the base should also be noted.

vi. *Surface.*—The surface should be examined with a lens, and the presence or absence of hairs, as well as the character of the latter, glandular or otherwise, determined; this particular is of very great importance.

vii. *Glands.*—Many leaves contain internal oil-glands or oil-cells which can usually be seen as translucent points when the leaf is examined with a lens by transmitted light. In very thick leaves they may be difficult to see, and it may be necessary to examine a section under the microscope.

viii. *Texture.*—Papery, leathery, fleshy, &c.

ix. *Colour.*—Greyish green, dark green, &c.

x. *Odour.*—This is best determined by crushing the leaf between the fingers and smelling it.

xi. *Taste.*—Often very characteristic.

xii. *Petiole.*—Presence or absence, length, &c.

## TEA

### (Folia Theæ)

**Source, &c.**—The tea shrub, *Camellia Thea*,[1] Link (N.O. *Ternstræmiaceæ*), has been cultivated from time immemorial in China and Japan, and more recently in Assam, Ceylon, Java, &c.

In the manufacture of black tea the leaves are subjected to processes that vary somewhat in the different countries. In Assam where much of the tea consumed in England is produced, the leaf-buds together with two or three of the youngest leaves are collected and spread upon trays until they have lost their turgidity. The 'withered' leaves are then rolled between two flat surfaces worked by machinery, during which many of the cells of the mesophyll are broken, and their sap squeezed out and allowed to come into contact with the air; at the same time the leaves acquire the twist characteristic of ordinary tea. They are next 'fermented' by exposing them on mats to the air; during this process, which is conducted at a temperature of 35°–40°, the colour changes from green to yellow and finally to coppery red, and the tea acquires its well-known odour. The fermented leaves are dried in a current of hot air, and graded by sifting by which a partial separation into leaf-bud, smaller, and larger leaves is effected.

The changes that take place during the fermentation are due to the action of an enzyme or more probably mixture of enzymes termed thease. Thease is very active at 54°, but is destroyed by a temperature of 76°–77°; it oxidises part of the tannin, converting it into a reddish brown, insoluble substance, while at the same time traces of volatile oil are produced and a bitter principle present in the fresh leaves is rendered insoluble. The quality of the tea produced depends largely upon the skill with which this process is conducted.

In the manufacture of green tea the leaves are subjected to a process of roasting in pans heated by direct fire, in which they are kept continually moving; they are then cooled, rolled into balls, and allowed to ferment. In this case the preliminary roasting probably destroys some at least of the various ferments of which the thease is composed, the tannin is not oxidised, and the leaves retain their green colour more or less unchanged.

[1] Specific names are written with a small initial letter, except in the following cases:

(*a*) When the specific name was formerly a generic name, e.g. *Camellia Thea.*

(*b*) When the specific name is the vernacular name, e.g. *Pilocarpus Jaborandi.*

(*c*) When the specific name is the proper name of a person used as a noun or as an adjective, e.g. *Garcinia Hanburyi*; *Aloe Hanburiana.*

(*d*) When the specific name is a noun, e.g. *Eucalyptus Globulus.*

**Description.**—The full-grown tea leaf is from 5 to 10 cm. long, dark green in colour, lanceolate or elliptical in outline and blunt at the apex, tapering at the base into a short stalk; the margin is distantly and shortly serrate, the serrations terminating in characteristic, glandular teeth which readily break off and are often absent from mature leaves. When quite young the leaves are covered with silky hairs, but as they mature these are lost, and the surface becomes glabrous. This difference is readily observed in commercial tea, the bud still bearing numerous hairs ('flowery' Pekoe), while the larger leaves are glabrous or nearly so (Congou).

**Microscopical characters.**—Fragments of the leaves may be readily recognised under the microscope by the small cells of the upper epidermis, those of the lower epidermis being rather larger and accompanied by numerous large stomata and long, thick-walled, one-celled simple hairs bent near the base; in the mesophyll, especially of older leaves, large, elongated, branching, thick-walled, sclerenchymatous idioblasts occur.[1]

**Constituents.**—The principal constituents of tea are caffeine and tannin. It contains in addition traces of theobromine, theophylline and volatile oil.

Caffeine or trimethylxanthine, $C_5H(CH_3)_3N_4O_2,H_2O$, an alkaloid obtainable in colourless silky crystals melting at 237°, was isolated from coffee by Runge in 1820 and from tea by Oudry in 1827. It occurs also in maté (the leaves of *Ilex paraguayensis*, Lambert, which are largely used in the Argentine Republic as tea is in this country), in cola seeds, and in guarana. Tea contains from 1 to 5 per cent. (usually 3 to 4) of it apparently in combination with tannin, the combination being decomposed by water and the caffeine liberated. The tannin varies from 10 to 24 per cent. But the commercial value of tea is not determined by the percentage of caffeine or of tannin contained in it, but by a combined consideration of several factors, such as the appearance, the size of the leaves as indicating their age, the presence of 'tip' (unexpanded leaf-bud), and the taste of the infusion.

Theobromine, or dimethylxanthine, $C_5H_2(CH_3)_2N_4O_2$, and its isomer theophylline are closely allied to caffeine, but occur in small proportion only.

Caffeine, theobromine, and theophylline are derivatives of xanthin, $C_5H_4N_4O_2$, which, with a number of other substances, is regarded as containing the purin group, $C_5H_4N_4$. Tea, coffee, and other drugs containing caffeine, theobromine, or a similar purin derivative form therefore a natural group the members of which are allied by the similarity of their chief constituents.

The following are the most important drugs belonging to the purin group:

*Tea*, the leaves of *C. Thea*, containing caffeine (1 to 5 per cent.), theobromine, and theophylline.

*Coffee*, the seeds of *Coffea arabica*, Linné (N.O. *Rubiaceæ*), containing caffeine (0·4 to 1·5 per cent.).

*Cola Seeds*, the seeds of *Cola vera*, Schumann (N.O. *Sterculiaceæ*), containing caffeine (2 to 2·5 per cent.) and traces of theobromine.

---

[1] Compare Greenish, *Foods and Drugs*, p. 122; Greenish and Collin, *Anatomical Atlas*, p. 89.

*Maté*, the leaves of *Ilex paraguayensis* (N.O. *Ilicineæ*), containing caffeine (0·2 to 2·0 per cent.).

*Guarana*, the crushed seeds of *Paullinia Cupana*, Humboldt, Bonpland, and Kunth (N.O. *Sapindaceæ*), containing caffeine (2·5 to 5 per cent.).

*Cocoa*, the seeds of *Theobroma Cacao*, Linné (N.O. *Sterculiaceæ*), containing theobromine (2 per cent.).

**Assay.**—Shake 6 gm. of the dried but not powdered tea with 120 gm. of chloroform for 15 minutes, add 6 c.c. of solution of ammonia, shake frequently for half an hour and set aside till clear Filter 100 gm. ( = 5 gm. of tea) into a flask, distil off the chloroform, add 3 c.c. of absolute alcohol, and dry on a water-bath. To the residue add 10 c.c. of 60 per cent. alcohol in which the caffeine is readily soluble, then 20 c.c. of water; shake vigorously, filter through a small filter, wash the residue and filter, evaporate, dry at 95°, and weigh. The weight of the residue multiplied by 20 gives the percentage of total caffeine.

**Adulterations.**—Tea has been adulterated with foreign leaves, as well as with exhausted tea leaves that have been rolled and dried. For details concerning the detection of adulterations in tea, the student should refer to Allen's 'Commercial Organic Analysis,' Vol. III., Part II.

**Uses.**—The chief pharmaceutical use of tea is as a source of caffeine which has a marked stimulant action on the nervous system and heart, and is also diuretic but less powerfully so than theobromine.

**Allied Drugs.**—*Abyssinian*, *Arabian* or *African tea* consists of the leaves of *Catha edulis* Forskal (N.O. *Celastrineæ*); it contains the alkaloids cathine, cathinine and cathidine.

*Bush tea* is the leaves of various species of *Cyclopia* (N.O. *Leguminosæ*); it is used in S. Africa.

*Marsh tea* is the leaves of *Ledum palustre*, Linné (N.O. *Ericaceæ*).

*Kaporie* tea is the leaves of *Epilobium angustifolium*, Linné (N.O. *Onagraceæ*).

None of these contains caffeine.

## COCA LEAVES

### (Folia Cocæ)

**Source, &c.**—The coca shrub, *Erythroxylum Coca*, Lamarck, and *E. truxillense*, Rusby (N.O. *Lineæ*) has been cultivated in Peru and Bolivia for so long that the plant is not known in the wild state. It is cultivated also in Java and formerly to a limited extent in Ceylon. The dried leaves are in almost daily use in Peru and Bolivia as a nerve-stimulant; mixed with lime or with the ashes of a species of *Chenopodium* and chewed they impart unusual powers of endurance, allaying the cravings of hunger and the feeling of fatigue. On the steep sides of the valleys in the spurs of the Andes the coca is cultivated in large quantities. The leaves are picked twice in the year, or sometimes oftener, and dried. Only a very small proportion of the coca produced is exported; by far the greater part is consumed by the natives, whilst a considerable quantity is used in Callao for the manufacture of cocaine, the crude alkaloid being exported instead of the leaves.

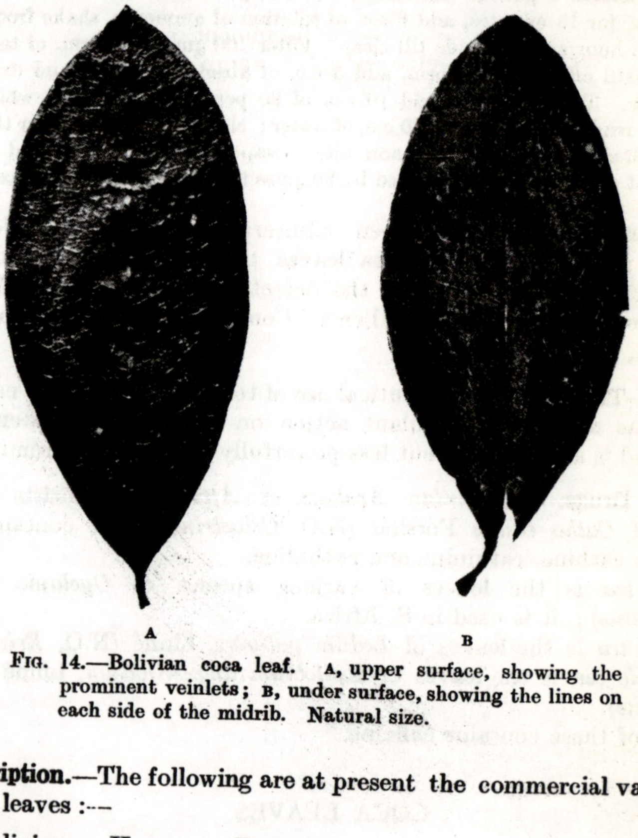


**Fig. 14.—Bolivian coca leaf.** A, upper surface, showing the prominent veinlets; B, under surface, showing the lines on each side of the midrib. Natural size.

**Description.**—The following are at present the commercial varieties of coca leaves :—

1. Bolivian or Huanuco (*E. Coca*).
2. Peruvian or Truxillo (*E. truxillense*).
3. Java (*E. truxillense*).

*Bolivian* coca leaves are usually fairly intact. They are brownish green in colour, and oval in outline, varying usually from 4 to 8 cm. in length and from 2·5 to 4 cm. in breadth. Both surfaces are glabrous, and the lateral veins and veinlets are distinctly prominent. The margin is entire, and the lamina tapers towards both base and apex; the latter is acute, the midrib projecting in the form of a minute horny point (apiculus) which, however, is frequently broken off.

The midrib is reddish brown, depressed on the upper surface and surmounted by a raised ridge; on the under surface on either side of the midrib a distinct, raised, curved line runs from the base to the apex of the leaf. This line is formed in the young leaf, each half of which is inrolled (involute) in the bud and forms a ridge closely applied to a similar ridge on the other half. Although the leaf is not thick it is by no means fragile, the lateral veins and veinlets being comparatively strong, and hence prominent on the upper surface (fig. 15). When a transverse section is examined under the microscope most of the epidermal cells of the under surface will be seen to project in the form of small papillæ; in surface view these appear as circular rings.[1] This peculiarity, although exhibited by other species of

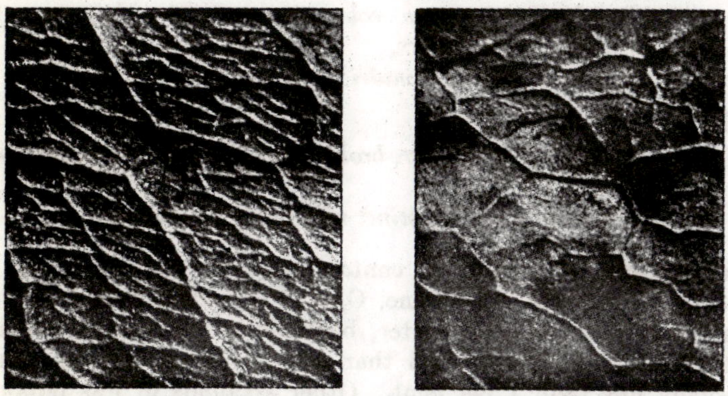

A B

FIG. 15.—A, under surface of Bolivian coca leaf; B, of Peruvian coca leaf, showing the difference in the veinlets. Magnified.

*Erythroxylum*, is so unusual as to render it of considerable diagnostic importance. The leaves have a faint but characteristic odour. The taste is slightly bitter, followed by a feeling of numbness in the mouth and throat. The small, oblong-ovoid, dark brown, pointed fruits are occasionally found in the drug.

*Peruvian* or *Truxillo* coca leaves are rather smaller than the Bolivian; they are pale green in colour and are more fragile, hence they are usually more or less broken. On the under surface the two curved lines are much less distinct; on the upper surface the ridge above the midrib is less marked; on the under surface the veinlets are less prominent and the midrib green in colour. These leaves frequently contain an admixture of the carefully picked and dried flowers of a species of *Inga* easily recognised by their yellowish brown, tubular, hairy calyx and numerous deep red filaments forming a plume; this is an intentional addition, made apparently with the view of

[1] Compare Greenish and Collin, *Anatomical Atlas*, p. 63.

improving the coca, and is not an adulteration. The pale, reddish brown fruits of the coca are also occasionally to be found.

The plant that yields this variety of coca has been distinguished as *E. truxillense*, Rusby.

*Java coca* is exported in the form of coarse powder; it does not reach the English market, but is used in Holland and Germany for the extraction of cocaine.

The student should observe in both Bolivian and Peruvian coca leaves:

(*a*) The *curved lines on the under surface of the leaf*,
(*b*) The *minute horny apiculus*,
(*c*) The *characteristic odour and taste*;

in the Bolivian leaves:

(*d*) The *brownish green colour*, *prominent veinlets*, and *less broken appearance*,
(*e*) The *ridge above the midrib*;

in the Peruvian leaves:

(*f*) The *pale green colour*, *broken appearance*, and *less prominent veinlets*,
(*g*) The *absence of a distinct ridge over the midrib*.

**Constituents.**—Coca leaves contain several alkaloids, the most important of which is cocaine, $C_{17}H_{21}NO_4$, which forms colourless monoclinic prisms with a bitter, benumbing taste. Bolivian leaves contain more of this alkaloid than the Peruvian, but the quantity is usually less than 1 per cent. Other alkaloids of less importance are cinnamyl-cocaine and $\alpha$- and $\beta$-truxilline (isotropylcocaine, cocamine, isococamine); these are frequently present in Peruvian leaves in larger quantities than cocaine. All these alkaloids are easily hydrolysed, and they all yield, together with other products, a crystalline alkaloid, ecgonine, from which by suitable treatment cocaine can be regenerated. Thus cocaine, when hydrolysed, yields methyl alcohol, benzoic acid, and ecgonine; cinnamyl-cocaine yields methyl alcohol, cinnamic acid, and ecgonine; and truxilline yields methyl alcohol, truxillic acid, and ecgonine. Ecgonine is very closely allied to tropine; $\alpha$-truxilline is a powerful heart-poison. (See under 'Belladonna-Leaves.')

In leaves imported from Java benzoyl-pseudotropine or tropacocaine has been found; this alkaloid yields, when hydrolysed, benzoic acid and pseudotropine; the latter is isomeric with tropine. This variety of coca contains about 0·7 per cent. of total alkaloid; it also contains four yellow, crystalline glucosides.

Coca leaves contain, in addition, cocatannic acid.

**Assay.**—Moisten 5 gm. of the finely powdered leaves with 2 per cent. solution of ammonia; pack in a long, narrow percolator, percolate with ammoniated

**ether until 100 c.c. have been collected or the drug is exhausted. Shake out with 2 per cent. hydrochloric acid till exhausted. Mix the acid solutions, wash once with ether, make alkaline with ammonia, and shake out with ether. Allow the mixed ethereal solutions to evaporate, dry at 100° and weigh.**

**Uses.**—Coca is a stimulant tonic and restorative, and is used during convalescence. Cocaine hydrochloride, when administered hypodermically, or applied to an exposed mucous surface, rapidly paralyses the sensory nerves and thus produces local anæsthesia. It is therefore of great value and much used in minor surgical operations of the eye, nose, ear, &c.

## BUCHU LEAVES

### (Folia Buchu)

**Source, &c.**—The official buchu leaves are obtained from *Barosma betulina*, Bartling & Wendland (N.O. *Rutaceæ*), a small shrubby plant indigenous to Cape Colony. The drug, the use of which appears

Fig. 16.—Buchu leaves (*B. betulina*), showing the shape, margin, and recurved apex. Natural size.

to have been learnt from the Hottentots, was introduced in 1821 but this was probably not derived from the species now official. The branches, collected while the plant is flowering and fruiting, are dried, and beaten to separate the leaves which are exported from Cape Town.

**Description.**—The leaves of *Barosma betulina*, commercially known as 'short' buchu, average from 1 to 2 cm. in length, and are of a pale green colour. They have a very characteristic, rhomboid-obovate outline, and a blunt, strongly recurved apex. They are rigid and brittle when quite dry, but cartilaginous when moist. The surface is glabrous or very nearly so, short hairs being often present on the midrib near the base; the upper surface presents small wart-like prominences due to the elevation of the epidermis by subjacent oil-glands; the lower surface is finely wrinkled. The margin is provided with numerous, minute, sharp teeth. When examined with a lens by transmitted light the lamina exhibits scattered oil-glands, one being situated at the base of each marginal indentation.

Not unfrequently the small flowers with five whitish petals and the brownish fruits with five carpels may be found in the drug.

The leaves have a characteristic odour recalling peppermint, and a strong aromatic taste.

The student should carefully observe

(*a*) The *rhomboidal outline* and *recurved apex*,

(*b*) The *oil-glands* and their distribution,

(*c*) The *characteristic odour and taste* ;

and should compare the official variety with the other commercial buchu leaves mentioned below, as well as with

*Bearberry leaves*, which are rounded at the apex, spathulate in outline, and destitute of oil-glands.

**Constituents.**—The principal constituents of buchu leaves are volatile oil and mucilage, the former being contained in the oil-glands,

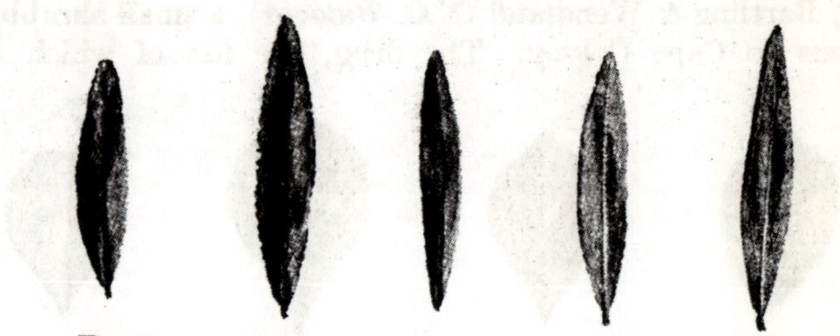

FIG. 17.—Buchu leaves (*B. serratifolia*). Natural size.

whilst the latter is deposited on the inner walls of the epidermal cells. They contain in addition a yellow crystalline substance, hesperidin, which is found in several other Rutaceous plants, but does not markedly contribute to the physiological action of the drug ; this substance forms sphero-crystals in the epidermal cells. The volatile oil (1·3 to 2 per cent.) deposits about 30 per cent. of crystalline diosphenol, $C_{10}H_{16}O_2$, when it is cooled. Diosphenol has antiseptic properties, and is considered by some to be the most important constituent of buchu ; its absence from the variety known as long buchu has led to the exclusion of the latter from the British Pharmacopœia. The oil also contains a ketone, probably identical with menthone, to which the peppermint-like odour is due.

Diosmin, which has been reported as a constituent of buchu, is probably impure hesperidin.

**Varieties.**—*Barosma serratifolia*, Willdenow.—The leaves of this plant, which, like *B. betulina*, is also a native of South Africa, are imported in considerable quantity, and known commercially as 'long buchu.' They are about 2·5 to 4 cm. long and linear-lanceolate in outline. The margin is serrate, and the apex distinctly truncate.

The leaves contain oil-glands similar to those of short buchu, one being distinctly visible in the truncate apex when examined with a lens. In odour and taste they resemble short buchu; they contain mucilage and about 1 per cent. of volatile oil, which, however, as previously mentioned, does not yield diosphenol when cooled.

*Barosma crenulata*, Hooker.—The leaves of this plant are also imported from the Cape of Good Hope, but in smaller quantity than either of the preceding varieties of the drug. They are rather broader than the long buchu, varying in outline from lanceolate to oval-oblong. The margin is minutely serrate, and the apex is blunt but not recurved (as in *B. betulina*). In odour and taste they resemble the official leaves; they yield about 1·6 per cent. of volatile oil, which also contains diosphenol.

**Substitutes.**—*Empleurum serrulatum*, Aiton; resemble long buchu under which name they have been offered for sale; apex acute

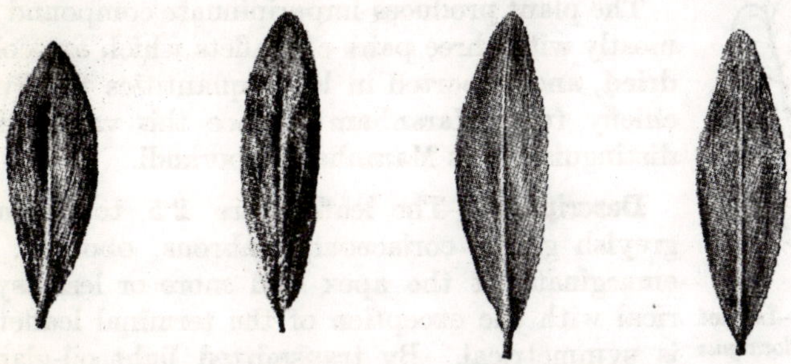

FIG. 18.—Buchu leaves (*B. crenulata*). Natural size.

without oil gland, colour usually yellowish green, taste bitterish, odour different.

*Agathosma Cerefolium*, Bartling and Wendland, *A. microphylla*, Meyer, *A. variabilis*, Sonder, have a distinct anise odour; *A. chortopila*, Ecklon and Zeyher, has a cummin odour.

*B. Eckloniana*, Bartling; rather broader than *B. crenulata*, more distantly crenate-serrate, apex very obtuse and without gland.

*B. venusta*, Ecklon and Zeyher; small, obovate or oblanceolate.

*B. pulchella*, Bartling and Zeyher; 7–12 mm. long, 5–8 mm. broad, ovate or ovate-lanceolate, margin thickened, apex obtuse, taste different.

*B. ericifolia*, Andr., *B. succulenta*, Thunberg; leaves heath-like, linear.

*B. lanceolata*, Sonder; rather narrower than *B. crenulata*, margin entire, recurved.

*Psoralea obliqua*, Meyer; lamina oblique, apiculus recurved, veins hairy.

*Adenandra fragrans*, Rœmer and Schultes; oblong, obtuse, caraway odour.

**Uses.**—Buchu is regarded as possessing a tonic and diuretic action; it is used in inflammatory conditions of the urinary tract.

## JABORANDI LEAVES

### (Folia Jaborandi)

**Source, &c.**—The name jaborandi is applied in South America, especially in Brazil, to a number of plants (belonging chiefly to the natural orders *Rutaceæ* and *Piperaceæ*) possessing salivant and sudorific properties. The leaflets of several species of *Pilocarpus* (N.O. *Rutaceæ*) have been imported as jaborandi but the variety at present almost exclusively in commerce is obtained from *P. microphyllus*, Stapf.

Fig. 19.—Leaflet of *Pilocarpus microphyllus*. (Holmes.)

The plant produces imparipinnate compound leaves, mostly with three pairs of leaflets which are collected, dried, and exported in large quantities to Liverpool, chiefly from Maranham; hence this variety is often distinguished as Maranham jaborandi.

**Description.**—The leaflets are 2·5 to 4 cm. long, greyish green, coriaceous, glabrous, obovate, deeply emarginate at the apex and more or less asymmetrical with the exception of the terminal leaflet which is symmetrical. By transmitted light oil-glands are visible under a lens. Odour characteristic but not powerful; taste aromatic and pungent, a copious flow of saliva being produced. The rachis, if present, is slightly winged.

The student should carefully observe

(*a*) The *presence of oil-glands*,
(*b*) The *emarginate apex*,
(*c*) The *asymmetry of the leaflet*.

**Constituents.**—Jaborandi leaves contain volatile oil (about 0·5 per cent.) resin and the alkaloids pilocarpine, isopilocarpine and pilosine (carpiline). Of these alkaloids pilocarpine, $C_{11}H_{16}N_2O_2$, is the most important and the one upon which the therapeutical value of the leaves almost entirely depends. The proportion present is usually about 0·5 per cent.

Pilocarpine is crystalline (m. pt. 34°) and dextrorotatory; 0·01 gm. produces powerful sweating and salivation. Heating with alcoholic potash converts it into the racemic form, isopilocarpine, which has about one-tenth of the activity.

Pilosine occurs in very small proportion only.

Pilocarpidine occurs in the leaves of *P. Jaborandi*, Holmes (see below); jaborine is non-existent. The therapeutical action of pilocarpine is not modified by the addition of isopilocarpine, pilosine, pilocarpidine or the resin; hence pilocarpine can well replace the drug for medicinal use (Jowett).

**Assay.**—The total quantity of alkaloid in jaborandi may be determined by the following process: Moisten 10 gm. of the leaves in fine powder with 2 c.c. of solution of ammonia and 3 c.c. of chloroform, pack in a small percolator and percolate to exhaustion with ammoniacal chloroform. Shake out with diluted sulphuric acid, make alkaline, and shake out again with chloroform. Evaporate the chloroformic solution to dryness, dissolve the residue in 7 c.c. N/10 sulphuric acid and titrate back with N/50 solution of potassium hydroxide, using cochineal as indicator. Each c.c. of the acid indicates 0·02 gm. of alkaloid.

**Uses.**—Jaborandi stimulates the nerve endings to the salivary, sweat, gastric, and other secreting glands, and induces therefore profuse salivation and perspiration. It is chiefly given as a powerful and rapid diaphoretic, and appears to be of most service in renal disease, eliminating both water and urea. Pilocarpine is used in ophthalmic surgery to contract the pupil of the eye; it is antagonistic to atropine.

**Substitutes and Varieties.**—*Swartzia sp.* (N.O. *Leguminosæ*); brownish green, glossy, midrib minutely hairy; very short, hairy petiolule, veinlets pellucid; about 3 cm. long but some only 0·5 long; substituted for *P. microphyllus*.

*P. Jaborandi*, Holmes; dull brownish green, 6–10 cm. long, oblong-lanceolate, margin revolute, apex emarginate, unequal at the base, upper surface glabrous with prominent lateral veinlets, under surface sometimes slightly hairy. Formerly official but now seldom imported; contain pilocarpine (about 0·5 per cent.), isopilocarpine and pilocarpidine.

*P. pennatifolius*, Lemaire; Paraguay Jaborandi; greyish green, pale and less coriaceous than *P. Jaborandi*; veinlets not prominent on the upper surface; usually equal at the base. Contain about 0·25 per cent. of alkaloid.

*P. selloanus*, Engler; imported from Rio Janeiro, closely resemble the above, but are more obovate.

*P. trachylophus*, Holmes; smaller than those of either *P. Jaborandi* or *P. pennatifolius*; dark olive green on the upper, yellowish green on the under surface, which is clothed with short curved hairs, similar ones being found scattered on the upper surface; contain about 0·4 per cent. of total alkaloid (Paul and Cownley, 1896), of which only 0·02 per cent. is possibly pilocarpine.

*P. spicatus* (?); Aracati Jaborandi; resemble bay leaves in size and shape; upper surface polished, veins scarcely visible, outline lanceolate, and the petiole short and twisted.

*P. racemosus*, Vahl; Guadeloupe Jaborandi; ovate, attaining 17 cm. or more in length; contain from 0·6 to 1·0 per cent. of total alkaloid, about one half of which appears to be pilocarpine.

*Piper Jaborandi*, Velley; large, thin, grey, papery leaves tapering towards both base and apex; usually mixed with the stems, which are swollen at the nodes.

## ALEXANDRIAN SENNA LEAVES
### (Senna Alexandrina)

**Source, &c.**—Alexandrian senna consists of the leaflets of *Cassia acutifolia*, Delile (N.O. *Leguminosæ*), a small shrub from 1 to 1·5 metres in height, indigenous to and cultivated in the middle and upper Nile territories, and cultivated also in India. The medicinal value of the pods and leaves of the plant was known to the Arabian physicians of the tenth and eleventh centuries, through whom European physicians probably became acquainted with the drug.

Alexandrian senna is collected by Bedouins, chiefly from wild plants between Suakin and Kassala, the shrubs being cut down and dried. The drug is either cleaned in the Soudan or sent in the crude

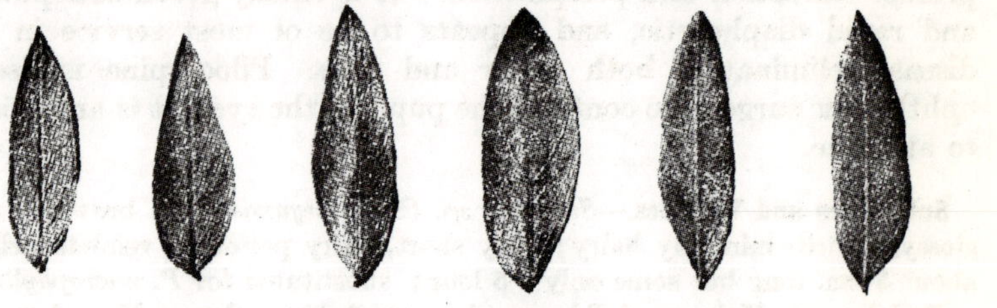

FIG. 20.—Alexandrian Senna leaves. Natural size.

state to Cairo, Alexandria or Suez and there cleaned and sorted from stalks, stones, etc. It is exported from Alexandria.

**Description.**—The plant produces a paripinnate compound leaf about 10 cm. in length. The leaflets average about 2·5 cm. in length, seldom exceeding 4 cm., and, when dried, are of a pale greyish green colour, thin, and brittle, lanceolate to ovate-lanceolate in outline, with an entire margin and acute, mucronate apex. They are decidedly unequal at the base, and on the under surface the veins are distinct. When examined with a lens both surfaces of the leaf are seen to be pubescent, small hairs being distinctly visible, especially near the veins.

The leaves frequently appear in commerce in a more or less broken condition due to their brittle, papery texture. They curl slightly as they dry, and, being loosely packed, retain this appearance ; Indian senna leaves, on the other hand, being pressed into bales, are commonly flat and bear other evidence of the pressure to which they have been subjected.

Alexandrian senna has a faint but characteristic odour and a mucilaginous, mawkish, unpleasant taste.

**Microscopical Characters.**—The epidermal cells of both surfaces have straight walls and many contain mucilage; the stomata are surrounded by two cells with their long axes parallel to the ostiole; the hairs are one-celled, thick-walled, warty, and often curved. Both surfaces possess long, narrow palisade cells. The midrib contains abundant pericyclic fibres, abutting on which are cells containing prismatic crystals of calcium oxalate, cluster crystals of which occur in the mesophyll.

The student should observe

(*a*) The *predominating asymmetrical*, *lanceolate* or *ovate-lanceolate shape*,

(*b*) The *veins distinct on the under surface*,

(*c*) *Both surfaces more or less distinctly pubescent*;

and should compare the leaves with

(i) *Mecca senna leaves*, which are elongated lanceolate,

(ii) *Indian senna leaves*, which are yellowish green in colour, less conspicuously asymmetrical, and less pubescent.

FIG. 21.—Obovate Senna leaves (*C. obovata*). Natural size.

**Constituents.**—According to the very exhaustive researches of Tutin the chief constituents of Alexandrian senna are rhein, aloe-emodin, kaempferol and isorhamnetin, all four substances occurring both free and in the form of glucosides; myricyl alcohol and a phytosterolin (phytosterol glucoside) are also present. The leaves contain in addition mucilage, calcium oxalate, resin, and amorphous glucosidic material.

Aloe-emodin, $C_{14}H_5O_2(OH)_2CH_2OH$, brownish green needles, is hydroxymethyldihydroxyanthraquinone and is the primary alcohol corresponding to chrysophanic acid (better termed chrysophanol) which is dihydroxymethylanthraquinone. Rhein, $C_{14}H_5O_2(OH)_2COOH$, orange needles, is the corresponding carboxylic acid and may be formed by the oxidation of aloe-emodin.. Both aloe-emodin and rhein exist in the free state and in the form of glucosides.

Kæmpferol, $C_{15}H_6O_2(OH)_4$, is 1 : 3 : 4-trihydroxyflavonol (bright yellow needles), and is yielded together with dextrose by the hydrolysis of kæmpferine, $C_{27}H_{30}O_{16},6H_2O$, which is also present in senna.

Isorhamnetin, $C_6H_{12}O_7$, yellow needles, was first obtained from yellow wallflowers.

Of these constituents aloe-emodin and its glucoside appear to be the only purgative ones. The total amount of methylanthraquinone derivatives present in senna has been variously estimated at 1·0 to 4·0 per cent. a proportion that is not consistent with the purgative action of the drug and the true active constituent is apparently still unknown.

Recent investigations have revealed the presence of oxymethylanthraquinones in a number of laxative drugs, which, therefore, form a natural group allied by the similarity of their constituents. The following are the most important members of the group:

*Senna*, which contains aloe-emodin, and rhein.

*Cascara Sagrada*, the bark of *Rhamnus Purshianus*, de Candolle (N.O. *Rhamnaceæ*), which contains emodin and frangula-emodin.

*Alderbuckthorn Bark*, from *Rhamnus Frangula*, Linné, which contains frangulin and frangula-emodin.

*Rhubarb*, the rhizome of *Rheum officinale*, Baillon, &c. (N.O. *Polygonaceæ*), which contains or yields chrysophanic acid, emodin, aloe-emodin, and rhein.

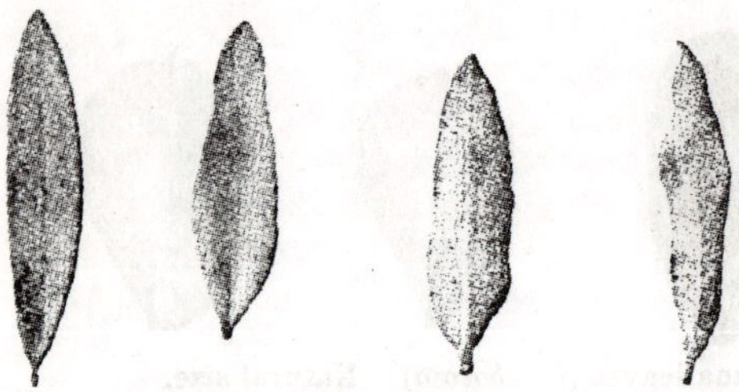


FIG. 22.—Argel leaves. A, showing the shape of the leaf; B, C, D, showing the curled appearance of the dried leaves. Natural size.

*Aloes*, from various species of *Aloe* (N.O. *Liliaceæ*), which contains and yields aloe-emodin.

*Cassia Pulp*, an aqueous extract of the fruit of *Cassia Fistula*, Linné, which contains unidentified oxymethylanthraquinones.

**Uses.**—Senna stimulates the muscular coat of the intestine and produces purgation, which is not followed, as is commonly the case, by constipation; it is therefore one of the most useful of purgatives, especially in cases of habitual constipation.

**Substitutes, &c.**—*Senna pods*, the fruits of *Cassia acutifolia*; occasionally found mixed with the senna, and imported as a separate article of commerce; very flat legumes. (See p. 100.)

*Cassia obovata* (dog senna), Upper Egypt; formerly highly valued as a drug, and cultivated in Italy and sometimes termed Italian senna; leaves broadly obovate, apex abruptly tapering venation, pinnate, distinct; constituents similar to those of senna; total oxymethylanthraquinones 3·8 per cent.; about equal to senna in activity. The leaves are sometimes broken up and mixed with broken Alexandrian senna, they may be recognised by the papillose cells of the lower epidermis.

*Argel leaves, Solenostemma Argel*, Hayne (N.O. *Asclepiadeæ*); resemble senna in colour and outline, but distinguished by their thick, rigid texture and peculiarly curled, curved, or twisted appearance;

surface finely wrinkled; veins not evident; leaf equal at the base; hairs three-celled; taste distinctly bitter; formerly regularly mixed with the senna but now of rare occurrence.

*Arabian, Mecca, or Bombay senna*, obtained from wild plants of *C. angustifolia*; collected in Southern Arabia and sometimes shipped *via* Bombay, or now more commonly by the Red Sea route, to London; elongated lanceolate, often discoloured and mixed with stalks; frequently mixed with Alexandrian senna, but may be distinguished by the shape; contain about 2·5 per cent. of total oxymethylanthraquinones.

Leaves of the following plants have also been imported as senna, or have occurred in commercial senna:

*Cassia holosericea*, Fresenius; smaller, more obtuse, hairy.

*Cassia montana*, Heyne; darker, rounded apex, dark network of veins.

*Tephrosia apollinea*, de Candolle; obovate-oblong, pubescent, emarginate lateral veins straight and parallel; fruits narrow-cylindrical; recently (1918) found in Alexandrian senna.

*Colutea arborescens*, Linné; green, very thin.

*Globularia Alypum*, Linné; spathulate, rounded apex, mucronate.

FIG. 23.—Arabian Senna leaves. Natural size.

*Coriaria myrtifolia*, Linné; ovate-lanceolate, greyish-green, two prominent lateral veins, conspicuous midrib.

## INDIAN SENNA LEAVES

### (Tinnevelly Senna, Senna Indica)

**Source, &c.**—Indian or Tinnevelly senna consists of the leaflets of *Cassia angustifolia*, Vahl, which is indigenous to southern Arabia, but is cultivated largely in southern India, especially in the district of Tinnevelly, in the extreme south-east. Here the plant attains an unusual luxuriance, and produces larger leaves than the Arabian wild plant. They are carefully collected, dried, pressed into bales, and exported from Tuticorin.

**Description.**—Tinnevelly senna is usually free from admixture either of foreign leaves or even of stalks and fruits. The leaves resemble Alexandrian senna rather closely, but are generally of a yellowish green rather than greyish green colour, a difference more noticeable in bulk than in single leaves. They attain a larger size than the Alexandrian, varying usually from 2·5 to 5 cm. in length. They differ also in being more uniformly lanceolate in shape, less conspicuously asymmetrical and less pubescent. They are somewhat firmer in texture than the Alexandrian, and are consequently less broken when they arrive, and, being exported in compressed bales, are usually flatter. There is also a slight but perceptible difference in the odour of the two varieties. The student should carefully compare these two varieties of senna. It is, however, quite possible to select from the two varieties exceptional leaves that are indistinguishable from one another. The student must be guided by the characters of the majority.

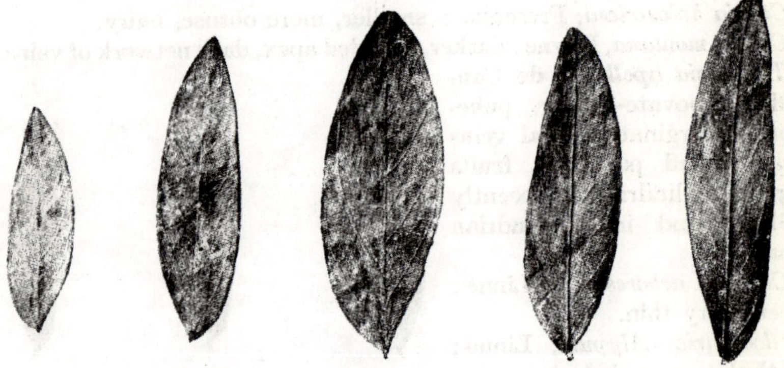

FIG. 24.—Tinnevelly Senna leaves. Natural size.

**Constituents.**—The constituents of Tinnevelly senna are identical with those of Alexandrian senna. Either variety may be used in making the official preparations of senna.

## CHERRY-LAUREL LEAVES

### (Folia Laurocerasi)

**Source, &c.**—The cherry-laurel, *Prunus Laurocerăsus*, Linné (N.O. *Rosaceæ*), is an evergreen shrub indigenous to Persia and Asia Minor but cultivated in most temperate regions. It was introduced into this country towards the end of the sixteenth century, but the poisonous property of the water distilled from the leaves was not known till 1731; the poisonous principle contained in it was identified as hydrocyanic acid soon after the discovery of the latter by Scheele in 1782.

**Description.**—Cherry-laurel leaves which are official in the fresh state only, average about 15 cm. in length by 5 cm. in breadth. The upper surface is dark green and glossy, the under surface paler. In outline they vary from oblong-lanceolate to nearly obovate, tapering towards both base and apex, the latter being shortly acute and recurved. The leaves are thick and coriaceous in texture, and are supported on short stout petioles. The margin is slightly recurved and is provided at intervals with short strong serrations. On the under surface of the leaf, near the base, and on either side of the midrib are from one to four yellowish depressed spots; these spots are the remains of glands in which a sugary substance is produced whilst the leaf is young. Both surfaces of the leaf are quite glabrous.

When fresh and entire, cherry-laurel leaves are almost inodorous; but when crushed, or even scratched with a needle, the young leaves evolve an odour recalling that of oil of bitter almond and hydrocyanic acid. Old leaves, or dried leaves crushed and moistened with water, yield a much less perceptible odour.

FIG. 25.—Cherry-laurel leaf. Under surface. Slightly reduced.

**Constituents.**—The principal constituent of cherry-laurel leaves is a glucoside. laurocerasin, first isolated by Lehmann (1885), and subsequently by Bourquelot and Hérissey, who further investigated it and called it prulaurasin, by which name it is now generally known. Prulaurasin, $C_{14}H_{17}NO_6$, has been obtained in colourless, odourless, bitter prisms (m.pt. 120°; O.R. = −53°·6). In contact with prunase (an enzyme also contained in the leaves) and water, prulaurasin is decomposed, yielding benzaldehyde (oil of bitter almonds), hydrocyanic acid, and dextrose.

$$\underset{\text{Prulaurasin}}{C_{14}H_{17}NO_6} + \underset{\text{Water}}{H_2O} = \underset{\text{Dextrose}}{C_6H_{12}O_6} + \underset{\text{Hydrocyanic acid}}{HCN} + \underset{\text{Benzaldehyde}}{C_6H_5CHO}.$$

As long as the leaves remain intact no decomposition takes place, for the emulsin is stored in the endodermis of the veins, whilst the prulaurasin is distributed through the parenchyma of the leaf. The two bodies cannot therefore come into contact with one another until the cells containing them are broken; directly this is effected the reaction takes place.

Fresh leaves from healthy plants yield on an average about 0·1 per cent. of hydrocyanic acid. In the autumn the yield is less than in the spring or early summer, the young leaves being especially rich in prulaurasin; newly unfolded leaves have yielded as much as 2·49 per cent, of hydrocyanic acid (Ven, 1898).

Prulaurasin resembles but is not identical with amygdalin. By acting upon amygdalin with yeast Fischer obtained an amygdonitrile glucoside, $C_6H_5.CH<{CN \atop O.C_6H_{11}O_5}$. This body is isomeric with prulaurasin, but differs in its optical rotation (—26°·9), and yields, when acted upon by hydrochloric acid, lævo-phenylglycollic acid (mandelic acid), $C_6H_5.CH<{OH \atop COOH}$. Under similar conditions prulaurasin yields inactive (racemic) phenylglycollic acid Hence prulaurasin is the glucoside of racemic phenylglycollic acid, and the amygdonitrile glucoside that of lævo-phenylglycollic acid. Sambunigrin, an allied glucoside contained in the leaves of the elder, is the glucoside of dextro-phenylglycollic acid. Both the amygdonitrile glucoside and sambunigrin are readily converted into prulaurasin by dilute alkalies.

Several other glucosides have been discovered (e.g. manihotoxin, lotusin, dhurrin, phaseolunatin, gynocardin, linamarin, corynocarpin) which under the influence of enzymes yield hydrocyanic acid; they are collectively termed 'cyanogenetic' glucosides.

**Uses.**—Cherry-laurel water is employed as a sedative, its action being due to the hydrocyanic acid it contains.

## WITCH-HAZEL LEAVES

### (Folia Hamamelidis)

**Source &c.**—The witch-hazel, *Hamamelis virginiana*, Linné (N.O. *Hamamelidĕæ*), is a common shrub in the United States and Canada. It attains a height of about 3 metres, and resembles the common hazel both in its leaves and in its fruit, which is edible. Both the fresh and dried leaves are official, the former being employed for the preparation of Liquor Hamamelidis. The bark is also used in medicine.

**Description.**—Commercial witch-hazel leaves are usually in a somewhat indifferent state of preservation, being frequently discoloured, broken, and pressed together into more or less compact masses. Well-preserved full-grown leaves are of a dark green or brownish green colour, and attain 15 cm. in length by 10 cm. in breadth. They are broadly oval or rounded-obovate, the lamina tapering somewhat towards the base, where it is decidedly oblique, usually cordate, and provided with a short petiole; the apex is acuminate, but is often imperfect and then frequently obtuse. The margin is coarsely crenate or even sinuate. On the under surface the midrib is prominent, and the lateral veins, which are also very distinct, branch from it at an acute angle, and run straight to and terminate in the crenations of the margin. In the angles thus formed and on the veins, hairs, which will be seen under the lens to have a characteristic branching form, are usually to be found; they are more frequent on young leaves, very young leaves being brown in colour and densely hairy. The leaves have only a slight odour, but a decidedly astringent and somewhat bitter taste.

The student should observe

(*a*) The *sinuate margin*,
(*b*) The *lateral veins running straight to the margin*,
(*c*) The *branching hairs*.

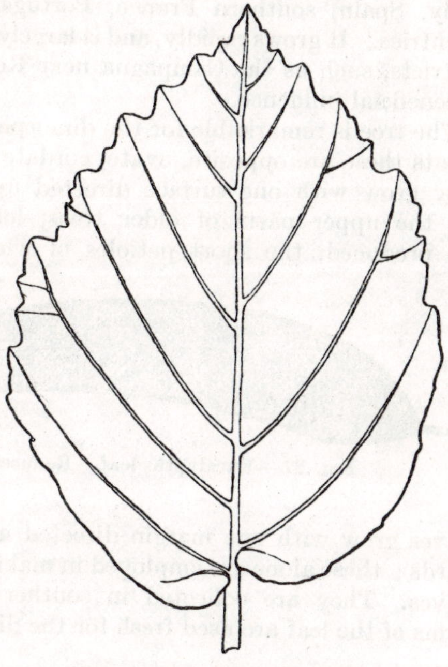

FIG. 26.—Witch-hazel leaf.

**Constituents.**—Witch-hazel leaves have not yet been subjected to careful analysis. They contain gallic acid, tannin, a bitter principle, and a trace of volatile oil. By distilling the fresh or dry leaves with water or dilute alcohol, a distillate is obtained possessing a distinct aroma, different from that of the leaves themselves, due probably to some product of decomposition. The official Liquor Hamamelidis, which is made by macerating the fresh leaves with dilute alcohol and distilling, contains, in addition to the aromatic body alluded to, a trace of protocatechuic acid.

Hamamelin is a mixture of substances obtained by extracting the leaves, or sometimes the bark, with strong alcohol and evaporating the tincture thus obtained.

The autumnal leaves are said to contain most tannin and to be distinguishable by the hairs, which have thicker walls, are yellow in colour, and devoid of contents (Cooley, 1900).

**Uses.**—Witch-hazel leaves are astringent and hæmostatic; they are useful as a local application in hæmorrhage from the nose, &c.

## EUCALYPTUS LEAVES

### (Folia Eucalypti)

**Source, &c.**—Eucalyptus leaves are obtained from *Eucalyptus Globŭlus*, Labillardier (N.O. *Myrtaceæ*), the ordinary 'blue gum' tree of Victoria and Tasmania. This tree, which is one of the largest known, attains a height of over 100 metres; it is cultivated in Italy, Spain, southern France, Portugal, Algeria, and other warm countries. It grows rapidly, and is largely planted in unhealthy marshy districts, such as the Campagna near Rome, upon which it exercises a beneficial influence.

The tree is remarkable for the dimorphism of its leaves. On young plants these are opposite, ovate, cordate at the base, and sessile, and they grow with one surface directed upwards and one downwards. On the upper parts of older trees, longer, scimitar-shaped leaves are produced, the short petioles of which are twisted so that the

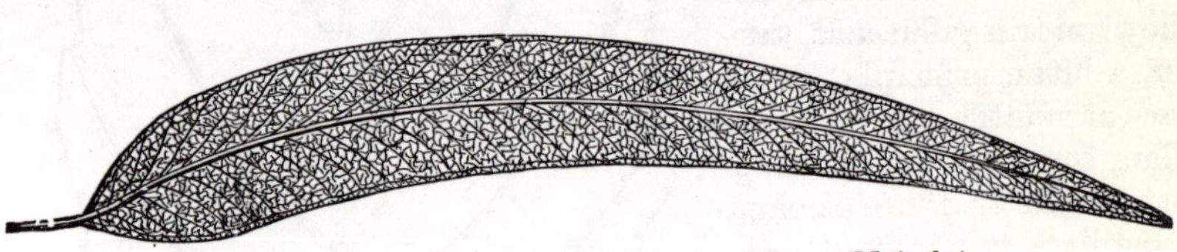

FIG. 27.—Eucalyptus leaf. Reduced in size. (Maisch.)

leaves grow with one margin directed upwards and the other downwards; these alone are employed in making preparations of eucalyptus leaves. They are collected in southern Europe and dried. Both forms of the leaf are used fresh for the distillation of the volatile oil.

**Description.**—Eucalyptus leaves are well characterised by their great length (up to 30 cm.) and narrow, ensiform outline. They taper gradually towards the apex, but narrow rather abruptly at the base into a short, twisted petiole. The lamina is thick, coriaceous, and, when quite dry, brittle. The margin is entire and somewhat thickened; the midrib is not prominent on either surface, and the lateral veins, most of which leave the midrib at an acute angle, anastomose near the margin to a continuous line. The leaves are quite glabrous, but distinctly punctate from the presence of numerous oil-glands situated in the mesophyll; these are best observed by examining the surface with a lens whilst the leaf is held against a strong light. The surfaces are frequently marked with a number of minute warty brown spots (groups of cork cells).

The odour of fresh eucalyptus leaves is strong, camphoraceous, and characteristic; in the dry leaves it is less perceptible until they are crushed. The taste of the dry leaves is aromatic, pungent, and slightly bitter.

The student should observe

(a) The *ensiform outline*,
(b) The *coriaceous texture*,
(c) The *similarity between the two surfaces*.

**Constituents.**—Eucalyptus leaves contain, when fresh from 3 to 5 per cent. of volatile oil containing 50 or more per cent. of cineol (eucalyptol, cajeputol). They also contain tannin, a bitter principle which has not yet been investigated, and several resins, one of which is crystalline.

**Uses.**—Eucalyptus leaves are used as an astringent ; they have also been employed in the form of a cigarette for asthma. The volatile oil has antiseptic properties.

**Note.**—The official eucalyptus oil may be obtained from *E. Globulus*, *E. dumosa*, A. Cunningham, and other species. Much is imported from Australia. Citron-scented eucalyptus oil is obtained from *E. citriodora*, Hooker (Queensland).

## BEARBERRY LEAVES

### (Folia Uvæ Ursi)

**Source &c.**—The common bearberry, *Arctostaphylos Uva-ursi*, Sprengel (N.O. *Ericaceæ*), is a small procumbent ever-green shrub distributed throughout central and northern Europe and North America. It is indigenous to Great Britain, but is confined to Scotland, the north of England, and Ireland. The plant sends out branching stems that take root, and so forms small clumps. The drug was probably in use long ago in this country but was first introduced into the London Pharmacopœia in 1788.

**Description.**—Bearberry leaves are small, shining, coriaceous leaves, seldom measuring as much as 25 mm. in length by 12 mm. in breadth. The upper surface is dark green or yellowish green, the under surface paler. They are spathulate or obovate, the lamina being rounded at the apex, but tapering gradually towards the base to a short petiole. They are more or less rigid, and, when quite dry, brittle. The margin is entire, slightly revolute, and in young leaves ciliate with short hairs, but these are scarcely discernible in the drug. The veins and veinlets, are depressed on the upper surface, which thus assumes a chequered or wrinkled appearance ; the greyish green under surface is reticulately marked with somewhat darker veins, which are often slightly raised, but it does not show raised brown points when examined with a lens. The leaves have no marked odour, but are strongly astringent and somewhat bitter.

The student should observe

(*a*) The *spathulate outline*, *entire margin*, and *rounded apex*,

(*b*) The *veinlets depressed on the upper surface*,

(*c*) The *absence of brown points on the under surface*;

and should compare the leaves with

(i) *Buchu* leaves, which have a toothed margin,

(ii) The substitutes mentioned below.

FIG. 28.—Bearberry. Reduced to about one-fourth natural size. (Maisch.)

**Constituents.**—Bearberry leaves contain both tannin and gallic acid; an infusion of the leaves accordingly gives a bluish black precipitate with ferric salts. They also contain arbutin, methyl-arbutin, ursone, quercetin, and possibly myricetin.

Arbutin, $C_{12}H_{16}O_7,\frac{1}{2}H_2O$, crystallises in long white bitter needles melting at 168° and easily soluble in boiling water and in alcohol; when hydrolysed with dilute sulphuric acid (or with emulsin) it yields dextrose and hydroquinone. A similar decomposition takes place when arbutin is administered by the mouth, both arbutin and hydroquinone being excreted by the urine; in fact, the activity of bearberry leaves is said to be partly due to the stimulant and antiseptic

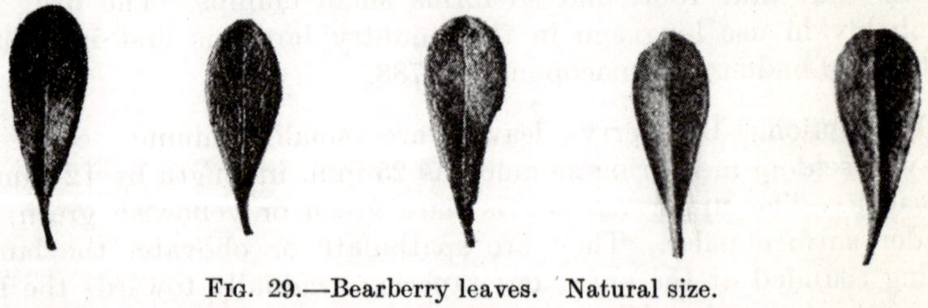

FIG. 29.—Bearberry leaves. Natural size.

properties of the latter substance. Arbutin has also been detected in several other Ericaceous plants.

Ursone, $C_{30}H_{48}O_3,2H_2O$, has been obtained in tasteless, colourless, odourless crystals.

Quercetin, $C_{15}H_{10}O_7,2H_2O$, is a yellow crystalline body which occurs in the fruits of *Rhamnus infectoria*, Linné, &c., and is also obtained when quercitrin, one of the constituents of quercitron bark (*Quercus discolor*, Aiton), is hydrolysed by boiling with dilute mineral acids.

Ericolin, an ill-defined amorphous substance said to occur in this and other Ericaceous plants, is probably a mixture.

**Substitutes.**—Bearberry leaves are seldom adulterated; the leaves of the box (*Buxus sempervirens*, Linné) and of the cowberry (*Vaccinium Vitis-idœa*, Linné) are reported to have been used for that purpose, but are easily distinguished, as the former are emarginate at the apex, and the latter have brown dots scattered over the under surface of the leaf.

**Uses.**—Bearberry leaves are used as a stimulant, diuretic, and antiseptic in diseases of the urino-genital tract; they resemble buchu in their action, but are more astringent.

## BELLADONNA LEAVES

### (Folia Belladonnæ)

**Source, &c.**—The deadly nightshade, or belladonna, *Atrŏpa Belladonna*, Linné (N.O. *Solanaceæ*), is a tall branching herb with a perennial root widely distributed over central and southern Europe; in England it is confined chiefly to the southern counties; it also widely cultivated not only in this country but also on the continent, in India, in the United States, etc.

The poisonous properties of belladonna were certainly known towards the end of the fifteenth century, probably much earlier. The leaves were introduced into the London Pharmacopœia of 1809, the root coming into use much later.

The fresh leaves and branches of the plant are used for preparing the juice and green extract of belladonna but only the dried leaves are official. Considerable quantities are dried in this country; much is also imported.

**Description.**—The stem of the belladonna plant divides usually a little above the ground into three large branches, each of which branches again freely. The lower leaves are solitary, but the upper are arranged in pairs, consisting of a large and a small leaf, the latter being really a leafy bract. The flowers are solitary and pendulous; they have a dull purplish, campanulate corolla and superior, two-celled, many-seeded ovary, which develops into a purplish black, pulpy fruit. The leaves vary from 8 to 20 cm. in length, They are ovate or broadly ovate, acute, entire, tapering towards the base, and glabrous or nearly so, except when quite young, when, in common with the young stems, they are softly pubescent. Both the midrib and the lateral veins are prominent on the under surface.

The dried leaves are thin, papery, and brittle; they are usually of a dull green colour, paler below. Their surface exhibits under the lens numerous whitish elevated points; these are caused by cells filled with calcium oxalate, which produce, when the tissue of the leaf dries and shrinks, the appearance described. Neither all specimens of the drug nor both surfaces exhibit this character equally well.

The leaves should be collected when the plant is in flower, as they are then richest in alkaloid. For the preparation of the extract the larger stalks should be rejected, as it has been shown that the extract

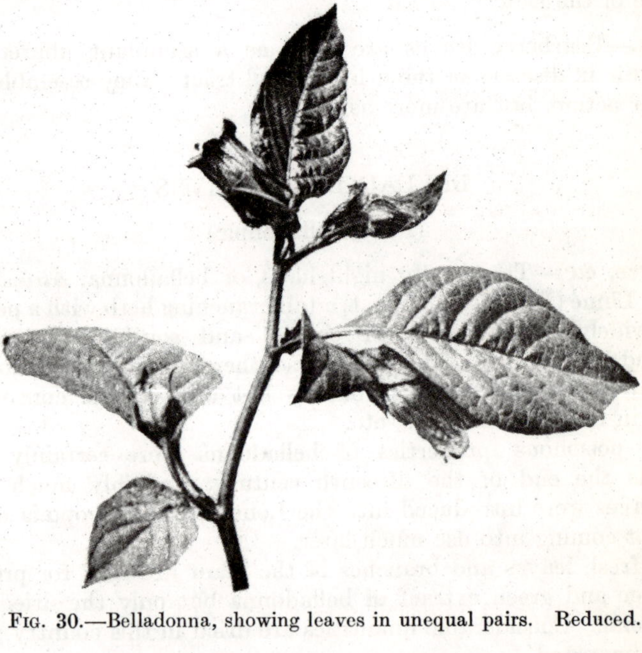

FIG. 30.—Belladonna, showing leaves in unequal pairs. Reduced.

prepared from them contains less alkaloid than that prepared from the leaves (Ranwez, 1897). For the same reason the dry drug should be as free as possible from stalk.

**Microscopical Characters.**—The transverse section of the midrib exhibits strands of bast both above and below the wood (bi-collateral bundles), a feature common in *Solanaceæ* and in a few other orders, but by no means of frequent occurrence, and hence of use in identifying the leaves. There are no bast-fibres or sclerenchymatous cells. Some of the cells of the mesophyll are filled with numerous, minute crystals of calcium oxalate, best seen when portions of the leaves are warmed with solution of chloral hydrate (5 parts) in water (2 parts), or with a mixture of cresol (2 parts) with glycerin (1 part), and examined under the microscope. This is an important character, as it enables these leaves to be distinguished from stramonium, henbane, foxglove, and many others. Each stoma is surrounded by three or four cells, one of which is smaller than the others; the epidermal cells have sinuous walls and striated cuticle.

The student should observe

(a) The *upper leaves in unequal pairs*,

(b) The *glabrous surface*,

(c) The *presence of whitish raised points* (on the dried leaves);

and should compare the dried leaves with

(i) *Stramonium leaves*, which curl and twist as they dry, have a characteristic odour and sinuate-dentate outline,

(ii) *Foxglove leaves*, which are hairy,

(iii) *Henbane leaves*, which are hairy.

**Constituents.**—The chief constituents of the drug are the alkaloids hyoscyamine and atropine. Good samples contain about 0·4 per cent. of total alkaloid, but as much as 1 per cent. has (exceptionally) been found. The alkaloids are contained in all parts of the plant; in the calices and young ovaries 0·79 per cent. has been found, in the ripe seeds 0·83 per cent., in the root 0·5 per cent., and in fresh fruit 0·12 per cent.

Belladonna leaves also contain a fluorescent substance, $\beta$-methylæsculetin (scopoletin, chrysatropic acid), and yield about 14 per cent. of ash.

Fig. 31.—Belladonna leaf. Under surface, showing the venation. Natural size.

Hyoscyamine, $C_{17}H_{23}NO_3$, forms colourless crystals melting at 108·5°; it is lævorotatory $[a]_D = -29°·97$), and is converted by heating to 110° into its isomer atropine. Atropine differs from hyoscyamine in having a higher melting-point, viz. 115°, and in being optically inactive; it is the racemic modification and can be resolved into laevo- and dextro-hyoscyamine, the latter being much less active than laevo-hyoscyamine. Both hyoscyamine and atropine yield tropine and tropic acid when hydrolysed by heating with baryta water. Hyoscyamine crystallises less readily than atropine, is more soluble in water and dilute alcohol, and yields with gold chloride an aurichloride ($C_{17}H_{23}NO_3$,HCl,$AuCl_3$), crystallising in brilliant yellow scales, melting at 160°, whereas the corresponding atropine gold salt crystallises in dull warty prisms melting at 136°. Other alkaloids (belladonnine, apoatropine) may possibly also be present, but it is more probable that these are formed from hyoscyamine during the process of extraction. Hyoscyamine is also found in *Hyoscyamus niger*, in *Hyoscyamus muticus* (see p. 53), in *Datura Stramonium*, as well as in other species of *Datura*, and is often accompanied by scopolamine (hyoscine) (compare 'Belladonna Root'). For the distribution of the alkaloid in the various parts of the plant compare 'Stramonium Leaves.'

Duboisine, the alkaloid from the Australian *Duboisia myoporoides* appears to consist of hyoscyamine or scopolamine according to the variety of drug from which it is obtained.

**Adulterations.**—(1) The leaves of *Scopŏla carniōlica*, Jacquin, a Solanaceous plant widely distributed in Hungary; they closely resemble belladonna leaves, but are more lanceolate, more translucent, brighter in colour, and with more distinct veinlets; the crystal cells are less numerous, hairs are absent, and stomata occur on the under surface only; they are usually accompanied by the fruit which is a nearly spherical pyxis almost completely covered by the papery calyx, that of belladonna being a berry; they contain hyoscyamine and scopolamine (0·5 per cent.)

(2) The leaves of *Phytolacca decandra*, Linné; these may be distinguished by containing acicular crystals of calcium oxalate, and by the epidermal cells which are elongated and straight-walled.

(3) The leaves of *Ailanthus glandulosa*, Desfontaines (N.O. *Simarubeæ*); these have cluster-crystals of calcium oxalate along the veins.

Other leaves have also been found in the commercial drug, but the microscopical characters of the true drug generally suffice for its identification.

**Uses.**—Belladonna acts as a local anæsthetic and anodyne, and is used externally to relieve pain. Internally, it is given to check the sweating in phthisis, as a sedative to the respiratory nerves, to relieve spasmodic cough, and in numerous other cases.

## STRAMONIUM LEAVES

### (Folia Stramonii)

**Source, &c.**—The thornapple, *Datūra Stramonium*, Linné (N.O. *Solanaceæ*), is a large bushy annual attaining about 1 metre in height, a native probably of the shores of the Caspian Sea, but found commonly on waste ground throughout the temperate and warmer regions, abundant in South Africa, and met with occasionally in England, escaped from cultivation. It is cultivated in this country, but our supply of the drug is derived largely from Germany, France, and Hungary. The introduction of stramonium into medicine is due chiefly to the exertions of Störck in the latter half of the eighteenth century. The leaves only are official, but usually the young shoots are collected when the plant is in flower and dried.

**Description.**—The stem of the thornapple is stout and erect, branching repeatedly, and producing in the forks of the branches a leaf and a single flower with a tubular calyx and large, white, funnel-shaped corolla. This flower is succeeded by an erect, spiny capsule about the size of a walnut. In addition to mature and young leaves the drug therefore frequently contains the remains of the flower with a tubular calyx and yellowish corolla, or of the young, shrivelled, bristly fruit.

The dried leaves are usually much shrivelled and wrinkled; they are of a dark greyish green colour, especially on the upper surface, ovate in outline, petiolate, and often unequal at the base. They attain 20 cm. or more in length, and are characterised by the very coarse pointed teeth which impart to the margin a sinuate-dentate outline; the apex of the leaves is shortly acuminate. On the under surface the midrib is distinct though not very prominent; the lateral veins leave it at an angle of about 45° and divide when near the margin, one branch passing into the pointed lobe of the leaf, whilst the other anastomoses with other, tertiary, veins. When quite young the leaves are covered with stout curved hairs, but as they reach maturity these fall off, and full-grown leaves are glabrous or nearly so. The odour of the drug, though not strong, is disagreeable and characteristic; the taste is unpleasantly bitter.

FIG. 32.—Thornapple leaf. Under surface, showing the venation. Natural size.

**Microscopical Characters.**—The distinctive characters of stramonium leaves are the abundant cluster crystals of calcium oxalate in the mesophyll; the hairs though not numerous, are also characteristic, being either simple, uniserial, three- to five-celled, and warty, or small, shortly stalked, and glandular. The leaves possess in addition typical Solanaceous stomata and bicollateral bundles devoid of bast-fibres.

The student should carefully observe

(*a*) The *curled and twisted appearance of the dried leaves*,
(*b*) The *angle made by the lateral veins with the midrib*,
(*c*) The occasional presence of the remains of the *flower and fruit*,
(*d*) The *characteristic odour*, and,
(*e*) After soaking in water, the *sinuate-dentate outline* (unless the leaves are too much broken);

and should compare them with

(i) *Belladonna leaves* (see p. 43),
(ii) *Henbane leaves* (see p. 49),
(iii) *Foxglove leaves* (see p. 53).

**Constituents.**—Stramonium leaves contain the same alkaloids as belladonna leaves, but in somewhat smaller proportion, the average of commercial samples being about 0·22 per cent.; the percentage may, however, rise to 0·4 or exceptionally to 0·7 per cent.; South African leaves 0·54 per cent., Egyptian 0·35 per cent (hyoscyamine only). Daturine was the name given to the mixture of alkaloids originally extracted from the drug.

The alkaloids in the leaf are localised chiefly in the epidermis, particularly the upper, and in the bast parenchyma of the veins, the midrib containing more than the petiole, and both being much richer than the leaf. Hence the practice sometimes followed of rubbing the leaves through a coarse sieve (laminating), and rejecting the midribs and larger veins should be discontinued, especially as thereby the difficulty of identifying the leaves is increased. The main stem contains but little alkaloid, and therefore should not be present in the drug.

**Assay.**—The drug may be assayed by the process official for belladonna leaves.

**Adulterations.**—*Xanthium strumarium* (N.O. *Compositæ*); short, conical hairs containing cystoliths, uniserial, simple hairs and globular, multicellular, glandular hairs; no cluster crystals of calcium oxalate.

*Xanthium macrocarpum*, de Candolle; said to possess three forms of hairs, two protective and one glandular, one of the former being short and lignified; cluster-crystals smaller and less numerous; transverse section of the midrib shows six bundles (*D. Stramonium* has only one).

*Carthamus helenioides*, Desfontaines; epidermal cells large, walls straight, cuticle striated; large, pluricellular protective hairs, small glandular hairs arising from the whole surface of an epidermal cell (in *D. Stramonium* from a small spot); no cluster-crystals;. well-developed secreting ducts.

*Chenopodium hybridum*, Linné; cluster-crystals abundant; epidermal cells small, walls nearly straight; occasional hairs with slender pedicel and large, bladdery, water-storing terminal cell.

*Solanum nigrum*, Linné; hairs resemble stramonium; contains no cluster-crystals.

**Uses.**—Stramonium leaves resemble belladonna in their action; they are, however, almost exclusively used in the treatment of spasmodic affections of the respiratory organs.

## DATURA LEAVES

### (Folia Daturæ)

**Source, &c.**—The official Datura leaves are the dried leaves of *D. fastuosa*, Linné, var. *alba*, Nees, and also of *D. Metel*, Linné, annual plants indigenous to India where the leaves are used as an equivalent of belladonna and thornapple. They are brownish or yellowish green, ovate, acuminate, sinuate-dentate, often unequal at the base, up to 20 cm. long and 12·5 wide, with long petioles. The odour is characteristic and unpleasant; taste bitter. The epidermis bears scattered simple hairs and stalked glandular hairs.

They contain about 0·5 per cent. of alkaloid, chiefly scopolamine (hyoscine) with traces of hyoscyamine and atropine. They serve as a commercial source of scopolamine.

**Note.**—*Datura Tatula,* Linné, is a variety of *D. Stramonium* with a violet corolla and purplish veins; the leaves are used together with stramonium leaves for making asthma cigarettes.

## HENBANE LEAVES

### (Folia Hyoscyami)

**Source, &c.**—The common henbane, *Hyoscyamus niger*, Linné (N.O. *Solanaceæ*), is an erect herb attaining a height of about 1 metre, distributed over the whole of Europe and extending to Persia and India. In England it is found chiefly on waste places near buildings; it is cultivated in this country for medicinal use, but much is imported from Germany and Russia. The medicinal use of the plant dates from very remote ages. It was well known to the Anglo-Saxons in the tenth and eleventh centuries, but subsequently fell into disuse. It was omitted from the London Pharmacopœia in 1746, but restored in 1809 chiefly by the influence of Störck, who also introduced stramonium.

Two varieties of the plant are known, an annual and a binneial, the distinction between them being by no means well marked.

Biennial henbane produces in the first year of its growth simply a rosette of large stalked leaves attaining 30 cm. in length, some of which are frequently collected and form the drug known commercially as 'first biennial henbane.' In the second year the plant sends up a large branching stem attaining a height of about 1·5 metres, which flowers, ripens its seeds, and dies. The smaller branches of this plant with the leaves and flowers constitute the drug from which the green extract of henbane and juice of henbane should be prepared; the flowering tops are dried and sold; the leaves separated from the stalks and dried constitute in part the official drug.

Fig. 33.—Henbane (*H. niger*). Flowering top of a large second year's plant of biennial Henbane. Reduced.

Annual henbane is a much smaller plant than the biennial. The stem does not branch as that of the biennial does, nor are the leaves so large or so deeply incised; the corolla is paler in colour and less deeply veined with purple. This variety flowers in the first year of its growth, ripens its seeds, and dies. The dried leaves are included in the official description, but the commercial drug frequently contains a large proportion of the stem.

Four commercial varieties of henbane leaves must therefore be distinguished, viz.:

'*First biennial*,' consisting of the leaves of the first year's growth of the biennial variety.

'*Second biennial*,' or '*biennial*,' consisting of the flowering tops of the second year's growth of the biennial variety.

'*Annual*,' consisting of the stem with leaves and flowers of the annual variety.

'*Official*' (*B.P. 1914*), consisting of the leaves collected from the flowering plant, whether biennial or annual.

**Description.**—As previously mentioned, in the second year of its growth the biennial variety of the plant produces a tall, stout, branching stem with leaves and flowers.

The leaves vary considerably in size. The lower attain as much as 25 cm. in length, and are stalked ; the upper are smaller and sessile. They are pale green in colour, and, especially when fresh, soft and unpleasantly clammy or sticky to the touch. This peculiarity is due to the soft hairs which are particularly abundant near the veins on the under surface ; these hairs possess glandular heads which secrete a resinous substance.

FIG. 34.—Annual Henbane. Small specimen. Reduced.

In outline the leaves vary from nearly ovate to elongated triangular ; the margin is coarsely dentate or even pinnatifid, and the midrib broad ; the stems are rounded, and bear glandular hairs like those of the leaf.

The flowers, which are usually crowded together, arise from the axils of large, hairy, leafy bracts ; they possess a hairy, urceolate calyx and a yellowish, gamopetalous corolla deeply veined with purple. The fruit is a two-celled pyxis containing numerous seeds.

The dried flowering tops are commonly found in irregular rounded or flattened masses about 2·5 to 5 cm. in diameter, in which the coarsely dentate hairy bracts, the yellowish corolla, with deep purple veins, and two-celled ovary with numerous ovules can easily be identified.

The fresh plant has a strong heavy odour, which is less perceptible in the dry drug ; but the taste of the latter is more distinctly bitter than that of the former.

**Microscopical Characters.**—The epidermis closely resembles that of belladonna and of stramonium, but bears numerous, very long, uniserial, multicellular hairs, some of which are simple, but the majority terminate in oval multicellular glands; the mesophyll contains calcium oxalate chiefly in the form of prismatic, often twin, crystals; in some specimens small cluster crystals are fairly numerous, and sandy crystals may occur in or near the midrib; the bundles are bicollateral, and there are no pericyclic or bast fibres.

The student should observe

(*a*) The *hairy leaves with sinuate-dentate outline and broad midrib*,
(*b*) The *purple veins of the corolla*,
(*c*) The *characteristic fruit* (or ovary).

**Constituents.**—The principal alkaloid in henbane leaves is hyoscyamine, but it appears to be accompanied by small proportions of atropine and scopolamine (hyoscine). The drug contains much less total alkaloid than either belladonna or stramonium, viz. from 0·045 to 0·14 or, exceptionally, 0·2 per cent. (of the dry drug). The cultivated plant contains about the same proportion as the wild, but the leaves contain more than the large stems and the extract made from the leaves, more than that made from the stem. The petiole is, however, rather richer than the lamina. Henbane should yield about 8 to 12 per cent. of ash, but commercial samples sometimes afford much more.

**Assay.**—Henbane may be assayed by the process official for belladonna leaves, using 25 gm. of the powdered drug instead of 10.

**Uses.**—The action and uses of henbane closely resemble those of belladonna and stramonium, but the drug is distinctly weaker. The extract has a decided laxative and carminative effect, whilst the tincture has a more marked action on the urinary organs.

**Varieties.**—*The first year's leaves* of biennial henbane may be recognised by their being longer and relatively narrower, stalked, and free from stem and flower. They are equal in activity to the leaves from flowering plants, and there seems to be no good reason why they should not be used. They are sometimes sold for annual henbane.

*Annual henbane* is distinguished by its slender simple stem, smaller leaves, and paler corolla, with less distinct purple venation. As in this case the entire plant is usually cut and dried, portions of the stem as well as leaf and flower are often found in the drug. Much imported henbane is of this variety, and it often arrives in very poor condition, probably mixed with other species of *Hyoscyamus*. Such henbane contains notably less alkaloid than either the first or second biennial, viz. about 0·03 per cent. English annual henbane consists of such plants as happen under favourable conditions to flower in the first year; such plants are much stronger than the foreign, and being more carefully dried are richer in alkaloid.

*H. muticus*, Linné, Egyptian henbane (Egypt, India); leaves entire or toothed, petiolate; calyx striated, pubescent; corolla yellow or nearly white; pyxis cylindrical not urn-shaped. The commercial drug is imported in large quantity from Egypt, often with a considerable proportion of stout, yellowish stalk; used as source of hyoscyamine (0·2 to 1·0 per cent.).

## FOXGLOVE LEAVES

### (Folia Digitalis)

**Source, &c.**—The purple foxglove, *Digitalis purpūrea*, Linné (N.O. *Scrophularineæ*), is a handsome biennial herb, widely distributed throughout Europe and common in England, where it occurs wild and is also cultivated as a garden plant, as well as for medicinal use. Much of the commercial drug was formerly imported from Germany, but latterly (1915–1918) large quantities have been collected and dried in this country. It appears to have been long used as a domestic medicine; it was introduced into the London Pharmacopœia as long ago as 1650, although it did not come into frequent use until about a century later.

The foxglove produces, like henbane, in the first year a rosette of leaves, but no aerial stem; in the second year a tall, erect, usually simple stem that may attain a height of 2 metres or more, and bear numerous flowers. The latter are well characterised by their crimson bell-shaped corolla, with darker spots on the inner part of the mouth; the ovary is conical and contains two cells with numerous ovules; the stamens are didynamous.

The plant flowers during the summer months, and the leaves for official use are directed to be gathered from plants commencing to flower, when the presence of the flowers precludes any possibility of the leaves of other plants being collected by mistake. They should be completely dried immediately after collection and preserved quite dry in air-tight containers, when they will retain their activity unimpaired; air-dry leaves are said to deteriorate rapidly but on this point the evidence is conflicting.

**Description.**—Foxglove leaves vary usually from 10 to 30 cm. in length, but may attain as much as 15 cm. in breadth; in shape they vary from broadly ovate to lanceolate, those on the upper part of the stem being the narrower. Towards the base of the leaf the lamina is contracted and passes into a winged petiole of varying length, down which the lower lateral veins are usually decurrent, the petioles of the lower leaves being longer than those of the upper. The upper surface is dull green in colour and bears numerous short hairs, the under surface paler and more or less densely pubescent, the hairs being seen under the lens to be simple and unbranched. The midrib is prominent on the under surface ; the majority of the lateral veins leave the midrib at a rather acute angle (especially in the narrower leaves) and gradually curve round towards the apex, passing into smaller ramifications near the margin. The latter is crenate or irregularly crenate-dentate, the apex blunt or subacute. Each crenation terminates in a small cartilaginous point which is brown or black in old leaves but colourless in young ; a veinlet enters each crenation and spreads like a brush, a marginal veinlet approaching from either side, the whole forming a well-marked character of the foxglove leaf. The odour of the fresh leaves is unpleasant, and the taste of both fresh and dried leaves disagreeably bitter.

FIG. 35.—Foxglove leaf. Under surface, showing the lower veins decurrent in the petiole. About two-thirds natural size.

Although the leaves are directed to be collected from flowering plants, it is not possible to distinguish these accurately from those of the first year ; generally speaking, the latter are narrower and have longer petioles, whilst the biennial leaves are usually broader and have shorter petioles. The separation is, however, not material, as the first and second year's leaves are about equal in activity.

**Microscopical Characters.**—Under epidermis composed of cells with wavy walls and smooth cuticle ; stomata numerous and small ; hairs abundant and of two kinds, either simple, uniserial, three to five or more cells, thin-walled, often warty, and frequently collapsed ; or small, stalked and glandular. Midrib normal in structure and free from bast or pericylic fibres. No crystals of calcium oxalate in any part of the leaf.

The student should direct his attention particularly to

(*a*) The *crenate margin*,
(*b*) The *winged petiole with decurrent veins*,

(*c*) The *simple unbranched hairs*,
(*d*) The *course taken by the lateral veins*;

and should compare these leaves with

(i) *Matico leaves*, in which the veinlets are depressed on the upper surface, dividing it into small squares,
(ii) The possible substitutions mentioned below.

**Constituents.**—Although the constituents of foxglove have repeatedly been the subject of research, our knowledge of them remains incomplete. The activity of the leaves appears to be due chiefly to digitoxin, digitalein (gitalin, Kraft) and digitophyllin; other constituents are saponins, digitoflavone (luteolin), an irritant resin (digitalic acid), and a very active oxydase.

Digitoxin, $C_{34}H_{54}O_{11}$, is a well-defined, colourless, odourless, crystalline, bitter substance, insoluble in water, but nevertheless passing into solution in appreciable quantity when foxglove leaves are infused in that menstruum. It is the most toxic of the active constituents of the leaves, and is cumulative in action, being apparently fixed by the muscles of the heart. It is hydrolysed by dilute mineral acids yielding digitoxigenin, $C_{22}H_{32}O_4$, and a sugar, digitoxose, $C_6H_{12}O_4$. It may be identified by Keller's reaction, which consists in dissolving it in glacial acetic acid, adding a drop of ferric chloride solution, and then gently a stratum of sulphuric acid; the upper part of the latter is coloured red, whilst above this an indigo blue band gradually appears.

Dry foxgloves leaves contain from 0·2 to 0·3 per cent. of digitoxin, the first year's leaves containing as much as the second. Wild plants contain more than cultivated, but the quantity rapidly diminishes towards the end of the flowering stage.

Digitalein (0·3 to 0·9 per cent.), a purified form of which is known as gitalin, is an amorphous, in hydrated form crystalline, glucoside soluble in water; it has a marked action on the heart and as it is not cumulative it is believed to be more valuable therapeutically than digitoxin. It passes readily into the infusion. By heating it changes into anhydrogitalin and by hydrolysis it yields anhydrogitaligenin and digitoxose.

*Digitophyllin*, crystallising in plates melting at 232°, is probably methyl-digitoxin and gives the same colour reaction as digitoxin.

*Digitoflavone*, (luteolin) is a yellow colouring matter allied to quercetin.

The saponins appear to comprise digitsaponin, amorphous, soluble in water and gitin, crystalline and insoluble in water. The digitsaponin accelerates the heart-beat but is non-cumulative.

The active constituents of foxglove seeds are:

Digitoxin (see above).

Digitalin, a crystalline, water-soluble, active glucoside not identical with digitalein.

Digitonin, a crystalline saponin resembling gitin but not identical with it.

The digitalins of commerce appear to be variable mixtures:

Homolle's digitalin is amorphous and a mixture of digitoxin and digitalin.

Nativelle's digitalin is crystalline and chiefly digitoxin.

German digitalin is amorphous and said to consist chiefly of digitalein.

Digitalinum verum is amorphous and said to be the best variety of digitalin for medicinal use.

Several other species of *Digitalis* have been examined; all were found to be toxic.

**Uses.**—Foxglove increases the activity of muscular tissue, especially that of the heart and arterioles, and is employed in most forms of cardiac failure. Digitoxin is cumulative, and the action of preparations of foxglove must therefore be watched.

**Assay.**—Since the therapeutical activity of foxglove leaves is due to more than one glucoside, it is obviously impossible to assay the leaves by determining one only of these, *e.g.* digitoxin. No method of determining digitalein or digitalin is known. Hence chemical methods of assay fail, and recourse must be had to the physiological assay (compare p. 168).

**Adulterations.**—*Leaves of D. Thapsi*, Linné, imported from Spain; greyish green or yellowish green, lamina less decurrent, hairs very numerous, long, curling. Said to be more toxic than *D. purpurea.*

*Mullein leaves* (*Verbascum Thapsus*, Linné); woolly; hairs branched.

*Comfrey leaves* (*Symphytum officinale*, Linné); lanceolate or ovate; isolated stiff hairs.

*Primrose leaves* (*Primula vulgaris*, Hudson); nearly spathulate; lateral veins straight, dividing near the margin.

*Ploughman's spikenard* (*Inula Conyza*, de Candolle); margin either entire or dentate, with horny points to the teeth.

*Elecampane leaves* (*Inula Helenium*, Linné); lower lateral veins not decurrent.

## LAUREL LEAVES

### (Folia Lauri)

**Source, &c.**—The true laurel, bay or bay laurel, *Laurus nobilĭs*, Linné (N.O. *Laurinĕæ*), is a small evergreen tree commonly cultivated in this country.

**Description.**—Laurel leaves are of a shining, green colour on the upper surface, paler below, coriaceous in texture, lanceolate and acuminate in outline, attaining 10 cm. or more in length and shortly stalked, with an entire, wavy margin, the apex being often acute but sometimes blunt. They are quite glabrous, with the exception of a few hairs often present on the under surface in the angles made by the lateral veins with the midrib. The upper surface appears finely shagreened; the lower is reticulated, an appearance due to the prominence of the network of minute veinlets. When crushed they emit an aromatic odour, due to the volatile oil contained in oil-cells situated in the mesophyll of the leaves. These oil-cells are with difficulty visible, even when the leaf is held against a strong light and examined with a lens. The taste is aromatic and bitter.

The student should note

(*a*) The *coriaceous texture*,
(*b*) The *entire, wavy margin* and *acute apex*,
(*c*) The *characteristic odour* ;

and should be careful not to confuse these leaves with *cherry*-laurel leaves.

**Constituents.**—The leaves contain from 1 to 3 per cent. of volatile oil, consisting chiefly of cineol (50 per cent.), eugenol, geraniol, and terpenes. This oil should not be confounded with that of *Pimenta acris*, Wight, also known as oil of bay, and from which bay rum is made. The latter oil is distilled in the West Indies.

**Uses.**—Laurel leaves are aromatic and stimulant. but are now seldom employed medicinally.

## MATICO LEAVES

### (Folia Maticæ)

**Source &c.**—Matico leaves were defined in the British Pharmacopœia of 1885, as the leaves of *Piper angustifolium*, Ruiz and Pavon (N.O. *Piperaceæ*), a climbing plant distributed over the north of South America, extending into Bolivia. The name matico appears, however, to be applied to a number of other Piperaceous plants, and the commercial drug is obtained from several species of *Piper* (see below). It has been much used in Peru as a styptic, and also for venereal disease, the leaves, together with occasional stalks and fruits, being exported in bales.

**Description.**—Matico leaves reach this country usually in brittle compressed masses of a dull, dark, greyish green or yellowish green colour. The leaves, which are very brittle, can easily be separated after soaking in water, spread out, and examined. Those of *P. angustifolium* are lanceolate in outline, about 10 to 15 cm. long and 2·5 to 4 cm. broad, and taper gradually to an acute apex ; they have short stalks, and are cordate and very unequal at the base, the lamina of one side extending over the petiole so as to conceal it. The margin is entire and revolute. On the upper surface the midrib and the network of lateral veins are so deeply depressed as to divide the surface into small raised squares about 1 mm. in diameter. On the under surface the midrib, lateral veins, and veinlets are nearly equally prominent, and both the veins and veinlets, as well as the interneural depressions, are covered with short, shaggy hairs. The latter are less prominent on the upper surface, but the leaves are not all equally hairy.

Both stalks and fruits are frequently found mixed with the leaves; the former are characterised by the swollen nodes which distinguish Piperaceous plants; the latter are long, slender, cylindrical spikes.

The leaves have a slight, aromatic odour and an aromatic, camphoraceous, somewhat bitter taste. Although the mesophyll contains oil-cells, they cannot easily be seen even with a strong lens if the leaf is very hairy.

The student should observe

(a) The *veinlets depressed on the upper surface, prominent on the under surface*,
(b) The *shaggy hairs*,
(c) The *characteristic taste*;

and should compare the leaves with

*Foxglove leaves*, in which the veinlets are not depressed and the taste not aromatic.

**Constituents.**—Matico leaves contain volatile oil (1 to 3·5 per cent.), tannin, crystalline artanthic acid, and a bitter principle which has not yet been satisfactorily examined. The volatile oil has the sp. gr. 0·93 to 0·99 and deposits on cooling matico camphor.

**Varieties.**—Leaves of the following species of *Piper* have been identified in commercial matico. They differ from those of *P. angustifolium* in appéarance, and vary also in the quantity and composition of the volatile oil:

*P. lineatum*, Ruiz and Pavon; volatile oil contains no matico camphor.

*P. camphoriferum*, de Candolle; volatile oil contains matico camphor and borneol.

*P. angustifolium*, var. *β-ossanum*; volatile oil possibly contains matico camphor and borneol.

*P. acutifolium*, var. *subverbascifolium*; volatile oil contains pinene and dill-apiol.

**Uses.**—Matico is used as an aromatic astringent in inflammatory conditions of the urinary passages.

## BETEL

### Betel.

**Source, &c.**—Betel leaves are obtained from *Piper Betle*, Linné (N.O. *Piperaceæ*), a shrub indigenous to and cultivated in India and the Malay Archipelago. The leaves are collected, put into baskets, and pressed under heavy weights, after which they are dried. They occur in commerce either loose or tied into small packets.

**Description.**—Betel leaves are brownish, broadly ovate, about 15 cm. long and 10 cm. broad, thin and brittle ; they are acuminate at the apex, unequally cordate at the base ; lateral veins well marked, 5 to 7 in number, curving from base to apex. Under the microscope abundant, rounded oil cells filled with a brownish secretion are conspicuous. Taste warm, aromatic.

**Constituents.**—The drug yields from 0·2 to 1 per cent. of volatile oil, the principal constituent of which is betel-phenol (about 70 per cent.), a substance isomeric with eugenol. Chavicol and cadinene have also been found in the oil.

**Uses.**—Fresh betel leaves are very extensively used in India and the Malay Archipelago as a masticatory and slight stimulant, commonly in conjunction with areca nut, lime, catechu, and cloves or other spice ; they are said to lose much of their volatile oil (and activity) by drying.

## LEAVES IN LESS FREQUENT USE

**Aconite Leaves.**—The leaves of *Aconitum Napellus*, Linné, (N.O. *Ranunculaceæ*). Leaves stalked, roundish in general outline, divided down to the stalk into three main segments, which are again subdivided into nearly opposite, linear, acute tapering segments ; dark green on the upper surface, glabrous ; much broken in the commercial drug, which often contains the zygomorphous flowers with dark blue, helmet-shaped sepals. No odour, taste numbing. Contain from 0·12 to 0·96 per cent. of total alkaloid, the toxic constituent of which is aconitine (compare ' Aconiti Radix ').

**Paraguay Tea or Maté.**—The leaves of *Ilex paraguayensis*, Hooker (N.O *Ilicineæ*), Brazil and Argentina. Ovate or oblong-lanceolate, 5 to 15 cm. long, coriaceous, distantly crenate-serrate, nearly glabrous ; cells of upper epidermis polygonal, with thick, striated cuticle ; stomata on under surface only, each surrounded by four or five cells ; numerous prismatic and cluster crystals of calcium oxalate ; taste bitterish and astringent ; usually imported in coarse powder ; contains 0·2 to 2 per cent. of caffeine and 10 to 16 per cent. of tannin.

**Damiana.**—Leaves of *Turnera diffusa*, Willdenow, var. *aphrodisiaca*, Urban, and probably other species of *Turnera* (N.O. *Turneraceæ*), herbs indigenous to south-western Texas and Mexico. Light green, 10 to 25 mm. long, 5 to 10 mm. wide ; broadly lanceolate, shortly petiolate ; margin dentate, with three to six teeth on either side ; veins prominent on under surface. Under the microscope long, unicellular hairs and cluster crystals of calcium oxalate ; odour and taste aromatic. Contain volatile oil, 0·5 to 1·0 per cent., amorphous bitter principle (damianin), resin, tannin. Used as an aphrodisiac but efficacy doubtful.

**Wintergreen Leaves.**—The leaves of *Gaultheria procumbens*, Linné, (N.O. *Ericaceæ*), United States. Leaves shortly stalked, elliptical or obovate, glabrous, green, brownish- or reddish-green, distantly serrate, each tooth with a bristle astringent and aromatic. Yield 0·7 per cent. of volatile oil (methyl salicylate) which is not contained in oil-cells but is formed from the glucoside gaultherin by the action of the enzyme gaultherase; contain, in addition to gaultherin, arbutin and tannin.

**Tylophora Leaves.**—The leaves of *Tylophora asthmatica*, Wight and Arnott (N.O. *Asclepiadeæ*), India. Leaves yellowish-brown, 5 to 12·5 cm. long, broadly ovate, cordate at base, shortly acuminate, margin entire, upper surface glabrous, under surface with slender, three- to five-celled hairs; little odour or taste. Contain alkaloid tylophorine; said to contain an emetic principle. Used as an expectorant and emetic in place of ipecacuanha.

**Eriodictyon** (Yerba Santa).—Leaves of *Eriodictyon californicum*, Gr en (N.O. *Hydrophyllaceæ*), California. Oblong lanceolate, 5 to 15 cm. long, 1 to 3 cm. broad, acute; margin more or less incurved, entire or irregularly serrate; upper surface yellowish green, smooth, covered with brownish resin; under surface whitish or yellowish white, reticulated and densely tomentose; odour and taste aromatic. Contain eriodictyol and four allied substances all of phenolic nature, volatile oil and resin. Used as a bitter tonic and expectorant.

**Duboisia Leaves.**—The leaves of *Duboisia myoporoides*, Robert Brown (N.O. *Solanaceæ*), Australia, New Caledonia; from them duboisine sulphate is obtained, a mixture of the sulphates chiefly of hyoscyamine and hyoscine.

**Adhatoda.**—The leaves of *Adhatoda vasica*, Nees (N.O. *Acanthaceæ*), an Indian shrub. Leaves 10 to 15 cm. long, about 4 cm. broad, entire, lanceolate, shortly petiolate, tapering at base and apex; brownish green, with characteristic odour and taste; the epidermis bears one- to three-celled, simple, warty hairs and small hairs with quadricellular secreting gland; below the epidermis there are cystoliths. Contains vasicine (alkaloid), volatile oil, and possibly a second alkaloid; used as an expectorant.

**Boldo Leaves.**—The leaves of *Peumus Boldus*, Molina, (N.O. *Monimiaceæ*), Chili. Leaves ovate or elliptical, 4 to 8 cm. long, shortly petiolate, greyish-green, coriaceous, brittle; margin entire, slightly revolute; both surfaces with numerous emergences each crowned with a group of one-celled, thick-walled hairs (easily broken off); in the mesophyll numerous oil-cells; aromatic odour; pungent, camphoraceous, bitter taste. Contain volatile oil, boldine (alkaloid), and boldoglucin (glucoside). Used as a diuretic and liver stimulant.

## SECTION II

## FLOWERS

Under this heading are grouped together a number of drugs that consist either of entire flowers or inflorescences, or of parts only of flowers, collected at varying stages of their development.

In some instances the student will find it necessary to soften the drug in water and spread out such parts as the shrivelled petals with a camel's hair brush in order to examine their shape. In other cases a sharp penknife and good lens must be used in order to observe the characteristic details alluded to in the text. For a description of the morphology of the flower and an enumeration of the various forms of inflorescence a text-book of botany must be consulted.

### RED POPPY PETALS

(Petala Rhœados, Flores Rhœados)

**Source, &c.**—The red or field poppy, *Papaver Rhœas*, Linné (N.O. *Papaveraceæ*), is a common herb, doubtfully indigenous to England, but abundant in cornfields and waste places throughout Europe and long used as a medicine. It is the commonest British poppy, and is distinguished by its rich scarlet petals and glabrous, nearly globular fruit.

The long-headed poppy, *P. dubium*, Linné, is also common, but is generally smaller, more slender, and possesses a capsule often twice as long as it is broad.

**Description.**—The two hairy sepals of the bud fall off as the four delicate crumpled petals expand. The latter are of a bright scarlet colour, with a short, dark violet claw, they are smooth and shining, broadly elliptical, with an entire margin. They have, when fresh, an unpleasant heavy odour and slightly bitter taste. By drying, the bright scarlet colour changes to a dingy violet, and the petals, which are used on account of the colouring matter they contain, are therefore official in the fresh state only.

**Constituents.**—The colour of the petals is due, in part at least, to mekocyanin, probably present in combination with an acid (see p. 65). The presence of meconic acid, an important constituent of opium, has not yet been definitely ascertained. In the juice of the capsules and herb Hesse found the alkaloid rhœadine, which is also contained in opium ; and it is worth noting that this alkaloid, when decomposed with hydrochloric acid, yields a red colouring matter of intense colorific power. Dieterich (1888) reported the drug to contain 0·1 to 0·7 per cent. of morphine, but according to Hesse and others it is quite free from morphine, Dieterich's alkaloid being rhœadine.

The petals of the *P. dubium* contain a toxic alkaloid, aporeine, resembling thebaine in its action, and they should therefore be rejected.

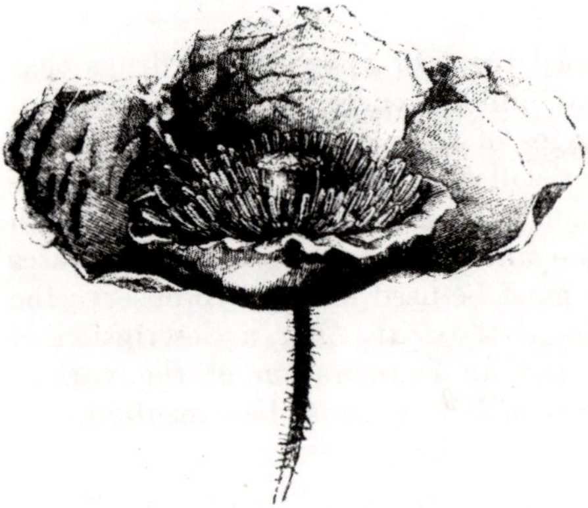

FIG. 36.—Red Poppy. (Planchon and Collin.)

**Use.**—Red poppy petals are employed solely as a colouring agent, chiefly in the form of syrup.

## KOUSSO

(Cusso ; Cousso)

**Source, &c.**—The tree yielding this drug, *Brayēra anthelmintica*, Kunth (N.O. *Rosaceæ*), is a native of north-eastern Africa. It is cultivated in Abyssinia, being commonly planted by the natives near their villages, and is in general use as a remedy for intestinal worms, from which they suffer severely. Bruce became acquainted with it in the course of his travels through Africa (about 1770).

The panicles of pistillate flowers are collected after fertilisation and dried. They are usually packed into cylindrical rolls (hanks) about 30 to 60 cm. in length and 5 to 8 cm. in diameter, bound round with a flexible monocotyledonous stem. The staminate inflorescences, which are sometimes borne on the same, sometimes on different trees, are also collected and the flowers stripped from them, but they are comparatively valueless and are not official.

**Description.**—The panicles attain as much as 60 cm. in length, and are of a decided, though dull, reddish or mauve colour (red kousso). The main axis is stout, covered with shaggy brown hairs, and branches repeatedly, forming a sympodial branch system. The branches spring from the axils of large bracts, and are more or less thickly covered with similar hairs and also with minute glands, the latter appearing under the lens as a brownish powder adhering to the surface. The flowers are very numerous and shortly stalked. Each bears on its pedicel two rounded bracts, and consists originally of two whorls of greenish sepals, a caducous white corolla, abortive stamens, and two monocarpellary ovaries enclosed in the tube of the calyx. After fertilisation the inner sepals bend over the young fruit and shrivel; the outer grow larger and become deeply veined with purple. Only one of the two ovaries arrives at maturity. In the drug the most conspicuous part of the flower is the outer whorl of reddish, veined sepals; in its centre may be found the inner sepals bending over the immature fruit; of corolla and abortive stamens no trace is to be found.

Kousso has no marked odour, but possesses a bitter, acrid taste.

The student should soak a little kousso in water and, when thoroughly softened, examine it with the help of lens and dissecting needles. The two whorls of sepals and the young fruit can then be distinguished without difficulty.

The drug is easily identified, but the student should particularly observe

(*a*) The *inner whorl of sepals bending over the young fruit*,
(*b*) The *enlarged, veined, outer whorl of sepals*,
(*c*) The *reddish colour of the drug*,

as these characters serve to distinguish the pistillate flowers, which alone are official, from the staminate.

**Constituents.**—The most important constituent of kousso is kosotoxin (Leichsenring, 1894), a highly active, amorphous, yellowish substance, of which 0·004 gm. is sufficient to kill a frog. Protokosin and kosidin are inactive, colourless, crystalline substances; *a*- and *β*-kosin are inactive, yellowish, crystalline substances and appear to be decomposition products, it being uncertain whether they occur preformed in the drug. Kousso also contains tannin and resin, and yields about 5 per cent. of ash.

Kosotoxin is insoluble in water but easily soluble in alcohol, ether, acetone, chloroform, &c., as well as in solutions of alkaline carbonates. Caustic alkalies convert it into kosin. By the action of zinc dust and caustic soda kosin and trimethylphloroglucin are obtained; this reaction indicates a close relationship with the vermifuge constituents of male fern and kamala, which under similar conditions yield trimethylphloroglucin and dimethylphloroglucin respectively.

Protokosin occurs in colourless needles melting at 176°, *a*-kosin, in yellowish needles melting at 161°, and *β*-kosin, in dark yellow prisms melting at 121°.

**Substitutions &c.**—Under the name of 'loose kousso' the flowers stripped from the panicles and dried are sometimes imported. They arrive usually in more or less fragmentary condition, and frequently contain a considerable admixture of staminate flowers. These may

FIG. 37.—*Brayera anthelmintica.* *A*, flowering branch, three-fourths natural size (after Berg and Schmidt). *B* and *C*, staminate flower closed and open, magnified. *D*, pistillate flower, magnified. *E*, the same, cut longitudinally, magnified: *b*, outer, *k*, inner whorl of sepals; *c*, corolla. (Luerssen.)

be easily distinguished by their greenish colour, small outer sepals densely covered with short hairs, and fertile stamens; they are often unexpanded. They are said to possess emetic properties, and to be a much less active anthelmintic than the pistillate flowers; the Pharmacopœia, by describing the drug in panicles, properly excludes the use of loose kousso.

**Use.**—Kousso is used as an anthelmintic, tapeworms being readily killed by it. In large doses it produces nausea, vomiting, and colic. It is commonly administered in the form of an infusion, the dregs being swallowed, but probably an ethereal or ether-alcohol extract would be preferable.

## RED-ROSE PETALS

(Petala Rosæ Gallicæ)

**Source, &c.**—The red or Provins rose, *Rosa gallica*, Linné (N.O. *Rosaceæ*), is probably indigenous to southern Europe, but has been cultivated as a garden plant in numerous varieties everywhere. For medicinal use the red rose is grown in England (Oxfordshire, Derbyshire, &c.), in the south of France, near Hamburg, &c.

The drug is collected by plucking the whole of the unexpanded petals from the calyx. The little masses of petals are then either used fresh for the preparation of the confection, or dried, in which state they are officially employed for making the infusion; in the latter case they are often gently sifted to remove the stamens. The petals of the red rose are obovate in outline, velvety, and of a deep purplish red colour, with a paler claw.

**Description.**—When collected as described they remain united in small conical masses of a deep purplish red colour, paler towards the base; these, however, are often broken up into the separate petals for convenience in use. They possess a delicate rose-like aroma and slightly astringent taste.

**Constituents.**—Red-rose petals contain a trace of volatile oil, gallic and possibly quercitannic acids. The colour is due to the glucoside cyanin (microcrystalline, dark brown powder, 2 per cent.) which is probably combined with an acid; when present as a potassium salt (as in cornflowers) cyanin produces a dark blue colour. The crystalline yellow substance obtained by Naylor and Chappel (1904) was probably a decomposition product of cyanin.

**Substitutes.**—Red-rose petals should have attached a portion of the paler base of the petal. Artificially coloured petals, may be recognised by their uniform dark reddish colour.

**Uses.**—The petals are slightly astringent; they are chiefly used medicinally, in the form of acid infusion, as an agreeable astringent vehicle or as a colouring agent.

**Allied Drugs.**—*Rosa centifolia*, Linné, largely cultivated as a garden plant, has pale red or pink petals formerly official for the production of rose water.

*Rosa damascena*, Linné, largely cultivated in European Turkey, Bulgaria, the south of France, &c., yields otto of rose and is the rose from which rose water is officially prepared.

## CLOVES

### (Caryophylla)

**Source, &c.**—The clove tree, *Eugenia caryophyllata*, Thunberg (N.O. *Myrtaceæ*), is a handsome evergreen tree and a native of the Molucca Islands, where, as well as on the neighbouring islands, it was formerly extensively cultivated.

Although the spice was known in China about 220 B.C. and in Europe in the fourth century, the Clove Islands were not discovered till 1504. They passed into the hands of the Portuguese and then into those of the Dutch, who unsuccessfully attempted to monopolise the trade in cloves and confine the tree to the Moluccas. The French succeeded in introducing the plant into Mauritius and Réunion, whence it was brought to Cayenne and to Zanzibar. On the latter island and its neighbour, Pemba, the clove tree is now extensively cultivated, and these two islands furnish the bulk of the world's supply, the remainder being obtained from Penang, Amboyna, Madagascar, &c.

The inflorescence of the plant is a compound raceme, on the ultimate ramifications of which the flowers are borne. When quite young the buds are white; as they develop the lower portion assumes a green and finally a crimson colour. The buds are then collected, before the white corolla expands, and dried in the sun, during which the crimson colour changes to a dark reddish brown. Sometimes the whole inflorescence is collected, or sometimes the buds are knocked off with bamboos. The buds are finally separated from peduncles, which are exported separately under the name of 'clove stalks.' The ripe fruits are also occasionally collected; they are known in commerce as 'mother cloves" (anthophylli).

**Description.**—The cloves of commerce are therefore the dried flower-buds of the tree. Each of them consists of a nearly cylindrical, dark reddish-brown portion, slightly tapering at the base, which is sometimes regarded as a gynophore, sometimes as a fleshy calyx-tube, but is perhaps most correctly interpreted as the solid lower portion of the ovary (Tschirch), crowned by four, thick, divergent calyx-teeth of a similar colour, from the centre of which arise four, paler, brown, unexpanded, imbricated petals. After soaking in water for twenty-four hours the petals can be removed, and they will be found to enclose a large number of stamens bending over a stiff erect style arising from a depression in the centre of a small disc. Just below the disc is the two-celled ovary with its numerous ovules; it can be found by cutting the clove either longitudinally or transversely.

The lower part of the ovary is solid and fleshy, spongy near the centre. It contains, especially near the periphery, a large number of oil-glands, visible, when the transverse section is examined under the lens, as dark shining points or small cavities. Similar glands can be seen both in the calyx-teeth and petals; in the latter they appear as translucent dots by transmitted light.

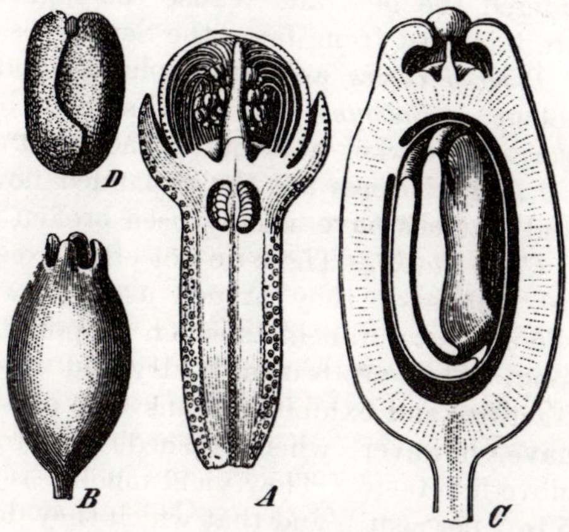

FIG. 38.—Clove (*Eugenia caryophyllata*). *A*, clove cut vertically, showing calyx, corolla, stamens, pistil, and ovules; near the margin oil-glands; magnified. *B*, fruit (mother clove), natural size. *C*, the same, cut vertically and magnified. *D*, embryo, natural size. (Luerssen.)

Cloves are strongly aromatic and have a pungent aromatic taste. Good cloves should be plump and heavy, have a bright, reddish-brown colour, sink in water, and exude oil when indented with the finger-nail.

**Constituents.**—Cloves contain a large quantity of volatile oil, a considerable proportion (13 per cent.—Peabody, 1895) of tannin, which has been identified as gallotannic acid, and a colourless, odourless, crystalline substance, caryophyllin. They yield from 5 to 7 per cent. of ash.

The most important of these constituents is the volatile oil, and the value of the drug is determined chiefly by the amount of oil that it contains, good cloves yielding from 15 to 20 per cent.

The chief constituent of the oil (not less than 85 per cent.) is eugenol, $C_{10}H_{12}O_2$, a colourless liquid with an odour of cloves, boiling at 253°; a terpene (caryophyllene), aceteugenol, α-methylfurfural, dimethylfurfural, methyl salicylate and other bodies are also present. The amount of eugenol present can be approximately determined by shaking a measured quantity of the oil with

5 per cent. solution of potassium hydroxide with which the eugenol forms a water-soluble compound; the caryophyllene which floats on the surface can be measured by suitable means and deducted from the volume of oil used, the difference being eugenol. Specific gravity 1·047 to 1·065. The oil is largely used for the production of vanillin.

**Uses.**—Cloves are used as an agreeable aromatic stimulant, antispasmodic, and carminative, properties that are due to the volatile oil they contain.

**Varieties, &c.**—Although Zanzibar supplies the bulk of the cloves imported, those from Penang, Amboyna and Madagascar are considered the best and realise the highest prices. Smaller quantities are imported from Java, the Seychelles, Ceylon, &c.

*Penang cloves* are large, plump, and of a bright, reddish-brown colour; *Amboyna cloves* are similar but rather smaller; *Zanzibar cloves* are darker in colour, leaner, and smaller still.

'*Blown*' *cloves* are the expanded flowers from which both corolla and stamens have usually been broken off (forming clove dust).

*Clove Stalks.*—These do not often exceed 3·5 cm. in length or 3 mm. in thickness; they branch usually twice or thrice trichotomously, the ultimate branchlets which support the flowers being about 3 mm. long. They are brownish, dry and woody; they break with a short fracture, and exhibit in transverse section but few oil-glands. They have, however, when crushed, an aromatic odour and a pungent clove-like taste. They yield much less volatile oil than cloves (about 5 to 7 per cent.), and that which they do yield is less aromatic. They are said to be used for adulterating powdered cloves, a sophistication easily determined by microscopical examination (they contain isodiametric sclerenchymatous cells[1] which do not occur in cloves) and by the amount of ash yielded by the drug, good cloves affording not more than 7 per cent.

*Mother cloves* ('anthophylli').—Of the numerous ovules contained in the ovary of the clove, only one arrives at maturity. After fertilisation it rapidly increases in size, pushes the other ovules and surrounding tissue aside, and forces its way into the lower part of the ovary (compare fig. 38, *C*). The ripe fruits are ovoid brown berries and about 12 mm. long. They contain much less oil than cloves, and are said to be used for adulterating powdered cloves; their presence can easily be detected, as they contain starch, from which cloves are free.

*Exhausted cloves*, i.e. cloves which have been deprived of their volatile oil by distillation, are darker, yield no oil when indented with the nail and float in water.

[1] Compare Greenish and Collin, *Anatomical Atlas*, p. 96.

## ELDER FLOWERS

### (Flores Sambuci)

**Source, &c.**—The common elder, *Sambucus nigra*, Linné (N.O. *Caprifoliaceæ*), is indigenous and common in England, and is distributed throughout the whole of central and southern Europe. It flowers in the early summer, producing large polychasial cymes, about 15 cm. in diameter, of small white flowers. The entire inflorescences are collected and allowed to remain in heaps for a few hours, during which they become slightly heated and the flowers drop off from the stalks. The latter are removed by sifting, and the flowers are either dried or preserved by mixing them with common salt. By this 'pickling' the rather disagreeable odour of the fresh flowers is gradually changed to a pleasant fragrance. Either fresh or pickled flowers may be used for the preparation of elder-flower water, but that made from the fresh flowers has a distinct and unpleasant odour, which it is said to lose when redistilled after having been kept for some weeks. The fresh flowers, infused in melted lard, yield elder-flower ointment.

**Description.**—The flower consists of a three-celled, three-seeded, inferior ovary, five small green calyx-teeth, a white, rotate, monopetalous corolla with five ovate or rounded lobes and a very short tube in which five stamens with short filaments and yellow anthers are inserted. The dried flowers, which are so shrivelled that their details are quite obscured, have a dingy brownish yellow colour and faint odour.

The student should examine the drug by boiling a little for a few moments in water and pouring on a plate to cool; in addition to the flowers, numerous portions of the stalks (pedicels), as well as occasional buds and immature fruits, may be readily seen under a lens.

He should observe

(*a*) The *five-lobed* monopetalous corolla,

(*b*) The *yellow anthers* of the stamens.

**Constituents.**—Elder flowers contain about 0·32 per cent. of volatile oil, possessing the odour of the flowers in a high degree. It is obtained by distilling the fresh flowers with water, saturating the distillate with salt and shaking it with ether. On evaporating the ethereal solution the oil is obtained as a yellowish buttery mass.

**Substitutes, &c.**—The flowers of various small composite plants (e.g. *Achillea Millefolium*, Linné) are said to have been used for adulterating elder flowers. The latter can be easily distinguished by the characters detailed above.

The flowers of the dwarf elder (*Sambucus Ebulus*, Linné), which, however, is a comparatively rare plant, are distinguished by their dark red anthers.

**Use.**—Elder flowers are chiefly used for the preparation of the distilled water.

**Notes.**—The fresh ripe fruits contain tyrosin ; the leaves and bark an alkaloid, sambucine, and a purgative principle. The leaves also contain the cyanogenetic glucoside sambunigrin, the glucoside of lævo-phenylglycollic acid (compare p. 38). Green elder ointment is prepared by infusing the leaves in a melted mixture of lard and suet.

## CHAMOMILE FLOWERS

### (Flores Anthemidis)

**Source, &c.**—The common or Roman chamomile, *Anthemis nobilis*, Linné (N.O. *Compositæ*), is a small creeping perennial plant with

A B

FIG. 39.—Single chamomile. A, entire ; B, cut vertically, showing the solid, conical receptacle. Natural size. (Bentley and Trimen.)

shortly ascending, leafy, flowering branches, bearing terminal white-rayed flowerheads. It is common on waste grounds in this country, and is cultivated for medicinal use both in England (Hampshire, &c.) and in Belgium, France, and Saxony. Formerly large quantities were produced near Mitcham, but at present there is but little grown in Surrey, and that little is used chiefly for distillation. The flowers have been used for centuries as a domestic medicine.

The inflorescence of the wild plant is a capitulum surrounded by two or three rows of overlapping bracts ; the disc-florets are yellow, tubular, closely packed on an elongated conical receptacle and surrounded by a single row of ray-florets with white ligulate corollas. Such chamomile flowers, which are properly designated 'single,' are collected to some extent in Scotland, but they are not official.

Under certain circumstances the flower may become more or less double. The yellow tubular corollas of the disc-florets become more or less completely white and ligulate, and the flowerhead is converted into a hemispherical mass of white ligulate florets. These are 'double' chamomiles, and form the bulk of the imported, cultivated flowers ;

intermediate forms, or semi-double flowers, in which the conversion of tubular into ligulate corollas has been only partially effected, may, however, frequently be found.

The entire flowerhead is collected and dried.

**Description.**—The official chamomiles are limited to the 'double' or semi-double flowerheads obtained from cultivated plants, those imported from Belgium, France, and Saxony being preferred by the drug trade on account of their handsome appearance, whilst English flowers yield more volatile oil when distilled.

The flowerheads are hemispherical in shape, 10 to 20 mm. in diameter, and white or nearly white in colour, becoming yellow or buff-coloured when kept. The involucre surrounding each flowerhead consists of two or three rows of overlapping bracts with membranous margins, and is almost entirely concealed in the drug by the reflexed outer ligulate florets. The latter are pistillate, the yellow tubular central florets (if present) being hermaphrodite. The corolla of the ligulate florets is white, rather narrow, and terminated by three teeth; towards the base it is contracted to a short tube, immediately below which is the small, nearly smooth ovary. The calyx is completely adherent to the ovary, and there is no pappus. If the florets are all carefully plucked from the flowerhead, a number of blunt, narrow, concave, scaly bracts (paleæ) with membranous wings will be found standing on a conical receptacle; on cutting a flowerhead longitudinally through the centre each floret will be seen to spring from the axil of a bract, and the conical receptacle will be found to be solid (or occasionally lacunous). When closely examined with a powerful lens, the lower part of the corolla may be seen to be sprinkled with minute, yellowish, shining oil-glands.

Fig. 40. — Double chamomile. Natural size. (Bentley and Trimen.)

Chamomiles have an aromatic odour and an aromatic, intensely bitter taste.

The student should carefully strip the florets from a flowerhead, and observe

(*a*) The *presence and shape of the paleæ*;

and should also cut a flowerhead longitudinally through the centre, and note

(*b*) The *solid, elongated, conical receptacle.*

He should further compare these with those of the flowers of the wild chamomile, German chamomile (*Matricaria Chamomilla*, Linné), and of the feverfew (*Chrysanthemum Parthenium*, Bernhardi), which are alluded to below. He should also soak a few ligulate florets in water, spread them out on a dark surface and examine the three-toothed apex.

**Constituents.**—The principal constituents of chamomile flowers are the volatile oil (0·8 to 1·0 per cent.) and the bitter principle, anthemic acid.

The volatile oil is blue when freshly distilled, but becomes greenish or brownish yellow on keeping; it consists chiefly of the esters of the isomeric acids angelic and tiglic, with butyl and amyl alcohols;

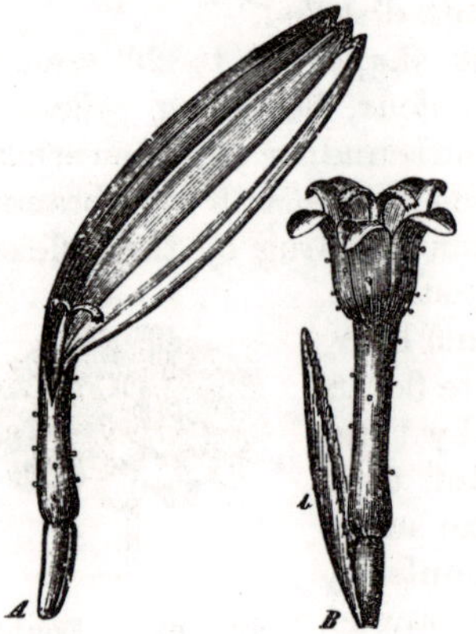


FIG. 41.—Chamomile (*Anthemis nobilis*). *A*, ray-floret, magnified 4 diam. *B*, disc-floret with bract (*b*), showing external glands; more highly magnified. (Luerssen.)

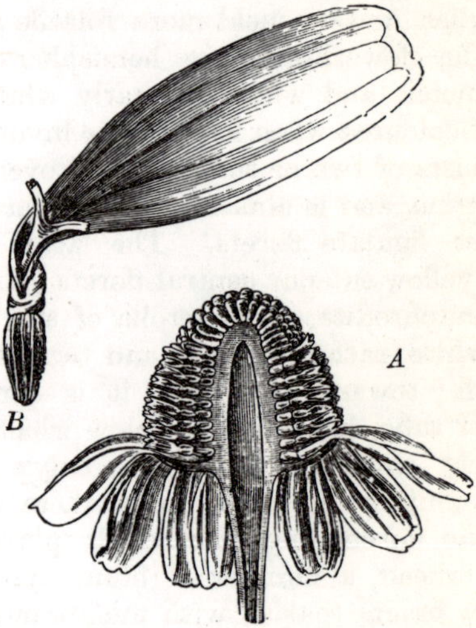


FIG. 42.—German chamomile (*Matricaria Chamomilla*). *A*, flowerhead cut vertically. *B*, ray-floret, magnified. (Moeller.)

it also contains an alcohol, anthemol, and a crystalline hydrocarbon, anthemene.

Anthemic acid is an intensely bitter, crystalline glucoside easily hydrolysed with total loss of bitterness, prolonged boiling in water being sufficient to produce the reaction.

The flowers also contain a crystalline phytosterin, anthesterin, melting at 221°, wax, fatty oil, glucose, &c. They yield about 5 per cent. of ash.

**Uses.**—Chamomile flowers possess aromatic, bitter stomachic properties; the oil is occasionally administered in pills as a carminative.

**Substitutes, &c.**—The characters given above will easily distinguish the true or Roman chamomile from the following, which somewhat resemble it:

*Matricaria Chamomilla*, Linné, German Chamomile.—The flowerheads are smaller, single, and have a hollow conical receptacle devoid of paleæ. They are sometimes sold as 'single chamomiles.'

*Chrysanthemum Parthenium*, Bernhardi, Feverfew.—The cultivated plant has double flowerheads, resembling those of the chamomile. The receptacle is flat; paleæ may be present or absent, according to the variety; if present they are acute.

## ARNICA FLOWERS

### (Flores Arnicæ)

**Source, &c.**—Arnica flowers are the flowerheads of *Arnica montana*, Linné (N.O. *Compositæ*), a small plant with creeping perennial rhizome, indigenous to central Europe. In the warmer districts it is common in the meadows on the lower mountain spurs; in the more northern districts it grows in the valleys. It produces large, solitary, orange-yellow flowerheads, not unlike yellow marguerites, measuring about 7 cm. in diameter.

The flowerheads are usually collected entire and dried; the receptacle being especially liable to be attacked by insects, this, together with the involucre, is sometimes separated, and the drug then consists of the ligulate and tubular florets but this variety is not official.

**Description.**—The drug as met with in English commerce commonly consists of the entire flowerheads, the most conspicuous parts of which are the green involucre and whitish, bristly pappus; the dark yellow ligulate corolla, so prominent in the fresh flower, shrivels so much as to become quite inconspicuous.

The involucre consists of two rows of dark green, acutely pointed, linear-lanceolate, hairy bracts. The florets of the ray are pistillate, and possess long, narrow, dark yellow ligulate corollas containing about seven to nine veins and terminated by three teeth.

The disc-florets are numerous, and have a long, dark yellow, tubular corolla which bears on its outer surface minute glands, visible under a strong lens; similar glands are found also on the ligulate ray-florets. The fruit, which is more or less shrivelled, is elongated, straight, covered with appressed hairs, and surmounted by a single ring of stiff, whitish, barbed bristles (pappus); the hairs consist of two long, narrow, contiguous cells which are divergent at the apex and pointed, but this character can scarcely be seen under a lens.

After the florets have been removed, the receptacle, which is about 8 mm. in diameter and arched, exhibits a corresponding number of depressions, each of which bears in its centre a stiff, dark bristle, and is surrounded by short, bristly hairs, the latter being, however, sometimes absent.

Arnica flowers have a slight pleasant odour, and bitter, rather acrid taste.

The student should particularly note

(*a*) The *narrow, ligulate corollas, with from seven to nine veins and three teeth,*

(*b*) The *straight, bristly fruit* and *abundant, whitish, bristly pappus,*

FIG. 43.—Arnica flower. Natural size.
(*Pharmaceutical Journal.*)

(*c*) The *linear-lanceolate, hairy, involucral bracts,*

(*d*) The *characteristic twin hairs on the fruit.*

**Constituents.**—Arnica flowers contain traces of volatile oil and a bitter principle, arnicin, which has been obtained in minute, yellow, deliquescent crystals. The flowers are said to contain more arnicin than the rhizome. The drug also contains tannin, resin, yellow colouring matter and a phytosterin, arnisterin.

**Uses.**—Preparations of arnica flowers applied to the skin appear to increase the activity of the circulation, and the tincture, diluted with water, is used for application to the skin ; internally they are stimulant and irritant, but they are now seldom administered.

**Substitutes, &c.**—Of the numerous other composite flowers reported to have been mixed with or sold for arnica, the following may be mentioned.

*Anthemis tinctoria*, Linné; fruits without pappus; paleæ on receptacle.

*Calendula officinalis*, Linné; ligulate corolla with four veins; fruit without pappus.

*Inula britannica*, Linné; ligulate corolla with four veins; pappus not bristly. Arnica from the Italian Alps is frequently derived from *I. britannica*.

*Doronicum pardalianches*, Linné; ligulate corolla with four veins; no pappus.

*Taraxacum* sp.; *Scorzonera* sp.; all florets with 5-toothed ligulate corollas.

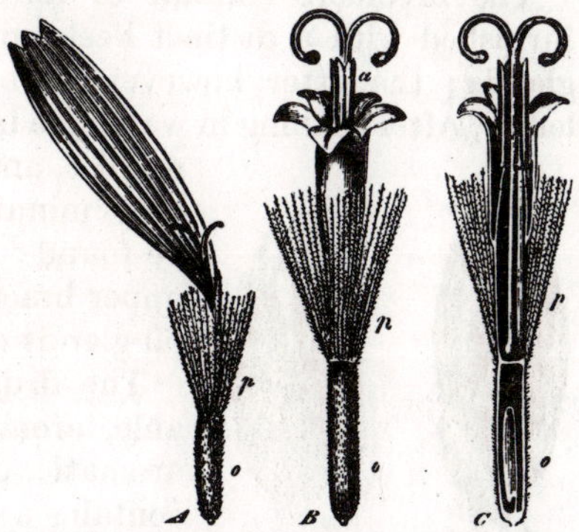


FIG. 44.—Arnica flowers. *A*, ray-floret. *B*, disc-floret. *C*, the same, cut vertically: *o*, ovary; *p*, pappus; *a*, anthers. All magnified. (Luerssen.)

## WORMSEED

(Santonica, Semen Cinæ, Semen Contra)

**Source, &c.** — Wormseed consists of the small, unexpanded flower-heads of *Artemisia maritima*, var. *a-Stechmanniana*, Besser (N.O. *Compositæ*). The species has, in numerous varieties, an extremely wide distribution, from the Bay of Biscay to Chinese Mongolia. The variety from which part at least of the drug is obtained, and which, according to some botanists, should be considered a distinct species (*A. Cina*, Berg), grows in enormous quantities in the deserts of the Kirghiz in Turkestan, especially near the town of Chimkent (which lies to the north of Taschkent and some seventy miles east of the Sir Daria river) where factories have been erected, in which large quantities of santonin are produced from the wormseed collected in the vicinity. Comparatively little of the crude drug is now exported.

The plant is small and woody, with numerous, erect branches about 40 cm. long, on which very small flowerheads are borne. These are stripped from the stems before they expand, and dried. They are collected in July and August by Kirghiz and other tribes and brought to Chimkent.

Wormseed has long been used as an anthelmintic; it was employed in Italy under the name of semenzina (diminutive of *semenza* = seed) under the belief that it consisted of small seeds. From the word semenzina is derived the name 'semen cinæ,' by which the drug is often known; semen contra (another of its names) is an abbreviation of 'semen contra vermes.' The drug appears at first sight to consist of a number of small, brownish, ridged seeds the true nature of which becomes apparent when they are dissected.

**Description.**—The flowerheads are of a greenish yellow colour, but turn brown by drying and keeping. They are only about 1·5 mm. long, elongated ovoid in shape, and somewhat angular; their surface is shining and glabrous, or at most only slightly hairy.

The involucre consists of imbricated ovate or lanceolate bracts, furnished with a distinct keel, on each side of which are shining oil-glands; the latter, however, are not easily seen even under a powerful lens. After soaking in water the bracts can be removed with dissecting needles, and in the centre from three to six very minute, unexpanded, tubular florets will be found; they are completely enclosed by the upper bracts of the involucre, and bear minute oil-glands on the lower portion of the corolla.

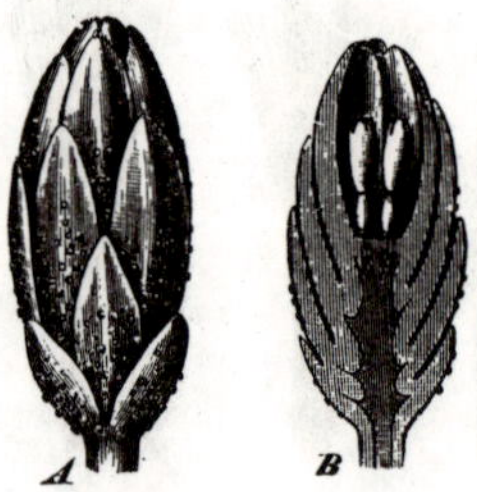

FIG. 45.—Flowerhead of *Artemisia*. *A*, entire, showing imbricated bracts with external glands: *B*, cut longitudinally, showing florets. Magnified. (Luerssen.)

The drug exhales, when crushed, an agreeable, aromatic odour, and possesses a bitter, aromatic, camphoraceous taste. It frequently contains a considerable admixture of fragments of the leaves and very slender flower stalks.

The student should soak the wormseed in water for twenty-four hours, and then dissect a flowerhead with the aid of the dissecting needles and a lens. He should note

(*a*) The *imbricated, keeled, glabrous bracts*,
(*b*) The *minute florets enclosed within them*,
(*c*) The *characteristic odour and taste.*

**Constituents.**—Wormseed contains a volatile oil and two crystalline principles, viz. santonin, to which the anthelmintic property of the drug is due and artemisin. The santonin attains its maximum (2·3 to 3·6 per cent.—Ehlinger, 1885) in July and August. After flowering it rapidly disappears.

Santonin, $C_{15}H_{18}O_3$, forms colourless bitter crystals that are very slightly soluble in water but unite with alkalies, forming soluble salts of monobasic santonic acid, $C_{15}H_{20}O_4$. Exposed to light, santonin assumes a yellow colour (photosantonin).

Santonin is extracted from the flowerheads by treating them with milk of lime, the santonin being converted into soluble calcium santonate. From the filtrate excess of calcium is removed by a current of carbon dioxide, and the calcium salt is converted into the sodium salt by means of sodium carbonate. The liquid is filtered warm and the santonin separated by the addition of sulphuric acid. It is purified by decolorisation with charcoal and recrystallisation from hot alcohol.

**Substitutes, &c.**—From time to time a variety of wormseed containing little or no santonin appears on the market; it may be recognised by the more or less hairy nature of the bracts.

**Use.**—Wormseed is now seldom administered, but santonin is often employed as an anthelmintic for round worms which it rapidly expels; it has less effect upon thread-worms and no action on tape worms. It produces remarkable disturbances of vision, objects appearing first blue and then yellow, and the absorbed santonin renders the urine intensely yellow if acid or purplish if alkaline.

Certain other species of *Artemisia* (e.g. *A. gallica*, Willdenow) have anthelmintic properties. American wormseed (*A. anthelmintica*) is cultivated in California.

## CALENDULA

### (Flores Calendulæ, Marigold Florets)

**Source, &c.**—The marigold, *Calendula officinalis*, Linné (N.O. *Compositæ*), is commonly cultivated in numerous varieties as a garden plant. The flower head is about 5 cm. in diameter, yellow to deep reddish orange in colour, with numerous barren disc-florets and one or more rows of fertile ligulate ray-florets. The latter are collected when the flower is fully expanded and dried.

**Description.**—The drug consists almost entirely of the ligulate corollas of the ray-florets, about 2·5 cm. in length, enclosing in a short tube the remains of the style and two stigmas. The limb of the corolla, the tube of which is hairy externally, is terminated by three teeth, and exhibits, when examined with a lens, four principal veins. (Compare fig. 47 *B.*)

The drug has a somewhat aromatic odour and a distinctly bitter taste.

The student should soften some calendula in water, spread the florets out, and examine them with a lens, noting

(*a*) The *three teeth of the corolla*,
(*b*) The *four principal veins*;

and should compare them with the ligulate florets of arnica (which have about seven to nine veins) and of dandelion (which have five teeth). (Compare figs. 44 *A*, and 47 *B.*)

**Constituents.**—Calendula contains traces of volatile oil, a bitter principle, and calendulin, the latter being a tasteless substance swelling in water (Geiger, 1818).

**Uses.**—Calendula is used chiefly in the form of the tincture diluted with water as an application to bruises to promote the absorption of effused blood.

## INSECT FLOWERS (Dalmatian)

### (Flores Pyrethri, Pyrethrum Flowers)

**Source, &c.**—Dalmatian insect flowers are the unexpanded flowerheads of *Chrysanthemum cinerariæfolium*, Visiani (N.O. *Compositæ*), a native of Dalmatia, Herzegovina, and Montenegro, and cultivated there as well as in Japan and California.

The flowers of several composite plants have long been known to possess the remarkable property of stupefying flies and other small insects. None, however, appears to act so energetically as insect flowers, and these are most active if collected when they are fully developed, but before they expand. They are then known commercially as 'closed' flowers, 'half-closed' and 'open' flowers being collected at more advanced stages of development. They retain their insecticidal properties indefinitely, even in the state of dry powder. The leaves have been shown to be destitute of insecticidal properties.

**Description.**—The closed flowerheads of commerce are of a dull brownish yellow or greyish brown colour and about 10 mm. in diameter. The bracts of the involucre, which are arranged in two or three rows, are yellowish or greyish in colour, lanceolate, hairy, and membranous at the margin. In the closed flowers they are erect, but as the flowers expand they bend outwards, the capitulum assuming a flattened hemispherical shape. There is only one row of ray-florets, with brownish or whitish ligulate corollas. The disc-florets, which are numerous, have comparatively short yellow corollas. The fruit is longer than the corolla, club-shaped, and provided with five ribs that project so strongly as to make it appear almost winged. Both the corolla and the fruit are sprinkled with yellow shining oil-glands. After the corolla has been removed, the calyx may be seen in the form of a raised membranous ring crowning the fruit. The receptacle is naked and nearly flat.

'Closed' flowers are sub-globular or sub-conical in shape, and usually bear the yellowish, shrivelled, ligulate corollas; 'half-closed' flowers are nearly hemispherical in shape, and the corollas of the ray-florets are more spreading; 'open' flowers are still flatter, and most of the ligulate corollas have been broken off; they are often devoid of the tubular corollas of the disc-florets also, and then present a reticulated appearance due to the membranous calices crowning the closely packed fruits.

Insect flowers possess a bitter, acrid taste; the odour is aromatic, but not strong.

The student should observe

(*a*) The *sub-globular or sub-conical shape of the 'closed' flowers*,
(*b*) The *yellowish colour of the bracts*,

(*c*) The *short corolla and membranous calyx of the tubular florets*,
(*d*) The *five prominent ribs of the fruit.*

**Constituents.**—Dalmatian insect flowers contain up to 1·25 per cent. of volatile oil, but the toxic principle is a yellow syrupy substance, pyrethrone, which is an ester yielding by saponification pyrethrol. Pyretol and pyrethrotoxic acid, also given as the active constituent, appear to consist essentially of pyrethrone. It is not volatile, and the flowers do not lose their activity when exposed to the air.

Substances of alkaloidal and glucosidal nature have also been isolated, but our knowledge of them is very imperfect.

Good insect powder should stupefy common house-flies kept near it within a minute ; less active powders may take as much as twenty minutes to effect this. It should yield about 8 or 9 per cent. of ash and about 10 per cent. of moisture. The quality of the powder is also indicated by the ethereal extract, good closed flowers affording from 7·5 to 10·5 per cent. of yellow extract ; half-closed flowers yield from 6 to 7 per cent., and open flowers from 5 to 6 per cent., while the presence of much stem or leaf is indicated by the greenish colour of the extract.

**Substitutes, &c.**—*Chrysanthemumc occineum*, Willdenow, (*C. roseum*, Weber et Mohr), a native of the Caucasus and northern Persia, yields the Persian (or Caucasian) insect flowers, which were formerly more commonly used than the Dalmatian. The importation has now, however, practically ceased.

The flowerheads are distinguished from the Dalmatian by the dark, nearly black, colour of the involucral bracts, by the rose colour of the ray-florets, and by the ten-ribbed fruit. They are said to be less active than the Dalmatian.

Insect powder may be adulterated with powdered leaves and stems, with the powder of other composite flowers, with lead chromate, &c. ; powdered quassia, pepper, aloes, euphorbium, etc., are said to have been found in it.

**Use.**—Powdered insect flowers and also a tincture prepared from them are used to stupefy small insects.

## LAVENDER FLOWERS

(Flores Lavandulæ)

**Source &c.**—The common lavender, *Lavandŭla vera*, de Candolle (N.O. *Labiatæ*), is a small under-shrub indigenous to southern France, Italy, and Spain, but cultivated in this country as a garden plant, as well as on a large scale for its aromatic flowers. It is also extensively cultivated in southern France.

The bulk of the flowers produced are used for the distillation of the volatile oil. The spikes are cut with a small hook about 15 to 30 cm. below the flowers. They are then thrown into large stills, water added, and distillation conducted over a naked fire or by steam, the distillate being received in a separating can, in which the oil that is carried over is retained.

**Description.**—The inflorescence of the lavender is a terminal spike, on which the flowers are arranged in small verticillasters, each of which arises from the axil of a rhomboidal bract. The calyx is tubular and ribbed, bluish-violet in colour, five-toothed, and hairy, shining oil-glands being visible with a lens amongst the hairs. The bulk of the oil yielded by the flowers is contained in these glands on the calyx. The corolla is bilabiate and of a beautiful bluish violet colour.

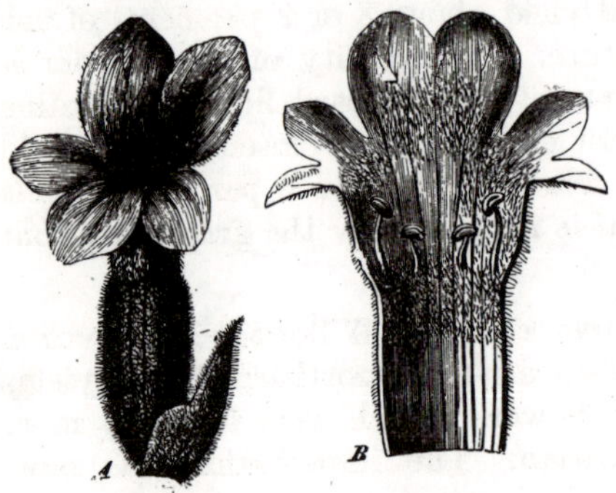


FIG. 46.—*Lavandula vera.* *A*, entire flower, showing bract and hairy calyx, magnified. *B*, the corolla opened vertically, magnified. (Luerssen.)

The student should observe

(*a*) The *hairy*, *ribbed*, *tubular calyx*,

(*b*) The *rhomboidal bracts supporting the flowers*.

**Constituents.**—The principal constituent is the volatile oil, of which the fresh flowers yield about 0·5 per cent.

The chief constituents of the volatile oil are linalool and linalyl acetate; cineol (in English oil), pinene, limonene, and other bodies are also present. The oil should have a sp. gr. of 0·885 to 0·900 and an optical rotation $-5°$ to $-10°$. English oil contains from 7 to 11 per cent. of esters, foreign oil not less than 30.

**Substitutes, &c.**—*Lavandula spica*, de Candolle, is distinguished from the true lavender by the linear bracts, spathulate leaves and more compressed inflorescence. The oil (oil of spike) is less fragrant than that of the true lavender, and slightly dextrorotatory.

## SAFFRON

### (Crocus)

**Source, &c.**—The saffron crocus, *Crocus sativus*, Linné (N.O. *Iridĕæ*), has been cultivated for so many centuries that its native country is no longer known. It was well known to the Greeks and Romans, who used it as a medicine, as a dye, and as a flavouring agent. It was cultivated in Spain in the tenth century, and was subsequently introduced into France and England; it is, however, no longer grown in this country.

At the present time Spain produces the bulk of European saffron. On the plains towards the east and south-east the saffron crocus is grown extensively, the saffron being exported from Valencia and Alicante. A little is also produced in France, near Pithiviers, about fifty miles south of Paris, as well as in Greece and Persia.

The plant produces in the autumn usually one or two pale purplish violet flowers, not unlike the ordinary garden crocus. The long pale yellow style terminates in three deep red elongated stigmas (fig. 47 *A*), which protrude from the flower and are pendulous. The whole flower is collected in the morning, spread out on mats to dry partially; then the stigmas are removed and dried in sieves over a low fire. This forms the hay saffron of commerce, which must necessarily be a valuable drug, since it takes about 750 fully developed stigmas to make one gramme of saffron.

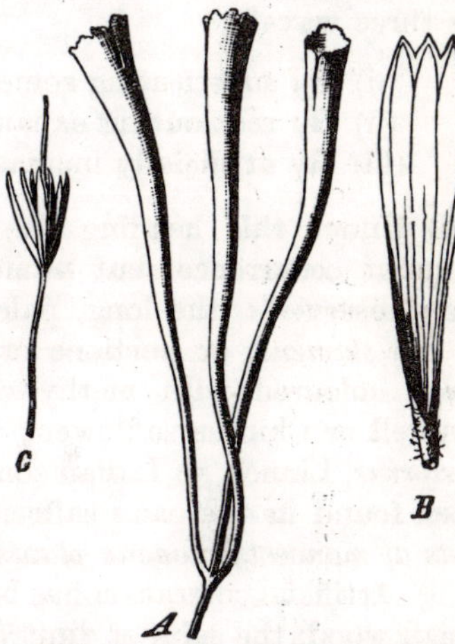


FIG. 47.—*A*, three stigmas of Saffron with a portion of the style, magnified 2 diam. *B*, Calendula floret, magnified 1½ diam. *C*, Safflower floret, magnified 1½ diam. (Vogl.)

**Description.**—Hay saffron forms a loosely matted mass of dark, reddish brown, flattened stigmas with a strong, characteristic odour and bitterish taste. When fresh it is unctuous to the touch and glossy, but after keeping, it becomes dull and brittle.

Thrown on the surface of water the dry stigmas rapidly expand, and their form can be easily studied. At the same time the water surrounding them slowly assumes a deep yellow colour. Each stigma is about 25 mm. in length, and has the shape of a long tube, narrow at the base, where it joins the style, but broadening towards the upper extremity, where it is slit on the inner side. The mouth of the tube is irregularly notched. The stigmas are either single or attached in threes to a short portion of the pale yellow style.

The characteristic shape of the saffron stigmas is of the greatest service in the detection of adulteration; the student should therefore soak a little saffron in water and examine it carefully with a lens.

**Constituents.**—Saffron contains a trace of volatile oil, a bitter principle, picrocrocin, and red colouring matter, polychroite (also called crocin).

The volatile oil consists principally of a terpene, probably associated with a small quantity of an oxygenated aromatic constituent.

Picrocrocin is a colourless crystalline glucoside yielding by hydrolysis crocose (a reducing sugar) and a terpene with an odour recalling saffron.

Polychroite is closely related to carotin and like carotin colours concentrated sulphuric acid deep blue. It has been obtained as a ruby-red, amorphous, glucosidal substance yielding by hydrolysis a new red colouring matter, crocetin (also called crocin), volatile oil (?) and sugar. Crocetin is acid and forms crystalline salts with ammonia, brucine, &c. Saffron apparently contains a second red colouring matter, insoluble in water but soluble in alcohol.

**Substitutes, &c.**—The high price of the drug has naturally been, even during the time of Dioscorides and Pliny, a great inducement to adulteration. This has been affected in one (or more) of the following three ways :

(i) By substituting some substance for the saffron stigmas,
(ii) By recolouring exhausted saffron,
(iii) By artificially increasing the weight of genuine saffron.

(i) Under this heading the following may be mentioned as of frequent occurrence, but numerous others have from time to time been observed : the long, pale coloured *styles* of the saffron crocus, or the *stamens*, or portions of the *perianth* of the same ; *calendula florets* coloured with methyl-orange and sold under the names of feminell or Chinese safflower ; the florets of the *safflower* (*Carthamus tinctorius*, Linné), an Indian composite plant, largely used as a dye and often found in the cake saffron of commerce ; the slender *stems and roots of monocotyledonous plants.*

(ii) Artificial coloration has been effected with aniline dyes, logwood, Brazil wood, the salts of dinitrocresylic acid (Victoria yellow, Victoria orange), &c. Genuine saffron, when thrown on to the surface of water, should slowly surround itself with a circle of yellow (not orange yellow, red or pink) liquid ; it should yield but little colour to ether or petroleum spirit, in which many artificial colours are readily soluble. Animal charcoal almost instantaneously decolorises an infusion of saffron, whereas aniline colours are much less readily removed. An infusion of 0·1 gm. of the suspected sample in 100 c.c. of water should agree with a similar infusion of genuine saffron, both in depth and tint of colour. A solution of chromic acid may also be used as a standard for comparison ; 0·1 gm. of saffron should impart as deep a colour to 50 c.c. of water as 0·275 gm. of chromic anhydride.

(iii) Saffron may be artificially weighted in a variety of ways. Vegetable or mineral oil, which, in addition to increasing the weight, improves the appearance, may be detected by pressing the saffron between thin sheets of paper ; a greasy stain indicates the presence of oil. Saffron should not contain more than 12·5 per cent. of moisture or yield more than 7 per cent. of ash. The presence of glycerin, ammonium nitrate, and of other substances soluble in water, but leaving no ash when incinerated, may be detected by ascertaining the amount of aqueous extract yielded by the saffron ; it should not exceed 51 per cent.

**Note.**—Cape Saffron, now seldom imported, consists of the flowers of a Scrophulariaceous shrub, *Lypēria atropurpūrea*, Bentham, a native of South Africa ; it contains a yellow colouring matter, but could scarcely be mistaken for saffron.

Cake saffron ('croci placentæ,' 'crocus in placenta') commonly consists of safflower florets made into cakes with an adhesive sugary substance. The structure of the florets is easily seen when a little of the drug is soaked in water

**Uses.**—Although chiefly employed as a colouring agent, saffron has been regarded as stimulant, antispasmodic, and emmenagogue.

## LILY OF THE VALLEY FLOWERS

### (Flores Convallariæ)

**Source, &c.**—The lily of the valley, *Convallaria majālis*, Linné (N.O. *Liliaceæ*), is a small herbaceous plant, with perennial creeping rhizome, widely distributed over Europe and indigenous in England, where it occurs in woods or thickets, being much more abundant in some countries than in others.

The plant produces two broadly elliptical leaves and a flowering scape, bearing in the axils of small bracts pedicellate, campanulate flowers, forming a unilateral raceme of fragrant, white drooping flowers. The root, leaves, and flowers have all been used in medicine, but the part now usually employed is the flowers. The entire inflorescence is collected and dried, during which process the white flowers assume a brownish yellow tinge and the fragrant odour almost entirely disappears.

**Description.**—The drug consists of the slender scape, bearing from three to eight brownish yellow campanulate flowers. The perianth has six, recurved teeth, and bears on its inner surface six large anthers ; the ovary is superior and three-celled. It possesses a slight agreeable odour and bitter taste.

**Constituents.**—Two crystalline glucosides, convallamarin and convallarin have been isolated from the lily of the valley.

Convallamarin, $C_{23}H_{44}O_{12}$, has been obtained as a crystalline powder readily soluble in water and in alcohol, but only slightly in ether ; it yields a violet coloration with sulphuric acid and may be hydrolysed into dextrose and convallamaretin ; its action is that of a cardiac tonic and diuretic.

Convallarin, $C_{34}H_{62}O_{11}$, crystallises in prisms, soluble in alcohol and slightly soluble in water, the aqueous solution frothing when shaken ; it has a purgative action. According to Kobert (1915) an acid and a neutral convallarin exist, both possessing the typical saponin action on blood (see p. 254).

**Use.**—Lily of the valley flowers are occasionally used as a cardiac tonic in the place of foxglove.

## SECTION III

## FRUITS

Fruits are generally classified into true, spurious, and collective fruits. A true fruit may be defined as the mature ovary of a single flower. A spurious fruit is one in the formation of which other structures have taken part, while a collective fruit consists of the mature carpels of several flowers united with the bracts and floral envelopes.

True fruits may be further subdivided according as they are indehiscent or dehiscent, monocarpellary or polycarpellary.

The following classification of the most important fruits may be useful to the student for reference :

### A. TRUE FRUITS

#### 1. Indehiscent Monocarpellary Fruits

(a) *The Achene*, a superior one-celled, one-seeded fruit with a dry, indehiscent, closely applied but separable pericarp (true fruits of the fig).

(b) *The Caryopsis*, a superior one-celled, one-seeded, fruit with a dry, indehiscent, inseparable pericarp (Graminaceous fruits).

(c) *The Drupe*, a superior one-celled, one- or two-seeded, indehiscent fruit with a fleshy or pulpy mesocarp and hard endocarp (prune, pimento, Cocculus indicus).

#### 2. Indehiscent Polycarpellary Fruits

(a) *The Nut*, a dry, hard, indehiscent, one-celled, one- or two-seeded fruit more or less enclosed in a cupule.

(b) *The Schizocarp*, an inferior dry, indehiscent fruit, splitting as it ripens into its component carpels, which remain attached to the carpophore (Umbelliferous fruits are two-celled schizocarps).

(c) *The Berry*, an indehiscent, one or more celled, many-seeded, pulpy fruit (orange, lemon, bitter apple, pomegranate, &c.).

### 3. Dehiscent Monocarpellary Fruits

(*a*) *The Follicle*, a superior one-celled, one- or many-seeded fruit dehiscing by one suture only (star anise).

(*b*) *The Legume*, a superior one-celled, one- or many-seeded fruit dehiscing by both sutures (senna pod, tamarind).

### 4. Dehiscent Polycarpellary Fruits

*The Capsule*, a superior one or more celled, many-seeded, dry, dehiscent fruit (poppy head). Fruits which differ from the capsule in being inferior are often termed capsular (cardamom); the siliqua, a superior, spuriously two-celled, many-seeded, long, narrow fruit dehiscing by two valves from below upwards (mustard) and the silicula, which differs from the siliqua only in being short and broad, are varieties of the capsule.

## B. SPURIOUS FRUITS

*The Pome*, an inferior, indehiscent, two or more celled, few-seeded, fleshy fruit, the flesh being generally considered to be the enlarged calyx-tube (quince).

## C. COLLECTIVE FRUITS

(*a*) *The Cone*, an elongated fruit composed of a number of indurated scales, each of which bears one or more seeds on its inner surface.

(*b*) *The Galbulus* differs from the cone in being more or less rounded; the scales are sometimes fleshy (juniper berry).

(*c*) *The Strobile* resembles the cone, but the scales are membranous, and the seeds are contained in carpels (hop).

(*d*) *The Syconus* is formed of an enlarged hollow and more or less succulent receptacle which bears a number of separate flowers on its inner surface (fig).

## STAR ANISE FRUIT

### (Fructus Anisi Stellati)

**Source, &c.**—Star anise fruit is the ripe fruit of *Illicium verum*, Hooker filius (N.O. *Magnoliaceæ*), a small tree indigenous to the southern and south-western provinces of China. Only a small proportion of the harvest is exported, the bulk being used in China for the distillation of the volatile oil.

**Description.**—Star anise fruit is apocarpous, and consists normally of eight one-seeded ovaries. In the flower these are erect, but as the fruit ripens they bend outwards, and finally radiate from a central axis; the pericarp becomes dark brown and woody, and dehisces by the now fully exposed ventral suture, disclosing a single, shining brown seed.

The carpels measure about 1·5 cm. in length; they are boat-shaped, and usually bluntly beaked at the apex, but nearly flat at the base, where they are attached to a short, central column proceeding from a curved peduncle. They are reddish brown and woody; externally irregularly wrinkled, internally paler, smoother, and glossy. The seed is reddish brown, smooth, shining, hard, ovoid, and slightly compressed. The hilum is conspicuous as an oval depression at one extremity. The hard, brittle seed-coats enclose a large, soft, oily kernel. Both the pericarp and the kernel have an agreeable aromatic odour and a sweet spicy taste.

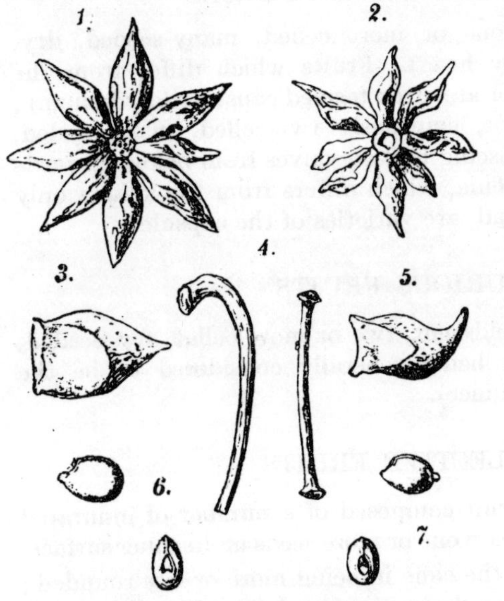

Fig. 48.—Fruits of *Illicium verum* and *I. religiosum*. 1, entire fruit of *I. verum*; 3, a single carpel; 4, stalk (left); 6, seed; 2, entire fruit of *I. religiosum*; 4, stalk (right); 5, single carpel; 7, seed. (Vogl.)

The student should particularly observe

(a) The *size, regular appearance, and blunt beaks of the carpels,*

(b) The *peduncle, which is curved near the fruit,*

(c) The *spicy odour and taste*;

and should compare these fruits with the Japanese star anise (see below).

**Constituents.**—The chief constituent of the fruit is the volatile oil contained in the pericarp of the fruit (about 5 per cent.) and in the kernel of the seed (about 2·5 per cent.).

The oil is distilled in large quantities in crude native stills by the peasants of Langson, in southern China, and brought to the ports of Hai-fong and Hong Kong for exportation. It is scarcely distinguishable from that of *Pimpinella Anisum*, Linné, and may be substituted for it. Its chief constituent is anethol, $C_{10}H_{12}O$ (80 to 90 per cent.).

**Substitutes, &c.**—Japanese star anise, also called sikimi or shikimi fruits (*Illicium religiosum*, Siebold). These fruits find their way to the Indian and occasionally to the London market. As they are toxic they must be carefully distinguished from the Chinese. The following characters will suffice: The fruits are less regularly developed, the carpels usually more wrinkled, the beak more acute and commonly directed upwards; the ventral suture is usually more open, and the peduncle, to which the carpels seldom remain attached, is straight. The Japanese fruits have a balsamic, not anise-like odour, and a disagreeable, bitterish taste; the taste and odour are the best characters by which to distinguish the genuine from the false, as they can be applied to fragments only of the fruit.

From the seeds of Japanese star anise fruit Eykmann (1881) obtained 52 per cent. of fixed oil and a poisonous crystalline principle, sikimin, soluble in hot water, alcohol, and chloroform; the leaves and fruits yielded volatile oil, sikimic (schikimic) acid, and sikimipicrin.

**Uses.**—The oil is employed as a carminative and as a flavouring agent, especially in cough mixtures, as it is supposed to possess a special action on the bronchial mucous surfaces.

## COCCULUS INDICUS

### (Levant Berries, Fishberries, Fructus Cocculi)

**Source, &c.**—The fruits commonly known as Cocculus indicus are produced by *Anamirta paniculata*, Colebrooke (N.O. *Menispermaceæ*), a tall, woody, climbing shrub indigenous to Eastern India and the Malay Archipelago. They derive the name of Levant berries, by which they are sometimes known, from the fact of their having been formerly brought from India by way of Alexandria and the ports of the Levant.

The plant produces a pendulous panicle of flowers; the ovaries of the pistillate flowers are apocarpus, each carpel being gibbous and developing into a drupaceous fruit containing a single seed. The fruits are collected when ripe, and dried; they are exported chiefly from Bombay and Madras.

**Description.**—Cocculus indicus of commerce consists of small dark brown or nearly black fruits, about 12 mm. in length. They are more or less distinctly reniform in shape, one side being flattened or even slightly concave, whilst the other is boldly arched. On the former the small scar left by the stalk can usually be distinguished, and near it is a minute prominence, the apex of the fruit. The gibbous dorsal surface of the carpel from which the fruit is formed develops much more rapidly than the ventral, and the apex of the fruit thus remains near to the base, the dorsal surface becoming conspicuously arched. The pericarp is rough and finely wrinkled, and although thin is hard and woody. It encloses a single, oily seed—so deeply hollowed out as to be cup-shaped. This cup-shaped hollow is completely filled by two parallel, lenticular ingrowths of the endocarp and mesocarp, which can easily be seen by cutting a fruit longitudinally through the median line of the carpel and removing the two halves of the seed, or less clearly by cutting the fruit transversely. The seed exhibits a crescent-shaped section when cut either longitudinally or transversely. Cocculus indicus has no odour ; the pericarp is tasteless, but the seed is very bitter.

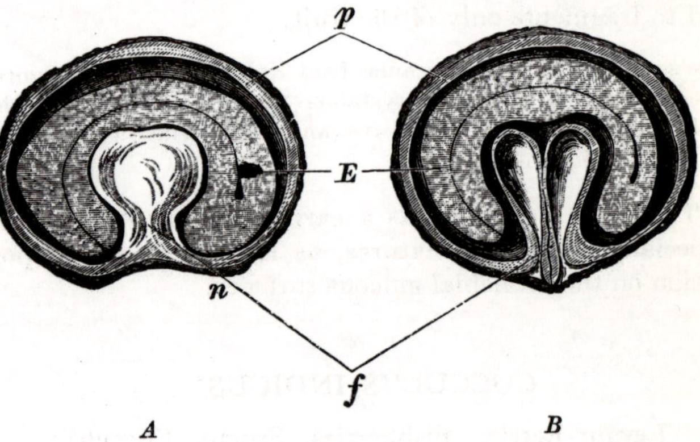


FIG. 49.—*Cocculus indicus.* *A*, vertical ; *B*, transverse section ; *p*, pericarp ; *n*, base of fruit ; *E*, seed ; *f*, fold of pericarp. Magnified. (Moeller.)

The student should observe

(*a*) The *sub-reniform shape of the fruit*,

and should cut it longitudinally and transversely, noting

(*b*) The *shape of the seed*,

(*c*) The characteristic *ingrowths from the pericarp*.

He should further notice that the seed is bitter, but the pericarp is almost tasteless.

**Constituents.**—The seed contains from 1 to 1·5 per cent. of an intensely bitter, crystalline principle, picrotoxin, accompanied by a crystalline but tasteless body, cocculin, and a large proportion of fat. Picrotoxin, $C_{45}H_{50}O_{19}$ (m.p. 199°), contains no nitrogen and is therefore not alkaloidal, nor does it possess glucosidal properties. It is readily separable into toxic picrotoxinin, $C_{15}H_{16}O_6$, and non-toxic picrotin, $C_{15}H_{18}O_7$, but its further constitution is as yet unknown.

From the pericarp, which is tasteless, two alkaloids, menispermine and paramenispermine (Pelletier and Couerbe, 1833) have been isolated but they require re-investigation.

**Uses.**—Cocculus indicus is now used almost exclusively for the preparation of picrotoxin, which is a powerful convulsive poison; it has been given internally to check the night-sweating of phthisis; and has also been employed to destroy pediculi. The power possessed by the fruits, when thrown into water, of stupefying fish has long been known, and is due to the picrotoxin contained in the seed. So susceptible are fish to the influence of picrotoxin, that they have been used as a means of detecting its presence. A number of other plants, however, share this property with Cocculus indicus.

## POPPY CAPSULES

### (Poppy Heads, Fructus Papaveris)

**Source, &c.**—The opium poppy, *Papaver somniferum*, Linné (N.O. *Papaveraceæ*), is probably a native of Asia Minor, but is now cultivated in many warm or temperate countries both as a garden plant and for the sake of its fruits and seeds. The plant is an erect herbaceous annual; it varies very much in the colour of the petals, as well as in the shape of the fruit and colour of the seeds. In England a variety with pale flowers and whitish seeds is cultivated for medicinal use.

The fruits are of a pale glaucous green when young, and exude when wounded a bitter, white, milky juice (latex); as they ripen they change to yellowish brown, and are then cut from the stems. In Germany the unripe fruits are considered to be more active than the ripe.

**Description.**—Poppy heads vary very much both in shape and size. Some varieties are ovoid, others are nearly globular, others again depressed both at the summit and base, the latter variety attaining 8 cm. or more in diameter. The fruit is pale yellowish brown, often marked with darker spots, glabrous, and crowned with the persistent remains of twelve to fifteen stellate, sessile stigmas; below, it is contracted into a neck which is swollen just above the point of attachment to the peduncle and marked there with the scars of the petals and sepals. Cut transversely the fruit is seen to be unilocular, but formed by the union of as many carpels as there are stigmas. From the inner surface of the thin, brittle pericarp, yellowish, membranous placentas, corresponding in number to the carpels, project into the cavity of the fruit.

The seeds, which for the most part lie loose in the fruit, are minute and very numerous. Under a lens they may be seen to be reniform in shape and covered with distinct, delicate reticulations. They vary in colour from whitish to slate (the latter being known as maw seed), and contain an oily endosperm.

Poppy capsules have no odour; the seeds have an oily taste, but the pericarp is distinctly bitter.

The student should observe

(*a*) The *sessile stellate stigmas*,

(*b*) The *membranous placentas*,

(*c*) The *reniform shape of the seed* and the *reticulations on its surface*.

**Constituents.**—Poppy capsules contain the principal constituents of opium, the most important of which is the crystalline alkaloid morphine. Ripe capsules have been found to contain 0·018 (Malin, 1906), 0·16 (Dieterich), and 0·28 (Paul and Cownley) per cent. of morphine; unripe 0·06 (Malin) and 0·086 (Dieterich) per cent.; there is, therefore, some uncertainty as to the actual and the relative medicinal value of ripe and unripe fruits.

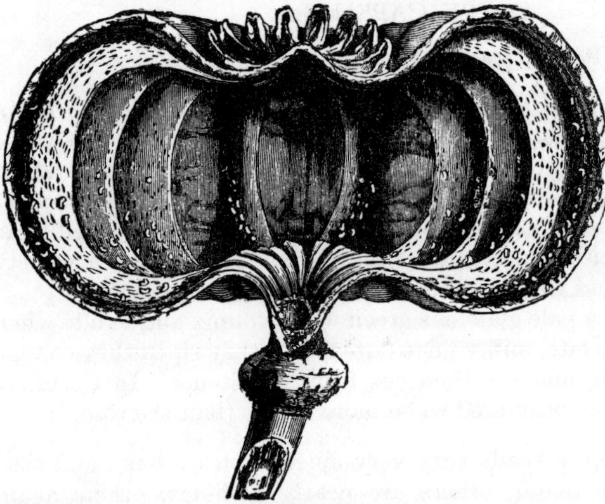

Fig. 50.—Poppy capsule, cut vertically. (Planchon and Collin.)

Fig. 51. — Poppy seed. Magnified. (Bentley and Trimen.)

Poppy capsules also contain meconic acid, an organic acid found only in the latex of the opium poppy. Although this acid is not an active constituent from a therapeutic point of view, the detection of its presence is often important as indicating a preparation of poppy capsules or of opium.

The seeds are free from morphine, but are said to contain traces of narcotine and amorphous alkaloid; the principal constituent in them is the pale yellow fixed oil (see 'Poppy Oil'); it is a drying oil and is used by artists as well as for culinary and various technical purposes. They occasionally (Russian seeds) contain henbane fruits.

**Uses.**—The action of poppy capsules is the same as that of opium, but much weaker. The warm decoction is a favourite anodyne fomentation. The extract and syrup are uncertain remedies, and preparations of opium are in every respect preferable.

## BAEL FRUIT

### (Indian Bael, Fructus Belæ)

**Source, &c.**—Indian bael is the fruit of *Ægle Marmĕlos*, Correa (N.O. *Rutaceæ*), a tree attaining a height of 12 metres and growing both wild and cultivated throughout the entire Indian peninsula. Being a sacred tree the Hindus plant it near their temples. It became known to the Portuguese as a remedy for dysentery when they occupied the eastern shores of India, the pulp of the half ripe fruit being eaten while fresh. It was introduced into European medicine about the middle of the last century, but the dried fruit appears to be much less efficacious than the fresh.

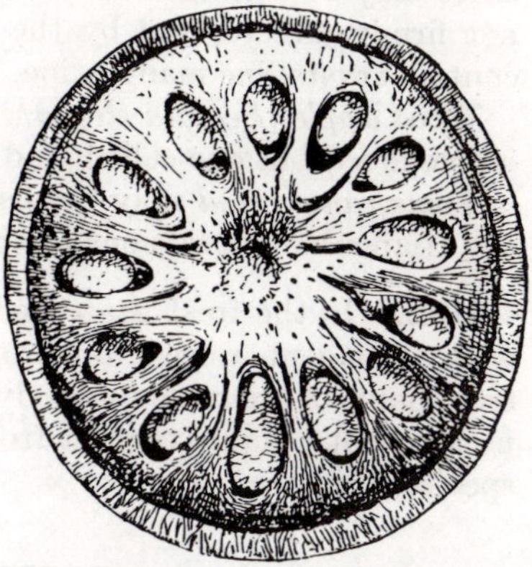

FIG. 52.—Bael fruit. Transverse section of a small specimen. Natural size. (Holmes.)

**Description.**—The fruit is ovoid or rounded, and about the size of an orange, although sometimes rather larger. Externally it is yellowish brown, smooth, or slightly granular and hard, and bears a circular scar at the point of attachment of the peduncle. The entire fruit is too hard to cut with a knife, but may be sawn across, and will then be found to consist of a reddish woody rind about 1·5 mm. thick, enclosing from ten to fifteen carpels, each containing several hairy seeds embedded in a transparent yellowish or red mucilage. The dried pulp, which is mucilaginous and aromatic when fresh, is hard, and varies in colour from pale to dark red ; it frequently breaks away from the rind during the drying, leaving only a thin layer attached to it. Even the dried fruit has an agreeable odour and mucilaginous, sometimes also aromatic, taste.

The fruit is frequently imported in dried quarters or in transverse slices. The latter have an appearance similar to that of the transverse section of the entire fruit, but the pulp usually adheres firmly to the rind and has a darker colour externally, being paler within. Preference should be given to unripe fruits, or the slices cut from them, and these may be recognised by the small undeveloped seeds they contain ; they are also less aromatic than the ripe fruits. In this respect the entire fruits are often inferior to the slices.

The student should observe

(*a*) The comparatively *smooth hard rind* to which (in the sliced fruit) the *reddish pulp firmly adheres*,

(*b*) The *numerous carpels*,

(*c*) The *hairy seeds embedded in mucilage*.

**Constituents.**—As far as is known the principal constituent of bael fruit is mucilage. Traces of tannin are present in the ripe but not in the unripe fruit.

**Substitutes.**—Several substitutes for bael have been met with, viz.: *Mangosteen fruits* (*Garcinia Magnostāna*, Linné, N.O. *Guttiferæ*); these may be distinguished by the darker rind to which the pulp does not firmly adhere, and by the wedge-shaped, radiate stigmas; they contain crystalline mangostine, tannin, and resin.

*Wood apple* (*Feronia elephantum*, Correa, N.O. *Rutaceæ*).—The fruit is five-lobed but one-celled, and has a rough exterior.

*Pomegranate rind* may be distinguished by its astringent taste and the triangular impressions of the seeds.

**Uses.**—In the fresh state Indian bael is a pleasant refreshing fruit with astringent, refrigerant properties which render it valuable in the treatment of diarrhœa and dysentery. As imported it is probably useless, but a liquid extract from the *fresh* fruit appears to possess its specific effects.

## BITTER ORANGE

### (Seville Orange, Fructus Aurantii)

**Source, &c.**—The bitter or Seville orange is the fruit of *Citrus Aurantium*, var. *Bigaradia*, Hooker filius (*C. Aurantium*, Linné; *C. vulgaris*, Risso, N.O. *Rutaceæ*), a small tree, probably a native of north-eastern India, but cultivated in most warm countries. In Europe it is grown in the countries bordering on the Mediterranean, whither it was probably brought by the Arabs, as it was unknown to the Greeks and Romans. The fruit is collected before it is quite ripe, packed in boxes and exported, the ripening being completed during the voyage. Bitter oranges are shipped chiefly from southern Spain (Seville and Malaga) and from Sicily (Messina and Palermo), Seville oranges being considered to be the best for medicinal use.

In addition to the fresh fruit the dried peel is also largely imported from Malta, Spain, and Morocco.

Indian orange peel (*Cortex Aurantii Indicus*) is obtained from varieties of *Citrus Aurantium*, Linné, grown in India and Ceylon; it closely resembles bitter-orange peel and may be employed in its place in India and the Eastern divisions of the Empire.

**Description.**—The ovary of the orange tree is superior and polycarpellary; it is composed of eight (or sometimes more) carpels, each containing two rows of seeds with axile placentation. During the growth of the ovary the loculi increase in size, and the outer wall (pericarp) in thickness. Into the loculi there grow from the inner epidermis of the pericarp numerous hair-like processes which fill with juice and completely occupy the loculi; they form the pulp contained

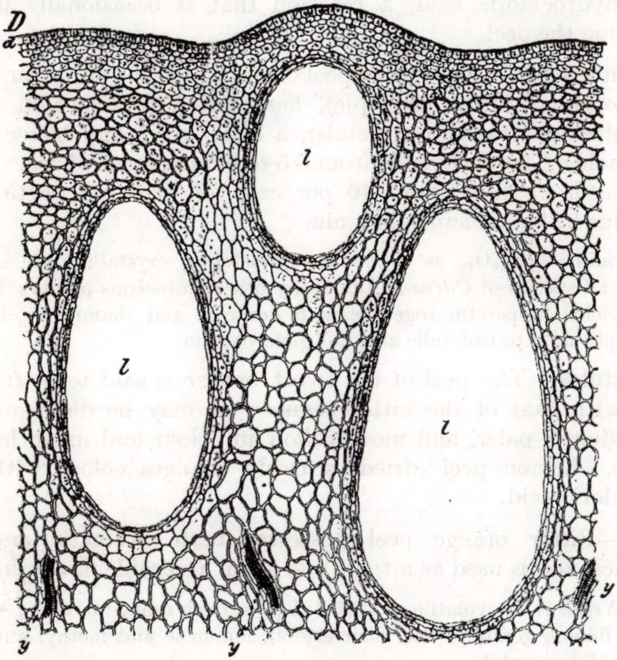


FIG. 53.—Orange peel. *D*, transverse section, enlarged 100 diam.; *l*, oil-glands. (Berg.)

in the 'quarters' of the orange, the thin membrane enclosing each quarter being the endocarp. The fruit is a variety of berry sometimes termed 'hesperidium.'

The bitter orange may be distinguished from the sweet orange by its rougher and usually darker rind and sour, bitter pulp. The rind, which is agreeably aromatic and has also a bitter taste, shows, when cut transversely, a narrow yellow outer part corresponding to the epicarp, in which large oil-glands are discernible with the naked eye, and an inner white portion corresponding to the mesocarp.

The peel is used both fresh and dried. In cutting it from the fruit care should be taken not to rupture the oil-glands more than is necessary, for it is to the volatile oil they contain that the pleasant aroma of the peel is due ; at the same time too much of the white 'zest' should not be removed, as that is lacking in bitterness. It may readily be dried in a warm room but large quantities of the dried peel are imported, especially from Malta ; in the latter case it is generally in the form of very narrow strips cut by machinery ('gelatin cut'), or in wider spiral strips or in four 'quarters.' English cut peel is generally rather thick, but has a fine colour and aroma. The transverse section assumes a dark green colour when moistened with strong hydrochloric acid, a reaction that is occasionally useful in identifying the peel.

**Constituents.**—Bitter orange peel contains volatile oil, aurantiamarin (an amorphous bitter principle), hesperidin, isohesperidin, hesperic acid (colourless, tasteless crystals), a bitter resin and bitter aurantiamaric acid The peel yields from 3·5 to 6·5 per cent. of ash.

The seeds contain about 40 per cent. of fixed oil, pectin and a crystalline bitter principle, limonin.

Hesperidin, $C_{22}H_{26}O_{12}$, is a colourless, tasteless, crystalline glucoside that occurs in all species of *Citrus* and in many other Rutaceous plants ; by hydrolysis it yields hesperetin together with dextrose and rhamnose ; hesperetin may be split up into isoferulic acid and phloroglucin.

**Substitutes.**—The peel of the sweet orange is said to be frequently mixed with that of the bitter orange ; it may be distinguished by being thinner, paler, and more yellow in colour and much less bitter in taste. Lemon peel (dried) scarcely changes colour with strong hydrochloric acid.

**Uses.**—Bitter orange peel possesses both aromatic and bitter properties, and is used as a tonic and as an agreeable flavouring agent.

*Oil of Neroli* is the volatile oil distilled from fresh orange flowers ; chief constituents linalol, geraniol (and their esters), limonene and methyl anthranilate (ortho-amidobenzoate).

*Orange Flower Water* is a saturated aqueous solution of the volatile oil obtained during the distillation ; the residue contains the bitter principle limonin (naringin).

*Oil of Orange* is the volatile oil obtained from the rind of the bitter orange ; chief constituents *d*-limonene, citral, citronellal, decyl alcohol, methyl anthranilate.

*Oil of Portugal*, the volatile oil from the rind of the sweet orange, has similar constituents.

*Oil of Petit Grain*, the volatile oil obtained originally from the immature fruits but now from the leaves and twigs, has also a very similar composition.

## LEMON

### (Fructus Limonis)

**Source, &c.**—The lemon is the fruit of *Citrus Medica*, Linné, var. *β-Limonum*, Hooker filius (N.O. *Rutaceæ*), a small tree which, like the orange, is probably a native of northern India. It is cultivated in all countries bordering on the Mediterranean, especially in Sicily and southern Italy, in Spain and Portugal, and in the Riviera.

The fruits are gathered whilst they are still green; the finest are wrapped in paper and exported in cases of 200 (Murcia lemons) to 360 (Messina lemons); less sightly fruits are packed in barrels and preserved with salt water. The finest lemons are those imported from Sicily (Messina and Palermo); those from Murcia are also of high quality, while Naples and Malaga lemons are less esteemed.

**Description.**—The lemon resembles the orange in development and in structure, but it is easily distinguished by its more elongated, ovoid shape and pale yellow colour; at the apex it is crowned by a distinct nipple. The pulp has a strongly but agreeably acid taste; the peel is aromatic and bitter.

The following parts of the fruit are official: the fresh peel, the juice, and the volatile oil obtained from the peel.

1. *Lemon peel* is pale yellow and more or less rough on the outer surface, whitish on the inner. The transverse section shows numerous large oil-glands embedded in the tissue. The peel has a strong fragrant odour and aromatic, bitter taste. It contains volatile oil and hesperidin, and is used chiefly as an agreeable flavouring agent.

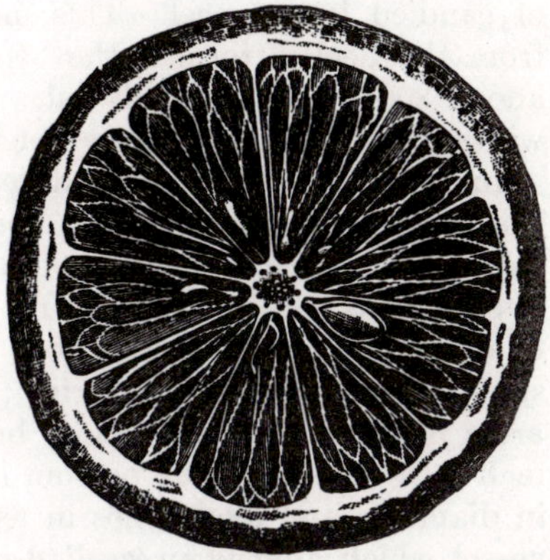

FIG. 54.—Lemon. Transverse section. (Planchon and Collin.)

2. *Lemon juice.*—Fresh lemons yield about 30 per cent. of juice, which, for pharmaceutical use, should be pressed from the fresh fruit. It is a turbid yellowish liquid with a characteristic odour and acid taste. The principal constituent is citric acid, of which it contains from 6·7 to 8·6 per cent. (30 to 40 grains in each fluid ounce). The amount of citric acid in the juice is largest in lemons imported in December and January, and smallest in August, both the fruit and the juice itself gradually diminishing in acidity when kept.

Large quantities of lemon juice are pressed in Sicily from the pulp that is left in the production of the volatile oil, the residual cake being used as cattle food. The juice is concentrated to a specific gravity of 1·233 to 1·235 and exported, chiefly to England, for the manufacture of citric acid; or the citric acid is precipitated as calcium citrate from which the citric acid is subsequently regenerated.

3. *Oil of Lemon.*—Various methods are used for the production of

oil (essence) of lemon, but the following appears to be largely adopted in Sicily, whence most of the oil is exported :—

The lemons are first cut lengthwise into three or four pieces, the pulp removed and pressed, and the peel set aside in a cool place for some hours. The workman holds in his left hand a medium-sized sponge, smaller ones being placed between the fingers. With the right hand he takes a piece of peel and squeezes it in such a way as to break the oil-glands and discharge their contents on the sponges. When sufficient has accumulated the sponges are pressed ; the liquid thus obtained separates on standing into a lower watery stratum and an upper clear oily layer which can be poured off. The oil thus obtained is filtered and exported in coppers of varying sizes. This hand-pressed peel is then pickled in brine and sold to the manufacturers of candied lemon peel. This method of separating the volatile oil from the peel is termed the 'sponge' process ; 1,000 lemons yield about 800 gm. of volatile oil. That which is obtained during the winter from lemons that are not fully ripe is considered the best.

In the south of France a different procedure is adopted, of which the following description has been given : 'The object being to set free and to collect the oil contained in the vesicles of the peel, an apparatus is employed which may be thus described : a stout saucer or shallow basin of pewter, about 25 cm. in diameter, with a lip on one side for convenience of pouring ; fixed in the bottom of this saucer are a number of stout, sharp, brass pins standing up about half an inch ; the centre of the bottom is deepened into a tube about an inch in diameter and five inches in length, closed at its lower end. This vessel, which is called an *écuelle à piquer*, has therefore some resemblance to a shallow, dish-shaped funnel, the tube of which is closed below.

'The workman takes a lemon in the hand and rubs it over the sharp pins, turning it round so that the oil-vessels of the entire surface may be punctured. The essential oil which is thus liberated is received in the saucer, whence it flows down into the tube ; and as this latter becomes filled it is poured into another vessel that it may separate from the turbid aqueous liquid that accompanies it. It is finally filtered, and is then known as *Essence de Citron au Zeste*' (Flückiger and Hanbury).

A machine (scorzetta), constructed on a similar principle, has been devised by which the scarification of the lemons is much more rapidly effected than is possible by the tedious hand process, and by which at the same time an increased yield of oil is obtained ; it has not, however, as yet been generally adopted.

Oil of lemon is also prepared by distilling the peel with water, but this distilled essence is inferior in fragrance to that prepared by either of the foregoing processes.

Oil of lemon is a pale yellow liquid of sp. gr. 0·857 to 0·860 and optical rotation $+ 58°$ to $+ 64°$, consisting of about 90 per cent. of the terpene limonene and about from 4 to 6 per cent. of the aldehyde citral, $C_{10}H_{16}O$, the latter being the chief odorous constituent of the oil. Other odorous constituents are citronellal, geranyl acetate, linalyl acetate, octyl and nonyl aldehydes, &c. 'Terpeneless' essence of lemon is prepared by distilling off the terpene *in vacuo*, and consists mainly of citral (50 per cent.) together with the above-mentioned esters and aldehydes.

**Allied Drugs.**—*Bergamot.*—The fruit of *Citrus Bergamia*, Risso, cultivated in southern Italy and Sicily. The volatile oil is obtained by rotating the fruits in a machine against sharp copper points, collecting and filtering the mixture of volatile oil and cell sap and decanting. The pulp is pressed and the juice used for the production of calcium citrate or citric acid. The oil contains linalyl acetate (34 to 40 per cent.), *d*-limonene and bergaptene.

*Lime.*—The fruit of *Citrus Medica*, var. *acida*, Brandis (West Indian Lime) or of *Citrus Limetta*, Risso (Italian Lime). The volatile oil of the former has an odour of citronella and contains citral and limonene ; that of the latter an odour of bergamot and has a composition similar to bergamot oil but contains less linalyl acetate (26 per cent.). The pulp of both contains citric acid (about 7 per cent.).

## RAISINS

### (Uvæ Passæ)

**Source, &c.**—The grape vine, *Vitis vinifera*, Linné (N.O. *Ampelidĕæ*) is cultivated in numerous varieties in most of the warmer temperate regions, particularly in central and southern Europe, California, and Australia.

When the grapes are ripe most of the leaves are stripped from the vine, and the stalk of the bunch is cut half through or twisted ; the grapes are then allowed to dry by the heat of the sun. This process, which occupies two or three weeks, is often hastened by cutting the bunch off when partially dried, sterilising the fruits by dipping them into a boiling alkaline solution and drying on wicker trays in the sun.

**Description.**—The ovary of the grape vine is superior and two-celled, each cell containing two ovules. As the fruit reaches maturity the interior becomes pulpy, the dissepiment disappears, and the ripe fruit may be regarded as a superior berry. It is usually ovoid or nearly globular in shape, and varies in colour according to the variety. The dried fruits are of a purplish black or brownish colour, and more or less shrivelled.

**Constituents.**—Raisins contain in the pulp dextrose (grape sugar) and acid potassium tartrate, both of which may be found crystallised in small granular masses in old raisins. The skin and the seeds contain tannin, and the latter a fixed oil in addition.

**Varieties**—*Muscatels*: these are considered the best and are exported from Malaga and Valencia; the clusters of purplish black fruits are usually packed in boxes in layers.

*Valencias*: these are also exported from Valencia, but are generally picked from their stalks; they are of a rather dark golden brown colour.

*Turkish*: a light brown, elongated raisin.

*Sultanas*: a seedless variety grown in Turkey, Greece, and Persia; exported chiefly from Smyrna.

*Currants*: a small seedless variety grown on the Ionian Islands and in the Grecian archipelago.

**Uses.**—Raisins possess demulcent, refreshing, and nutrient properties, and are also slightly laxative.

## CASSIA PODS

(Purging Cassia, Fructus Cassiæ Fistulæ)

**Source, &c.**—Cassia pods are the ripe fruits of *Cassia Fistŭla*, Linné (N.O. *Leguminosæ*), a tree of moderate size, indigenous to India, and often cultivated as an ornamental plant (Indian laburnum). The tree bears a pendulous raceme of fragrant flowers, each with a one-celled many-seeded ovary, which develops into a long, leguminous fruit. As the latter ripens the seeds become separated from one another by the formation of numerous thin, transverse, spurious dissepiments, and the fruit, which was originally one-celled, becomes spuriously many-celled. It differs from a typical legume in being indehiscent as well as many-celled, and may be described therefore as a many-celled indehiscent legume.

**Description.**—The ripe pods, as met with in commerce, are long, nearly straight, cylindrical, dark chocolate-brown fruits attaining 45 to 60 cm. in length and 20 to 25 mm. in thickness. The surface, which appears smooth and shining to the naked eye, is seen under a lens to be marked with minute transverse striations; both the dorsal and ventral sutures are evident but not prominent. To one end of the fruit a short stalk is attached; the other terminates very abruptly in a short point.

The pericarp, although thin, is hard and woody. The interior of the fruit is divided into a number of compartments by transverse dissepiments placed about a quarter of an inch apart. Each compartment contains a single seed attached to the ventral suture by a long, dark, thread-like funiculus. A thin layer of nearly black firm pulp, which in the fresh fruit is soft and fills the compartment, adheres to each side of the membranous dissepiments. In very dry pods the pulp is often so much contracted that the seeds lie loose in the cavity and rattle when shaken; these are less esteemed.

The seeds are flattened-ovoid, reddish brown, smooth, and extremely hard. When cut transversely they exhibit a curved yellow embryo obliquely crossing a whitish horny endosperm. The pulp has a sweetish taste and a somewhat sickly odour.

The pods are exported from the West Indies (Dominica) and from Java. From the West Indies they arrive in cylindrical baskets made of plaited split canes.

The student should observe

(*a*) The *smooth surface and long cylindrical shape,*

(*b*) The *spurious dissepiments with adhering pulp,*

(*c*) The *odour* and *taste.*

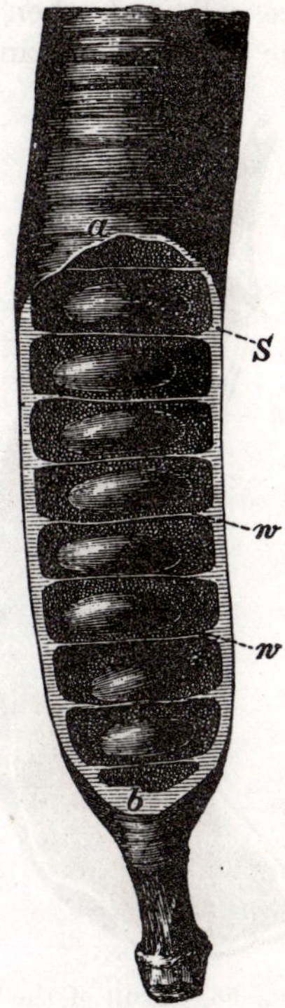


FIG. 55.—*Cassia Fistula.* Lower portion of a pod partly opened to show the seeds. *S*, seed; *w*, transverse dissepiments; *b*, pericarp of fruit. (Moeller, after Wiesner.)

**Constituents.**—The pulp, which is the only official part of the fruit, is separated by crushing the fruits, macerating them with water, straining the liquid, and evaporating it to a soft extract. It contains about 50 per cent. of sugar and also oxymethylanthraquinones (not yet investigated); both of these probably contribute to its laxative action.

**Substitutes.**—Pods of *Cassia grandis*, Linné (horse cassia). Longer, thicker and heavier than those of *C. Fistula*; laterally compressed; surface rough; one prominent ridge on the dorsal and two on the ventral suture; odour of pulp disagreeable, taste bitter and astringent.

Pods of *C. moschata*, Humboldt, Bonpland and Kunth. Smaller and narrower than those of *C. Fistula*; pulp paler, odour musk-like.

**Uses.**—Cassia pulp is laxative, but is seldom used except as an ingredient in confection of senna.

## TAMARINDS

### (Tamarindus, Fructus Tamarindi)

**Source, &c.**—The tamarind tree, *Tamarindus indica*, Linné (N.O. *Leguminosæ*), a large and handsome tree indigenous to tropical Africa is, cultivated throughout India and the West Indies, where it forms a valuable article of diet. The Arabian name of the fruit (Tamare Hindi, or Indian date) would indicate that the Arabians became acquainted with it from the Hindus, and probably introduced it into Europe.

The tree produces an erect raceme of flowers succeeded by indehiscent legumes from 5 to 20 cm. in length. The epicarp of the legumes is rough, brownish, firm, but easily broken and separated from the pulp; the latter (mesocarp) is juicy and acid, and traversed by a number of stout branching fibrovascular bundles. The seeds are few in number and enclosed in a leathery membrane (endocarp).

In the West Indies tamarinds, exported from Barbados and Antigua, are collected when ripe, and prepared for the market by removing the epicarp, packing them in barrels, and preserving them with hot syrup. In India the pulp is simply pressed into cakes, with or without the addition of 10 per cent. of salt as a preservative. Small quantities of tamarinds are also exported from other countries (Egypt, &c.).

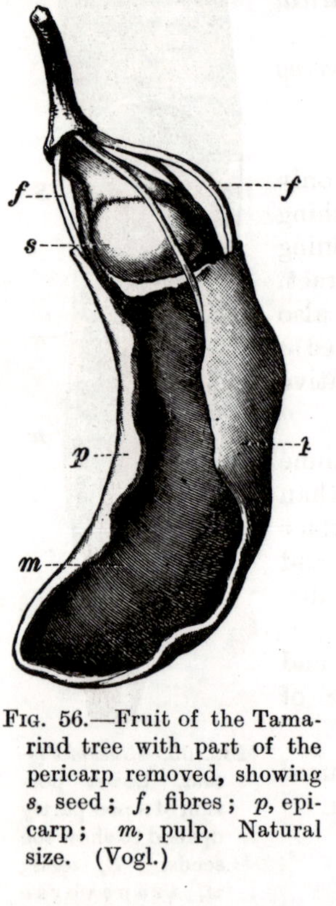


FIG. 56.—Fruit of the Tamarind tree with part of the pericarp removed, showing *s*, seed; *f*, fibres; *p*, epicarp; *m*, pulp. Natural size. (Vogl.)

**Description.**—The official tamarinds are the fruits preserved with syrup, as above described. They form a reddish brown. moist, sugary mass, in which the branching fibres and hard, reddish brown, obscurely quadrangular seeds enclosed in the leathery endocarp can readily be found. They have an agreeable odour and a sweet and acid taste.

From India tamarinds are sent in the form of firm, black cakes consisting of the pulp of the fruit, together with the fibres, seeds, and small portions of the epicarp. Although not official it is frequently employed, and yields a good pulp. The cakes have scarcely any odour, but a rather strongly acid taste.

**Constituents.**—Tamarind pulp, though varying in composition, contains about 10 per cent. of free tartaric acid, about 8 per cent. of acid potassium tartrate, and from 25 to 40 per cent. of invert sugar. The total acidity varies from 11 to 16 per cent. West Indian tamarinds contain much added sugar and proportionately less acid.

**Uses.**—Tamarinds are a pleasant acid refrigerant and gentle laxative.

## SENNA PODS

### (Senna Legumes, Sennæ Fructus)

**Source, &c.**—Senna pods are the dried ripe fruits of *Cassia acutifolia*, Delile (Alexandrian senna pods), and of *C. angustifolia*, Vahl (Indian senna pods) (see "Senna").

**Description.**—The pods are brown or greenish brown or pale green and brownish in the centre over the seeds, broadly oblong, very flat and rounded at each extremity and very slightly mucronate at the apex. They are about 5 cm. long and 2 to 2·5 cm. wide and contain several pale greyish green, hard, flat, irregularly reticulated, obovate-cuneate seeds. Indian senna pods are usually rather darker than Alexandrian and slightly narrower and the remains of the base of the style are rather more distinct.

**Constituents.**—The constituents are presumably the same as those of senna.

**Uses.**—Senna pods are used as a laxative and are considered to be more certain than the leaves and to cause less griping. The liquid extract made from the fruits (with seeds) appears to cause more pain than one made from the pericarps alone; the reason for this is not known.

## PRUNES
### (Prunum)

**Source, &c.**—The prune, or French plum, is the dried fruit of *Prunus domestica*, Linné, var. *Juliana*, de Candolle (N.O. *Rosaceæ*), a variety of the plum tree cultivated in France, especially in the valley of the Loire.

The plums are collected when ripe, and dried by exposing them to the sun for a day or two, after which the drying is completed in an oven or drying apparatus, the temperature being raised in stages from 45° to 90°. A more rapid process consists in drying in a special 'evaporator' on wicker trays.

**Description.**—Prunes are of irregular flattened-ovoid shape, and about 3 cm. long. The surface is shrivelled and nearly black; the pulp, which should be soft and fleshy, is dark-coloured and surrounds a hard oval flattened stone; the latter is broadly rounded at one end and provided at the other with a shallow, slightly oblique depression. They have an agreeable odour and sweet fruity taste.

The seeds contain about 45 per cent. of fixed oil and also amygdalin (see p. 163).

**Varieties.**—The best prunes are the French from the Loire districts, but California now exports large quantities of excellent fruit, and smaller amounts come from Bosnia, Spain, and Portugal.

**Constituents.**—The pulp contains about 40 per cent. of dextrose, about 2 per cent. of vegetable acids (malic, tartaric, &c.), and about 30 per cent. of water, the composition varying with the season, degree of ripeness, &c.

**Uses.**—The prune is nutritive, demulcent, and slightly laxative; it is frequently ordered as an article of diet in habitual constipation.

## HIPS

(Fructus Rosæ Caninæ ; Fructus Cynosbati)

**Source, &c.**—Hips are the fresh, ripe fruits of the dog rose, *Rosa canina*, Linné (N.O. *Rosaceæ*), and other indigenous allied species.

**Description.**—The fruit of the dog rose consists of the hollowed thalamus on the inner surface of which a number of achenes are borne, each provided with a style and stigma, the whole forming a spurious fruit distinguished by the name 'cynarrhodon.'

The ripe fruit is ovoid, smooth and shining, and scarlet or red in colour. It is crowned with five calyx-teeth, beyond which a dense tuft of hair-like styles slightly protrudes. Cut longitudinally it is seen to consist of a fleshy, deeply concave receptacle, on the inner surface of which are borne a number of hairs as well as small, very hard, hairy achines. The fleshy receptacle has an agreeable acid taste, and is separated for pharmaceutical use from the achenes, &c., by beating the fruits in a mortar and rubbing the pulp through a hair-sieve.

The only other wild rose sufficiently common to serve as a source of the fruit is the field rose, *R. arvensis*, Hudson ; it is distinguished from the dog rose by its trailing habit and globular fruits, from which the styles protrude in the form of a distinct column.

The student should observe

(*a*) The *fleshy hollow receptacle*,
(*b*) The *small hard hairy achenes* within it.

**Constituents.**—The pulp contains citric and malic acids (3 per cent.), invert sugar (11 to 15 per cent.), and tannin (2 per cent.).

**Uses.**—Confection of hips is occasionally used as a pill excipient, or mixed with water to form an agreeable acid cough linctus.

## MYROBALANS

(Myrobalanum ; Black Chebulic Myrobalans)

**Source, &c.**—The official myrobalans, which are properly termed black chebulic myrobalans, are the dried immature fruits of *Terminalia Chebula*, Retzius (N.O. *Combretaceæ*), a large tree abundant in northern India. The fruits are collected at varying stages of development. If collected when about the size of a raisin and dried they constitute the official drug ; if allowed to grow to the size of a walnut they form the valuable tanning material known commercially as chebulic myrobalans or simply myrobalans.

**Description.**—Black chebulic myrobalans are dark brown or nearly black in colour, ovoid or fusiform in shape and from 10 to 30 mm. or more in length and 5 to 15 mm. in width; they are much shrivelled longitudinally, hard and brittle; the fracture is dark and shiny and in the centre there is often a small cavity; they have no odour but a very astringent taste.

**Constituents.**—The fruits contain from 20 to 30 per cent. of a mixture of gallic acid and tannic acid, apparently derived from an organic acid, chebulinic acid; they contain also a greenish oleo-resin which has been termed myrobalanin.

**Uses.**—Notwithstanding the large proportion of tannin they contain black chebulic myrobalans, taken internally, act as an efficient purgative. They are also used externally as an astringent in the place of galls.

**Note.**—Commercial myrobalans are the fruits collected when about the size of a walnut and dried; they are oval, pointed at each end, brownish or greenish yellow in colour and ribbed. They contain about 40 per cent. of tannin and are imported in large quantities for use in tanneries.

## PIMENTO

### (Allspice, Jamaica Pepper, Fructus Pimentæ)

**Source, &c.**—The pimento tree, *Pimenta officinalis*, Lindley (N.O. *Myrtaceæ*), is a handsome tree indigenous to and common in the West Indies, and found also in Central America and Venezuela. It is cultivated, especially in Jamaica, in plantations known as pimento walks.

The tree bears large panicles of fragrant, white flowers with two-celled inferior ovaries, each cell containing a single ovule. The flowers are succeeded by small fruits; as soon as these have attained their full size, but before they ripen, the whole panicle is collected and dried in the sun, during which the green colour changes to a reddish brown. The stalks are then separated and the fruits are ready for packing. Had they been allowed to ripen they would have become dark purple and filled with a sweet pulp, but would have lost much of their aroma; hence they are collected whilst unripe.

Pimento berries appear to have been long known to the Mexicans, who used them, together with vanilla, for flavouring chocolate; through them the Spaniards became acquainted with the spice towards the end of the sixteenth century, and introduced it into Europe, where great quantities were consumed. It was supposed to possess the combined flavour of cloves and other spices; hence its name of 'allspice.' It is also called Jamaica pepper.

**Description.**—Pimento berries are small, nearly globular, reddish-brown fruits from 5 to 8 mm. in diameter. They have a rough surface, and are crowned with the remains of the calyx in the form of a raised ring, in which the four calyx-teeth are scarcely discernible; in the centre of this ring are the short remains of the style. The pericarp is thin, woody, and brittle. At the base a small scar indicates the point of attachment of the pedicel.

Cut transversely, the fruit is seen to be two-celled, each cell containing a single flattened or irregularly plano-convex, nearly black seed, within which there is a dark coiled embryo but no endosperm.

The thin pericarp contains a large number of oil-glands embedded in it; many are situated just below the outer epidermis, and elevate it at intervals, thus producing the roughness of the pericarp; these glands are just visible in a transverse section when examined with a lens. The seeds also contain oil-glands, but are much less aromatic than the pericarp.

The odour of pimento berries is agreeably aromatic; it is especially perceptible when the fruits are crushed. The taste is warm and aromatic, resembling, but distinct from, that of cloves.

The student should observe

(*a*) The *remains of the calyx crowning the fruit*,
(*b*) The *two cells, each containing a single seed*;

and should compare these fruits with

(i) *Cubebs*, which are one-celled, dark greyish black or greyish brown in colour, and reticulated on the surface,
(ii) *Black pepper*, which is one-celled and one-seeded.

**Constituents.**—Pimento berries contain from 3 to 4·5 per cent. of volatile oil (sp. gr. 1·040 to 1·055; O.R. = — 2°) consisting principally of eugenol (about 65 per cent.), which is also the chief constituent of oil of cloves. They contain, further, a notable quantity of tannin and yield from 2·5 to 5 per cent. of ash.

**Substitutes.**—Dark (nearly ripe) fruits have been coloured with bole or brown ochre and passed off as genuine; the fraud may be detected by boiling them for a few moments with hydrochloric acid, filtering, and testing with potassium ferrocyanide, when at most a bluish green colour should be produced.

The fruits of *Pimenta acris*, Wight (compare p. 57), may be distinguished by their five calyx teeth. Mexican allspice, the fruits of *Eugenia Tabasco*, G. Don, are larger than the genuine, but neither so dark nor so aromatic. Both are of rare occurrence.

**Uses.**—Pimento is used as a flavouring agent and as an aromatic stimulant, resembling cloves in its action.

## POMEGRANATE RIND

### (Cortex Fructus Granati)

**Source, &c.**—The pomegranate tree, *Punica Granātum*, Linné (N.O. *Lythrariĕæ*), is a shrub or small tree indigenous to north-western India, but cultivated in the warmer parts of the temperate regions, especially in the countries bordering on the Mediterranean. More than one variety is known, but that producing crimson flowers is the commonest. The flowers are succeeded by handsome reddish fruits, about the size of a large orange, much esteemed for their agreeable juicy contents. The leaves and flowers, as well as the fruits, were employed in medicine by the ancients; the Romans used the peel of the fruit for tanning, and in the south of Europe it still serves the same purpose.

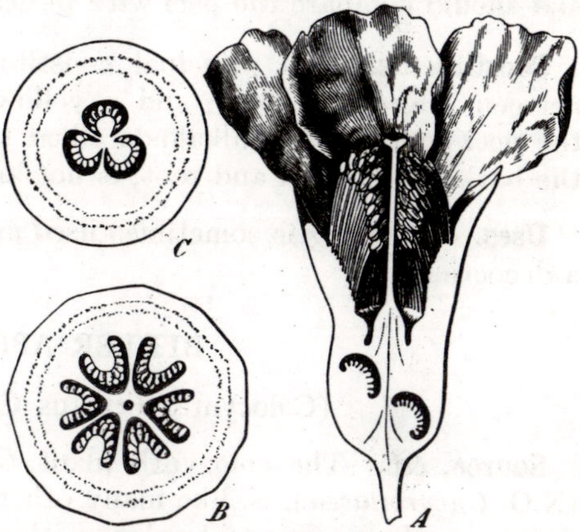

FIG. 57.—Pomegranate (*Punica Granatum*). *A*, flower cut vertically; *B*, transverse section of the upper part of the ovary, showing six loculi, magnified; *C*, transverse section of lower part, showing three loculi, magnified. (Luerssen.)

**Description.**—The pomegranate is a large handsome fruit of sub-globular shape and brownish yellow colour, passing into dull crimson red. It is crowned with a large, deeply five-toothed, tubular, coriaceous calyx, within which are the remains of the stamens and style. The pericarp, the outer surface of which is granular in appearance, is about 1·5 mm. thick, hard, and coriaceous. The large cavity is divided by thin, membranous dissepiments into a number of cells, each of which is completely filled with numerous seeds.

When cut longitudinally the fruit is seen to be divided by a membrane into an upper and lower portion. Transverse sections show the former to be divided by radiating dissepiments into six cavities, in each of which the placentation of the seeds is parietal, whilst in the lower portion from three to five cavities are irregularly arranged. The seeds are about 12 mm. long, subpyramidal in shape, and contain a sweetish acidulous juice in a translucent outer portion. The fruit is a berry sometimes distinguished by the special name 'balausta.'

The peel of the fruit, separated from the seeds, was formerly much used as an astringent. It occurs in thin curved fragments, granular and brownish yellow or reddish on the outer surface, uniformly yellowish brown within, and there exhibiting distinct, shallow, angular depressions left by the seeds ; to some fragments the remains of the coriaceous calyx are attached. It breaks with a short granular fracture, has little odour, but a powerfully astringent taste.

The student should observe

(a) The *astringent taste*,
(b) The *impressions of the seeds*,
(c) The occasional *fragments of the superior calyx* ;

and should compare the peel with pieces of Indian bael.

**Constituents.**—The principal constituent is gallotannic acid (28 per cent., Kramer, 1883), and a yellow colouring matter ; whether the peel contains any alkaloids allied to, or identical with, those of the bark of the stem and root, is not known.

**Uses.**—The drug is sometimes used as an astringent in the form of a decoction.

## BITTER APPLE

### (Colocynth, Fructus Colocynthidis)

**Source, &c.**—The colocynth plant, *Citrullus Colocynthis*, Schrader (N.O. *Cucurbitaceæ*), is, like many other members of the same natural order, a scabrous, prostrate plant with a perennial root. It enjoys an extensive distribution throughout northern Africa, Syria, and north-western India. In the African and Egyptian deserts it frequently occurs in enormous quantities ; it is cultivated in Syria, Spain, Cyprus, and the north-western provinces of India.

The remarkable properties of this fruit must necessarily have attracted attention very early. Dioscorides and Pliny were both acquainted with it, and so also were the Arabian physicians, by whom probably its use was introduced into Europe.

The fruit, which is about the size of an apple, is collected when ripe, dried and then freed from the thin rind by peeling with a sharp knife, the peeled fruits being often termed 'apples.' Persian colocynth is peeled before drying and consequently shrinks considerably. Mogadore colocynth is imported unpeeled and commonly used for filling the show-jars in the pharmacist's window.

**Description.**—The fruit of the colocynth plant is, when young, three-celled and bears numerous ovules attached to axile placentas. The placentas, however, grow from the centre towards the circumference of the fruit, and when they have nearly reached the pericarp divide in two, each half curving inwards and bearing numerous seeds on its margins. During these changes the carpellary walls disappear, and the fruit, originally three-celled, becomes spuriously one-celled. The pulp consists largely of the fleshy placentas; they usually split in a radial direction throughout the greater part of their length, and in the dried fruit they are deeply fissured.

As seen in commerce, colocynth fruits form more or less broken white balls, 5 cm. in diameter and light in weight. The rind has generally been carefully peeled off, and the outer surface is formed by part of the thin, whitish mesocarp, which, however, has in places itself been removed, disclosing the yellowish or nearly brown seeds, or the

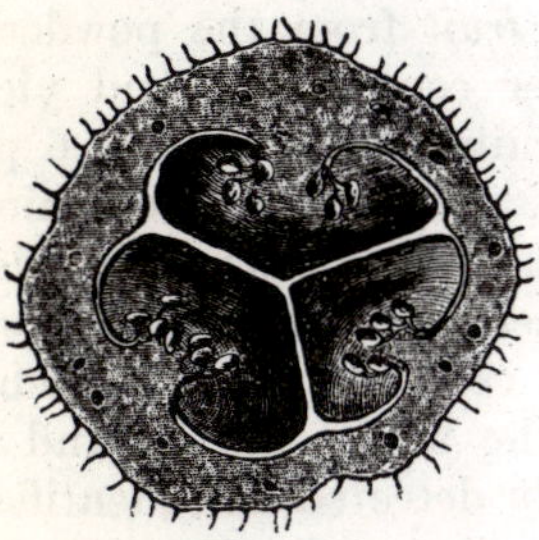
A

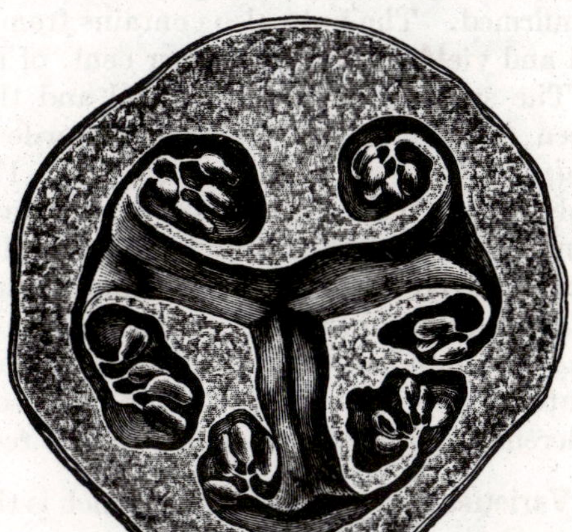
B

FIG. 58.—Colocynth fruit. *A*, transverse section of young fruit, showing the placentas dividing near the margin and curving inwards; *B*, ripe fruit, showing each placenta fissured. (Moeller.)

fleshy placentas. If a fruit is cut transversely the placentas will generally exhibit radiating fissures dividing the fruit into three portions, in each of which, near the periphery, a half of each placenta curves inwards and bears on its inner margin several vertical rows of seeds. These number from 200 to 300, and constitute about three-quarters of the weight of the fruit; they vary in colour from yellowish white to dark brown according to the degree of ripeness. They have a flattened ovoid shape, possess a hard seed-coat, and contain a small oily kernel which, however, strange as it may seem, forms, when properly prepared, a valuable addition to the scanty diet of certain tribes of Arabs.

For medicinal use the light, whitish, spongy, pith-like pulp, the bulk of which is furnished by the placentas, should be freed from the seeds, as the pulp alone is official. It possesses a slight odour, but an intensely bitter taste, the seeds when quite free from pulp being almost tasteless.

Much of the colocynth at present imported consists of the pulp free from the seeds; for the latter some use, possibly the production of fixed oil, has apparently been found.

**Constituents.**—The chief constituents of colocynth pulp appear to be an alkaloid producing very drastic purgation even in small doses, and amorphous resins soluble in ether and chloroform which also are powerful purgatives. Other constituents are a crystalline alcohol, citrullol, and $\alpha$-elaterin; neither of these are purgative; $\alpha$-elaterin is also present in elaterium (see 'Elaterium'). The presence of a crystallin glucoside, colocynthin, formerly reported, has not been confirmed. The pulp also contains from 1·0 to 1·3 per cent. of fixed oil and yields from 7 to 13 per cent. of ash; it contains no starch.

The small amount of fixed oil and the large amount of ash have been utilised to distinguish the powdered *fruit* from the powdered *pulp*. The seeds contain from 15 to 17 per cent. of oil, and yield only 2·5 to 3·0 per cent. of ash. Powdered fruits yield from 4 to 6 per cent. of ash and about 10 per cent. fixed oil. Microscopical examination of the powder yields more definite results. Colocynth pulp consists of very large thin-walled parenchymatous cells with occasional fibro-vascular bundles, accompanied by tubular cells but no sclerenchymatous cells are present in it, whereas the seed-coats abound in sclerenchymatous tissue which can be readily detected and identified.

**Varieties.**—*Turkey colocynth*, which is the most esteemed, is imported from Syria, Cyprus, &c.; the fruits are carefully peeled, nearly white, and contain a larger proportion of pulp. During and since the war large quantities of very similar Sudan colocynth have been imported.

*Spanish colocynth* is less sightly, often discoloured, and contains less pulp.

*Persian colocynth* occurs in small, shrunken balls with little pulp; it is now seldom imported.

*Mogadore colocynth* is occasionally imported; it is unpeeled.

**Uses.**—Colocynth is a gastro-intestinal stimulant or irritant and one of the most powerful of the official purgatives, acting as a hydragogue cathartic. It is employed as an occasional purgative to produce free evacuation of the bowels in bilious derangements or chronic constipation, but as it causes griping is seldom prescribed alone.

## CARAWAY FRUITS

### (Fructus Carui)

**Source, &c.**—Caraways are the ripe fruits of *Carum Carvi*, Linné (N.O. *Umbelliferæ*), an erect biennial herb distributed over central and northern Europe. The plant is found in Britain, apparently wild, but possibly only naturalised. It is cultivated principally in Holland, but Sweden, Norway, Russia, Germany, and Morocco also produce caraways, whilst a small quantity is grown in England. When the fruit ripens, the plant is cut and the caraways are separated by thrashing.

**Description.**—The ovary of the caraway is inferior and two-celled; as it ripens it develops into a schizocarp, that is, a fruit which separates into its component carpels by their splitting away from the central axis (carpophore), to which, however, they remain attached. Each complete fruit is termed a cremocarp (a variety of schizocarp), and

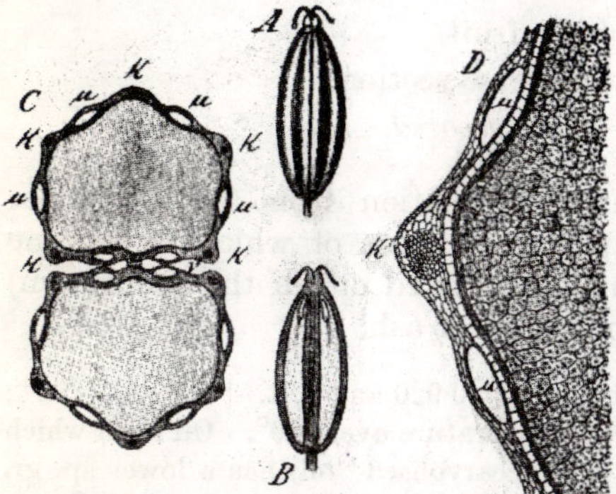


FIG. 59.—Caraway fruit. *A*, entire fruit, side view magnified 3 diam. *B*, longitudinal section, magnified 3 diam. *C*, transverse section: $\mu$, vittæ; $\kappa$, ridges, magnified 14 diam. *D*, portion of the same, further enlarged; $\iota$, pericarp; $\tau$, seedcoat. (Berg.)

FIG. 60—Caraway fruit. Mericarp, magnified. 6 diam.

each half-fruit a mericarp, the mutually apposed inner faces of which are the commissures or commissural surfaces.

The cremocarp of the caraway easily separates into its constituent mericarps, and the commercial drug consists almost entirely of separate mericarps, about 4 to 6 mm. long, very narrow, tapering at each end, and slightly curved. They are quite glabrous, brown, and traversed from base to apex by five narrow raised yellow ridges or 'costæ' (fig. 59 *C* and *D*, $\kappa$). Each of these (primary) ridges contains a fibro-vascular bundle, and this distinguishes them from other (secondary) ridges which some Umbelliferous fruits (coriander) possess. In caraway fruits no secondary ridges are present. In the depressions ('intercostal regions') between these ridges, embedded in the pericarp of the fruit and extending from base to apex, lie elongated oil-glands (vittæ, fig. 59 *C* and *D*, $\mu$). In the caraway there are four such vittæ on the dorsal and two on the commissural surface of each mericarp. Very many Umbelliferous fruits contain six vittæ, but some contain more (anise), whilst others contain none at all (hemlock). They are generally best seen in transverse section under a strong lens, when they appear as dark brown minute spots or cavities (compare fennel, in which they are conspicuous).

The transverse section exhibits also a narrow pericarp and a large oily endosperm, the commissural surface of which is not grooved. The small embryo lies near the apex of the fruit. (Compare fig. 59.)

Caraways exhale, when crushed, an agreeable aromatic odour, and possess a pleasant aromatic taste.

The student should observe

(*a*) The *glabrous surface* of the fruit,
(*b*) The *six vittæ* on the transverse section,
(*c*) The *endosperm, which is not grooved.*

**Constituents.**—Caraways yield by distillation from 3·5 to 7·0 per cent. of volatile oil, the principal constituent of which is carvone (50 to 60 per cent.). They also contain fixed oil (in the endosperm) and yield about 6 (not over 9) per cent. of ash.

The volatile oil should have a sp. gr. 0·910 to 0·920 and O.R. $+ 75°$ to $+ 82°$; at least 50 per cent. should distil at a temperature over 200°. Oil from which part of the carvone has been removed ('decarvolised' oil) has a lower sp. gr. lower optical rotation, and yields less than 50 per cent. boiling above 200°.

**Varieties.**—*Dutch*: Holland produces and exports far larger quantities of caraways than any other country; they are of medium size, dark colour, and clean appearance, and yield from 4·0 to 6·5 per cent. of oil.

*English*: these are produced in small quantity only; they are similar to the Dutch but brighter in colour.

**Adulteration.**—Caraways and other Umbelliferous fruits used in the production of volatile oil are distilled either ground or entire; in the former case the yield is increased, and the exhausted fruits are used as cattle food; entire fruits yield less oil, and the exhausted drug after drying may be mixed with the genuine fruits. These exhausted fruits may be recognised by their dark colour, deficiency in aroma, and low yield of aqueous extract; they often also sink at once in water (fennel), whereas the genuine float for a considerable time.

**Uses.**—Caraways, or the volatile oil obtained from them, are extensively used as an aromatic carminative.

## HEMLOCK FRUITS

### (Fructus Conii)

**Source, &c.**—The common or spotted hemlock, *Cŏnīum*[1] *maculatum*, Linné (N.O. *Umbelliferæ*), is a biennial plant widely spread throughout temperate Europe and generally distributed over Great Britain. It is distinguished by its smooth hollow stem with purple spots, the general and partial involucres on the inflorescence, its glabrous, decompound leaves and ovoid fruit without vittæ but with deeply grooved endosperm (compare also the details given under 'Hemlock Herb'). This plant was in all probability the one employed by the Greeks in the preparation of poisonous draughts. It was used in Anglo-Saxon medicine, but has latterly lost much of its reputation owing to the uncertain action of its preparations. Researches have indicated the reason of this and suggested a remedy.

The fruits should be gathered from wild plants when full-grown, but before they ripen—that is, before the colour changes from green to yellow—and thoroughly dried; if carefully preserved quite dry they will long retain their activity unimpaired.

**Description.**—Hemlock fruits, gathered when unripe, possess, after drying, a distinct greenish grey colour, which changes by long keeping to yellowish grey. They are small (about 3 mm. long), broadly ovoid and slightly compressed laterally. They are crowned with small stylopods bearing the remains of the stigmas. In the commercial drug the mericarps are usually separate and freed from the stalks. Each mericarp is glabrous and exhibits five paler sharply prominent primary ridges, which, from the presence of small protuberances, are irregularly crenate and wavy, a character not equally well shown by all fruits, and more conspicuous in the fresh than in the dried. Cut transversely and examined with a strong lens, the endosperm exhibits a deep furrow or groove on its commissural surface (fig. 61, *C*). This furrow, being filled with the tissue of the pericarp, is not visible on the commissural surface of the mericarp, but only on the transverse section; its presence should be carefully noted, as it indicates that the fruits are derived from a plant belonging to the suborder *Campylospermeæ*, and distinguishes them at once from anise, fennel, &c., that have an endosperm nearly flat on its commissural surface (suborder *Orthospermeæ*), and from coriander and other fruits that have an endosperm strongly curved in both radial and transverse section (suborder *Cœlospermeæ*). The pericarp contains no vittæ, and this is also an important character, but one

[1] This pronunciation, though well established, is not strictly correct. The word is derived from the Greek κώνειον, hemlock, and should be pronounced Cōnīum.

that can be ascertained only by subjecting a transverse section to examination under the microscope.

The fruit has scarcely any odour or taste, but when to the crushed fruit a solution of caustic potash is added a strong mouse-like odour is developed.

The student should observe

(*a*) The *glabrous surface*,
(*b*) The *irregular crenate ridges*,
(*c*) The *grooved endosperm*, and, if possible, under the microscope, the *absence of vittæ* ;

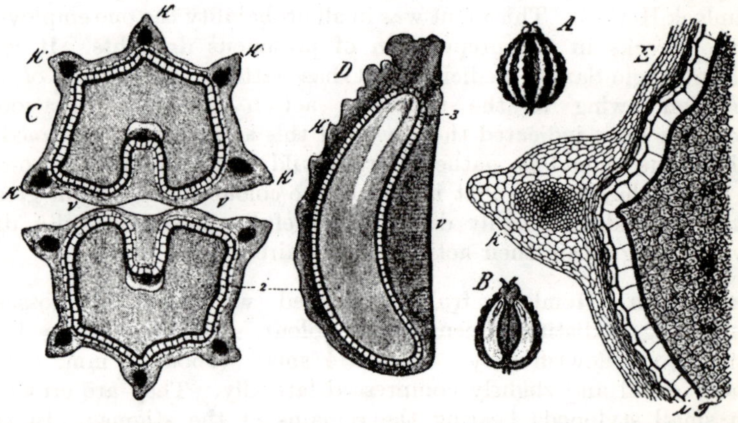


FIG. 61.—Hemlock fruit. *A*, side view of entire fruit. *B*, longitudinal section of the same, showing endosperm and embryo, magnified 3 diam. *D*, half of the same, further enlarged ; 3, embryo. *C*, transverse section : *ν*, commissural surface ; *κ*, ridges ; 2, endosperm magnified 14 diam. *E*, portion of the same, further enlarged ; *ι*, pericarp ; *τ*, seed-coat. (Berg.)

and should compare the fruits with anise fruits, which are distinguished by the short bristly hairs with which they are provided, by the presence of numerous vittæ, and by the endosperm, which is not grooved, the latter being the most definite character visible under a lens.

**Constituents.**—The principal constituent of hemlock fruits is the poisonous, volatile, liquid alkaloid coniine, of which they may contain, when collected at the proper time and dried, as much as 2·77 per cent. (Farr and Wright, 1904), the average being 1·65 per cent. This alkaloid is the source of the strong odour produced when hemlock fruits are crushed and mixed with caustic potash, the alkali liberating the alkaloid which was previously combined with acids.

The proportion of coniine present attains its maximum when the fruits are full or nearly full grown, but before the colour changes from green to yellow ; during this change (the ripening of the fruit) the proportion of coniine rapidly diminishes. Commercial fruits yield from 0·5 to 1·3 per cent. of coniine.

It is accompanied in the fruit by small quantities of methylconiine, which is oily and resembles coniine, and by conhydrine, which is colourless and crystalline. Ethyl piperidine and pseudoconhydrine are also present.

Pure coniine, $C_8H_{17}N$, is $\alpha$-propyl-piperidine. It is a colourless, oily, very poisonous liquid with an extremely unpleasant odour. It boils at 163·5° without decomposition in an atmosphere free from oxygen; but exposed to the air, even at ordinary temperatures, it rapidly assumes a brown colour. It yields crystalline non-volatile salts. It may be produced from $\alpha$-picoline, $C_5H_4(CH_3)N$, one of the constituents of coal tar, by heating it to 250° for 10 hours with acetic aldehyde, by which it is converted into $\alpha$-allyl pyridine $C_5H_4(C_3H_5)N$; this by reduction with sodium yields coniine $C_5H_9(C_3H_7)NH$. Coniine thus prepared is, however, optically inactive, whereas natural coniine is dextrorotatory; it can be separated into the dextro- and lævo-rotatory alkaloids by inducing crystallisation of the tartrate by means of a crystal of natural coniine tartrate.

**Assay.**—Percolate 10 gm. of the finely produced fruit with 70 per cent. alcohol until exhausted, add 20 c.c. N/10 sulphuric acid and evaporate off the alcohol. Transfer to a separator, wash twice with chloroform, make alkaline with solution of potash and shake out with three successive 5 c.c. of chloroform. Transfer the alkaloids to acidified water and back to chloroform. Repeat this purification twice. Finally shake the chloroformic solution of the alkaloids with three drops of fuming hydrochloric acid, wash the separator with a little alcohol, evaporate, dry at 90° and weigh. The weight of the alkaloidal hydrochlorides multiplied by 0·777 gives the weight of alkaloid in 10 gm. of the fruit.

**Uses.**—Hemlock is used in spasmodic and convulsive diseases, such as tetanus, chorea, and epilepsy; in asthma, whooping-cough, and spasmodic affections of the larynx. Coniine depresses the medulla and motor nerve endings, excessive doses resulting in death from respiratory paralysis.

## CORIANDER FRUITS

### (Fructus Coriandri)

**Source, &c.**—The coriander, *Coriandrum sativum*, Linné (N.O. *Umbelliferæ*), is an erect herbaceous annual that has become naturalised throughout temperate Europe. It is cultivated principally in Russia, Thuringia, Moravia, and Hungary, as well as in northern Africa, Malta, and India. The whole plant, and especially the unripe fruit, is characterised by a strong disagreeable odour, whence the name coriander (from the Greek *κόρις*, a bug). As the fruit ripens, this gives place to an agreeable aroma, and the plant is then cut and thrashed. It has been cultivated for many centuries for use as a spice.

**Description.**—The two mericarps of which the fruit consists remain firmly united by their margins, and enclose between them a small cavity, the cremocarp being nearly globular in shape. The fruit is of small size, averaging only about 5 mm. in diameter, of a uniform brownish yellow colour, and quite glabrous. It is crowned by the remains of the calyx-teeth and styles, and bears on each mericarp five inconspicuous wavy primary ridges (containing fibro-vascular bundles) and four more conspicuous, straight, secondary ridges alternating with the primary. Both transverse and radial sections show a curved

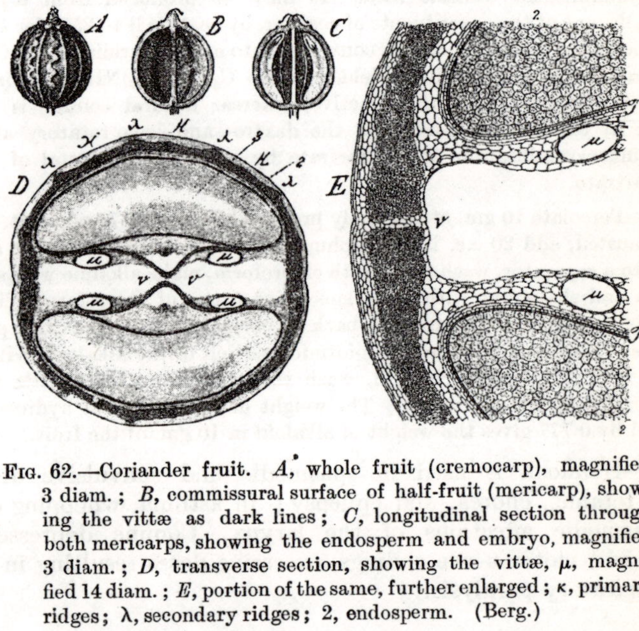

Fig. 62.—Coriander fruit. *A*, whole fruit (cremocarp), magnified 3 diam.; *B*, commissural surface of half-fruit (mericarp), showing the vittæ as dark lines; *C*, longitudinal section through both mericarps, showing the endosperm and embryo, magnified 3 diam.; *D*, transverse section, showing the vittæ, $\mu$, magnified 14 diam.; *E*, portion of the same, further enlarged; $\kappa$, primary ridges; $\lambda$, secondary ridges; 2, endosperm. (Berg.)

endosperm (suborder *Cœlospermeæ*), and on the former only two vittæ, both of which are situated on the commissural surface, can be discerned. In addition to these, however, the young ovary contains vittæ on the dorsal surface; as the fruit ripens these become compressed and break down into tangentially elongated cavities, the outer portion of the pericarp being finally thrown off.

The odour of the bruised fruit is aromatic, and the taste agreeable and spicy.

The student should observe

(*a*) The *firmly united mericarps*,

(*b*) The *wavy primary* and *straight secondary ridges*,

(*c*) The *two vittæ on each commissural surface*.

**Constituents.**—Coriander fruits of good quality yield from 0·8 to 1·0 per cent. of volatile oil (sp. gr. 0·870 to 0·885, O.R. + 8° to + 14° chief constituent 90 per cent. of the alcohol linalol). That distilled from unripe fruit has a fetid odour, which, however, disappears on keeping.

**Varieties.**—*English*: these are grown in Essex, and said to have the finest flavour, though not specially rich in oil.

*Russian and German*: these are the richest in oil, yielding up to 1·0 per cent.

*Mogadore*, which are the largest and brightest of all, but contain only 0·2 to 0·3 per cent. of oil.

*Bombay*, which are distinguished by their oval shape; they yield only 0·15 to 0·2 per cent. of oil.

**Uses.**—The fruit and the oil distilled from it are used as aromatic carminatives.

## CUMMIN FRUIT

### (Fructus Cumini)

**Source, &c.**—The cummin, *Cumīnum Cymīnum*, Linné (N.O. *Umbelliferæ*), is a small annual plant indigenous to the upper Nile territory, and cultivated in Morocco, Sicily, Malta, Syria, and India. The fruits were well known to the ancients, and were much used in Europe in the Middle Ages as a culinary spice.

The plants are cut down when the fruits are ripe, and thrashed.

**Description.**—Cummin fruits are brown, elongated oval, tapering towards both base and apex, and somewhat laterally compressed. In the commercial drug the mericarps are sometimes united and attached to a short stalk; sometimes they are free. Each mericarp is nearly straight, about 6 mm. in length, and furnished with five yellow, smooth or scabrous primary ridges. In the depressions between the primary ridges are secondary ridges bearing short bristly hairs.

The transverse section of a mericarp exhibits an oily endosperm, which is not deeply grooved, and six vittæ—four on the dorsal surface below the bristly secondary ridges and two on the commissural.

Cummin fruits differ from caraways in being nearly straight instead of curved (as caraways usually are), and in being bristly instead of smooth. The odour and taste, though similar to those of caraways, are by no means so agreeable.

The student should observe

(*a*) The *straight mericarp*,
(*b*) The *bristles on the secondary ridges*,
(*c*) The *characteristic odour*.

**Constituents.**—Cummin fruits yield from 3 to 4 per cent. of *volatile oil* (sp. gr. 0·972 ; O.R. + 25·5° ; chief constituent cuminic aldehyde).

**Uses.**—The fruits have been used as a stimulant and carminative ; they are now chiefly employed in veterinary medicine.

*Persian cummin* is probably derived from a species of *Carum* ; fruits smaller than cummin or caraway fruits ; odour similar to cummin ; oil contains no carvone.

## FENNEL FRUITS

### (Fructus Fœniculi)

**Source, &c.**—Fennel, *Fœniculum capillăceum*, Gilibert (*F. vulgare*, Miller ; N.O. *Umbelliferæ*), is apparently indigenous to the shores of the Mediterranean, extending eastwards, but is cultivated for medicinal use in the south of France, in Saxony and Würtemberg, in Russia and Galicia, and also in India, Japan, &c. Like other aromatic Umbelliferous fruits, fennel was well known to the ancients, and was largely used in Europe during the Middle Ages. For medicinal use Saxon, Russian, Galician, or Roumanian fruits are to be preferred, as it has been shown that they yield most volatile oil, and that the latter contains fenchone (see below).

**Description.**—Fennel fruits occur in several commercial varieties, varying considerably in size and appearance. Saxon fruits, which may be regarded as the best, are of a greenish or yellowish brown colour, and oblong in shape, varying from 8 to 10 mm. in length, and 3 to 4 mm. in width. The mericarps frequently remain united and attached to a pedicel. They are glabrous, and bear five paler, very prominent, primary ridges. In a transverse section four large vittæ can be distinguished by the naked eye on the dorsal surface, and two on the commissural surface of each mericarp ; the endosperm is dark in colour, oily, and not deeply grooved. They have an aromatic odour, recalling anise, and a sweet, camphoraceous taste.

The student should observe

(*a*) The *very prominent ridges*,
(*b*) The *large vittæ*,
(*c*) The *characteristic odour and taste*.

**Constituents.**—The best varieties of fennel (Saxon, Galician, and Russian) yield from 4 to 5 per cent. of volatile oil (sp. gr. 0·960 to 0·980 ; O.R. + 6° to + 12° ; solidifying-point 5° to 20°), the principal constituents of which are anethol ($C_{10}H_{12}O$, 50 to 60 per cent.) and fenchone ($C_{10}H_{16}O$, 18 to 20 per cent.).

Anethol is also the chief constituent of anise oil (compare p. 121).

Fenchone is a colourless liquid possessing a pungent, camphoraceous odour and taste; it probably contributes materially to the medicinal properties of the oil, hence only such varieties of fennel as contain a good proportion of fenchone are suitable for medicinal use.

**Varieties.**—The following are the chief commercial varieties of fennel fruits:

1. *Saxon*: as above described; they yield 4·7 per cent. of volatile oil containing 22 per cent. of fenchone.

2. *Russian*, *Galician*, and *Roumanian*: these closely resemble one another; from 4 to 6 mm. in length and 1 to 2 mm. in width;

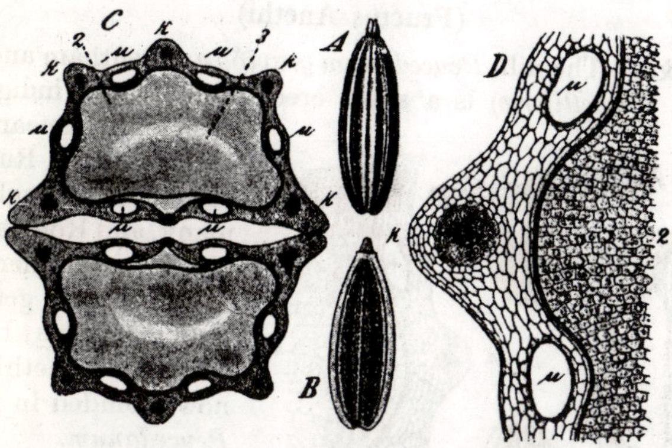

FIG. 63.—Fennel fruit. *A*, entire fruit, side view, magnified 3 diam. *B*, half-fruit, showing commissural surface, magnified 3 diam. *C*, transverse section; $\mu$, vittæ; $\kappa$, ridges; 2, endosperm; 3, embryo; magnified 14 diam. *D*, portion of the same, further enlarged. (Berg.)

yield of oil from 4 to 5 per cent. of which about 18 per cent. is fenchone; taste very camphoraceous.

3. *French Sweet* or *Roman*, 7 to 8 mm. long, 2 to 3 mm. wide, often arched; pale yellowish green with a sweet, anise taste. Yield of oil 2·1 per cent., free from fenchone.

4. *Indian*, 6 to 7 mm. long, brownish, stalky, with sweet anise taste. Yield of oil 0·72 per cent., containing 6·7 per cent. of fenchone.

5. *Japanese*, 3 to 4 mm. long, 2 to 3 mm. wide, ovoid, not curved, pale greenish brown in colour; taste camphoraceous and very sweet. Yield of oil 2·7 per cent., containing 10·2 per cent. of fenchone.

6. *French Bitter*, 4 to 5 mm. long, 2 mm. wide, scurfy in the furrows, ridges less prominent, and colour darker than the sweet.

**Adulteration.**—Fennel is said to be specially subject to admixture with exhausted fruits. These include the fruits partially exhausted of their oil by distillation in a current of alcohol vapour in liqueur-making, as well as the residues obtained after distillation with water or in a current of steam. Fruits exhausted by water or steam are darker, contain less oil, and sink at once in water, but those exhausted by alcohol vapour retain 1·0 to 2·0 per cent. of oil, and are but little altered in appearance; they acquire, however, a peculiar fusel-oil odour. Recoloured fennel may be detected by rubbing the fruits between the hands.

**Uses.**—Fennel is used as an agreeable aromatic and carminative.

## DILL FRUITS

### (Fructus Anethi)

**Source, &c.**—The dill, *Peucedănum gravēŏlens*, Bentham and Hooker filius (N.O. *Umbelliferæ*) is a short erect annual herb indigenous to the Mediterranean districts and southern Russia, but cultivated in England, Germany, and Roumania. The plant was formerly placed in a separate genus, *Anethum*, whence the name 'Fructus Anethi'; it is now included in the genus *Peucedanum*.

The drug has been in use in this country from very early times.

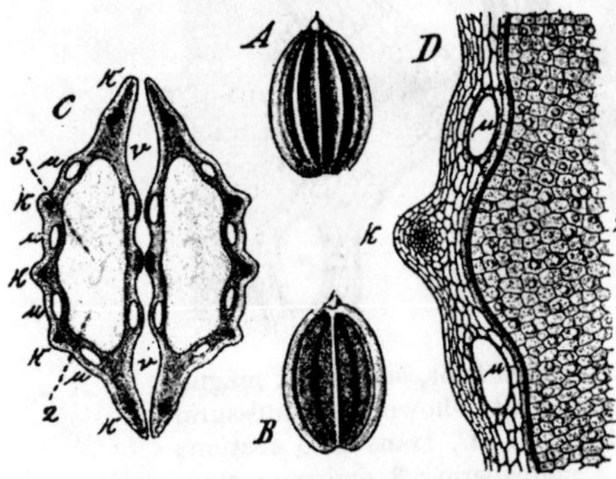


FIG. 64.—Dill fruit. *A*, entire fruit, side view, magnified 3 diam. *B*, commissural surface of mericarp, showing the vittæ as dark lines, magnified 3 diam. *C*, transverse section: $\mu$, vittæ; $\kappa$, ridges; $\nu$, commissural surface; 2, endosperm; magnified 14 diam. *D*, portion of the same, further enlarged. (Berg.)

**Description.**—In the commercial drug the two mericarps of which the fruit is composed are usually separate and free from the pedicel; they are broadly oval in outline, and so strongly dorsally compressed as to be nearly flat. They average about 4 mm. in length and 2·5 mm. in breadth.

Each mericarp is glabrous, brown, and traversed from base to apex by five primary ridges, of which the three dorsal are only slightly raised, and are brown, filiform, and inconspicuous; whilst the lateral are prolonged into thin, yellowish, membranous wings. The transverse section exhibits under the lens six vittæ, viz. four on the dorsal and two on the commissural surface; the endosperm is oily, and is not grooved on its commissural surface either in transverse or longitudinal section. Both the odour and taste are agreeably aromatic.

The student should observe

(*a*) The *very flat shape of the mericarp,*

(*b*) The *inconspicuous dorsal ridges*, and the *prominent lateral ones* (wings),

(*c*) The *mericarps, usually separate and free from the pedicel.*

**Constituents.**—Dill fruit contains from 3 to 4 per cent. of volatile oil (sp. gr. 0·900 to 0·915 ; O.R. + 70° to + 80°), consisting of carvone ($C_{14}H_{10}O$, 40 to 60 per cent.), a colourless liquid with strong dill odour, and limonene associated with phellandrene and other terpenes.

**Substitutes.**—Indian dill (*P. Sowa*, Kurz) ; imported in large quantities and used as a substitute for caraways when these are scarce ; mericarps frequently united and attached to a pedicel ;

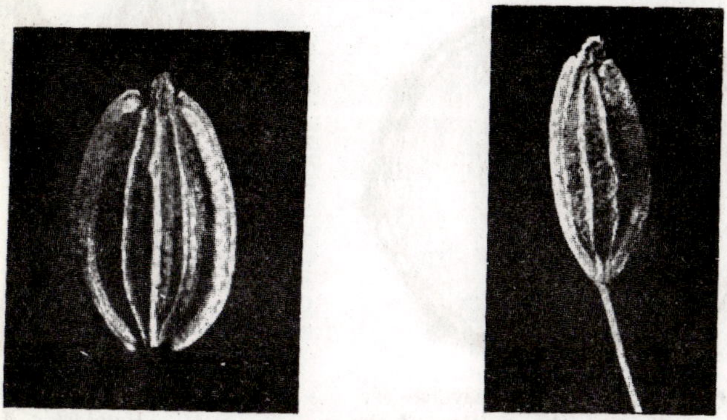

A B

FIG. 65.—Dill fruit. A, English ; B, Indian. Magnified 6 diam.

dorsal surface more convex than European dill, ridges paler and more conspicuous, mericarps narrower and less prominently winged. The volatile oil is different in composition (sp. gr. 0·948 to 0·968, optical rotation + 47°·5 to + 50°·5, 40 to 50 per cent. distils over 230°, contains dill-apiol) and the fruits should not be used in place of European dill.

**Uses.**—Dill is employed as an aromatic stimulant and carminative.

## ANISE FRUITS

### (Aniseed, Fructus Anisi)

**Source, &c.**—The anise, *Pimpinella Anisum*, Linné (N.O. *Umbelliferæ*), is an annual plant indigenous to Greece, Egypt, and Asia Minor, but cultivated largely in southern Russia, and also in Spain and Bulgaria. Aniseed is one of the oldest of medicines and spices ; it was cultivated by the ancient Egyptians, and was known to Dioscorides and Pliny; in this country it has been in use since the fourteenth century. The drug consists of the ripe fruits.

**Description.**—Anise fruits are greyish brown, about 5 mm. long, ovoid or pear-shaped, and somewhat compressed laterally. They are broad near the base, and taper gradually towards the apex, which is crowned with a stylopod and two short divergent styles. The mericarps usually remain united and attached to a pedicel longer than the fruit. Each mericap possesses five distinct but not prominent primary ridges which are usually slightly wavy, but do not exhibit the prominent crenations characteristic of hemlock fruits; the depressions between them are more or less distinctly bristly from the presence of short, stout hairs.

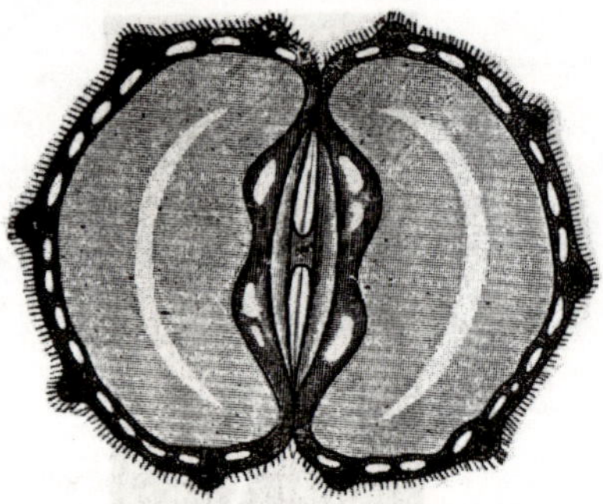

FIG. 66.—Anise fruit. Transverse section indicating the position of the vittæ, not all of which (about forty in each mericarp) are shown. Magnified. (Moeller.)

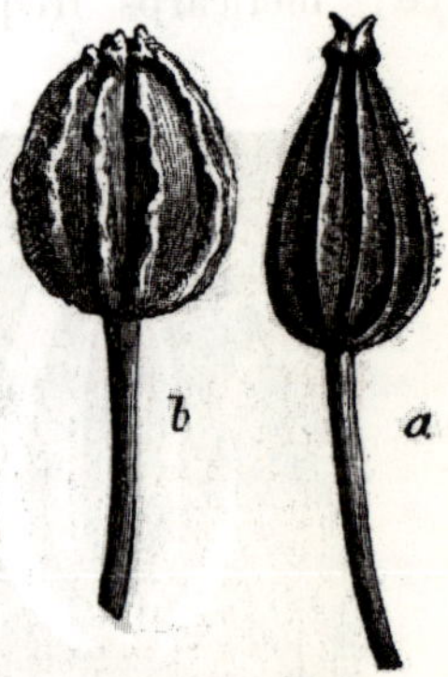


FIG. 67.—*a*, Anise fruit. *b*, Hemlock fruit. Magnified. (Vogl.)

The vittæ branch repeatedly, and the transverse section taken from the middle of the fruit exhibits under the microscope from thirty to forty such branches in each mericarp; these, however, are so small that they are scarcely visible under a lens. The endosperm is slightly concave on the commissural surface, but is not deeply grooved.

Anise fruits possess a sweet aromatic taste, and exhale, when crushed, an aromatic odour.

The student should observe

(*a*) The *short, stout, bristly hairs*, which are sometimes inconspicuous,
(*b*) The *united mericarps* and *the pedicel attached* to them,
(*c*) The *absence of any deep groove in the endosperm*,
(*d*) The *characteristic odour and taste*,
(*e*) The *absence of prominent crenations on the primary ridges*, which are themselves not conspicuous;

and should carefully compare the fruits with those of *Conium maculatum*, noting the absence of any deep groove in the endosperm, this being the best macroscopic diagnostic feature.

**Constituents.**—The fruits yield from 1·5 to 3·5 per cent. of volatile oil (sp. gr. 0·975 to 0·990 ; O.R. $-2°$ to $+1°$), of which anethol, present to the extent of about 90 per cent., is the principal aromatic constituent. Anethol, $C_{10}H_{12}O$, forms a white, crystalline solid melting at 22° ; it has a strong anise odour.

**Varieties.**—*Spanish*, exported from Alicante ; the fruits are distinguished by their large size (4 mm.), grey or brownish grey colour, and slender tapering shape ; they yield about 3 per cent. of oil.

*Russian* fruits are smaller, darker, and rather more ovoid in shape ; they are exported in very large quantities, and are the variety generally used for distillation.

*Italian* anise fruits are frequently contaminated with hemlock fruits, which may be identified by their glabrous surface, by the irregularly crenate ridges, by the absence of the pedicel, and (best of all) by the deep groove in the endosperm.

**Note.**—Anise fruits are frequently adulterated with dried, sifted earth ; the ash should not exceed 11 per cent. ; in adulterated anise it may rise to 25 per cent.

**Uses.**—Anise is employed as an aromatic and carminative.

## EMBELIA

(Embelia)

**Source, &c.**—Embelia is the dried fruit of *Embelia Ribes*, Burmann filius, and also of *Embelia robusta*, Roxburgh (N.O. *Myrsineæ*). The former is a large climbing shrub abundant in the hilly parts of India, and the latter a very common Indian shrub or small tree.

**Description.**—These plants produce large bunches of small fruits which are collected when ripe and dried. They are then nearly globular, about the size of a small pea and dull red or nearly black. They are superior, minutely beaked and often attached to a 5-partite calyx and slender pedicel. The commercial fruits are usually derived from *E. robusta* and have a striated surface, those from *E. Ribes* being warty or wrinkled and resembling black pepper (in which they have actually been detected). Within the brittle pericarp is a single, globular seed surrounded by a delicate membrane. The seed itself is reddish, has a cavity at the base and is marked with lighter spots which become fainter on long soaking in water and in which minute crystals are visible under a strong lens. The endosperm is horny and slightly ruminate. The fruit has a somewhat astringent and aromatic taste.

**Constituents.**—The principal constituent of the fruit is embelic acid which crystallises in golden yellow, lamellar crystals, insoluble in water, but soluble in alcohol and ether and yielding wine-red solutions with alkalies. In addition volatile oil (trace), fixed oil, resin, tannin and an alkaloid, christembine, are present.

**Uses.**—Both the fruit and the ammonium salt of embelic acid are used as a tænicide.

**Note.**—The drug is sharply characterised by the bluish violet colour obtained by extracting the powder with ether and shaking the ethereal solution with dilute ammonia.

## CAPSICUM FRUITS

### (Cayenne Pepper, Chillies, Fructus Capsici)

**Source, &c.**—The capsicum fruit official in the British Pharmacopœia is derived from *Capsicum minimum*, Roxburgh (N.O. *Solanaceæ*), a small erect shrub with spreading branches, producing oblong-conical fruits commercially known as chillies, the fruits of *C. annuum* being termed capsicums. It is apparently a native of Southern India but is cultivated in Eastern and Western Africa, in South America, &c. The plant shows a great disposition to vary and the chillies and capsicums of commerce vary exceedingly in their size shape, and pungency. In Hungary *C. tetragōnum*, Miller, and *C. annuum*, var. *Szegedinense* are widely grown, and produce the large fruits with a short peduncle and conspicuous green calyx known as paprika. In Italy and Spain *C. grossum*, Linné (*C. annuum*, Linné, var. *grossum*) is cultivated, but in tropical countries *C. minimum* and *C. frutescens*, Linné, are most largely cultivated; these yield smaller and more pungent fruits, those of *C. minimum* being alone official.

The fruits are collected when ripe, and dried.

**Description.**—The fruits of *C. minimum* have, when fresh, a scarlet colour which changes on drying to a dull orange red. They are oblong-conical in shape, and obtuse at the apex; they vary usually from 12 to 20 mm. in length, and do not exceed 7 mm. in diameter. They are superior, and not unfrequently remain attached to a small, inconspicuous, five-toothed, inferior calyx, and straight, slender peduncle, which is as long as, or rather longer than, the fruit itself. The pericarp is somewhat flattened, shrunken, and shrivelled. It is quite glabrous and shining, thin and leathery in texture, and more or less translucent.

Cut transversely, the fruit is seen to consist of two cells separated by a thin, reddish, membranous dissepiment. Each cell contains from five to ten small, flat, nearly circular, whitish seeds with a characteristic, thickened margin; they were originally attached to the dissepiment, but are frequently found loose in the cavity.

Capsicum fruits have a characteristic but not powerful odour, and an extremely fiery pungent taste. The latter resides principally in the membranous dissepiment that divides the fruit into two cells.

The student should observe

(a) The *shape and size of fruit and peduncle*,
(b) The *dissepiment* and its *pungent taste*,
(c) The *shape of the seeds*.

**Constituents.**—The most important constituent of capsicum fruit is the pungent principle, capsaicin (0·02 per cent.), first isolated by Thresh (1876) in colourless, odourless crystals. It also contains

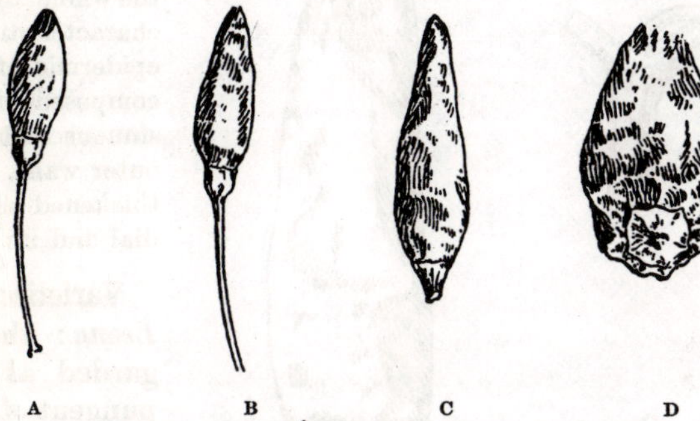

FIG. 68.—A, Zanzibar; B, Sierra Leone; C, Japanese chillies; D, Bombay capsicums. Natural size. (*Chemist and Druggist.*)

a minute quantity of a liquid alkaloid which is not pungent, a fixed oil, and red colouring matter; the seeds may contain traces of starch.

Capsicum fruits yield about 5 (not over 7) per cent. of ash and from 20 to 25 per cent. of alcoholic extract, known in commerce as capsicin; the same name has also been applied to an ethereal extract as well as to a soft red substance extracted by ether from an alcoholic extract. All of these contain capsaicin associated with fixed oil, colouring matter, &c.

Capsaicin (capsacutin) is secreted by the epidermis of the dissepiments of the fruit, between the outer cell wall and the cuticle.

Thresh assigned to it the formula $C_9H_{18}O_2$, which has been corrected to $C_{18}H_{27}NO_3$. The crystals melt at 65° and at higher temperatures are volatile, the vapour being extremely irritating; they are phenolic in character, dissolving in dilute solution of potassium hydroxide, but being reprecipitated by a current of carbon dioxide. The pungency is not affected by the alkali but is destroyed by oxidation with potassium bichromate or permanganate. Much interest is attached to this substance, as it is one of the very few pungent principles that have been isolated in a crystalline form, most of them being obtainable only in the state of oily liquids.

**Microscopical Characters.**—For the complete identification of the fruit of *C. minimum* the following histological characters are necessary in addition to the macroscopical characters detailed above. The outer epidermis of the pericarp consists of rectangular cells with a delicately striated cuticle and straight, thick, yellow, sparsely pitted walls abutting on parenchymatous tissue, the cells of which contain droplets of red oil and have thin cellulose walls. The inner epidermis of the pericarp exhibits groups of sclerenchymatous cells alternating with bands of thin-walled parenchyma, the whole forming a very characteristic tissue. The epidermis of the seed is composed of very large sinuous cells with thin outer walls, but strongly thickened and pitted radial and inner walls.

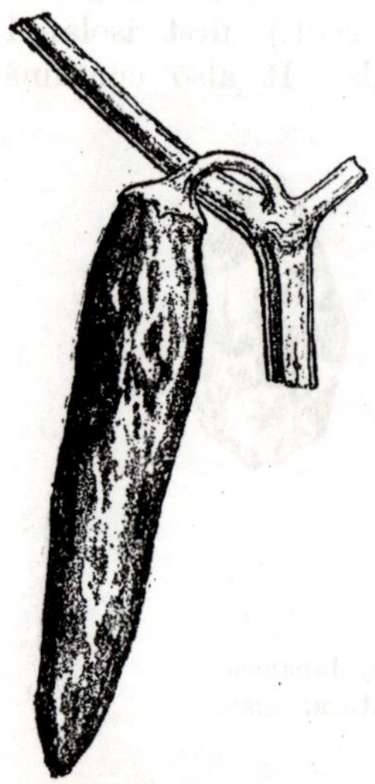

FIG. 69.—*Capsicum annuum*, fruit. Natural size. (Bentley and Trimen.)

FIG. 70.—Natal capsicum fruit. Natural size. (*Chemist and Druggist.*)

**Varieties.** — *Sierra Leone*: these are regarded as the most pungent of all. They have the macroscopical and microscopical characters detailed above. The pod is rather slender, bright in colour, with the stalk occasionally attached.

*Nyassaland* closely resemble Sierra Leone, but are rather brighter and more free from stalk.

*Zanzibar* are usually duller in colour, more stalky, and the pod rather shorter and broader.

*Japanese*: these are distinguished by their very bright, reddish colour and freedom from stalk; the cells of the epidermis of the pericarp have a *smooth* (not striated) *cuticle, strongly thickened walls and radiate lumen*; the cells of the single-layered hypoderma have rather thick, pitted, *cuticularised* walls. These fruits are less pungent than the African, but are valued for their very bright colour. According to Holmes the small Japanese chillies of commerce are probably derived from *C. frutescens*, Linné; the larger from a form of *C. minimum*, Roxburgh.

*Bombay capsicums*, the fruits of *C. annuum*, Linné. They vary considerably in size and shape from nearly globular to oblong; they are larger than the fruits of *C. minimum*, and have a less pungent taste. The stalk is usually bent, the calyx larger, the pericarp more leathery, and the dissepiment does not extend throughout the entire length of the fruit. The cells of the epidermis are *polygonal*, larger than those of *C. minimum* and with numerous pits, while the hypoderma consists of *several layers of cuticularised collenchymatous cells.*

*Natal capsicums*: these are much larger, averaging about 8 cm. in length, and are distinguished by their beautiful, transparent red pericarp.

*Paprika*, which is largely grown and used in Hungary, is derived from *C. tetragonum*, Miller, or from *C. annuum*, var. *Szegedinense*; the fruits are large and usually more or less distinctly tetragonal in shape.

*Bird Pepper*, the powder of which is given to canaries to improve the colour of their plumage is derived from *C. annuum*, var. *grossum*, Sendtn. (Spanish) or *C. annuum*, var. *Szegedinense* (Hungarian); it is free from pungency.

**Uses.**—Cayenne pepper is applied externally as a stimulant and counter-irritant; internally it is used as a pungent stomachic carminative and stimulant, to dispel flatulence and rouse the appetite.

## LAUREL BERRIES

### (Bay Berries, Fructus Lauri)

**Source, &c.**—Laurel berries are the ripe fruits of *Laurus nobilis*, Linné (N.O. *Laurineæ*) (see p. 56).

**Description.**—The laurel produces one-celled, one-seeded, drupaceous fruits, which are collected when ripe, and dried. They are then nearly black in colour and ovoid in shape, about 15 mm. long, and slightly pointed at the apex. The surface is glabrous, shining, and coarsely wrinkled. The pericarp is thin and brittle, and encloses a single seed, the kernel of which lies loose in the cavity, the seed coats remaining closely adherent to the inner surface of the pericarp. The kernel is usually brownish yellow in colour, and easily separable into two large firm cotyledons; it has an aromatic odour and aromatic bitter taste, the pericarp being less aromatic but much more bitter.

The student should observe

(*a*) The *ovoid shape*,
(*b*) The *loose kernel of the seed*,
(*c*) The *aromatic odour and taste*;

and should compare these fruits with Cocculus indicus, which are reniform in shape and destitute of aroma.

**Constituents.**—Laurel berries contain about 1 per cent. of an aromatic volatile oil and upwards of one-fourth of their weight of solid fat. The latter, separated by hot pressure, is the Oleum Lauri Expressum of commerce; when pure it has a dull green colour, granular consistence, and aromatic odour. The principal constituent is laurostearin (glyceryl laurate), the odour being due to the volatile oil, and the green colour to chlorophyll, both of which are simultaneously extracted.

The volatile oil consists of cineol (eucalyptol, 50 per cent.), accompanied by eugenol, pinene, geraniol, &c.

The nature of the bitter principle contained in the pericarp of the fruit is unknown.

**Uses.**—The expressed oil is sometimes used as a stimulant in veterinary practice.

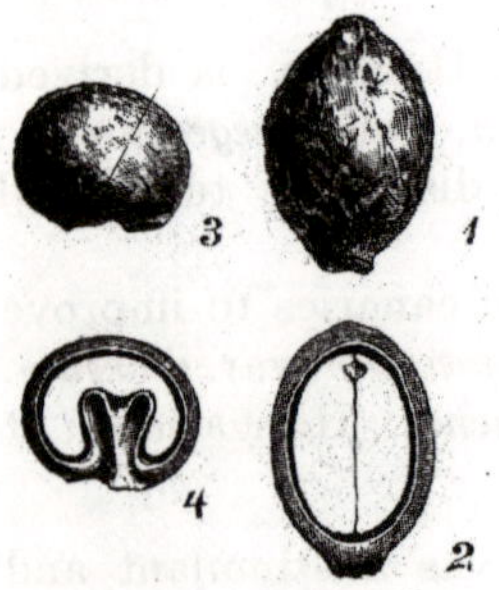

FIG. 71.—1 and 2, Laurel berries, whole and cut vertically. 3 and 4, *Cocculus indicus*, whole and cut transversely. Slightly magnified. (Vogl.)

## FIGS

### (Ficus)

**Source, &c.**—The fig tree, *Ficus Carica*, Linné (N.O. *Urticaceæ*), is a native of Persia and the surrounding countries. Figs have long constituted an important article of food in Eastern countries; they have been found provided as food for the dead in Egypt as early as 2400 years before the Christian era; from Egypt the tree was probably introduced into Greece and thence into southern Europe. It is now cultivated in most temperate and warm climates, and even ripens its fruit in England.

The fruit, which is often cauline, is produced by the abnormal growth of a lateral shoot, by which a green, hollow, pear-shaped body is produced, having a small aperture closed with bracts. On the inner walls of this hollow body numerous small flowers are borne; the wall itself is then fleshy and contains laticiferous vessels which, when wounded, discharge a milky latex. In this condition figs are inedible. As the fruit ripens the latex disappears, the fleshy wall fills with sugar and becomes pulpy, and the taste sweet and agreeable. They are then eaten in the fresh state, or are collected and dried in the sun, being sometimes previously sterilised by sulphuring or by dipping in a boiling alkaline solution. The dried fruits constitute the figs of commerce. If, without further manipulation, they are packed loose, retaining more or less their original shape, they are called 'natural' figs; but if subjected to a process of kneading by which they become supple and the skin translucent, they are called 'pulled' figs; the latter are usually packed into small boxes for exportation, and are considered the best. 'Pressed' or 'layered' figs are those which have been flattened by the pressure used in the packing.

**Description.**—The ordinary fig is too well known to need much description. The fruit, which is sometimes called a syconus, is a collective fruit formed from the enlarged, succulent, hollow, receptacle which bears on its inner wall very numerous, minute, one-seeded true fruits (achenes), commonly called the seeds.

**Constituents.**—The principal constituent of figs is sugar (dextrose), of which they may contain upwards of 50 per cent.

**Varieties.**—*Elémé* : exported from Smyrna either 'pulled' or 'layered,' or in wooden drums or barrels; they are distinguished by their thin skins and soft luscious pulp.

*Spanish* : from Malaga and Valencia, either 'layered' packed in boxes, or 'natural' packed in mats.

*Greek* : generally of small size and threaded on strings or packed in barrels; they have thicker, tougher skins, and less abundant pulp.

**Uses.**—Dried figs have agreeable demulcent, nutritive, and laxative properties, and are therefore sometimes prescribed as an article of diet in habitual constipation. They form one of the ingredients in confection of senna.

## HOPS

### (Lupulus, Humulus, Strobili Lupuli)

**Source, &c.**—The hop, *Humŭlus Lupŭlus*, Linné (N.O. *Urticaceæ*), is a scabrous climbing plant with perennial root, widely diffused over the whole of Europe and common in England, growing in hedges and thickets. It is largely cultivated in England, Germany, Russia, California, &c.

The hop is diœcious, but the pistillate plant only is cultivated, since from that alone the fruits (hops) can be obtained; these are preferred with undeveloped seeds, to ensure which the staminate plants are usually excluded.

The fruit is a cone-like collective fruit (strobile) consisting of leafy stipules and bracts borne on a zigzag axis. These are picked from the plant when fully developed, and dried; they are frequently exposed to fumes of burning sulphur (sulphur dioxide), by which the colour is preserved and change in the aroma is said to be hindered. They are sometimes dried loose, but usually they are pressed into compact bales known as 'pockets.'

**Description.**—The strobile of the hop averages about 3·5 cm. in length and is ovoid or rounded in shape, the most conspicuous part being the yellowish green imbricated membranous bracts and stipules. If these leafy organs are removed from the strobile, the axis will be seen to be hairy and have a zigzag course, bearing rudimentary branches on alternate sides. Some of the leafy bodies enfold at their bases minute fruits : these are the bracts (fig. 72, *b*) ; others, the stipules, do not. The fruits are minute achenes, and are partially surrounded by a perianth (fig. 72, *c*). Both the minute fruits and the bases of the bracts are sprinkled over with yellowish shining translucent glands, which contain volatile oil and constitute, when separated, the drug lupulin.

The odour of fresh hops is strong and characteristic, although scarcely agreeable ; the taste strongly aromatic and disagreeably bitter and acrid. By keeping, the aroma becomes less powerful but also less agreeable, the bracts and stipules turn yellow, and the glands acquire a brown colour.

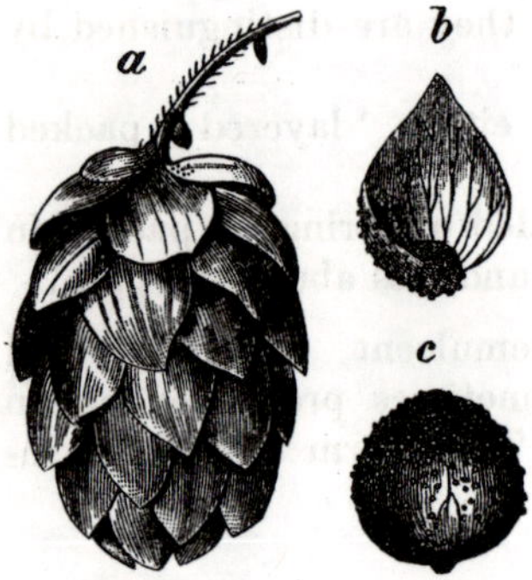


FIG. 72.—Hop. *a*, strobile of the Hop, natural size. *b*, bract enfolding at its base a small fruit, and showing lupulin glands, natural size. *c*, fruit magnified, showing lupulin glands. (Tschirch.)

The student should strip the bracts and stipules from the strobile and observe

(*a*) The *zigzag hairy axis*,
(*b*) The *bracts enclosing fruits and bearing glands*.

**Constituents.**—The composition of hops is very complex. They contain volatile oil, tannin, sugar, fatty acids, resins, &c.

The volatile oil (0·3 to 1·0 per cent.) consists chiefly of the terpene humulene.

The bitterness is due to a number of substances only one of which, humulol, has been obtained crystalline and pure. Some of these substances are soluble in water, others are constituents of the resin ; xanthohumol is an orange yellow, crystalline body. One of the constituent resins yields by oxidation valerianic acid, a reaction that explains the change in the odour when hops are kept.

**Uses.**—The volatile oil produces sedative and soporific effects, whilst the bitter substances are stomachic and tonic ; hops accordingly improve the appetite and promote sleep.

## CUBEBS

### (Fructus Cubebæ)

**Source, &c.**—Cubebs are the fruits of *Piper Cubēba*, Linné filius (N.O. *Piperaceæ*), a diœcious woody climber indigenous to Java, Sumatra, and Borneo, and apparently cultivated also in those islands,

FIG. 73.—*Piper Cubeba*. *A*, fruiting branch, natural size. *C*, portion of the same magnified, showing the stalk-like elongation of the pericarp. *B*, branch of staminate plant, natural size. *D*, bract and stamens, magnified. *E*, fruit cut to show the endosperm and perisperm, magnified. *F*, portion of same, more highly magnified, showing the embryo, *e*, endosperm, *a*, and perisperm, *p*. *G*, *Piper nigrum*; portion of flowering spike, magnified. (Luerssen.)

although exact information concerning the cultivation is difficult to obtain. The fruits were used as a spice, and as a medicine in the Middle Ages.

The pistillate inflorescence of the cubeb is a spike of sessile flowers. The young fruits are also sessile, but as they mature they become elevated on a slender stalk produced by an abnormal development of the pericarp of the fruit at its base (fig. 73, *C*). When the fruits are full-grown, but whilst they are still green and unripe, they are stripped from the rachis, bringing with them the stalk-like prolongation of the pericarp which remains permanently attached to them, whence the name 'tailed pepper' by which they are sometimes known. They are then dried in the sun, during which the green colour changes to a greyish black; they are bought up by Chinese traders and exported chiefly from Batavia to Amsterdam or from Singapore to London.

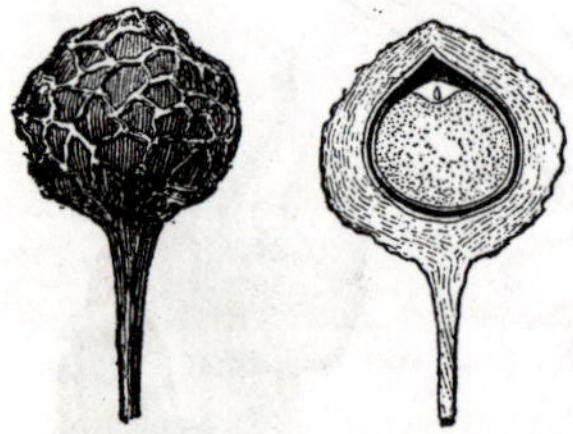


FIG. 74.—Cubeb. *A*, entire fruit, magnified. *B*, entire fruit, cut vertically, magnified. (Planchon and Collin.)

**Description.**—The commercial drug consists of nearly globular fruits, sometimes depressed at the base (immature fruits), measuring about 4 mm. in diameter, and usually of a greyish brown or nearly black colour. The pericarp is reticulately wrinkled (due to the shrinking as the fresh fruit dries) or, in very young fruits, shrivelled and abruptly prolonged at the base into a slender stalk about 6 mm. in length which is usually rounded or slightly flattened. The apex of the fruit bears the minute remains of three or four stigmas.

Within the pericarp, which is thin and brittle, is a single seed attached by the base; frequently only the dark shrunken remains of a partially developed seed are to be found. The fully developed seed is reddish brown in colour. The embryo is very small and embedded near the apex of the seed in a somewhat scanty endosperm surrounded by a copious, whitish perisperm, which constitutes the greater part of the seed.

Cubebs exhale, when crushed, a strong, characteristic, spicy odour, and possess a strong, spicy, somewhat bitter taste. The crushed fruit, sprinkled upon the surface of concentrated sulphuric acid, produces a crimson coloration. This reaction is an important one, as most of the substitutes for cubebs yield only a brownish red under the same conditions. It is especially valuable when taken in conjunction with the microscopical characters, for there is no substitute known that possesses an anatomical structure identical with that of genuine cubebs, and also yields the crimson colour with sulphuric acid (Hartwich, 1898).

**Microscopical Characters.**—Below the outer epidermis of the pericarp is an interrupted layer consisting of one or two rows of thick-walled sclerenchymatous cells. Distributed in the thin-walled parenchyma of the mesocarp are numerous large oil-cells. Within the inner epidermis of the pericarp is a second, uninterrupted layer of one or two rows of very thick-walled, radially elongated cells. The seed consists chiefly of perisperm, the cells of which are packed with small starch grains.

The student should observe

(*a*) The *slender stalk*, which is not easily detached,
(*b*) The *seed attached only by its base* to the pericarp,
(*c*) The *characteristic odour and taste*,
(*d*) The *reaction with sulphuric acid.*

**Constituents.**—Cubebs yield from 10 to 18 per cent, of volatile oil (sp. gr. 0·910 to 0·930 ; O.R. —25 ° to —40° ), which is contained in oil-cells both in the pericarp and perisperm ; similar oil-cells are contained also in the rachis, stem, leaves, &c., but the rachis, which is frequently found mixed with the cubebs, yields only about 1·7 per cent. of volatile oil.

Cubebs contain further an indifferent substance, cubebin, an acid resin, cubebic acid, and an indifferent resin. They afford about 6 (not over 8) per cent. of ash and yield to ether about 22 (not less than 20) per cent. of oleo-resin.

Cubebin has been obtained in colourless crystals, yielding with sulphuric acid a cherry-red colour ; it appears to be devoid of any remarkable physiological action.

Cubebic acid (0·96 per cent.) is white and amorphous ; it gives with sulphuric acid a crimson colour, and together with the indifferent resin (2·5 per cent.) produces purely diuretic effects. These two principles may therefore be regarded as active constituents of cubebs.

Cubeb-camphor ($C_{15}H_{24},H_2O$), found in old cubebs, is a crystalline but unstable hydrate of a sesquiterpene.

**Adulterations, &c.**—The exportation of cubebs has been subject to considerable variation, and during times of scarcity the drug has been liable to adulteration and substitution. Many other plants belonging either to the same or to other natural orders produce fruits resembling cubebs in appearance, a fact which has doubtless facilitated sophistication. Genuine cubebs can, however, be easily distinguished by the crimson colour they impart to sulphuric acid, taken in conjunction with the anatomical characters of the pericarp as above described.

The fact that the fruits on the spike mature in succession from base to apex necessitates the presence in the drug of some immature fruits, but too many of these, which may be recognised by their small size and shrivelled appearance, should not be present. Portions of the rachis also find their way into the commercial drug ; they too should not be present in large quantity, as they contain less cubebic acid and resin, and much less volatile oil.

The following fruits have from time to time been found mixed with or substituted for genuine cubebs :

*Rinoe badak*, greyish, 6 mm. in diameter, odour mace-like.

*Piper crassipes*, Korthals ; black, larger than genuine, stalk longer and curved, usually depressed at base.

*Piper ribesioides*, Wallich ; brown, larger than genuine, taste less aromatic and more bitter, sclerenchymatous cells in mesocarp.

*Piper mollissimum*, Blume ; large fruits, long stalks, no inner row of sclerenchymatous cells.

*Piper Lowong*, Blume ; fruit about 6 mm. in diameter, stalk 6 mm. long, no inner row of sclerenchymatous cells.

*Piper Clusii*, Casimir de Candolle ; small, grey ; stalk curved ; taste very peppery ; imported from the Congo under the name of Congo or African cubebs.

**Uses.**—Cubebs have a stimulant and antiseptic action on the mucous membrane of the genito-urinary organs and are also diuretic. They are chiefly used in gonorrhœa and affections of the bladder, sometimes also in chronic bronchitis, the active constituents of the drug leaving the body by the kidneys and urinary passages, the skin and the respiratory organs.

## BLACK PEPPER

### (Piper Nigrum)

**Source, &c.**—The pepper vine, *Piper nigrum*, Linné (N.O. *Piperaceæ*), is a perennial climbing plant indigenous to southern India, but spread by cultivation over the islands of the Malay Archipelago and to the West Indies and South America. It is now cultivated chiefly in Sumatra, Singapore, Johore, the islands of the Rioux-Lingga Archipelago, Penang and Siam. Probably no drug or spice has been so sought after, or has played so important a part, as pepper. It was apparently brought to Europe soon after the expedition of Alexander the Great to India. During the Middle Ages the trade in it was concentrated in Venice, and the desire to divert so profitable a commerce acted as a direct inducement to the Genoese and Portuguese to seek a sea route to India, which resulted in the discovery by Vasco da Gama in 1498 of the route round the Cape of Good Hope to India. At the present moment the centre of the pepper trade is Singapore, whence about one-half of the world's supply is exported (Tschirch, 1892).

The pepper vines are cultivated in gardens as hops are in England, but instead of poles being used, trees are planted to afford the necessary shade and support. The plant bears a pendulous spike of sessile flowers (fig. 73 *G*), which are succeeded by small drupaceous fruits; these do not become raised upon stalks as the cubebs do, but remain sessile. As soon as the lower berries on the spike begin to ripen, which is shown by their colour changing from green to red, the whole spike is picked and dried in the sun, during which the fruits turn black. Separated from the rachis and sorted they constitute the black pepper of commerce.

For the production of white pepper the fruits are allowed to become nearly ripe; they assume a crimson colour, and the mesocarp acquires a pulpy consistence. They are then either heaped and allowed to ferment, or soaked in water, until the outer part of the pericarp is loosened; from this they are separated by rubbing between the hands or treading under the feet, after which they are washed free from the pulp and dried. White pepper is therefore the nearly ripe fruit from which the outer part of the pericarp has been removed. A similar separation is easily effected from the black pepper of commerce after soaking in water.

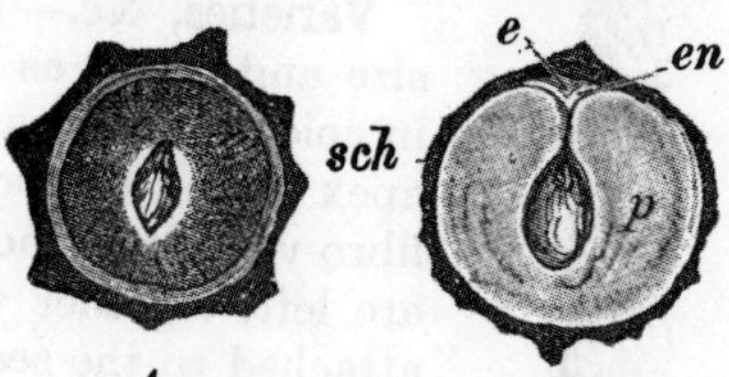


FIG. 75.—Black Pepper. *A*, transverse section. *B*, longitudinal section, showing pericarp, *sch*; perisperm, *p*; endosperm, *en*; and embryo, *e*. Magnified. (Tschirch.)

**Description.**—Black pepper consists of small dark brown or nearly black spherical fruits, measuring about 5 mm. in diameter, with a more or less regularly and deeply reticulate, wrinkled surface. At the apex the remains of the sessile stigmas can be traced; the base bears a scar indicating the point of attachment to the rachis, but, the fruits being sessile, there is no stalk.

If a fruit be cut in halves and examined with a lens, the pericarp will be seen to be thin and dark and completely filled with a seed to which it is closely adherent. The seed is globular and covered with a brown seed-coat. It contains a small endosperm, near the apex of the fruit, in which the minute embryo is situated; but the major part of the seed is composed of perisperm, which is yellowish and horny near the periphery, whitish, mealy, and frequently hollow in the centre.

Both pericarp and seed have an aromatic odour and pungent taste; they both contain oil-cells, visible under the microscope.

The student should observe

(*a*) The *absence of stalk*,
(*b*) The *seed completely filling the pericarp and adherent to it*.
(*c*) The *characteristic odour and taste*,
(*d*) In *white pepper the adherent inner part of the pericarp, with its fibro-vascular bundles*.

**Constituents.**—Both the seed and the pericarp contain numerous oil-cells in which a volatile oil (1·0 to 2·3 per cent.), a resin, and a crystalline alkaloid, piperine (5·0 to 8·25 per cent.), are contained. The volatile oil consists almost entirely of terpenes. The resin, which has been termed chavicin, and to which the pungency of pepper is chiefly due, has not been sufficiently examined. The alkaloid piperine, $C_{17}H_{19}NO_3$, forms colourless and odourless crystals, which are at first tasteless, but subsequently pungent; heated with alcoholic solution of potassium hydroxide it is split up into piperic acid and piperidine.

The seeds contain, further, a large quantity of starch in minute angular grains united into polygonal masses. Black pepper yields from 4·0 to 7·0 per cent. of ash.

FIG. 76.—Long Pepper. (Planchon and Collin.)

**Varieties, &c.**—*White Pepper* is of about the same size and shape as black. The surface is greyish white in colour, and nearly smooth. From the base to the apex there run about sixteen light lines; these are the fibro-vascular bundles that traverse the pericarp; they are left, together with the inner part of the pericarp, attached to the seed when the outer part of the fruit is removed. By gently scraping white pepper the adherent part of the pericarp can be removed and the dark brown seed disclosed; the latter possesses the characters above described. It contains from 4 to 6·5 per cent. of piperine.

• *Long Pepper.*—Long pepper is the dried unripe fruit of *Piper officinārum*, Casimir de Candolle, and is exported chiefly from Java. It consists of a large number of minute sessile fruits which, together with the bracts that support them, are crowded together on and partially embedded in an elongated axis so as to form a dense spike. Each spike is about 40 mm. long and 6 mm. thick; it is nearly cylindrical, tapering to a rounded apex, and often covered with a greyish powder. When washed free from this the spikes are seen to be reddish brown in colour, and the minute fruits are then more easily visible; they are arranged in a close spiral, and each bears the remains of the stigma at its apex. Cut transversely, the section of the spike shows eight or ten fruits with starchy perisperm arranged around a central axis.

The taste and odour resemble those of black pepper, but are not so strong. The active constituents are the same, but they are present in smaller proportion (volatile oil 1·0 per cent., piperine 6·2 per cent.).

The student should cut long pepper transversely and observe the minute fruits, which appear white and starchy in section.

**Uses.**—Applied externally pepper acts as a rubefacient, anodyne, and counter-irritant. It is given internally as a local stimulant, and as a stimulant to the urethra and rectum; it increases gastric secretion and improves the appetite; it is also occasionally used for hæmorrhoids and other diseases of the rectum.

## JUNIPER BERRIES

### (Fructus Juniperi)

**Source, &c.**—The juniper, *Junipĕrus communis*, Linné (N.O. *Coniferæ*), is a small diœcious evergreen shrub with linear spreading prickly leaves, indigenous to Great Britain, and widely distributed over Europe.

In the axils of certain of the upper leaves small rudimentary branches are produced, each of which bears a number of minute bracts;

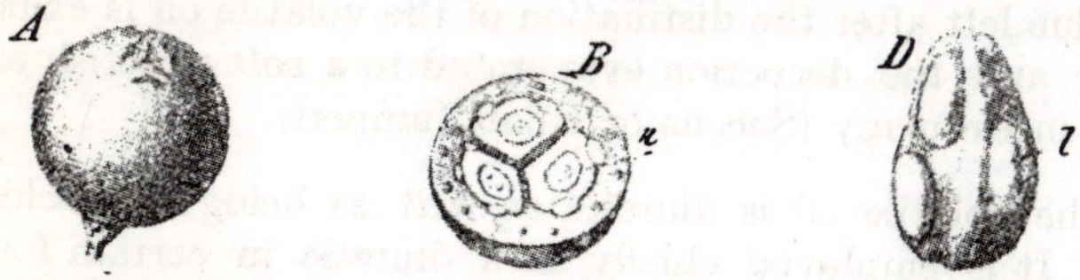


FIG. 77.—Juniper berry. *A*, entire fruit, magnified 3 diam. *B*, transverse section of the same; *κ*, seed. *D*, Seed, magnified 3 diam.; *l*, oil-gland. (Berg.)

in the axils of the uppermost whorl of three such bracts three ovules are produced, which, as in all gymnosperms, are not enclosed in an ovary. After fertilisation the three supporting bracts increase in size, become fleshy, grow round, and finally, with the exception of a minute canal at the apex, completely enclose the three ovules. At the same time the ovules develop to seeds and become surrounded with hard coats developed partly from the integuments, partly from the tissue of the enclosing bracts. The fruit ripens in the second year, and forms then a purple, berry-like fruit which is termed a 'galbulus.'

Juniper berries are collected in various parts of Europe, but especially in Italy and in Hungary.

**Description.**—Ripe juniper berries are of a dark purplish colour and nearly globular shape, measuring about 8 mm. in diameter. The apex bears a very distinct tri-radiate scar indicating the sutures of the three bracts by which the seeds are enclosed; at the base are (usually) six minute pointed bracts arranged in two whorls. The outer skin (corresponding to the epicarp) is thin and often covered with a bluish 'bloom'; the tissue corresponding to the mesocarp is loose and soft, of a yellowish brown colour, and contains embedded in it three hard triangular seeds. These usually bear, partly sunk in the hard tissue, large oil-glands filled with volatile oil, or, in old fruits, with yellowish transparent resin; they are often concealed by the soft surrounding tissue which adheres to the resinous secretion. Smaller oil-glands are present in the pulp of the fruit, but they are less conspicuous.

Juniper berries have an agreeable aromatic odour resembling turpentine, and a sweetish taste.

The student should observe

(a) The *scar at the apex*,
(b) The *minute bracts at the base*,
(c) The *three hard seeds with oil-glands* partly embedded in them.

**Constituents.**—Ripe Juniper berries yield about 1 per cent. of volatile oil (sp. gr. 0·862 to 0·890; O.R. $-$ 3° to $-$ 15°), consisting of pinene, cadinene, and juniper camphor, and contain about 33 per cent. of invert sugar.

The residue left after the distillation of the volatile oil is exhausted with water and the decoction evaporated to a soft sweetish extract much used in Germany (Succus or Roob Juniperi).

**Uses.**—The volatile oil is diuretic as well as being stomachic and stimulant. It is employed chiefly as a diuretic in certain forms of dropsy.

## VANILLA

### (Fructus Vanillæ)

**Source, &c.**—The vanilla plant, *Vanilla planifolia*, Andrews (N.O. *Orchideæ*), is a climbing plant growing wild in the moist woods on the eastern coast of Mexico. It is now largely cultivated in the islands Réunion, the Scychelles, Mauritius, Tahiti, Java, &c., where the climate resembles in temperature and humidity that of its native country. The genus includes many species, but *V. planifolia* is the chief vanilla-yielding one.

Like other members of the same natural order, the fertilisation of the Mexican wild plant is effected by insects; but when the plant is cultivated in other countries, in the absence of these insects, the fertilisation is accomplished by hand. This is effected by introducing a pointed stick into flower after flower, selecting the strongest. Before the fruits ripen, when the colour begins to change from yellowish green to brown, they are picked and subjected to a process of 'curing,' during which certain aromatic principles are formed and the fruits acquire the well-known vanilla odour. This process, the details of which vary considerably, consists essentially in slowly drying the fruits by exposure to the warmth of the sun or to artificial heat. The change in the constituents is probably due to the action of enzymes of which two, a hydrolysing enzyme and an oxydase have been detected, but the exact nature of the change or of the substances acted upon is not known. The cured fruits, which are of a dark brown or nearly black colour, are bound in bundles and packed in tins, in which they gradually become coated with minute crystals.

**Description.**—Commercial vanilla occurs in slender flexible stick-like pods about 15 to 20 cm. in length, and of a dark brown or nearly black colour. They have a flattened-cylindrical shape, due to the mutual pressure of the pods in the bundle, and taper towards both

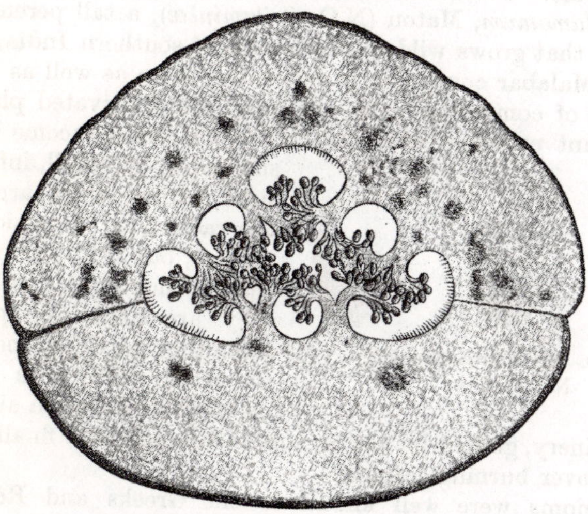

FIG. 78.—Vanilla. Transverse section of fruit, showing the placentas, seeds, and secreting hairs on inner epidermis of the pericarp. Magnified. (Moeller.)

base and apex. The surface is longitudinally wrinkled and frequently more or less covered with numerous minute glistening crystals (of vanillin). The fruits are one-celled and contain innumerable, minute, black seeds embedded in a dark-coloured, aromatic, balsamic fluid secreted by the cells of the inner epidermis of the pericarp, which are developed into short hair-like processes projecting into the cavity of the fruit. The drug has an extremely fragrant odour and an agreeable aromatic taste.

**Constituents.**—The principal aromatic constituent of vanilla pods is vanillin, though probably other aromatic substances are present; it is contained in the fluid secreted by the inner epidermis of the pericarp, which gradually permeates the whole fruit, and constitutes the crystalline deposit that slowly accumulates on the pods.

Vanillin, the aldehyde of methylprotocatechuic acid, can be obtained in colourless acicular crystals with a fragrant vanilla-like odour. It also occurs in balsam of Peru, balsam of Tolu, benzoin, and other drugs; it is now prepared in large quantities from eugenol, the principal constituent of clove oil.

## CARDAMOMS

### (Fructus Cardamomi)

**Source, &c.**—Cardamoms are the dried, nearly ripe fruits of *Elettaria Cardamomum*, Maton (N.O. *Scitaminĕæ*), a tall perennial, reed-like plant that grows wild in the forests of southern India, especially near the Malabar coast. It is cultivated there as well as in Ceylon, the fruits of commerce being obtained from cultivated plants.

The plant produces near the root a long loose raceme of flowers, succeeded by small, inferior, capsular fruits, which are cut from the rachis in succession as they mature, but before they are quite ripe; if they are left till quite ripe they are liable to spring open when they are dried and discharge the seeds. They are then dried on trays in the open air, trimmed by machinery, graded by sieves, sorted to colour and finally bleached in trays over burning sulphur.

FIG. 79.—Mysore Cardamoms. Natural size.

Cardamoms were well known to the Greeks and Romans, but probably not the species now under consideration.

**Description.**—Cardamom fruits differ considerably in size and shape. They vary usually from 1 to 2 cm. in length, the smallest variety being frequently nearly globular in shape, while the longer are ovate or oblong and more or less distinctly three-sided. They are of a pale buff or yellowish colour, and nearly smooth or longitudinally striated. The base is rounded and may bear the remains of the stalk; towards the apex they taper more or less abruptly, and are sometimes crowned by a short beak formed of the remains of the calyx. They are three-celled, and contain in each cell two rows of small seeds attached to axile placentas. From good plump cardamom fruits the seeds usually separate in a cohering mass from each cell. They are a dark reddish brown, about 3 mm. in length, irregularly angular, hard, and marked with transverse wrinkles, which are very distinct when examined with a lens; unripe seeds are paler in colour and not so plump. The thin, colourless, membranous aril that covers the seeds become more evident after soaking in water. The hilum is depressed, and a channel, extending on one side from the base to the apex of the seed, indicates the position of the raphe.

Cut transversely, the seed exhibits a thin, dark, seed-coat, a whitish perisperm grooved on one side, and in the centre a small, yellowish, translucent endosperm and minute embryo.

The seeds have a powerful aromatic odour and an agreeable, pungent, aromatic taste, but the pericarps possess neither aroma nor taste. The seeds of fruits which have partially opened are less aromatic, and such fruits ('splits') are less esteemed.

**Microscopical Characters.**—The most important tissues in cardamon seeds are (i) the epidermis, consisting of long, narrow, tapering cells with walls 3 to

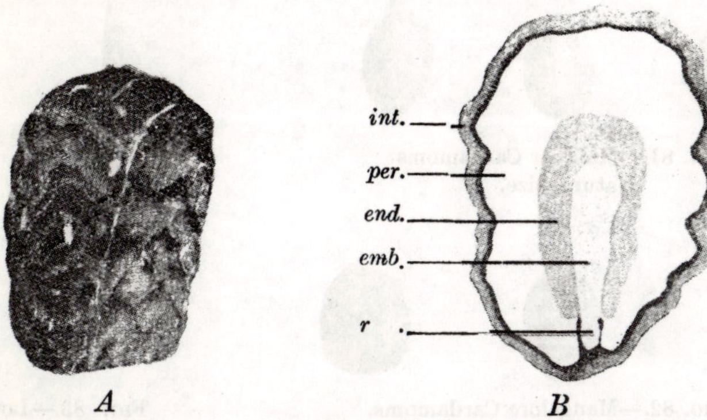


FIG. 80.—Cardamom seed. *A*, entire; *B*, longitudinal section; *int.*, integuments; *per.*, perisperm; *end.*, endosperm; *emb.*, embryo; *rad.*, radicle. Magnified 12 and 14 diam. respectively.

4 $\mu$ thick; (ii) a single layer of large, thin-walled oil-cells in the outer seed-coat; (iii) the inner seed-coat, consisting of dark brown, very strongly thickened, radially prismatic cells, each containing a small nodule of silica; (iv) the thin-walled parenchyma of the endosperm packed with minute starch grains and containing in the centre of each cell one or more small prismatic crystals of calcium oxalate.

The pericarps contain abundant empty parenchymatous cells, spiral vessels, and elongated sclerenchymatous cells and fibres, none of which occurs in the seed.[1]

The student should observe in the fruits

(*a*) The *pale colour* and *plump appearance*,
(*b*) The *abruptly tapering apex*;

and should compare them with Ceylon (wild) cardamoms (see below).

[1] Compare Greenish, *Food and Drugs*.

He should also observe in the seeds

(*a*) The *dark reddish brown colour* and *aromatic taste*,

(*b*) The *transverse wrinkles, depressed hilum and raphe*,

(*c*) The characters of the *section* ;

and should compare them with

(i) Seeds of *Ceylon cardamoms* (see below),

(ii) *Grains of Paradise*, which have a rich reddish brown colour, papillose surface, and thick fibrous funicle.

**Constituents.**—The principal constituent of the seeds is the volatile oil, of which they yield from 2 to 8 (average about 5) per cent. They also contain abundance of starch. Good ripe seeds yield about

FIG. 81.—Malabar Cardamoms. Natural size.

FIG. 82.—Mangalore Cardamoms. Natural size.

FIG. 83.—Long wild native Cardamoms. Natural size.

3·5 to 5·5 per cent. of ash, unripe seeds more ; ripe fruits from 4 to 7 per cent.

**Varieties.**—*Mysore* : although bearing the name of an Indian province, these are imported from Ceylon ; they constitute the chief commercial variety, and the one to which the official description most nearly applies. They are distinguished by their ovoid shape and cream coloured, nearly smooth surface.

*Malabar* : these are also imported from Ceylon ; they are shorter, plumper, and often not so smooth as Mysore cardamoms.

*Mangalore*, which closely resembles the Malabar variety, but are often rather larger, nearly globular, have a roughish, almost scurfy coat ; they are imported from India.

The seeds of these three varieties are practically indistinguishable from one another, though differing very slightly in flavour.

*Long wild natives* : these are the fruits of *E. Cardamomum*, var. *β major*, Smith, imported from Ceylon ; they are readily distinguished by their elongated shape, shrivelled appearance, and greyish brown colour; the seeds closely resemble the foregoing, but have a slightly bitterish taste, different odour, and rather thicker epidermal cell walls (4·5 to 6·0 μ; see above). They are used for the production of commercial cardamom oil, and also for making liqueurs and for flavouring.

Other varieties occasionally met with are:—

*Round or Cluster cardamom* (*Amomum Cardamomum*, Linné); fruits globose, about 12–15 mm. in diameter; seeds have strong camphoraceous taste (Java); frequently found in imported shelled seeds.

*Korarima cardamoms* (*Amomum Korarima*, Pereira); fruits ovate, pointed; seeds larger than genuine, reddish brown, striated; taste similar (Abyssinia).

*Bengal cardamoms* (*A. aromaticum*, Roxburgh); fruits large, winged; seeds about 3 mm. long; taste very aromatic, camphoraceous (India).

*Wild cardamoms* (*A. xanthioides*, Wallich); fruits spiny; seeds resemble genuine but flavour is different (Siam).

Powdered cardamom seeds may be distinguished from the powdered fruits by the absence of the tissues of the pericarp (see above).

**Uses.**—Cardamoms are employed as an aromatic carminative and as an agreeable flavouring agent.

## FRUITS IN LESS FREQUENT USE

*Annatto.* A colouring matter obtained from the seeds of *Bixa Orellana*, Linné (N.O. *Bixineæ*), a small tree indigenous to South America (Venezuela, Brazil) and cultivated in most tropical countries. The outer part of the shell of the seed consists of a soft, sticky, red, resinous mass. The capsular fruit is opened, the seeds removed and the annatto separated by vigorously stirring with water, pulping, drying the pulped mass and pressing it into cakes; or the seeds mixed with water may be allowed to ferment, the annatto suspended in the water separated by sieving, boiling, skimming off the scum, evaporating it, and forming it into cakes. Annatto from Brazil (Spanish annatto) has an agreeable odour; that from Cayenne (French annatto) has an unpleasant odour, due, it is said, to the addition of urine in order to keep the cakes moist. The chief constituent is bixin (minute, dark red crystals, soluble in alcohol, ether and alkalies, insoluble in water), fat, resin, &c. It is used to colour butter, cheese, wax, varnish, &c.

*Buckthorn Berries.* The fruits of *Rhamnus cathartica*, Linné (N.O. *Rhamneæ*). The fresh fruits are almost black, spherical up to 1 cm. in diameter, inferior, crowned with an eight-rayed calyx; mesocarp fleshy, greenish; each of the four carpels contains a single ovule but usually two or three only develop into hard seeds. The fruits are usually used in the fresh state for the expression of the juice (Succus Rhamni) which is made into syrup (Syrupus Rhamni) by the addition of sugar. The chief constituents are rhamno-emodin (Tschirch and Polasco), rhamno-carthartin (an emodin glucoside) and emodin-anthranol; they are accompanied by yellow colouring substances, rhamnoxanthin, xanthorhamnin, quercetin, rhamnocitrin, rhamnolutin, rhamnochrysin. The syrup is used as a purgative.

*Ajowan Fruits.* The fruits of *Carum copticum,* Bentham and Hooker filius (N.O. *Umbelliferæ*), India ; greyish-brown about 2 mm. long, compressed, with short protuberances ; odour of thymol. The volatile oil (3 to 4 per cent.) contains thymol (30 to 40 per cent.). Used as a source of thymol ; in India as a spice.

*Chenopodium* (American Wormseed). The fruits of *Chenopodium ambrosioides,* Linné and *C. ambrosioides,* var. *anthelminticum,* Gray (N.O. *Chenopodiaceæ*). Fruits very small, about 1·5 mm. in diameter, globular, greenish yellow or brown, depressed, enveloped by the persistent calyx ; in each a single, small brownish-black lenticular seed. Odour strong, recalling eucalyptus. Contains volatile oil (3 to 4 per cent.). Used as an anthelmintic.

*Parsley Fruits.* The fruits of *Carum Petroselinum,* Bentham and Hooker (N.O. *Umbelliferæ*). Fruits about 2 mm. long, 1 to 2 mm. thick, laterally compressed, ovate, greyish green or greenish brown ; odour and taste aromatic. Chief constituent volatile oil (2 to 5 per cent.) containing apiol ($C_{12}H_{14}O_4$) which separates in crystals on standing or on cooling (parsley camphor). Commercial apiol is a greenish viscous oily liquid obtained by exhausting the fruits with ether and distilling. Yellow apiol is a purified form of commercial apiol. Apiol is used in dysmenorrhœa and amenorrhœa.

*Cashew Nuts* (East Indian Almonds). The seeds of *Anacardium occidentale,* Linné (N.O. *Anacardiaceæ*), West Indies. Peduncle of fruit large, fleshy, pear-shaped, sweet, edible. Fruit reniform, up to 3 cm. long, brownish ; in the mesocarp dark brown, oily, very acrid and irritant secretion. Seed reniform, taste bland, edible. Chief constituents of the oily secretion cardol (yellowish oily, irritant and vesicant), anacardic acid, tannin ; in the seed 40 to 50 per cent. fixed oil. The fruits of *Semecarpus Anacardium,* Linné (N.O. *Anacardiaceæ*) India, are flattened cordate ; the mesocarp contains a similar secretion. Cardol exposed to the air gradually turns black and may be used to prepare an indelible ink, hence the fruits are sometimes called 'marking nuts.'

*Pearl Barley* (Hordeum Decorticatum). The dried fruit of *Hordeum distichon,* Linné (N.O. *Gramineæ*), from which the enclosing firmly attached paleæ, pericarp and seed-coats have been more or less completely removed by milling, the elongated barley grain being thereby reduced to a small rounded grain with a groove containing remains of the various coats on one side. Chief constituents are starch and proteids.

Malt is barley which has been induced to germinate by moistening it and has then been dried. It contains the enzyme diastase which is capable of converting gelatinised starch into dextrin and maltose. Taka-diastase is a similar enzyme obtained by growing the fungus *Eurotium oryzæ* on cooked rice.

## SECTION IV

## SEEDS

The seed is the body produced from the ovule as the result of fertilisation. It consists ordinarily of two seed-coats, an outer or testa and an inner or tegmen, enclosing a kernel composed of the embryo with one or more cotyledons and in many cases an endosperm. Sometimes an outgrowth (arillus) is developed from the funicle or from the placenta which may extend so as partially or even almost entirely to enclose the seed.

Most seeds may be readily distinguished from one another on account of the great diversity in their size, shape, surface, structure, and contents. Those with which the student will have to deal may be very minute (black mustard, lobelia, henbane) or of considerable size (calabar bean, cola seed) ; they may be spherical, ovoid, reniform, &c. ; the surface may exhibit a great variety in the sculpture of the outer wall, which may be pitted (mustard seed) or variously reticulate (poppy, henbane, lobelia), &c. Further, the epidermal cells may be prolonged into hairs (nux vomica, strophanthus), or contain mucilage (linseed, mustard seed), or offer other distinctive characters.

The internal structure of the seed-coats also possesses valuable diagnostic features, but these are only visible under the microscope. A section examined with a lens often exhibits the disposition of the cotyledons and radicle, and allows of the presence or absence of an endosperm being determined ; the endosperm itself varies considerably in its nature, being oily, mucilaginous, horny, ruminate, &c.

In examining small seeds the student should constantly employ a hand lens with the aid of which the surface should be carefully observed. Transverse sections cut with the aid of a sharp penknife or razor should also be scrutinised. Large, hard seeds such as nux vomica should be soaked in hot water till they are sufficiently softened to be cut.

### STAVESACRE SEEDS

### (Semina Staphisagriæ)

**Source, &c.**—Stavesacre, *Delphinium Staphisagria*, Linné (N.O. *Ranunculaceæ*), a stout erect herb attaining 4 feet in height, is indigenous to Asia Minor and southern Europe. The plant is cultivated in France and Italy, our supplies being derived chiefly from Trieste and from the south of Italy.

Stavesacre was well known to both the Greeks and the Romans. Dioscorides mentions it, and Pliny describes its use as a parasiticide. It continued to be extensively employed throughout the Middle Ages, but is now in much less demand.

The fruit consists of three follicles, in each of which a few seeds are closely packed; these are collected when ripe.

**Description.**—The seed, as seen in commerce, are about 6 mm. in length and rather less in breadth. They usually appear to be dark grey in colour, but when freed from the dust with which they are covered are seen to be dark brown, and then the characteristic markings on the surface of the seed become more evident. In shape they are usually irregularly or obscurely four-sided, one side being curved and larger than the others, which are nearly flat or even depressed. The surface of the seed is more or less uniformly reticulate and covered with minute papillæ visible under a lens. One end of the seed is usually more pointed than the other; near the pointed end the hilum is visible as a narrow line. By soaking a seed in water and cutting it transversely just below the hilum, the minute embryo may be found embedded in a large, whitish or yellowish oily endosperm.

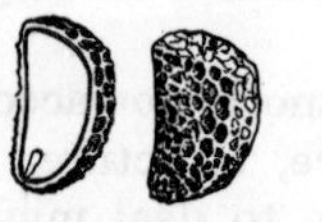

FIG. 84.—Stavesacre seed, entire and cut longitudinally, showing embryo. Magnified 2 diam. (Maisch.)

The seed-coat is nearly tasteless, but the endosperm is intensely bitter and acrid; the seeds have no marked odour.

The student should observe

(*a*) The *dull earthy colour of the seed*,
(*b*) The *characteristic obscurely quadrangular shape*,
(*c*) The *rough as well as pitted surface*,
(*d*) The *bitter, oily endosperm*.

**Constituents.**—The seeds contain several alkaloids (in all about 1 per cent.), the most important of which are delphinine, delphisine, and delphinoidine; less important are staphisagroine, of which traces only are present, and staphisagrine, which is probably a mixture.

Delphinine, $C_{31}H_{49}NO_7$, is crystalline and extremely poisonous. Delphisine is amorphous, but about twice as poisonous as delphinine, while delphinoidine, which is also amorphous, is much less toxic.

The seeds also contain from 30 to 35 per cent. of fixed oil, which may be extracted either by expression (expressed oil of stavesacre) or by solvents such as petroleum spirit, &c.; in either case the oil carries with it the greater part of the alkaloids, including practically all the delphinine; these may be removed from the oil by shaking it with an aqueous solution of tartaric acid. The drug affords from 10 to 13 per cent. of ash.

**Uses.**—Stavesacre seeds are extremely poisonous, delphinine and delphinoidine resembling aconitine in action, but being weaker; the seeds are used only as a parasiticide to kill pediculi, chiefly in the form of the official ointment, the expressed oil, the powdered seeds, or an acid aqueous extract containing the alkaloids. Delphinine has also been employed similarly to aconitine both internally and externally for neuralgia, &c.

**Note.**—*Delphinium Consolida,* Linné, contains three alkaloids apparently not identical with those of *D. Staphisagria. Delphinium Ajacis,* Linné, contains ajacine and ajaconine. The large size of stavesacre seeds distinguishes them from the seeds of other species of *Delphinium.*

## BLACK MUSTARD SEEDS

### (Semina Sinapis Nigræ)

**Source, &c.**—The black mustard plant, *Brassica sinapioides*, Roth (*B. nigra*, Koch, N.O. *Cruciferæ*), an erect annual plant attaining a height of 1 metre or more, is largely cultivated in Holland, England, Italy, Germany, and other countries. The fruits are smooth, erect, appressed siliquas, which dehisce, when ripe, by both sutures, disclosing about ten or twelve minute dark seeds. These are separated when ripe, and dried.

**Description.**—Black, or, as they are sometimes termed, brown or red, mustard seeds are of a dark reddish brown or greyish brown colour, sometimes nearly black, and are frequently partially covered with very thin, whitish scales (dried mucilage from the epidermis, probably the result of damp). They are nearly spherical or flattened-ovoid in shape and about 1 mm. in diameter. Under the lens the surface appears minutely pitted, and the hilum can be distinguished as a paler point.

Although the seeds are hard, the seed-coats are thin and brittle. The kernel is greenish yellow and oily, and consists of the two folded cotyledons embracing the small radicle. The position of cotyledons and radicle can be well seen by cutting a seed in halves midway between the hilum and the apex. After the seed has been soaked in water, during which it surrounds itself with mucilage, the seed-coats can be easily removed, disclosing the cotyledons folded over and incumbent upon the radicle; there is no endosperm, the seeds being exalbuminous.

Black mustard seeds, even when powdered, have no marked odour. The taste is at first bitter, but rapidly becomes strongly pungent, and although the dry seeds are almost inodorous they develop, when moistened with water, a volatile substance of extreme pungency that rapidly attacks both the nostrils and the eyes. This volatile substance is not, however, developed if the seeds have been previously thrown into boiling water.

The student should observe :

(a) The *minute size and spherical shape*,

(b) The *pitted surface*,

(c) The *incumbent and folded cotyledons*,

(d) The *pungent taste* ;

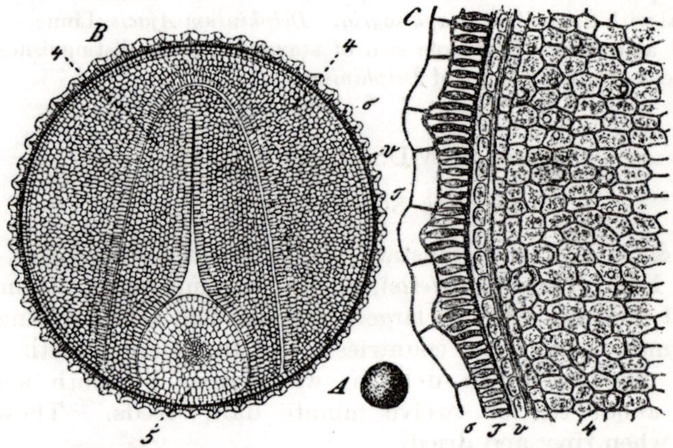


FIG. 85.—Black Mustard seed. *A*, entire seed, magnified 3 diam. *B*, transverse section, magnified 65 diam. : 4, 4, the two cotyledons ; 5, the radicle. *C*, portion of the same, further enlarged : σ, epidermis containing mucilage. (Berg.)

and should compare these seeds with colchicum seeds, which are larger, rougher, and harder than black mustard, and have a bitter, not pungent, taste.

**Constituents.**—Black mustard seeds contain in the kernels about 27 per cent. of fixed oil (compare Expressed Oil of Mustard) which can be extracted by crushing and pressing the seed. In the seed-coal (in the cells of the epidermis) there is mucilage, which dissolves when the seeds are soaked in water. The seeds contain, further, a small quantity of acid sinapine sulphate (compare p. 149), and two substances, sinigrin (about 4 per cent.) and an enzyme, myrosin, which, by interaction in the presence of water, yield the volatile pungent body previously referred to ; the latter is not a constituent of the seed, but is produced from two of its constituents under certain conditions. The assertion that black mustard seeds do not contain sufficient myrosin to decompose the whole of the sinigrin has been shown to be incorrect.

Sinigrin (also called potassium myronate), $C_{10}H_{18}NS_2O_{10}K$, may be crystallised from either alcohol or water. It may be extracted by first boiling the powdered seeds (not previously defatted) for a short time with 80 per cent. alcohol, by which the activity of the myrosin is destroyed; from the pressed and dried seeds cold water then extracts the sinigrin, which may be obtained by evaporating the aqueous solution in the presence of barium carbonate, extracting with alcohol, concentrating and crystallising (yield about 0·6 per cent.).

Sinigrin, when submitted to the action of myrosin in aqueous solution, decomposes into allyl isothiocyanate, acid potassium sulphate, and dextrose:

$$\underset{\text{Sinigrin}}{C_{10}H_{18}NS_2O_{10}K} = \underset{\text{Allyl isothiocyanate}}{C_3H_5NCS} + \underset{\text{Acid pot. sulph.}}{KHSO_4} + \underset{\text{Dextrose}}{C_6H_{12}O_6}.$$

This reaction, however, does not take place in the seed, as the myrosin and sinigrin are stored up in separate cells.

Allyl isothiocyanate, or volatile oil of mustard, is a mobile volatile liquid with an extremely pungent taste and odour, and is the body to which the pungent taste and odour of black mustard seeds (after crushing and moistening with water) are due. The power of the myrosin to effect this decomposition is destroyed by boiling water, and therefore the seeds that have been subjected to such treatment do not develop allyl isothiocyanate when treated with cold water.

The cake that is left after the fixed oil has been expressed from the seed contains both the sinigrin and the myrosin. From this cake volatile oil of mustard may be prepared by crushing, macerating in water for several hours, and distilling; a mobile liquid collects at the bottom of the distillate, and this, after redistillation, constitutes volatile oil of mustard. It consists of allyl isothiocyanate, associated with small quantities of other bodies such as allyl cyanide, $C_3H_5CN$, &c.

Allyl isothiocyanate can also be produced artificially by decomposing allyl iodide with potassium thiocyanate; allyl thiocyanate ($C_3H_5S.C{:}N$) is formed, which on distillation is converted into the isomeric, pungent allyl isothiocyanate, $C_3H_5N{:}C{:}S$ (allyl thiocarbimide)

$$\begin{matrix} C\equiv N \\ | \\ S\text{—}C_3H_5 \end{matrix} \quad = \quad \begin{matrix} C = S \\ \| \\ N\text{—}C_3H_5. \end{matrix}$$

The seeds yield about 0·7 to 1·3 per cent. of volatile oil, Dutch seeds being the best. Official volatile oil of mustard (s.g., 1·014 to 1·025, distils at 148°–156°) contains at least 92 per cent. of allyl isothiocyanate.

Many Cruciferous plants, and some belonging to other natural orders, yield under similar conditions a pungent volatile oil, thus horse-radish root yields allyl isothiocyanate; scurvy grass (*Cochlearia officinalis*, Linné) and nasturtium seed (*Tropæolum majus*, Linné), isobutyl isothiocyanate ($C_4H_9NCS$); radish (*Raphanus sativus*, Linné) and watercress (*Nasturtium officinale*, Linné), principally phenylethyl isothiocyanate ($C_8H_9NCS$); and cress (*Lepidium sativum*, Linné), benzyl isothiocyanate, ($C_7H_7NCS$).

Black mustard seeds also contain about 29 per cent. of proteid matter and yield from 4·2 to 5·7 per cent. of ash. Starch, which is present in the unripe seed, is not found in any of the cells of the ripe seed.

**Uses.**—Applied externally, black mustard acts as a rubefacient and counter-irritant; this effect is followed by loss of sensibility in the part, and consequently relief from previous pain. Prolonged action may result in vesication. Internally, mustard is used as a condiment, and in full doses as an emetic.

**Varieties.**—*Indian mustard* (*B. juncea*, Hooker and Thoms) is widely cultivated in Southern Russia and India; the seeds resemble black mustard but are rather larger and browner in colour; the volatile oil is believed to contain about 40 per cent. of allyl isothiocyanate, and 50 per cent. of crotonyl isothiocyanate, $C_4H_7NCS$; they are sometimes sold as black mustard seeds.

## WHITE MUSTARD SEEDS

### (Semina Sinapis Albæ)

**Source, &c.**—*Brassica alba*, Boissier (N.O. *Cruciferæ*), the white mustard, is cultivated like the black mustard, which it closely resembles, but not to so large an extent. It differs from the black mustard in producing more or less horizontal hairy fruits, those of the black mustard being erect, appressed, and smooth. Each fruit contains from four to six seeds.

**Description.**—White mustard seeds are yellow in colour, nearly spherical in shape and about 2 mm. in diameter. The seed-coat is very minutely pitted, the pits being so small that the seed appears smooth until examined with a lens. Internally the seed is yellow, and the oily kernel consists, as that of the black mustard seed does, of the two folded cotyledons embracing the small radicle; it becomes coated with mucilage when soaked in water, and can afterwards be easily deprived of its seed-coats.

White mustard seeds, either whole or powdered, are free from pungent odour, even when triturated with water. They have, nevertheless, a pungent taste.

The student should soak some white mustard seeds in water, remove the seed-coats, and observe the cotyledons and radicle, noting that the seed is exalbuminous; he should also crush the seeds, moisten them with water, and note that the taste is pungent, but the odour is not. He should also observe:

(a) The *minutely pitted surface*,
(b) The *incumbent, folded cotyledons*.

**Constituents.**—White mustard seeds contain a fixed oil (about 30 per cent.), mucilage (in the epidermis of the seed-coat), and proteids (about 25 per cent.). Starch is not present in the ripe seeds, which yield about 4 per cent. of ash.

They contain, in addition, a crystalline glucoside, sinalbin, and the same enzyme as is found in the black mustard seed—viz. myrosin. Sinalbin is readily soluble in water and in boiling alcohol, but only very sparingly in cold alcohol; it assumes an intense yellow colour when acted upon by alkalies. Under the influence of myrosin, and in the presence of water, it yields acid sinapine sulphate, dextrose, and acrinyl isothiocyanate. The decomposition may be represented by the following equation:

$$\underset{\text{Sinalbin}}{C_{30}H_{44}N_2S_2O_{16}} = \underset{\text{Acrinyl isothiocyanate}}{C_7H_7O.N{:}C{:}S} + \underset{\text{Dextrose}}{C_6H_{12}O_6} + \underset{\text{Acid sinapine sulphate}}{C_{16}H_{24}NO_5HSO_4}.$$

Of these three substances, acrinyl isothiocyanate is a yellow oily liquid with a pungent taste and powerful rubefacient action, but as it is not volatile it is destitute of pungent odour or pungent effect on the eyes. In this particular the pungent principle obtained from white mustard differs essentially from that yielded by black, and since this principle is not volatile, it is evident that volatile (or essential) oil of mustard cannot be obtained from white mustard seeds. It is very remarkable, considering the close relationship of the two plants and similarity in other constituents, that white mustard should contain no sinigrin and black no sinalbin.

Sinapine is an alkaloid which is so unstable that it has not yet been isolated; the acid sulphate and other salts are crystalline and assume a bright yellow colour in contact with alkalies.

**Uses.**—White mustard possesses rubefacient and vesicant properties similar to those of black mustard.

## LINSEED

(Flax Seed, Semina Lini)

**Source, &c.**—The flax plant, *Linum usitatissimum*, Linné (N.O. *Linĕæ*), is a tall, erect annual, spread by cultivation over all temperate and tropical regions. Not only has the plant been known and cultivated for so many centuries that its geographical origin cannot be identified, but the use of its fibre can be traced back to the thirteenth or fourteenth century before the Christian era. Flax seeds, as well as cloth woven from flax, have been found in Egyptian tombs, and the process of weaving is depicted on their buildings. The seeds were used as a food; the medicinal use of the mucilage and the value of the oil contained in them appear not to have been known till later.

The flax plant bears a small globular capsular fruit containing ten seeds; these are separated when ripe. Argentina, Russia, Canada, India, the United States and Holland furnish the principal supplies.

**Description.**—The seeds are commonly dark brown in colour, smooth and shining. They are elongated ovoid, flattened, with an acute edge; from 4 to 6 mm. long. One end is rounded, but the other has an oblique point, on one side of which, just below the apex, is a slight depression. In this depression both hilum and micropyle are situated, the seed being anatropous. The surface of the seed is glossy and minutely pitted. A transverse section exhibits a yellowish white oily kernel consisting of two large cotyledons surrounded by a narrow endosperm (not always easily seen). Soaked in water the seeds lose their glossy appearance and become covered with mucilage. They have but a slight odour, and an oily, mucilaginous taste.

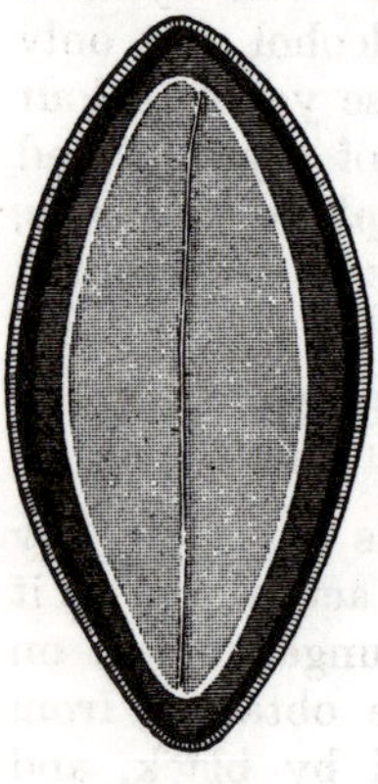

FIG. 86.—Linseed. Transverse section, showing the seed-coats, endosperm, and cotyledons. Magnified. (Moeller).

Both in colour and size linseed exhibits notable variations according to its source. As a general rule, warm climates yield larger and paler seeds than cold climates.

The student should observe

(*a*) The *glossy pitted surface,*

(*b*) The *mucilage extracted by water,*

(*c*) In the transverse section the *two oily cotyledons* and *narrow endosperm* (under a powerful lens).

**Constituents.**—Linseed contains from 30 to 40 per cent. of fixed oil, about three-fourths of which can be extracted by pressure. The epidermis of the seed-coat contains mucilage (6 per cent.), which, together with the proteids (25 per cent.) present in the seed and part of the oil, are left in the cake obtained when the seeds are pressed. The presence of oil, proteid, and carbohydrate (mucilage) in the cake renders it valuable as a cattle food. Unripe seeds contain numerous small starch grains, but the ripe seeds are free from starch. The seeds also contain a cyanogenetic glucoside, linamarin, identical with phaseolunatin (compare p. 38).

The fixed oil (Oleum Lini, B.P.) consists chiefly of linolein, a mixture of the glyceryl esters of linolic, linolenic, and isolinolenic acids, the acids themselves being sometimes designated as linoleic acid; it also contains similar esters of oleic, stearic, palmitic, and myristic acids. On exposure to the air it slowly ardens to a varnish, a change due to the oxidation of the linolein. The oil is haracterised by its high specific gravity (0·930 to 0·940) and high iodine value 1ot under 170). See also 'Linseed Oil.'

The mucilage consists of a mixture of hexanes and pentanes corresponding o the formula $2(C_6H_{10}O_5),2(C_5H_8O_4)$; it yields by hydrolysis the sugars galactose, extrose, arabinose, and xylose.

*Crushed linseed*, Lini Semina Contusa, B.P., consists of the seeds reduced o a coarse powder without being deprived of any part of their constituents. t should, when mixed with water, have a bland, not pungent (cruciferous seeds) r rancid (stale linseed) odour. It should yield not less than 30 per cent. of il to carbon disulphide (indicating the absence of ground cake left after emoval of part of the oil), and the oil thus extracted should respond to the fficial tests for Oleum Lini. It should not give the characteristic reactions with the tests for starch, nor leave, when incinerated, more than 5 per cent. of ash (absence of added starch and undue proportion of mineral matter). The presence of starch may be conveniently detected by microscopical examination or by applying the usual iodine test to a cooled decoction of the crushed linseed previously freed from oil by treatment with ether or carbon disulphide.

**Uses.**—Crushed linseed is used externally, in the form of a poultice, to convey heat and moisture to certain parts; the entire seeds are used to make a demulcent infusion containing a large quantity of mucilage.

**Note.**—Linseed usually contains foreign (weed) seeds, the proportion of which should not exceed 4 per cent.; these weed seeds often afford a guide to the geographical source of the linseed.

## COLA SEEDS

### (Semina Colæ)

**Source, &c.**—Cola seeds, sometimes called Cola or Kola nuts, Gooroo nuts, or Bissy nuts, are obtained from *Cola vera*, Schumann (N.O. *Sterculiaceæ*), a large and handsome tree resembling in habit the Spanish chestnut. It is a native of tropical Africa, growing wild in Sierra Leone, North Ashanti, near the sources of the Niger River, &c., but is cultivated in other tropical countries, such as the West Indies, Brazil, Java, &c., our supplies being derived either from the west coast of Africa or from the West Indies.

The woody capsular fruit of the tree contains from five to fifteen large white or crimson seeds which are removed and deprived of their seed-coats, the kernels only being used. These are chewed whilst still fresh, either before or after germination, and have been highly valued by the negroes for many centuries for their stimulating properties, in which they resemble tea, coffee, cocoa, &c.

Large quantities of the seeds are collected and consumed by the natives, who also carry on a considerable trade in them. Packed in baskets with the leaves of the cola tree they can be kept fresh, and in this state are brought chiefly from Lagos to Sokoto, Kano, and Timbuctoo, whence they are distributed to other parts of Africa. The fresh seeds are also occasionally exported, but more commonly the kernels are separated into the two large fleshy cotyledons and dried, during which the white or crimson colour changes to a dull reddish brown.

**Description.**—Dried cola seeds, as commonly seen in this country, consist of the kernels only of the seeds, sometimes entire, but more often separated into the two cotyledons. Externally, they have a dull dark brown or reddish brown colour ; internally they are usually somewhat paler. They are hard and solid, and exhibit, when cut, a perfectly uniform section, no mucilage glands being observable. They vary in length from about 2 to 5 cm., and are rather less in breadth and in thickness. In shape, too, they exhibit considerable differences, being frequently flat on one side and curved on the other, or wedge-shaped or irregularly six-sided. A shallow furrow encircles the kernel, dividing it into two cotyledons ; transverse to this furrow at one end of the seed a distinct cleft may be found, partially separating each cotyledon into two portions. Complete kernels may easily be separated into their constituent cotyledons, and the small radicle will be found towards the bottom of this transverse cleft.

Fresh cola seeds have a bitterish astringent taste, which is scarcely perceptible in the dry seed ; the latter are also destitute of any marked odour.

The student should observe

(*a*) The *two large fleshy cotylodons* and *small radicle*,
(*b*) The *absence of seed-coats*, the drug consisting of the *kernel only*.

**Constituents.**—The most important constituents of kola seeds are caffeine (1 to 2·5 per cent.), kolatin (0·75 per cent. mostly combined with the caffeine), and traces of theobromine. They also contain kolatein, an oxydase enzyme, fat, sugar, and abundance of starch.

Kolatin, $C_8H_8O_4$, is crystalline, slightly soluble in water, but readily in alcohol. During the drying of the seeds it is converted by the oxydase into kola-red, a substance allied to the phlobaphenes, which is therefore present in the dried drug to the exclusion of kolatin and imparts to it the characteristic colour. If, however, the seeds are boiled and the oxydase thus destroyed, the dried seeds retain the colour of the fresh and contain the kolatin. The fresh seeds may also be preserved by beating them into a pulp with an equal weight of loaf sugar.

Kolatein is also crystalline, soluble in alcohol and in hot water ; it has not yet been completely investigated.

Kolanin, a reputed constituent of the seeds, appears to be a mixture of kola-red and caffeine.

**Uses.**—Cola seeds have properties similar to those of tea, coffee, &c., and are used as a nerve stimulant. Kolatin increases the energy of the cardiac contractions, and as this substance is not present in he commercial drug, it having been converted into kola-red, the herapeutical action of the dried seeds is somewhat different from hat of the fresh; hence possibly the strong preference in Africa or the fresh seeds. The action of the dried *sterilised* seeds resembles hat of the fresh.

**Varieties.**—*C. acuminata*, Schott and Endlicher; Cameroon and Congo States; the seeds have three to five cotyledons; they are aten like the genuine, and are sometimes imported, but contain less affeine and are less esteemed.

*C. Ballayi*, Carnu; Gaboon; the seeds have six cotyledons and ontain but little caffeine.

*C. astrophora*, Warburg, the red cola of the Ashantis, always has ed seeds; *C. alba*, the white cola of the Ngaus, always has whitish eeds. *C. vera* is said to be a hybrid of these two species; its seeds re sometimes red, sometimes white.

Other seeds have from time to time been substituted for cola seeds, ut the genuine are easily distinguished by the characters given.

## COCOA SEEDS

### (Semina Theobromatis)

**Source, &c.**—The cocoa tree, *Theobroma Cacao*, Linné (N.O. *Sterculiaceæ*), is a native of tropical America, and is cultivated there s well as in other tropical countries, such as Java, Ceylon, the West ndies, &c.

The seeds had without doubt long been an important article of diet o the Mexicans when they were conquered by the Spaniards. Their se soon spread to Spain and thence over Europe.

The flowers are small, and spring directly from the trunk; they re succeeded by large orange or deep red fruits of the shape of a nall pointed vegetable marrow and about 15 or 20 cm. in length. Each fruit contains forty or fifty nearly colourless, fleshy seeds mbedded in a scanty, mucilaginous pulp. The seeds are separated nd packed in boxes, in which they undergo a process of fermentation, onsiderable heat, which, however, should not be allowed to exceed bout 42°, being developed; they are then dried in the sun. During hese processes the seeds acquire a reddish brown colour, and the aste, which at first is astringent and bitter, becomes mild and oily. Sometimes the seeds are simply freed from the pulp and dried in the un; they have then a more astringent and bitter taste and are less aluable.

**Description.**—Cocoa seeds, as they occur in commerce, are abou 2·5 cm. in length and of a flattened-ovoid shape. The seed-coat i reddish brown or chocolate brown in colour, thin and brittle. It ca easily be separated from the kernel, which consists mainly of tw irregularly folded, chocolate-coloured cotyledons; the latter are s brittle that they easily separate into small angular fragments (coco nibs of commerce).

**Constituents.**—Both the kernels and the shells contain the alkaloi theobromine, the former yielding about 2 per cent., the latter abou 1 per cent. The kernels contain, further, about half their weigh of solid fat, which is obtained as a by-product in the manufacture c cocoa essences by submitting the heated seeds to strong pressu (Oleum Theobromatis). The seeds also contain a trace of volatile oi

Theobromine or dimethylxanthine, $C_5H_2(CH_3)_2N_4O_2$, has been obtained a white crystalline powder very slightly soluble in cold water or alcohol, mo readily in boiling. It sublimes at 290°, and combines readily with alkali and alkaline earths to form salts. By introduction of the methyl group it converted into trimethylxanthine (caffeine).

Oleum Theobromatis is obtained as a by-product in the manufacture of coco and chocolate. The seeds are first roasted, then broken and the shell an embryo removed; the resulting cocoa nibs are ground in horizontal burr mi during which sufficient heat is developed to liquefy the fat. The mass (coc mass) is then pressed, the expressed oil filtered and run into moulds to set. I forms a hard, yellowish-white solid, melting at 30° to 33°, consisting chief of the glycerides of stearic (40 per cent.), palmitic, and oleic acids existin partly at least as mixed glycerides, *i.e.* glycerides in which two or more of the acids are combined with one glyceryl group. See also 'Oil of Theobroma.'

**Uses.**—Cocoa is largely used as a more nutritious and le stimulating beverage than tea or coffee. Its principal activ constituent, theobromine, has an action similar to that of caffein but while its effect on the nervous system is less than that of caffei its diuretic action is greater.

## GUARANA

### (Pasta Guarana)

**Source, &c.**—Guarana is prepared from the seeds of *Paullini Cupāna*, Humboldt, Bonpland, and Kunth (N.O. *Sapindaceæ*), a elegant climbing shrub indigenous to and common in Brazil, especial in the basins of the Amazon and its tributaries. It derives its nan from the Guarinis, an aboriginal tribe of Indians. The difficulty collecting the seeds from the wild plants, which, though common, a not easily accessible, has apparently led to their cultivation, th plants being trained to poles like hops. When the pods open discharge the ripe seeds they are collected, and the seeds, which resemble small horse-chestnuts in shape and colour, are separated by shelling. They are first washed and then roasted to loosen them from a papery shell, from which they are freed by beating. The broken kernels are made into a dough with water ; this is then divided into masses of varying size and shape, sometimes representing a fish or other animal, which are finally dried at a gentle heat by means of a slow fire. From the hard mass thus obtained portions are grated off with a large file and served in glasses of water, forming a refreshing drink.

**Description.**—Guarana commonly appears on the market in the form of extremely hard, heavy, sausage-shaped masses, varying from 10 to 30 cm. in length and from 2·5 to 4 cm. in thickness. The outer surface is dark chocolate brown in colour, and would be smooth and uniform were it not that small angular fragments, often of lighter colour than the rest, project slightly ; these fragments are evidently the larger pieces of the broken seed. The fractured surface, smoothed with a knife, is reddish in colour, and exhibits, like the outer surface, small paler irregular fragments embedded in a darker reddish mass, but no definite structure is discernible. The powder, in which form the drug is administered, is of a pale red colour ; it has a scarcely perceptible odour and slightly astringent, bitter taste.

The student should observe

(*a*) The extremely *hard heavy masses* in which the drug occurs,

(*b*) The presence of *small, paler fragments* embedded in them.

**Constituents.**—Guarana contains from 2·5 to nearly 5 per cent. of caffeine, together with a little catechu-tannic acid. There is abundance of starch present also, but only a little fat.

**Adulteration.**—The detection of foreign substances in such a drug as guarana is attended with considerable difficulty. The amount of caffeine should not be less than 2·5 per cent. Microscopical examination has shown the presence of the seed-coats, which therefore are only imperfectly separated, and frequently of foreign starches ; the latter, according to Schär (1897), are regularly present. Thoms (1894) found in guarana 8·63 per cent. of moisture, 1·68 per cent. of ash, and 2·68 per cent. of caffeine, the seeds themselves yielding closely concordant figures.

**Uses.**—Guarana is employed as a nervine stimulant in the same way that tea and coffee are, and produces similar effects. It has been long in common use in Brazil.

## TONCO BEANS

### (Semina Tonco)

**Source, &c.**—Tonco (tonka or tonquin) beans are the seeds of two species of Dipteryx (N.O. *Leguminosæ*), viz. *D. odorata*, Willdenow, and *D. oppositifolia*, Willdenow, both trees of considerable size, the former a native of Guiana, the latter of Brazil.

The tree produces an indehiscent drupaceous fruit about the size of an egg, with a fibrous pericarp containing a single brownish violet seed, about the size and shape of an almond. The fallen fruits are collected, split open and the seeds removed and dried on flat rocks. Sometimes they are placed upon the market without further treatment, but large quantities are brought from South America to Trinidad, where they are prepared for the European and American markets. This preparation consists in steeping them in rum, removing the excess and drying the seeds. By this treatment a white crystalline crust (of coumarin) is produced on the surface of the seeds; the latter may therefore occur 'black' or 'frosted'; when dry they are packed in cases or casks for shipment.

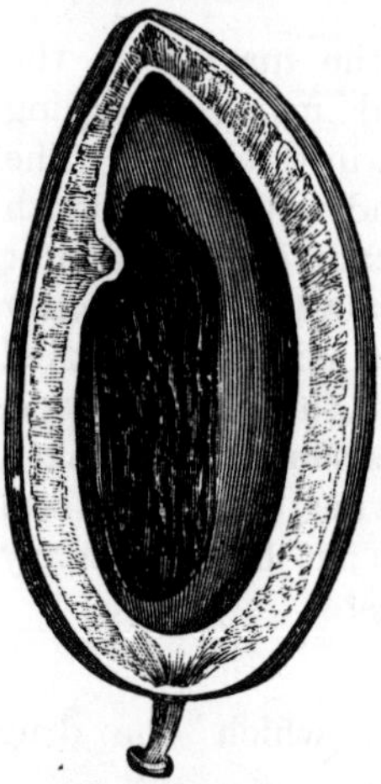

FIG. 87.—Tonco bean. Fruit cut vertically, showing the seed. (Planchon and Collin.)

**Description.**—Tonco beans closely resemble a Jordan almond in size and shape; they are usually rather longer, varying from 3 to 4 cm. in length, and differ from the almond in having a nearly black, coarsely wrinkled surface which, in the frosted seeds, is covered with minute whitish crystals. The beans are rounded at one extremity but terminate at the other in a broad, flat point, just below which on the obtuse margin of the seed the micropyle may easily be discerned as a brownish scar. Internally they are dark yellow, yellowish brown, or nearly black, and consist of two large oily cotyledons, without endosperm, enclosing a plumule with two folded leaves and a short thick radicle.

Tonco beans have a powerful and agreeably fragrant odour and an aromatic pungent taste.

**Constituents.**—The seeds owe their fragrance to coumarin, of which they may contain as much as 3 per cent.

Coumarin, coumaric or ortho-oxycinnamic anhydride, $C_6H_4(CH)_2OCO$, forms colourless, fragrant crystals melting at 67°. It may be prepared synthetically by heating together salicylic anhydride, acetic anhydride, and anhydrous sodium acetate. It has been isolated from a variety of plants belonging not only to *Leguminosæ*, but several other natural orders, especially *Gramineæ* and *Orchideæ*. Cherry wood (*Prunus Mahaleb*, Linné), woodruff (*Asperula odorata*, Linné), and melilot (*Melilotus officinalis*, Desvaux) owe their pleasant aroma to coumarin.

**Varieties.**—The chief varieties of tonco beans are the Angostura and the Pará, each of which may occur frosted or black ; the Angostura beans are the larger and more valuable.

**Uses.**—Tonco beans find their principal use in perfumery ; they are frequently mixed with vanilla beans in the preparation of extract of vanilla.

## CALABAR BEANS

### (Semina Physostigmatis)

**Source, &c.**—Calabar beans are the ripe seeds of *Physostigma venenosum*, Balfour (N.O. *Leguminosæ*), a woody climbing plant indigenous

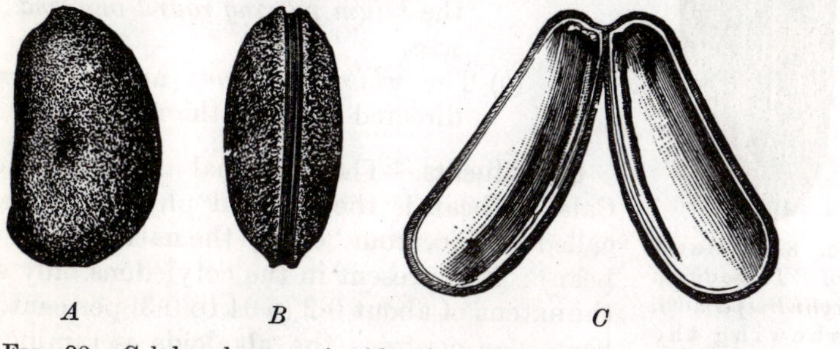

FIG. 88.—Calabar bean. *A*, side view, showing the sub-reniform shape. *B*, edge, showing the long hilum. *C*, seed split open, showing the concave cotyledons. (Maisch.)

to the west coast of Africa, especially near the mouths of the Old Calabar and Niger rivers. It ascends trees and, drooping down, bears pendulous racemes of flowers. These are succeeded by legumes about 15 cm. in length, in each of which two or three large seeds are contained. Calabar beans have long been used on the west coast of Africa as ' ordeal beans. They became known in England in 1840 ; their power of contracting the pupil of the eye was discovered by Fraser in 1862.

**Description.**—Calabar beans are dark reddish brown or chocolate-brown in colour and average about 25 mm. in length and 12 mm. in thickness. They are usually oblong-reniform in shape, being nearly flat or only slightly convex on one side, but boldly curved on the other. Along nearly the entire length of the curved side, and passing completely round one end of the seed, runs a broad, deep groove, the lips of which are thickened and paler in colour, and the bottom of which is black and bears a distinct brown furrow in its centre. This deep groove is the hilum, and in it may frequently be found portions of a white, papery funiculus; at one end a minute depression (the micropyle) can be detected.

The outer seed-coat (testa) is hard and thick, and appears nearly smooth to the naked eye, but under the lens is seen to be rather rough. After the seeds have been well soaked in water they can easily be split longitudinally, and exhibit then two firm white starchy cotyledons, which are curved so as to include between them a large lenticular cavity filled with air; this enables the seed to float when thrown on the surface of water. Near the micropyle is the small white radicle attached to one of the cotyledons; there is no endosperm.

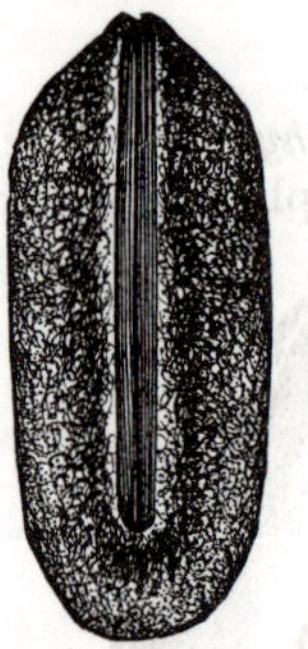

FIG. 89.—Seed of *Physostigma cylindrospermum*, showing the shorter hilum. (Maisch.)

The seeds have no marked odour or taste beyond those of an ordinary bean; they are nevertheless extremely poisonous.

The student should observe

(*a*) The *reniform* (not cylindrical) *shape*, and the *hilum passing round one end of the seed*,

(*b*) The *white cotyledons* and *small radicle* directed towards the micropyle.

**Constituents.**—The principal constituent of the Calabar bean is the alkaloid physostigmine (also called eserine, from 'esere,' the native name of the bean); it is present in the cotyledons only and to the extent of about 0·2 (0·04 to 0·3) per cent. The bean also contains the alkaloids eseramine (trace only), physovenine (Salway), and geneserine (Polonowski, 1915); stigmasterol, sitosterol, and the dihydric alcohols trifolianol and calabarol, and abundance of starch. The drug yields about 4 per cent. of ash.

*Physostigmine*, $C_{15}H_{21}N_3O_2$, forms colourless rhombic crystals melting at 105°; aqueous solutions of its salts rapidly become pink, due to conversion of the physostigmine into rubreserine; it powerfully contracts the pupil of the eye.

*Physovenine*, $C_{14}H_{18}O_3N_2$, has been obtained in colourless crystals, strongly mydriatic; traces only.

*Eseramine* crystallises in needles melting at 238°.

*Geneserine* yields a crystalline salicylate; the physiological action differs from that of physovenine.

*Calabarine* and *eseridine*, formerly considered constituents of the seed are now regarded as non-existent.

**Substitutes.**—*P. cylindrospermum*, Holmes; seed nearly cylindrical, hilum shorter; were imported in 1879 and are said to contain physostigmine.

*Mucuna urens*, de Candolle (horse-eye beans); brownish and rounded.

*Entada scandens*, Bentham (garbee beans); flattened, discoid, 5 cm. in diameter.

*Pentaclethra macrophylla*, Bentham, mussel-shaped; 7 cm. long, 5 cm. wide.

**Uses.**—Calabar beans are chiefly used as a source of the alkaloid physostigmine, which is much employed to produce contraction of the pupil of the eye. Both the drug and the alkaloid have been employed in tetanus, locomotor ataxy, and as an antidote in cases of strychnine poisoning; large doses produce an increase of blood-pressure, retardation of respiration, and finally death by asphyxia.

## BUTEA SEEDS

### (Buteæ Semina)

**Source, &c.**—Butea seeds are the seeds of *Butea frondosa*, Roxburgh (N.O. *Leguminosæ*), a tree indigenous to India.

**Description.**—The seeds are reniform in shape and very flat, from 25 to 38 mm. long, 16 to 25 mm. broad, and 1·5 to 2 mm. thick. Seed-coat dark reddish brown, thin, glossy, veined and wrinkled; hilum near the middle of the concave edge, conspicuous; cotyledons large, leafy and yellowish. They have a faint odour and slightly acrid and bitter taste.

**Constituents.**—Butea seeds contain fat (about 18 per cent.), proteids (about 19 per cent.), and sugar. No definite principle of therapeutic activity has been isolated.

**Uses.**—They are used as an aperient and anthelmintic, and are said to act as a rubefacient when pounded with lemon juice and applied to the skin.

## FENUGREEK SEEDS

### (Semina Fœni-græci)

**Source, &c.**—Fenugreek, *Trigonella Fœnum-græcum*, Linné (N.O. *Leguminosæ*), is an annual herb indigenous to the countries bordering on the eastern shores of the Mediterranean and largely cultivated in India, Egypt, and Morocco. It was well known to the ancients, who used the herb as cattle fodder and employed the seeds medicinally. In Egypt the seeds are roasted and eaten, and in India the young shoots form a favourite vegetable. In England the seeds are chiefly used in veterinary practice. They are contained in long, narrow, sickle-shaped pods, from which they are separated, when ripe, by thrashing. Our supplies are derived chiefly from Mogadore and Bombay.

**Description.**—Fenugreek seeds are small (about 5 mm. long), hard, and brownish yellow although the colour varies. They are flattened and have a very characteristic rhomboidal outline. Nearly in the centre of one of the long, narrow sides is a small depression in which both hilum and micropyle are situated, the former being distinctly visible as a whitish point; this depression is continued in the form of a furrow running diagonally across part of each of the adjoining sides, thus dividing the seed into two unequal lobes. If the seed is cut in a direction transverse to the side in which the hilum lies, so as to pass through both lobes of the seed (fig. 90, *B*), it will be found that the larger lobe contains two accumbent cotyledons—the smaller, the radicle; both are yellowish in colour, and surrounded by a darker, horny, translucent endosperm, which also separates the radicle from the cotyledons. Soaked in water the endosperm swells and yields mucilage to the surrounding liquid. Entire seeds macerated in warm water burst their seed-coats by the swelling of the mucilage, and disclose the structure of the seed. (Compare fig. 90, *A*.)

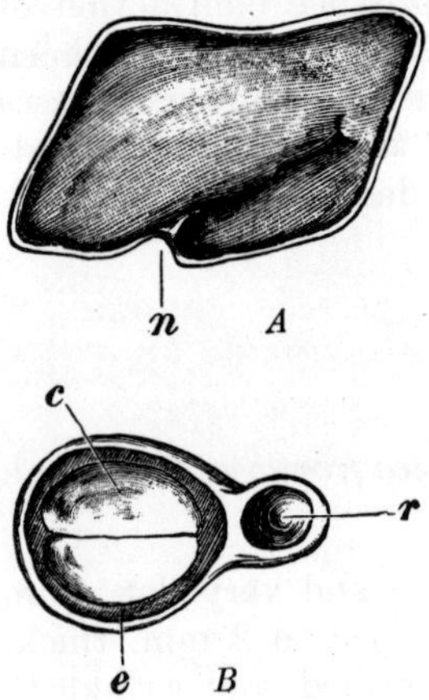


FIG. 90.—Fenugreek seed. *A*, vertical section, showing the radicle, one of the cotyledons, and hilum, *n*. *B*, transverse section, showing the radicle, *r*; both cotyledons, *c*; and endosperm, *e*. Magnified. (Moeller.)

The odour of fenugreek, especially if powdered, is strong and characteristic; the taste is disagreeable.

The student should observe.

(*a*) The characteristic *shape*,

(*b*) The *cotyledons*, *radicle*, and *endosperm* (in transverse section),

(*c*) The characteristic *odour*.

**Constituents.**—Fenugreek contains 28 per cent. of mucilage, which resides in the endosperm, not, as in the case of mustard and linseed, in the seed-coats; it yields by hydrolysis the sugars mannose and galactose. The drug contains, further, about 22 per cent. of proteids, 6 per cent. of fixed oil, and two alkaloids, trigonelline and choline, the latter being a frequent constituent of both animal and vegetable substances.

**Uses.**—The seeds are now used in veterinary medicine and occasionally as a spice in curry powders.

## SWEET ALMONDS

### (Amygdala Dulcis)

**Source, &c.**—The sweet almond is the seed of *Prunus Amygdalus*, Stokes, var. *dulcis*, Baillon (N.O. *Rosaceæ*). The tree is a native probably of Persia and Asia Minor, but is cultivated in all the countries that border on the Mediterranean, and produces ripe fruit even in the south of England. The seeds—almonds—as well as the oil pressed from them, were well known in Greece and Italy long before the Christian era; during the Middle Ages they became an important article of commerce in Central Europe.

The almond tree produces in early spring handsome pink flowers, which are succeeded by green, velvety, drupaceous fruits about the size of a plum, but differing from it in possessing a firm, felt-like mesocarp. As the fruit ripens the mesocarp gradually dries, splits, and falls away (or is easily removed), carrying the thin epicarp with it, and leaving the seed, enclosed in the endocarp or shell, attached to the tree. Sometimes almonds are exported enclosed in their endocarps (almonds in the shell), but more frequently the shells are broken and the seeds alone exported.

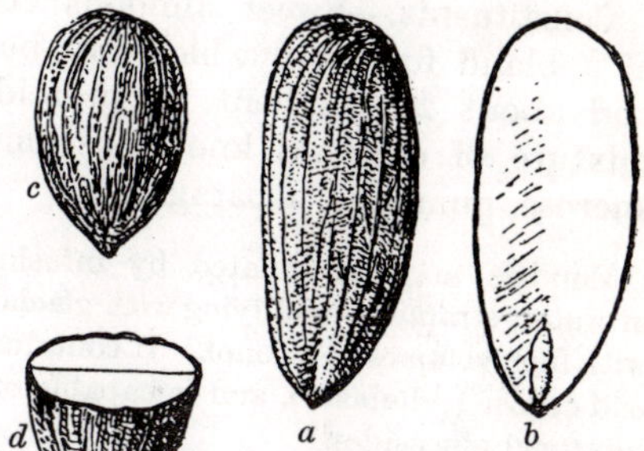


FIG. 91.—Almond: *a*, Sweet Almond; *b*, same, cut longitudinally. *c*, Bitter Almond; *d*, same, cut transversely. Natural size. (Holmes.)

Sicily and southern Italy are the chief almond-producing countries. Spain, Portugal, the south of France, the Balearic Islands, and Morocco also export considerable quantities.

**Description.**—The endocarp or shell of the almond is yellowish buff in colour and flattened-ovoid in shape, the outer surface being usually pitted with small holes; frequently it has a more or less fibrous nature. Sometimes it is thin and friable (soft-shell almonds), sometimes extremely hard and woody (hard-shell almonds). The seed is rounded at one end, pointed at the other, and covered with a thin, brown, scurfy seed-coat. The hilum is long, and situated on the acute edge of the seed near the pointed end; the raphé is distinguishable as a dark line running from the hilum to the broad end of the seed, where it terminates in a dark spot, the chalaza, from which a number of veins radiate.

After maceration in water the thin seed-coat is easily separated; the kernel consists of two, large plano-convex, oily cotyledons, enclosing a small plumule and radicle. The seed is exalbuminous, there being no endosperm. Sweet almonds have a bland nutty taste, and yield, when triturated with water, a white emulsion destitute of any marked odour. The latter character is important, as it is the only definite one by which the sweet almond can be distinguished from the bitter.

The student should observe

(a) The *elongated shape* (of the Jordan almond),
(b) The *thin seed-coats* and *two cotyledons without endosperm*,
(c) The *bland taste and odour* (of the emulsion).

**Constituents.**—Sweet almonds contain from 40 to 46 per cent. of a bland fixed oil, which can be obtained by pressing the seeds, and about 20 per cent. of proteids, amongst which is included a mixture of enzymes known as emulsin. They contain also a little sucrose, gum, and asparagin.

Emulsin may be isolated by infusing powdered blanched sweet almonds in water, straining, acidifying with glacial acetic acid, filtering, and precipitating with four volumes of alcohol. It contains lactase, $\beta$-glucosidase, $\beta$-galactosidase, and cellase (cellobiase), and is capable of hydrolysing a large number of (lævo-rotatory) glucosides.

Almond oil is pressed both from the sweet and the bitter almond. It consists chiefly of olein accompanied by a small proportion of linolein, &c.; it is characterised by its specific gravity (0·915 to 0·920), by the quantity of iodine it is capable of combining with (93 to 100 per cent.), by the melting-point of the mixed fat acids obtainable from it (not over 15°), and by the whitish colour when shaken with a cooled mixture of equal parts by weight of sulphuric acid, fuming nitric acid, and water. Drying oils such as cotton seed, &c., combine with larger proportions of iodine; sesame, ground nut, and certain other oils yield fat acids, melting at a much higher temperature. Apricot kernel oil gives a reddish colour with the nitric acid test.

**Varieties.**—*Jordan* (French 'jardin') almonds, exported from Malaga, constitute the official variety; they are characterised by their narrow, elongated shape and thin skin.

*Valencia*: these are broadly ovoid, shorter, and have a thicker, dusty brown, scurfy coat.

*Sicilian* and *Barbary*: both of these closely resemble the Valencia, but are rather smaller; they occasionally contain admixtures of bitter almonds.

*East Indian*: these are the seeds of the cashew nut *Anacardium occidentale*, Linné (N.O. *Anacardiaceæ*); the pericarp of the fruit contains an oily, vesicating liquid, cardol, but the seeds are edible.

**Uses.**—Sweet almonds are demulcent and nutritive. They are used as a non-starchy food for diabetic patients.

## BITTER ALMONDS

### (Amygdala Amara)

**Source, &c.**—The bitter almond tree (*Prunus Amygdalus*, Stokes, var. *amara*, Baillon) is indistinguishable from the sweet by any permanent botanical character, and enjoys the same geographical distribution, although it is not cultivated to so large an extent. It is remarkable, therefore, that the seeds should invariably contain a constituent—amygdalin—that is never present in the seed of the sweet almond tree. The poisonous properties of the water distilled from the bitter almond (and the cherry-laurel leaf), due to the presence and decomposition of a cyanogenetic glucoside, have long been known ; it was the discovery of hydrocyanic acid in it that led to the recognition of the poisonous nature of this acid, which, curiously enough, had up to then (1802) escaped observation.

Bitter almonds are imported chiefly from northern Africa, from Sicily, and from the south of France.

**Description.**—In form and appearance bitter almonds closely resemble Valencia almonds, but they are usually smaller and less regular. They have, however, a bitter taste, and yield with water an emulsion easily distinguished from that of the sweet almond by its characteristic odour.

**Constituents.**—Bitter almonds resemble the sweet in containing both a bland fixed oil and proteids, the former of which is obtained by crushing the almonds between horizontal grooved rollers and pressing in powerful hydraulic presses. But they differ essentially from the sweet in containing a colourless, crystalline glucoside, amygdalin (2·5 to 4 per cent.), which is not present in the sweet. This substance is left in the cake obtained after the oil has been expressed, and can be extracted from it by digestion with alcohol. It has a bitter taste and is odourless, but when an aqueous solution is mixed with an emulsion of sweet almonds the amygdalin is decomposed with production of benzaldehyde, hydrocyanic acid and dextrose.

$$\underset{\text{Amygdalin}}{C_{20}H_{27}NO_{11}} + 2H_2O = \underset{\text{Dextrose}}{2C_6H_{12}O_6} + \underset{\text{Benzaldehyde}}{C_6H_5COH} + \underset{\text{Hydrocyanic acid}}{HCN}$$

This change is effected by the emulsin contained in the sweet almond. Emulsin is also contained in the bitter almond itself, but, being localised in particular cells of the seed, is unable to act upon the amygdalin until the seeds are crushed and water added. When therefore bitter almonds (or cake) are crushed and mixed with water, the characteristic odours of benzaldehyde and hydrocyanic acid are developed. If, after standing a few hours, the mixture is subjected to distillation, an oily liquid of strong bitter almond odour is obtained, together with a quantity of watery distillate; the oil is volatile or essential oil of bitter almonds, and consists of benzaldehyde and hydrocyanic acid, partly in the free state and partly combined as benzaldehyde-cyanhydrin. It can be freed from hydrocyanic acid by shaking it with milk of lime; all the hydrocyanic acid, both free and combined, forms calcium cyanide. The addition of ferrous sulphate then converts the cyanide into calcium ferrocyanide, and the oil can be obtained free from hydrocyanic acid by redistilling in a current of steam. No benzaldehyde or hydrocyanic acid is developed by the sweet almond, because it contains no amygdalin.

Bitter almonds yield from 0·5 to 0·8 per cent. of volatile oil and about 0·25 per cent. of hydrocyanic acid.

Many other Rosaceous plants contain amygdalin, such as the peach, apricot, plum, &c., not only in the seed, but also in the young shoots and flower-buds.

Amygdalin is closely allied to, but not identical with, prulaurasin, and is also distinct from a similar glucoside that occurs in the bark of *Prunus serotina*, Ehrhart (compare also p. 252). Its hydrolysis is said to occur in three stages, each effected by a particular enzyme contained in the emulsin.

Benzaldehyde is also prepared from benzyl chloride, $C_6H_5.CH_2Cl$, or benzal chloride, $C_6H_5.CHCl_2$, and sold as oil of bitter almonds; it has a less pleasant odour than that obtained from bitter almonds.

**Uses.**—Bitter almonds are sedative, but as the poisonous hydrocyanic acid yielded by them varies in quantity they are unreliable. They are also employed for flavouring, but they should for a similar reason be used with caution.

**Substitutes.**—*Apricot kernels* contain constituents similar to those of the bitter almond. They are imported in large quantities from Syria and California and are often used by confectioners in the place of bitter almonds. The fixed oil expressed from them is commonly sold as 'Oleum Amygdalæ Persicum' or 'peach kernel oil.' From the cake an essential oil (0·6 to 1·0 per cent.) is distilled as from bitter almond cake.

## MELON PUMPKIN SEEDS

### (Cucurbitæ Semina Præparata)

**Source, &c.**—Melon pumpkin seeds are obtained from *Cucurbita maxima*, Duch. (N.O. *Cucurbitaceæ*), a native of the Levant but cultivated on the shores of the Mediterranean. For use in medicine they should not be more than one month old and should be deprived of their seed-coats.

**Description.**—The seeds are ovate, flat, nearly white, pitted and brittle; about 8 to 20 mm. long and 9 to 12 mm. broad; they have a flat ridge and shallow groove round the edge. The kernels consist of two white, fleshy, oily, easily separable cotyledons and a small radicle; they have a faint odour and only a slight taste.

**Constituents.**—Melon pumpkin seeds contain an acrid resin to which their activity has been ascribed and about 30 per cent. of a reddish fixed oil together with proteids, sugar and starch. Recent experiments have failed to show that either the resin or the oil possesses therapeutic activity (Power and Salway, 1910).

**Uses.**—As a tænicide.

## QUINCE SEEDS

### (Semina Cydoniæ)

**Source, &c.**—The quince is the fruit of *Pyrus Cydonia*, Linné (N.O. *Rosaceæ*), a small tree indigenous to Persia, but distributed by cultivation throughout central Europe and other warm countries; the seeds are imported chiefly from Cape Colony.

The fruit, which resembles a pear, contains five carpellary cavities, in each of which there are about twenty seeds closely packed in two vertical rows. These seeds are separated from the ripe fruit and dried; being coated with mucilage they adhere more or less firmly together.

**Description.**—Quince seeds bear a general resemblance to apple pips. They are of about the same size, and of a similar mahogany-brown colour. By the mutual pressure to which they are subjected in the fruit, quince seeds, however, are distinctly flattened on the two larger sides, whilst of the two narrow sides or edges one is obtuse and boldly arched, the other only slightly curved and often provided with a distinct acute ridge. They frequently adhere firmly to one another in small irregular masses or in two more or less regular rows, being cemented together by dry mucilage, which is visible in the form of whitish flakes on the surface of the seeds and in the interstices between them. This mucilage is derived from the cells of the epidermis of the seed-coat in which it is stored. (Compare fig. 92, *E.*)

The seeds are pointed at one end, where the hilum may be distinguished as a minute paler spot, and obtuse at the other (the chalazal extremity). Transverse sections through the seed, which is rather hard, exhibit two firm, yellowish white cotyledons with a very narrow endosperm. The kernel possesses a taste resembling that of bitter almonds, but much fainter. The seed-coats, when chewed, are simply mucilaginous.

The student should observe

(*a*) The *angular shape of the seed,*

(*b*) The *dry mucilage* which cements them together,

(*c*) The *taste of the kernel and seed-coats.*

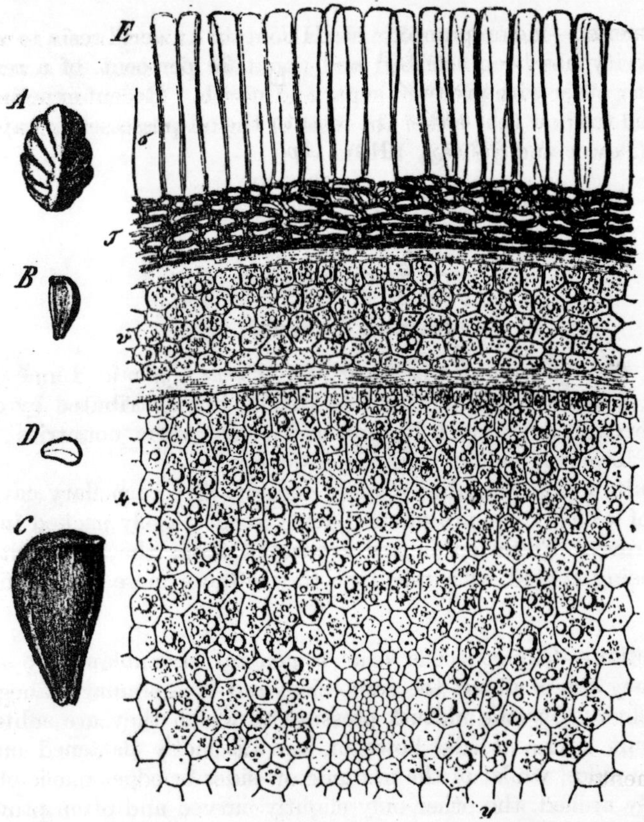

FIG. 92.—Quince seed. *A*, the seeds cemented together by mucilage, natural size. *B*, a single seed, natural size. *C*, the same, softened in water and magnified 3 diam. *D*, transverse section of *B*. *E*, portion of the same, magnified 190 diam.; *σ*, the epidermis, in which the mucilage is secreted; *v*, endosperm. (Berg.)

**Constituents.**—The principal constituent of quince seeds is the mucilage, of which they are said to yield as much as 20 per cent. It is contained in the cells of the outer epidermis of the seed-coat, and swells and dissolves when the seeds are soaked in water. The seeds also contain about 15 per cent. of fixed oil and probably a small proportion of amygdalin and of emulsin since they evolve an odour resembling that of bitter almonds when they are crushed and mixed with water.

**Uses.**—Quince seeds have been employed as a demulcent, but are not now much used medicinally in this country.

## STROPHANTHUS SEEDS

### (Semina Strophanthi)

**Source, &c.**—The official strophanthus seeds are obtained from *Strophanthus Kombé*, Oliver (N.O. *Apocynaceæ*), a climbing plant of considerable size, indigenous to eastern tropical Africa, near the Nyanza and Tanganyika lakes, the Shiré river, &c. An extract prepared from them (and possibly from the seeds of other species of *Strophanthus*) is used in Africa as an arrow poison. Specimens of this extract were sent to England in 1861–64 and recognised by Sharpey (1862) as a cardiac poison. The seeds were examined by Fraser (1885), who isolated the active principle strophanthin, and subsequently recommended the seeds as a substitute for foxglove leaves.

The fruit of the plant consists of two follicles about 30 cm. in length and 2·5 cm. in breadth, slightly narrowed at the base and tapering at the apex. Each follicle contains, closely packed together, a large number of seeds provided with long awns. The fruits are collected when ripe, and are sometimes exported after having been freed from their epicarp and fleshy mesocarp. More commonly the seeds, separated from the fruits and deprived of their awns, are sent into commerce, chiefly from Somba, Quilimane, Inhambane, Chinde, and other East African ports.

**Description.**—These very beautiful seeds are remarkable for the long plumose awn of white silky hairs that is attached to them. The integuments of the seed are prolonged at the apex to a slender brittle extension, which is terminated by a handsome feathery tuft of hairs about 5 cm. long and the same in breadth, the entire seed measuring about 10 cm. in length; these awns are commonly removed before the seeds are exported. The seeds as seen in commerce measure from 12 to 20 mm. in length and about 4 mm. in breadth. They are of an elongated oval shape, acuminate towards the apex, and narrowed towards the base, which is obtuse. They are flattened and provided on one side with a ridge running from the centre to the apex of the seed, and terminating in a broken point left by the removal of the awn. In colour they vary from greyish green to fawn; they are covered with satiny appressed hairs that impart to them a conspicuous sheen.

The seeds break easily; the transverse section exhibits a white oily kernel consisting of two straight cotyledons surrounded by a narrow endosperm. Moistened with 80 per cent. sulphuric acid the section assumes, at least in the endosperm but sometimes in the cotyledons also, a deep emerald green colour. The same reaction is afforded by the kernels from which the seed-coats have been removed after soaking in water. The seed-coats thus separated, warmed on a slide with a little saturated aqueous solution of choral hydrate and examined under the microscope, are seen to contain not more than an occasional crystal of calcium oxalate. The odour of the crushed seed is characteristic, though not very powerful; the taste is intensely bitter.

The student should observe

(a) The *colour of the seed* and the *satiny hairs*,
(b) The *straight cotyledons* and *narrow endosperm*,
(c) The *green* reaction *with sulphuric acid*.

**Constituents.**—The active constituent of strophanthus seeds is the glucoside strophanthin, of which they contain from 8 to 10 per cent. This substance, which is the source of the green reaction with sulphuric acid, is present in the endosperm, and frequently also in the cotyledons.

The seeds contain, in addition, about 30 per cent. of fixed oil, together with kombic acid, strophanthic acid, choline, and trigonelline. They yield about 4 per cent. of ash on incineration.

Strophanthin, $C_{40}H_{66}O_{19}$, forms colourless, intensely bitter, very hygroscopic crystals freely soluble in water and dilute alcohol, sparingly soluble in absolute alcohol, and almost insoluble in ether or chloroform. Warmed with a dilute mineral acid it is hydrolysed into strophanthidin and a sugar (strophanthobiosemethylether). Strophanthidin is also crystalline, but almost insoluble in water, and separates in crystalline form when strophanthin is hydrolysed.

It must, however, be observed that commercial strophanthin has been shown to vary considerably in toxicity and that the strophanthins prepared by different chemists from (as far as can be ascertained) the same variety of seed, viz. *S. Kombé*, have shown differences in the melting-point, in the hydrolysis, in the products yielded by the hydrolysis, and in other respects, so that they appeared to be different substances. It must, therefore, be admitted that the chemistry of this body has not yet been sufficiently investigated. Brown and Closson (1914) assert the presence of two strophanthins, a crystalline and an amorphous.

Strophanthic acid is an acid saponin (see p. 254).

**Assay.**—The chemical methods that have been proposed for the assay of strophanthus seed depend upon the extraction of the glucoside, and conversion into strophanthidin which is weighed and from which the strophanthin present is calculated. They are, however, all unsatisfactory. The drug may be standardised by the bio-chemical method which consists in determining the quantity of tincture which, after dilution with an equal volume of water, will kill a normal frog of about 25 gm. weight within an hour when injected into the dorsal lymph sac. The amount required has been designated a 'unit.' The seeds or their preparations may be compared in activity by determining experimentally the quantity that represents a unit.

**Uses.**—The action of strophanthus seed, which is due to the strophanthin they contain, resembles that of foxglove leaves. It raises the blood-pressure, is an efficient diuretic, and a powerful cardiac poison. It is not cumulative, and less liable than foxglove to produce gastro-intestinal irritation; hence is sometimes substituted for foxglove when this remedy has failed or disagreed.

**Varieties.**—The official strophanthus seed often occurs in commerce mixed with or replaced by the seeds of other species of *Strophanthus* which resemble them more or less closely. The following are the most important.

*S. hispidus*, de Candolle.—The seeds of this species are exported from West Africa; they are smaller than the official, although similar in shape, are brownish in colour, bear scattered hairs, and give with sulphuric acid a green reaction. They contain strophanthin (*h*-strophanthin; see below) said to be closely allied to, if not identical with, the amorphous strophanthin of Kombé seed.

*S. Courmontii*, Sacleux.—This is an East African species; the seeds usually have a brownish tinge, but often closely resemble the official, from which, however, they may be distinguished by their rather smaller size, lanceolate shape, less bitter taste, red reaction with sulphuric acid, and abundant prismatic calcium oxalate crystals in the seed-coat. The active constituent is unknown. The seeds are said to be about one-fourth as active as Kombé seeds.

*S. Nicholsoni*, Holmes.—The seeds of this species are whitish and woolly, the hairs project beyond the apex of the seed, and the reaction with sulphuric acid is red; they have been exported from Natal. The active constituent is unknown.

*S. gratus*, Franchet.—The seeds are brown and glabrous; they give a red reaction with sulphuric acid, and contain the glucoside ouabain, which was first isolated from the wood of *Acokanthera Schimperi*, Oliver, and is much more toxic than strophanthin. For this glucoside the distinctive name of '*g*-strophanthin' has been proposed (Thoms, 1900); its use has been advocated as a substitute for the official drug, as the seeds are readily obtainable and easily recognised.

*S. Emini*, Aschers.—The seeds are greyish green, but they contain cluster crystals of calcium oxalate, and give a red reaction with sulphuric acid. They contain the glucoside *e*-strophanthin (Thoms), which is distinct from the strophanthin of Kombé seed (*k*-strophanthin) and also from that of the seed of *S. hispidus* (*h*-strophanthin). Their action on the heart is said to be slight.

## NUX VOMICA

### (Semina Strychni, Nux Vomica)

**Source, &c.**—The nux vomica tree, *Strychnos Nux-vomica*, Linné (N.O. *Loganiaceæ*), is a small tree widely distributed in India, and occurring also in Ceylon, Siam, and northern Australia. Notwith-

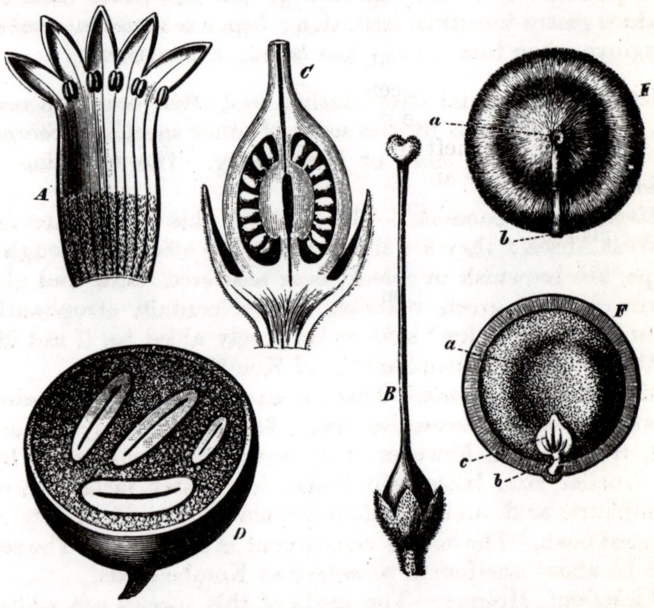


FIG. 93.—Nux Vomica. *A*, corolla opened, magnified. *B*, ovary with style, magnified. *C*, ovary cut longitudinally, more highly magnified. *D*, ripe fruit, cut transversely, showing seeds ; reduced to about one-half. *E*, seed, entire, natural size : *a*, hilum ; *b*, micropyle. *F*, the same, cut vertically : *a*, endosperm ; *b*, radicle ; *c*, cotyledon. (Luerssen.)

standing the wide distribution of the tree, and the poisonous properties of the seeds, the Hindus appear to have been unaware of its medicinal action. It was introduced into Europe in the sixteenth century, but was not much used in medicine, being chiefly employed to poison dogs, cats, crows, &c.

The ripe fruit, which resembles an orange, contains usually from three to five seeds embedded in a bitter whitish pulp. The seeds are washed free from adhering pulp, dried in the sun and sorted. They are collected extensively in the Eastern Ghat mountains and in the Travancore hills, and are exported from various Indian ports (Madras, Bombay, Cochin, Coconada, &c.).

**Description.**—The seeds have the shape of small discs, 20 to 25 mm. in diameter and about 4 mm. thick, of an ash-grey or greenish grey colour, and possessing a distinct sheen. They are usually not quite flat, being a little depressed on one side and arched on the other, or sometimes irregularly bent. They are covered with numerous, closely appressed hairs, radiating from the centre to the circumference, to which the satiny sheen of the seeds is due. The edge of the seed is sometimes rounded, sometimes acute, according to the variety; at one point on the margin there is a distinct prominence (micropyle) from which a raised line passes to the centre of the seed. This line does not exist in the fresh seed, but makes its appearance during the drying; it has frequently been mistaken for the raphé. The hilum is in the centre of either the raised or depressed surface, and may be recognised by the scar left by the funicle.

The dry seed is hard and horny, but softens when soaked in water; it can then easily be split into two thinner discs, exhibiting the small embryo, consisting of a radicle and two leafy cotyledons, embedded in a copious, grey, translucent, horny endosperm.

The seeds are almost odourless, but have a persistent, intensely bitter taste.

The student should not fail to observe

(*a*) The *characteristic shape* and *satiny appressed hairs*,
(*b*) The *horny endosperm* and *characteristic cotyledons*.
(*c*) The *bitter taste*.

**Microscopical Characters.**—The seed-coat is very thin and consists of a narrow layer of brown, collapsed parenchyma covered with an epidermis, the cells of which have developed into hairs. These are bent near the base and thus lie close to the surface of the seed. The base is thick-walled and marked with oblique pits; above it passes into a tubular hair bearing longitudinal rod-like thickenings. The endosperm consists of large cells with very thick walls which swell in boiling water or in solution of potassium hydroxide. These cells contain the alkaloids.

In powdered nux vomica the hairs are broken up into the thick-walled bases and rod-like fragments of the upper part.

**Constituents.**—The principal constituents of nux vomica are the two alkaloids, strychnine and brucine, in addition to which they contain a small percentage of a glucoside, loganin, which is much more abundant in the pulp, and an acid that has been termed igasuric acid, but is probably identical with caffeotannic acid. The alkaloid strychnicine which occurs in the leaves may also occur in minute quantity in the commercial drug (Boorsma, 1902). An alkaloid, struxine, has been obtained during the separation of strychnine from more or less decomposed Cochin-China seeds; it is probably not a normal constituent of the seeds.

The cell-wall of the endosperm, which is very thick, appears to consist of galactan and mannan, the former preponderating; it yields by hydrolysis galactose and mannose.

The seeds contain, further, about 3 per cent. of a viscous oil which appears to reside chiefly in the hairs. A trace of copper is also invariably present, and passes into solution when the seeds are exhausted with alcohol, giving rise to the green reaction often observed when the diluted tincture is mixed with ammonia.

The proportion of total alkaloid varies from 1·84 to 5·3 per cent., averaging in seeds of good quality about 2·8 per cent. This total alkaloid contains a variable proportion, ranging from 35 to 50 per cent. of strychnine, the average amount of strychnine in the d ug being 1·23 per cent., and of brucine 1·55 per cent. As the toxicity of brucine is only one-eighth that of strychnine, the drug and its preparations are assayed for strychnine and not for the total alkaloid present.

Strychnine, $C_{21}H_{22}N_2O_2$, crystallises in colourless, odourless prisms melting at 265°, and having an intensely bitter taste, though very slightly soluble in water (1 in 7000). The hydrochloride crystallises in needles which are soluble in water, but less soluble in water acidified with hydrochloric acid.

Brucine, $C_{23}H_{26}N_2O_4,4H_2O$, is also crystalline and more soluble in water than strychnine (1 in 320); melting-point 178°; is probably dimethoxystrychnine.

Caffeotannic acid (chlorogenic acid) is also found in the leaves and seeds of *Caffea arabica*, Linné, and in other plants; exposed to the air in the presence of ammonia it turns green (viridic acid).

**Action and Uses.**—Nux vomica is largely used as a bitter stomachic and tonic, its action being practically identical with (although slower than) that of the alkaloid strychnine. It stimulates the muscular coat of the intestine, increasing peristalsis, and hence is given in constipation from an atonic condition of the intestine. It increases the blood pressure, and is therefore valuable in certain cases of cardiac failure. In large doses the excitability of the motor nerve cells is so much increased that violent convulsions may occur; these involve the respiratory muscles, and death ensues from asphyxiation.

**Note.**—The seeds of *Strychnos potatorum*, Linné, and of *S. Nux-blanda*, Hill, have been offered as nux vomica. The former are smaller, thicker and free from bitterness; they are used in India for clearing turbid water. The latter closely resemble nux vomica, but may be distinguished by the small ridge on the edge; they are also free from bitterness (and consequently from strychnine and brucine).

## IGNATIUS BEANS

### (Semina Ignatii)

**Source, &c.**—Ignatius beans are the seeds of *Strychnos Ignatii*, Bergius (N.O. *Loganiaceæ*), a stout climbing plant with woody stem, indigenous to the southern Philippine Islands. It became known to Europeans through a Jesuit, Father Camellus, towards the end of the seventeenth century, and the seeds were called Fabæ Sancti Ignatii, in honour of Ignatius Loyola, the founder of the order. The large ovoid or nearly globose fruit contains about twelve seeds embedded in a pulp, from which they are separated and dried. When fresh they are covered with greyish satiny hairs, resembling those of the nux vomica; these, however, are very easily rubbed off, and in the commercial drug are mostly lost, together with much of the thin seed-coats.

**Description.**—Ignatius beans are dull, dark grey, irregularly ovoid and about 25 mm. long. Frequently there is one large curved side and three or four smaller flattish surfaces with rounded angles, but some seeds are irregularly sub-angular. Here and there are patches of the dull, ash-grey seed-coat, which under a lens is seen to be covered with appressed hairs, but these, unlike the hairs of nux

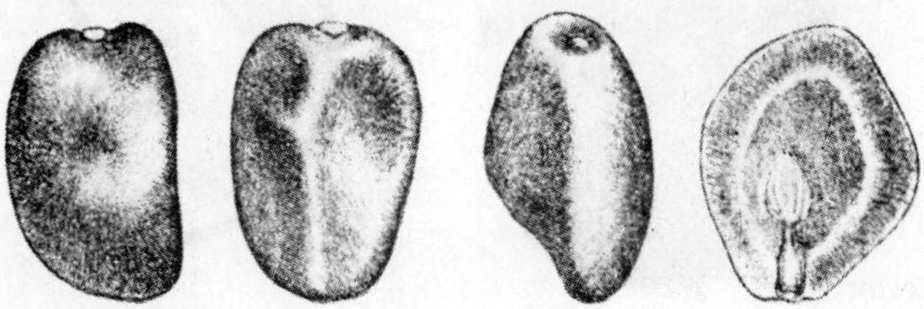

FIG. 94.—Ignatius Beans. Natural size. (Bentley and Trimen.)

vomica, are not so regularly arranged and are destitute of any satiny sheen. Usually, however, the seed-coat, which is very thin, has been removed by the friction of the seeds against one another, exposing the surface of the dark, translucent, horny endosperm. The hilum is generally distinguishable with ease at one extremity of the seed.

After soaking in water, the large endosperm can be divided into two portions, between which, in a cavity, lies the embryo with its small radicle and leafy cotyledons. The seeds are inodorous, but have an extremely bitter taste.

The student should observe

(*a*) The *irregular shape*,
(*b*) The *dark endosperm* with *patches of seed-coat*.

**Constituents.**—Ignatius beans contain strychnine and brucine to about the same extent as nux vomica, viz. 2·5 to 3 per cent. of total alkaloid, of which from 46 to 62 per cent. is strychnine.

**Uses.**—The drug possesses a medicinal action similar to that of nux vomica, over which it has no evident superiority.

## STRAMONIUM SEEDS

### (Thornapple Seeds, Semina Stramonii)

**Source, &c.**—The leaves of the thornapple, *Datura Stramonium*, Linné (N.O. *Solanaceæ*), have already been described (p. 46), and the attention of the student has been directed to the white flower and spiny fruit. The latter when ripe is about the size of a walnut, and dehisces septifragally, the four walls separating from apex to base. The fruit is two-celled in its upper part, but, from the presence of spurious dissepiments, four-celled near the base. It contains a large number of seeds, which are collected when ripe.

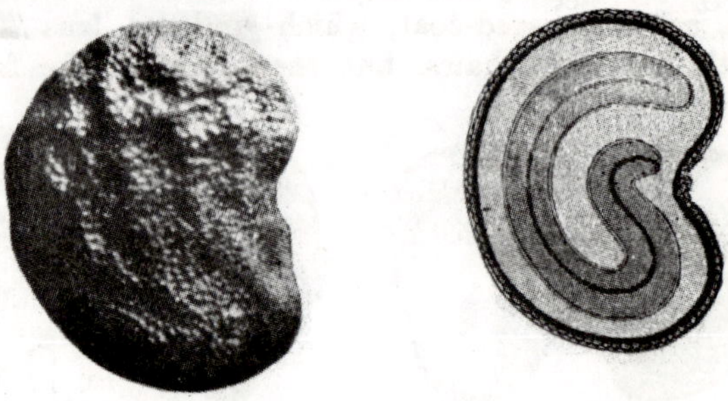

*A* *B*

FIG. 95.—Thornapple seed. *A*, showing the pitted surface ; *B*, longitudinal section, showing curved embryo. Magnified. (*B*, after Moeller.)

**Description.**—Stramonium seeds are dull dark brown or, more commonly, nearly black, flattened and distinctly reniform in outline, averaging about 3 mm. in length. The hilum is distinct in the form of a light spot on the concave edge. The seed-coat is marked with distinct, but not sharp, reticulate depressions, and, in addition, is minutely pitted.

If the seed is split parallel to one of the flattened sides, the crook-shaped embryo may be distinguished embedded in an oily endosperm ; in transverse section the embryo appears rounded, and will, owing to its curved shape, be cut at two, or possibly three, different points.

The odour of the seeds is scarely perceptible until they are crushed, then it becomes rather disagreeable ; the taste is bitterish and oily.

The student should observe

(*a*) The *size, reniform outline*, and *dark colour*,
(*b*) The *reticulated and pitted surface*,
(*c*) The *curved embryo* embedded in an *oily endosperm.*

**Constituents.**—In 1833, two years after the discovery of atropine, Geiger and Hesse extracted from stramonium seeds an alkaloid to which they gave the name of daturine. Planta proved this alkaloid to be atropine, and Schmidt subsequently showed that these seeds contain the alkaloid hyoscyamine, associated with a small proportion of atropine and scopolamine (hyoscine). Farr and Wright found English grown seed to yield from 0·16 to 0·28 (average 0·22 per cent.) of alkaloid, but as much as 0·5 per cent. has been recorded. In addition, the seeds contain from 15 to 30 per cent. of fixed oil containing daturic and other acids. The drug yields about 2 per cent. of ash.

**Uses.**—In this respect stramonium seeds resemble stramonium leaves. An extract of the seeds is given in spasmodic affections of the respiratory organs.

## HENBANE SEEDS

(Semina Hyoscyami)

**Source, &c.**—The henbane plant, *Hyoscyamus niger*, Linné (N.O. *Solanaceæ*), has been already alluded to (p. 49). The fruit of the plant is a small two-celled capsule, which dehisces transversely, the upper part separating from the lower like the lid of a box (pyxis). Within the fruit is a large number of minute seeds; these, separated when ripe, form the commercial drug.

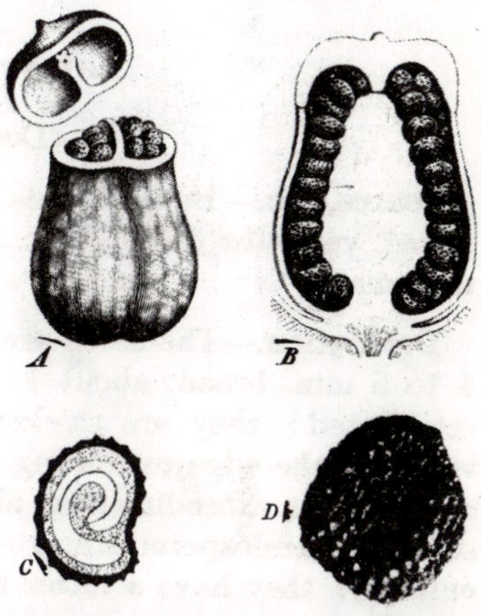


FIG. 96.—Henbane seed. *A*, entire fruit, showing dehiscence. *B*, the same cut vertically. *C*, seed, cut longitudinally. (Luerssen.) *D*, seed, showing reticulate surface. Magnified.

**Description.**—Henbane seeds are dark grey, about 1·5 mm. long, flattened and obscurely reniform in shape, being slightly pointed at one extremity (the hilum). Under a strong lens the surface, which appears dull to the naked eye, is seen to be marked with very characteristic reticulations. Cut longitudinally, parallel to the flat surface, the seed exhibits a coiled embryo embedded in an oily endosperm.

The seeds have no odour and only a slightly bitter taste.

The student should carefully observe

(*a*) The *small size* and *flattened sub-reniform shape*,

(*b*) The *reticulate surface*;

and should compare them with

(i) *Poppy seeds*, which are distinctly reniform, and have larger and shallower reticulations on the surface,

(ii) *Thornapple seeds*, which are much larger and darker,

**Constituents.**—The principal constituent of henbane seed is the alkaloid hyoscyamine with which is associated a small proportion of scopolamine (hyoscine).

The total amount of alkaloid is 0·058 per cent. (Ransom, 1891). In addition, the seed contains about 20 per cent. of fixed oil.

**Uses.**—Henbane seeds are comparatively seldom used in medicine. Thrown upon hot coals they form a domestic remedy for toothache the vapour that arises being allowed to enter the mouth. They have been used as a source of the alkaloid scopolamine.

## DATURA SEEDS

### (Daturæ Semina)

**Source, &c.**—Datura seeds are the dried seeds of *Datura fastuosa*, Linné, var. *alba*, Nees (N.O. *Solanaceæ*), an annual plant indigenous to India.

**Description.**—The seeds are yellowish brown in colour, ear-shaped, 4 to 5 mm. broad, about 1 mm. thick, flattened, finely pitted and reticulated; they are thickened towards the curved margin which is wavy, the edge exhibiting a triple ridge; the hilum is large and conspicuous, extending from about the middle to the acute end of the seed; the endosperm is narrow and translucent and encloses a curved embryo; they have a bitter taste but no odour.

**Constituents.**—The chief constituent of the seeds is the alkaloid scopolamine (hyoscine) of which they contain about 0·2 per cent., but traces of hyoscyamine and atropine are also present; they also contain resin and fixed oil.

**Uses.**—Used in India as an equivalent of stramonium seed.

## KALADANA

### (Kaladana; Pharbitis Seeds)

**Source, &c.**—Kaladana consists of the dried seeds of *Ipomœa hederacea*, Jacquin (N.O. *Convolvulaceæ*), a twining plant with a large, blue corolla resembling the major convolvulus of English gardens, common in the mountains of India. It was formerly called *Pharbitis Nil*, Choisy, whence the official synonym (Pharbitis is said to have reference to the colour of the corolla and Nil is the Hindustani for blue; kala-dana in Hindustani means 'black seed' and has been applied to several other seeds, giving rise to much confusion).

**Description.**—The seeds resemble in shape the 'quarter' of an orange, there being two flat sides meeting at an acute angle and an arched back. They are dull black, about 5 mm. long and 3 mm. wide; the hilum is distinct as a brown, slightly hairy, depressed spot. A transverse section shows plaited cotyledons in which small, slightly darker resin cells may be seen. The taste is at first not marked, but is subsequently acrid.

**Constituents.**—Kaladana contains about 14 per cent. of fatty oil, about 8 per cent. of resin and a little tannin, mucilage, &c. The resin, which is the most important constituent, is insoluble in ether and is probably identical with the ether-insoluble resin of jalap.

**Uses.**—The seeds are much used in India as a cheap and efficient laxative.

**Substitutes.**—According to Flückiger and Hanbury[1] a variety of kaladana closely resembling the official but about twice as large are used in India. Lately (1920) the seeds of *Ipomœa muricata*, Jacquin, have been imported as kaladana; these are imported into India from Persia and are largely used in Bombay for the same purpose; they resemble kaladana in shape but are less angular and dark brown in colour; they are also much larger, measuring 7 to 8 mm. in length and about 5 mm. in width; they are occasionally mixed with the seeds of *Acacia arabica*, Willdenow (N.O. *Leguminosæ*) which closely resemble them in size and colour, but are flattened-oval with a darker areola in the middle of each flat side. Another variety recently in commerce is smaller than the official, being only about 4 mm. long, dark grey, with a minutely but distinctly pitted surface; this variety may contain a small admixture of a nearly black Convolvulaceous seed with a smooth surface and also of the small, yellowish brown, flat seeds of *Crotalaria juncea*, Linné (N.O. *Leguminosæ*). The seeds of *Clitorea terneata*, Linné (N.O. *Leguminosæ*) have also been found in kaladana: they resemble vetch seeds and are mottled green and brown.

## ISPAGHULA

### (Ispaghula, Ispaghul, Spogel Seeds)

**Source, &c.**—Ispaghul consists of the dried seeds of *Plantago ovata*, Forskal (N.O. *Plantaginĕæ*), a herb widely distributed in the Punjab, Sind and Persia.

**Description.**—The seeds are about 2 to 3 mm. long and 1 to 1·5 mm. wide, boat-shaped, pale greyish brown in colour, usually with a pinkish tinge; on the convex surface there is a small, elongated, brown spot; on the concave surface is the hilum covered with a thin, whitish membrane; when soaked in water the seed coat swells and the seeds become surrounded with a transparent, colourless mucilage free from taste and odour. One part of the seeds with twenty of water forms a thick, tasteless jelly.

[1] *Pharmacographia*, p. 403.

12

**Constituents.**—The chief constituent is the mucilage which is contained in the cells of the epidermis; it swells and dissolves when the seeds are immersed in water.

**Uses.**—In dysentery and chronic diarrhœa; taken dry they absorb water in the intestine yielding a protective mucilage; they are also used as a poultice.

**Allied Drug.**—Psyllium seeds (flea seeds), the seeds of *Plantago Psyllium*, Linné, are similar but smaller, dark brown and shining; they are much used on the continent for diarrhœa.

## NUTMEGS

(Myristicæ Semina, Myristicæ, Nux Moschata)

**Source, &c.**—The nutmeg tree, *Myristica fragrans*, Van Houtte (N.O. *Myristicaceæ*), is indigenous to the Molucca Islands and a few neighbouring islands, as well as north-western New Guinea, but has been introduced into Penang, Sumatra, Malacca, Java, the West Indies, and Ceylon, nutmegs and mace being exported from the Malay Archipelago, the Straits Settlements, the West Indies, and Ceylon.

The use of the spice was introduced into Europe probably during the twelfth century. The Banda Islands, where they were produced, were discovered about 1506, and passed into the possession of the Portuguese, and finally of the Dutch, who, in this case as in that of cloves and cinnamon, made every endeavour to restrict the cultivation of the trees to the islands of Banda and Amboyna, and thus create a profitable monopoly. The nutmeg trees of adjacent islands were destroyed, and the nutmegs themselves soaked in a mixture of slaked lime and water to render them, it was said, incapable of germination, a precaution that was quite unnecessary, as the vitality of the seed is destroyed by the simple process of drying. For some time these efforts were successful, and the nutmeg trade remained in the hands of the Dutch; but eventually the trees were successfully introduced into Malacca, Ceylon, and Jamaica.

The fruit is a fleshy drupe resembling a small peach in size and shape. As it ripens the fleshy pericarp splits longitudinally and discloses a crimson, lobed arillus surrounding a brown seed (fig. 97). The fruits are collected, the pericarps removed, and the crimson arillus (mace) stripped off in a single piece (double blade) or in two halves (single blade), flattened, and dried, during which the crimson colour changes to a reddish yellow.

The seeds are then carefully dried, usually over a charcoal fire, a process that requires several weeks. When quite dry the kernel rattles in the thin, brittle, brown shell. The latter is broken and the kernel removed. They are frequently dusted over with slaked lime, or washed in milk of lime and dried, before they are exported; this protects them from the attacks of insects, to which they are

FIG. 97.—Nutmeg. *A*, fruiting branch of *Myristica fragrans*, showing fruit dehiscing. *B*, stamens of staminate flower; magnified. *C*, pistillate flower cut longitudinally; *p*, perianth; *g*, ovary; magnified. *D*, Nutmeg surrounded by the arillus (mace). *E*, the same cut longitudinally, showing the embryo, *e*. (Luerssen.)

otherwise very liable. Very probably the original 'liming' of nutmegs was intended to protect them from insects and not to destroy the vitality of the seeds, as has been often assumed (Tschirch, 1898). They are usually imported in cases which, on arrival in this country, are opened and the contents bulked. The nutmegs are then thrown on to a coarse iron riddle, the broken and otherwise damaged picked out and the sound ones sorted according to their size which is indicated by the number (65, 80, 110, &c.) required to make up a pound weight. Some of these are sold without further treatment, but many are limed (or re-limed), limed nutmegs being preferred in Holland and in the United States. The damaged nutmegs are sold for the production of the volatile and expressed oil.

**Description.**—Nutmegs are broadly ovoid in shape and about 2·5 cm. in length; they are usually of a greyish brown colour and marked with shallow reticulate furrows. The hilum lies in a little circular depression surrounded by a raised ring, and from it the raphe can usually be traced in a furrow extending to the chalaza at the apex. When examined with a powerful lens the surface is seen to be very finely pitted and marked with minute reddish points and larger dark reddish brown lines and irregularly elongated spots. The section exhibits dark, reddish brown, wavy lines alternating with pale brownish or greyish interspaces. The greater portion of the nutmeg consists of the ruminated albumen, the ruminations being produced by the infolding of part of the perisperm and deposition in its cells of dark colouring matter. These infoldings occur near the fibro-vascular bundles, and produce the depressed lines on the surface of the nutmeg corresponding to the branching bundles.

The cut surface easily yields oil when indented with the nail. The odour is strong and aromatic, the taste aromatic and bitterish.

**Constituents.**—The chief constituents of nutmegs are volatile oil (8 to 15 per cent.) and solid fat (about 40 per cent.); they contain in addition as reserve material amylodextrin, a substance intermediate between starch and dextrin.

The volatile oil (sp. gr. 0·870 to 0·925; O.R. + 13° to + 30°) consists chiefly of terpenes together with myristicin which possesses an intense odour of mace and passes over in the last portions of the distillate.

Expressed oil of nutmeg is a yellowish, very aromatic solid, melting at 25° to 43°, obtained from imperfect or broken nutmegs by hot pressure; it contains about 12 per cent. of the volatile oil together with the glycerides of myristic, palmitic, and oleic acids.

Amylodextrin occurs in granules of irregular shape which are coloured reddish brown by iodine (distinction from starch); it can be obtained in a crystalline form, and appears to be an intermediate product between starch and maltose (or dextrose).

Myristicin, $C_{11}H_{12}O_3$, is crystalline and toxic; it is more easily absorbed in the presence of the other constituents of nutmegs and is less toxic to lower animals than to human beings.

**Varieties.**—*Penang nutmegs* are broadly ovoid and very aromatic. *Singapore nutmegs* are more deeply and minutely wrinkled and frequently show marks of scorching.

*West Indian* nutmegs are somewhat elongated and frequently have dark marks on them.

All the above varieties may, however, be of small size, shrivelled, or otherwise defective.

**Substitutes.**—Of the other species of *Myristica* yielding seeds resembling nutmegs, only one, viz. *M. argentea*, Warburg, the Macassar or Papua nutmeg, yields an aromatic seed (wild, Papua, Macassar or long nutmeg ; exported from New Guinea) ; this is long, narrower and less aromatic than the official, and has a uniform, brown, scurfy surface and a distinctly acrid taste.

*Bombay nutmegs* (*M. malabarica*, Lamarck) are also long and narrow, but are destitute of aroma.

*Factitious nutmegs*, made from exhausted or damaged nutmegs mixed with mineral matter (clay) and pressed into moulds, have several times been detected ; they yield from 11 to 18 per cent. of ash, whereas genuine nutmegs do not afford over 4 per cent. ; they also yield less volatile oil (1·76 per cent.), and the section is not regularly reticulated.

**Uses.**—Nutmegs have stimulant and carminative properties ; in large doses they are toxic, producing convulsions, an action due to the myristicin contained in them. The expressed and volatile oils have been used externally in chronic rheumatism.

## MACE

*Mace* is the dried arillus of the nutmeg. It has its origin in a thickening of the funiculus extending to the outer integument of the seed near the exostome ; it is therefore intermediate in nature between an arillus and arillode. As it develops and surrounds the seed it divides into branching lobes which approach one another near the apex of the seed. It is carefully separated from the seed and dried, and then forms flattened lobed pieces about 25 mm. or rather more in length, somewhat less in breadth, and about 1 mm. thick. When soaked in water and restored to its original form, it is seen to be cup-shaped. It is of a dull reddish colour, translucent and brittle. The strong and fragrant odour and aromatic taste resemble those of nutmeg.

Mace contains from 4 to 15 per cent. of volatile oil which appears to be identical in all essential particulars with that obtained from the nutmeg.

**Varieties.**—*Bombay mace* (*M. malabarica*), considerable quantities of which are imported, is in longer, narrower pieces of dark red colour, dividing into numerous narrow lobes which are twisted together at the apex. It is devoid of aroma and valueless as a spice. The powder can be distinguished from that of genuine mace by the large amount of substances yielded to ether after exhaustion with petroleum spirit (30 per cent., as against 3·5 from genuine mace).

*Macassar* or *Papua mace* (*M. argentea*) is in dull, brownish fragments with a dusty surface ; the lobes are few, broad and widely separated, ultimately uniting to a compact cap ; the taste is distinctly acrid.

## CASTOR SEEDS

### (Castor-oil Seeds, Semina Ricini)

**Source, &c.**—The castor-oil plant, *Ricinus communis*, Linné (N.O. *Euphorbiaceæ*), is a native of India, but is diffused now over all tropical and subtropical countries. In India it may attain a height of 40 metres and be a perennial tree, but in cooler climates it is either a shrub or an annual herb. The plant, and with it the seed, is subject to much variation, the larger, arborescent forms yielding large seeds, the small, annual varieties small seeds.

The oil pressed from the seeds was well known to the Egyptians, Greeks, and Romans, and employed both as a medicine and illuminant, as well as for various technical purposes. It was used in Europe during the early Middle Ages, but subsequently fell into disuse, being re-introduced from the West Indies toward the end of the last century. The supplies of seeds came then chiefly from Jamaica, but the exports from India increased with great rapidity. The market is now chiefly supplied from India and South America, but considerable quantities of the seed are raised in other countries, as, for instance, in Italy.

**Description.**—The fruit of the plant is a three-celled, three-seeded, thorny capsule. The seeds are oblong in outline, somewhat flattened, and from about 8 to 15 mm. long. The seed-coat is quite smooth and glossy, and varies in colour from a greyish brown to a beautiful grey marbled with reddish brown or black spots and stripes. The dorsal surface is arched, the ventral nearly flat ; at one extremity is a prominent, usually pale, caruncle, from which a distinct line (raphe) runs along the ventral surface to the other extremity of the seed, where it terminates in a raised point (chalaza), branches and disappears. The caruncle can be easily removed, disclosing a dark spot (hilum) beneath.

The seed-coat is thin and brittle. The marbled outer layer, which, especially after soaking in water, can easily be scraped off, is succeeded by a hard, dark layer. Within is the kernel, consisting of a large, yellowish white, oily endosperm enclosing a small embryo with two papery cotyledons. Surrounding the kernel is a delicate silvery white membrane.

Castor seeds have an almost imperceptible odour and very slightly acrid taste.

The student should observe

(a) The *glossy, mottled seed-coat*,
(b) The *small caruncle*,
(c) The *oily endosperm* and *papery cotyledons* ;

and compare the seeds with croton seeds, which have a uniformly dull brown surface.

**Constituents.**—The most important medicinal constituent of castor seed is the fixed oil. It exists in the seed to the extent of about 50 per cent., and may be obtained by pressing the unshelled or shelled seed without heat (' cold drawn castor oil '). Usually the seeds are first graded to size, cracked between rollers, the shells removed by fanning and the kernels pressed ; the oil is filtered, steamed to coagulate proteids (see below) and again filtered. The cake still contains 8 to 10 per cent. of oil which can be recovered by extraction with benzene. Much is thus produced in Italy, Marseilles, London, Hull, Belgium, &c. Considerable quantities are also produced in India and elsewhere by pressing the seeds between hot plates, or by boiling the crushed seeds and skimming off the oil, but such oil has a dark colour and disagreeable odour, and is unsuitable for medicinal use.

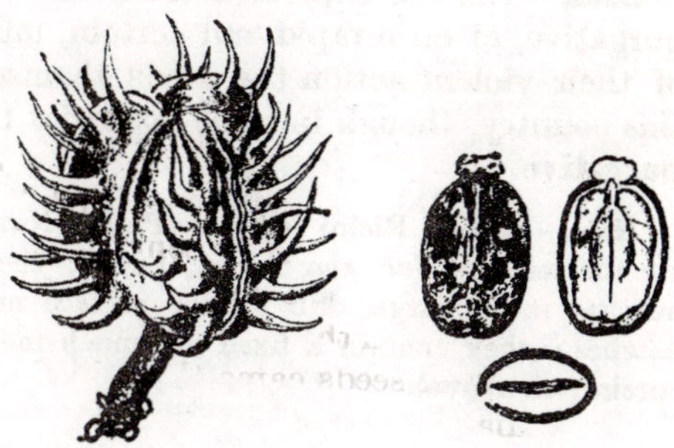

FIG. 98.—Fruit and seed of *Ricinus communis*. Natural size. (Bentley and Trimen.)

The cake left after the expression of the oil contains ricinine (Tuson, 1864), a crystalline principle melting at 201·5°, but present in small quantity only (about 0·2 per cent.), ricin (Stillmark, 1889), a toxin similar in nature to the bacterial toxins, and also (ripe seeds) a very active lipase (fat-splitting enzyme) and other enzymes.

Both castor seeds and the cake left after the expression of the oil act as violent purgatives, a property probably due to the ricin contained in them.

Castor oil consists chiefly of the glycerides of ricinoleic, isoricinoleic dihydroxystearic, and other acids. It owes its purgative action in all probability to the ricinoleic acid. It is characterised by its specific gravity (0·958 to 0·970), solubility in alcohol (1 in 3·5), insolubility in petroleum spirit (which, however, disappears if other vegetable oils are present) and by its high acetyl value (150).

It is much used for lubricating hot machinery, and for illuminating purposes. See also 'Castor Oil.'

Ricin is extremely poisonous, 0·002 of a milligramme per kilo. being fatal to rabbits. It is capable of producing an antitoxin (antiricin) in the body, and it is interesting to note that Ehrlich's researches on this property led to the preparation of antidiphtheritic and other sera (Müller, 1899). Ricin is probably identical with Bubnow and Dixon's ricinone (Finnemore and Deane, 1905); its activity is destroyed by heating to 100°. It forms with abrin (*Abrus precatorius*, Linné), crotin (*Croton Tiglium*, Linné), curcin (*Jatropha Curcas*, Linné), and two or three others a group of toxins (toxalbumoses, phytalbumoses) possessing similar properties; being insoluble in oil it is left in the cake when the seeds are pressed and renders the cake unfit for cattle food.

The lipase in the seeds has been turned to commercial use as a convenient means of splitting oils into glycerin and fat-acids; any lipase present in the expressed oil is rendered inactive by the steaming.

**Uses.**—The oil expressed from the seeds is valuable as a simple purgative, at once rapid and certain, mild and painless. On account of their violent action the seeds themselves are never employed in this country, though in some countries they are said to be a favourite purgative.

**Note.**—Semina Ricini Majores (Physic Nuts, Purging Nuts, Pignons d'Inde) are the seeds of *Jatropha Curcas*, Linné; they resemble castor seeds in shape, but are rather large, dull black; surface minutely rugose, with small white patches; they contain a fixed oil, much more purgative than castor oil, and curcin (see above).

## CROTON SEEDS

### (Semina Crotonis)

**Source, &c.**—Croton seeds are the seeds of *Croton Tiglium*, Linné (N.O. *Euphorbiaceæ*), a small tree indigenous to and cultivated in India. They were used medicinally in the seventeenth century, but fell into disuse owing probably to the violence and uncertainty of their action. The oil was introduced from India about 1819, and was found to be, in certain cases, a valuable cathartic.

The tree produces a three-celled, three-seeded capsular fruit resembling that of the castor plant, but devoid of spines. The seeds are exported, and the oil pressed from them in this country.

**Description.**—Croton seeds are of a dull cinnamon-brown colour and oblong outline; they are about 12 mm. in length, and resemble castor seeds in size and shape, though they are rather more angular in transverse section, the ventral and dorsal surfaces being separated by a prominent line. The caruncle which the seed possesses is easily detached, and is therefore seldom to be found in the drug; the hilum is less distinct than in castor seed, and from it the raphe runs along the ventral surface of the seed, terminating in a dark chalaza at the opposite extremity. The outer, dull brownish layer is easily removed, disclosing a hard dark coat; in many commercial specimens the friction of the seeds against one another has been sufficient partially to effect this, giving the seeds a mottled appearance.

The kernel is yellowish and oily, and consists of a large endosperm enclosing papery cotyledons and a small radicle.

The taste of the kernel, in ascertaining which great caution is necessary, is at first oily, but this is succeeded by an unpleasant acridity; the seeds have no marked odour.

The student should compare these seeds with castor seeds and observe

(*a*) The *dull brownish outer layer*,

(*b*) The *prominent line* separating the ventral from the dorsal surface.

**Constituents.**—Croton seeds contain about 50 per cent. of fixed oil which possesses violent cathartic and vesicant properties, due to a resinous constituent, croton-resin (Dunstan, 1895).

The seeds also contain the toxic albumoses croton-globulin and croton-albumin, which together are also known as crotin, and resemble ricin. Croton oil is brownish yellow and slightly fluorescent; it is soluble in less than its own volume of absolute alcohol, but on further addition of alcohol two layers are formed, the active constituent of the oil being contained in the alcoholic layer. The solubility appears to depend on the proportion of free acid present and to increase with the age of the oil. See also 'Croton Oil.'

**Uses.**—Croton oil is a powerful irritant, producing, when applied to the skin, a burning sensation and redness, followed by severe pustules; it is used, diluted, as a counter-irritant. Internally it is a very rapid, drastic cathartic, and is given in certain cases of apoplexy.

## GRAINS OF PARADISE

### (Guinea Grains, Grana Paradisi)

**Source, &c.**—Grains of paradise are the seeds of *Amomum Melegueta*, Roscoe (N.O. *Scitaminĕæ*), a herb attaining about 1·5 metres in height, indigenous to the west coast of Africa. These seeds were much esteemed as a spice in the twelfth and thirteenth centuries; the country from which they were derived being unknown, they were called 'grains of paradise.' At that time they were imported from Tripoli, whither they had made the land journey from West Africa. Subsequently they were brought direct from the west coast of Africa to Portugal by Portuguese traders. They are now imported from West Africa.

The plant produces an ovoid pointed fruit about 10 cm. long, containing a large number of small seeds.

**Description.**—Grains of paradise are small, hard, shining seeds of a rich reddish brown colour. They vary much in shape but are frequently sub-pyramidal with rounded or obtuse angles, and average about 3 mm. in length ; the surface is seen under a lens to be minutely but distinctly papillose. Attached to one extremity of the seed are usually the paler fibrous remains of the funiculus, which project in the form of a beak. Cut transversely near the hilum, they exhibit a copious, white, starchy perisperm surrounding a yellowish, horny endosperm in which a minute paler embryo is embedded. The longitudinal section also exhibits the perisperm, endosperm, and embryo, the radicle of the latter being directed towards the funiculus. The crushed seeds have a faintly aromatic odour, but the taste is intensely pungent, rivalling that of capsicum fruit.

Fig. 99.—Grain of Paradise. (Magnified.)

The student should observe

(*a*) The *rich, reddish brown colour*,
(*b*) The *papillose surface*,
(*c*) The *large projecting funiculus*,
(*d*) The *pungent taste* ;

and should compare them with cardamom seeds (see page 139).

**Constituents.**—Grains of paradise contain a little (0·3 per cent.) volatile oil and a yellowish extremely pungent, oily body, paradol which has not yet been obtained in a crystalline form. It resembles gingerol (see p. 378), but its pungency is not destroyed by boiling with 2 per cent. solution of caustic potash.

**Uses.**—The seeds possess stimulant properties, and were formerly employed as a condiment ; now they are chiefly used in veterinary medicine.

## COLCHICUM SEEDS

### (Semina Colchici)

**Source, &c.**—The meadow saffron, *Colchicum autumnale*, Linné (N.O. *Liliaceæ*), is widely distributed over Europe, and abundant in some parts of England in moist meadows and pastures. It produces in the autumn a conspicuous, reddish purple flower that springs from the side of an enlarged contracted stem (corm) several centimetres below the surface of the ground. The ovary is superior and lies at about the same depth in the ground. The leaves appear soon after the flower, and attain in the spring a length of 20 or 25 cm. The ovary is then raised to the surface by the elongation of the peduncle, after which the leaves wither. The fruit, a three-celled capsule, ripens in the summer, dehiscing septicidally and disclosing numerous seeds

**Fig. 100.**—*Colchicum autumnale.* *A*, flowering plant, much reduced. *B*, lower part of the same, natural size. *C*, corm, cut vertically: *k*, corm; *k'*, young corm for the next year; *k''*, bud destined to reproduce the plant after *k'* has developed. *D*, portion of the same, showing *h*, the leaves enveloping the corm. *E*, upper part of perianth, halved. *F*, ovary, *o*, with the three styles. *G*, ovary, magnified. *H*, fruit. *J*, the same, cut transversely, showing the seeds. *K*, seed, magnified. *L*, the same, cut to show the embryo, *e*. (Luerssen.)

which, when quite fresh, are pale in colour, but darken as they dry, becoming at the same time covered with a saccharine exudation, as much as 5 per cent. of glucose having been found on them. These were introduced into medicine about 1820 to replace the corm, which was considered uncertain in action.

**Description.**—Colchicum seeds are small, very hard, and of a dull dark reddish brown colour. They are about 2·5 mm. in diameter and nearly spherical, the remains of a thick funiculus rendering them somewhat pointed. Under a lens the surface is seen to be rough from the presence of minute pits. They are extremely hard and tough, and are difficult to cut until they have been soaked in water; the section exhibits a yellowish, oily endosperm, in which, near the margin and removed from the hilum, the minute embryo is embedded. They are odourless, but have an unpleasantly bitter taste.

The student should observe

(*a*) The *rough surface* and *hard nature* of the seed,
(*b*) The remains of the *thick funiculus*;

and should compare them with

(i) *Black mustard seeds*, which are smaller,
(ii) *Henbane seeds*, which are reniform in outline.

**Constituents.**—All parts of the plant contain the alkaloid, colchicine, which possesses weak basic properties. The seeds contain from 0·02 to 0·8 per cent., but a good sample of the drug should yield not less than 0·5 per cent. The seeds also contain a resin, colchicoresin, and about 6 per cent. of fixed oil. They yield about 3 per cent. of ash.

Colchicine, $C_{22}H_{25}NO_6$, crystallises in pale yellow needles melting at 155° to 157° and yields a crystalline compound, $C_{22}H_{25}NO_6.2CHCl_3$, with chloroform. The alkaloid is very soluble in cold water, alcohol, and chloroform, but only slightly in ether. Chloroform removes it from an acid as well as from an alkaline aqueous solution; boiled with dilute mineral acids it yields colchiceine and methyl alcohol. Colchicine has also been found in the root of *Gloriosa superba*, Linné.

**Assay.**—The seeds may be assayed by the process devised by Farr and Wright.[1]

**Uses.**—Colchicum is chiefly used to relieve the pain and inflammation and shorten the duration of acute gout and certain gouty affections. Its action depends upon the alkaloid colchicine. Colchicine has a marked action upon plain muscle, especially that of the intestine, producing diarrhœa and vomiting. In large doses it produces death from failure of the respiration.

**Adulteration.**—Colchicum seeds are said to be liable to adulteration by the fraudulent addition of glucose.

[1] *Yearbook of Pharmacy*, 1911, p. 23.

## CEVADILLA SEEDS

### (Sabadilla Seeds, Semina Cevadillæ)

**Source, &c.**—Cevadilla seeds are the ripe seeds of *Schœnocaulon officinale*, Asa Gray (N.O. *Liliaceæ*), a tall herbaceous plant growing on the lower mountain slopes near the eastern coast of Mexico, in Guatemala, and in Venezuela. At the time of the Spanish conquest the drug was known to the American Indians as a caustic application to wounds; it came into use in Europe much later as a parasiticide. The seeds are now chiefly used as the source of veratrine.

The plant produces a tall raceme of yellowish flowers, succeeded by small three-celled capsular fruits; as the fruit ripens it separates septicidally into three follicles, which dehisce by their ventral sutures. Each follicle contains from one to six seeds. Formerly the dried fruits were imported, but now chiefly the seeds freed from the thin, brown, papery pericarps.

FIG. 101.—Cevadilla seed. *a*, flower, magnified. *b*, stamen, magnified. *c*, fruit, magnified. *d*, *e*, fruit after dehiscence, natural size. *f*, *g*, *h*, *i*, seeds, natural size; *k*, *l*, *m*, enlarged; *n*, cut longitudinally. (Luerssen.)

**Description.**—Cevadilla seeds are long and narrow, glossy dark brown or nearly black and about 6 mm. long, tapering to an acute point. From mutual pressure in the fruits there is usually on one side a longitudinal depression with acute edges; for a similar reason the seeds are slightly curved; the surface is finely wrinkled.

The seeds are inodorous, but have an unpleasant, bitter, and acrid taste; the powder produces violent sneezing.

The student should observe

(*a*) The *dark colour*,

(*b*) The *long narrow shape* with *acute angles*,

(*c*) The *acrid taste*.

**Constituents.**—The seeds contain several alkaloids of which cevadine, $C_{32}H_{49}NO_9$ (also called crystalline veratrine), is the most important and the most toxic; it is easily hydrolysed by alkalies yielding cevine and methylcrotonic acid. Veratridine (also called veratrine) accompanies cevadine in the seeds; it is amorphous and yields veratric acid and verine when hydrolysed. Both cevadine and veratridine are toxic and sternutatory. Other alkaloids of less importance are sabadilline (cevadilline), sabadine, sabadinine, and sabatrine, the latter being said to be only a mixture.

Commercial veratrine is a mixture of these alkaloids and consists chiefly of cevadine and veratridine; the former may readily be obtained from it by dissolving in alcohol, adding water until the solution becomes faintly opalescent and setting aside to crystallise.

Keller (1895) found 4·25 per cent. of alkaloid in the seeds.

**Uses.**—Cevadilla seeds (and veratrine) act both internally and externally as a powerful irritant. This primary effect is followed by depression, and, when used externally, loss of sensibility; hence ointment of veratrine is used to relieve neuralgic pains, &c. Veratrine is employed also as a parasiticide, but is seldom administered internally.

## ARECA NUTS

### (Betel Nuts, Semina Arecæ)

**Source, &c.**—Areca nuts are the seeds of *Areca Catechu*, Linné (N.O. *Palmæ*), a handsome palm with a tall slender stem crowned by a number of large elegant leaves. The tree is widely cultivated throughout India and the whole of the islands of the Eastern Archipelago for its seeds, which are universally employed by the natives as a masticatory.

The fruits, which are orange yellow in colour when ripe, and of the size and shape of an egg, are borne in large numbers on a branching axis. The pericarp is fibrous and surrounds a single seed, from which it is easily separated. The seeds are usually boiled in water with the addition of a little lime and dried.

**Description.**—The areca nut has the shape of a short, bluntly rounded cone about 2·5 cm. long. It has a brownish colour, and the surface is marked with a network of paler depressed lines. Frequently portions of a hard, brittle, grey, silvery coat are attached more or less firmly to the seed; these are portions of the inner layer of the pericarp, and do not belong to the seed proper. The latter is hard, but can be cut with a knife, and then exhibits a marbled interior (*ruminate endosperm*), dark brown lines alternating with opal-white portions. Examination with the lens shows these lines to be slightly fissured. They are folds of the seed-coats, corresponding in position to the fibrovascular bundles that pass into them from the funiculus; they follow the branching of the bundles and produce the reticulate markings on the seed. The outer cells of the seed-coat fill with tannin, to which the brown colour is due, and contain alkaloid (Osenbrug, 1894).

The seed has little odour, but an astringent, slightly bitter taste.

The student should observe

(*a*) The *shape* of the seed,

(*b*) The *ruminate endosperm.*

**Constituents.**—Areca nuts contain several alkaloids together with tannin (15 per cent.) and fat (14 per cent.). The most important of the alkaloids, and the one to which the sialogogue and vermifuge properties of the seed are due, is arecoline (methylarecaidine). Other alkaloids are arecaidine, arecaine, guvacine, and choline.

Arecoline, $C_8H_{13}NO_2$ (boiling-point 220°), is a colourless oily liquid yielding crystallisable salts. Boiling with hydrochloric acid converts it into methyl chloride and crystalline arecaidine, from which arecoline can be regenerated. Arecaidine, $C_7H_{11}NO_2$, is methyltetrahydronicotinic acid and can be synthesised from trigonelline. Arecaine is methylguvacine and can be prepared by methylating guvacine.

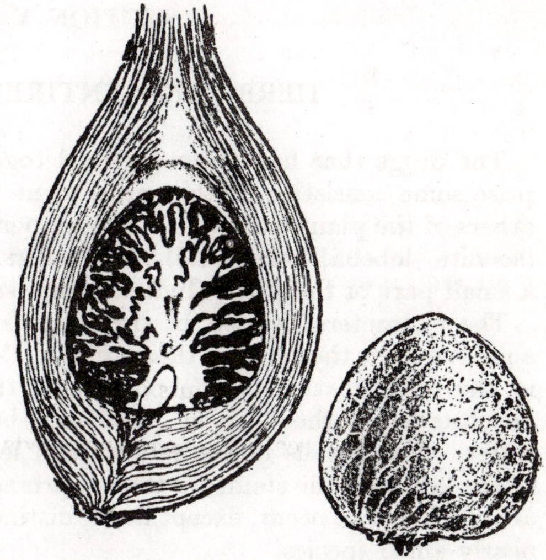

a b

FIG. 102.—Areca nut. *a*, vertical section of the fruit and seed, showing fibrous pericarp of the former and ruminate albumen of the latter. *b*, entire seed, natural size. (Bentley and Trimen.)

**Use.**—The powdered seeds are chiefly used in this country as a vermifuge for dogs.

## SEEDS IN LESS FREQUENT USE

**Jequirity Seeds.** (Prayer beads)—The seeds of *Abrus precatorius*, Linné (N.O. *Leguminosæ*), a common Indian and Brazilian shrub. The seeds are sharply characterised by their smooth, glossy surface and bright scarlet colour with a black patch at the hilum; ovoid or subglobular, 5 to 8 mm. long, 4 to 5 mm. broad. They contain a substance known as abrin which is a mixture of two toxic proteins, a paraglobulin which loses its activity at 75° to 80° and a phytalbumose which is destroyed at 85°. The action of abrin resembles that of snake venom. An infusion of the seeds has been applied to the eyes for opacities of the cornea but is dangerous; a paste of the seeds has been used for lupus.

Abrus root (Indian liquorice) contains a sweet substance resembling but not identical with glycyrrhizin, but it possesses toxic properties.

# SECTION V

## HERBS AND ENTIRE PLANTS

The drugs that have been grouped together under this term comprise some consisting of the entire plant (chiretta, carrageen, ergot), others of the plant from which a small portion only has been removed (aconite, lobelia, horehound), and others again that represent only a small part of the plant (Indian hemp, savin).

The characters utilised in the identification of the herbs are to a great extent the same as those that would be employed by botanists in the identification of the plant, and these characters can usually be observed if the drug be previously boiled for a few minutes in water. The details of the flowers and leaves, and the arrangement of the latter on the stem, are thus of primary importance. Difficulties are not likely to occur, except in the distinction of the drug from other nearly allied species.

Flowering herbs are commonly gathered when in flower, and dried in the shade. They should then be completely dried, and kept in airtight containers to obviate changes that are likely to ensue as the result of the action of enzymes. Such changes in some cases show themselves by changes in colour, &c., but in others they may affect and modify the active constituents without producing any visible alteration.

### ACONITE HERB

### (Herba Aconiti)

**Source, &c.**—The aconite, monkshood, or wolfsbane, *Aconitum Napellus*, Linné (N.O. *Ranunculaceæ*), is a perennial herb growing abundantly on the lower mountain slopes of Central Europe. It is cultivated in England as a garden plant as well as for medicinal use, and is found apparently wild in some localities, having probably escaped from cultivation. The drug has only recently been introduced into medicine, but the poisonous properties of aconite have long been known. Both the fresh leaves and flowering tops as well as the dried roots have been used, the latter alone being now official.

The plant should be cut when the flowers are beginning to expand, and the leaves and flowering tops separated from the larger stalks; they are used for the preparation of the juice and green extract of aconite.

**Description.**—The stem, which attains a height of about 1 metre or more, is upright, smooth, and usually simple, terminating in a leafy raceme of bluish flowers.

The lower leaves are petiolate, radiately veined and deeply pal-

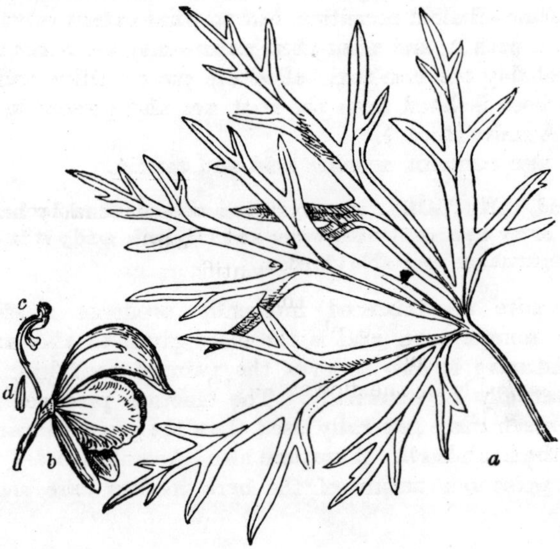


FIG. 103.—Aconite herb. *a*, leaf; *b*, flower; *c*, *d*, petals. Three-fourths natural size. (Holmes.)

matisected, the three primary divisions extending very nearly to the petiole; towards the upper part of the plant the petioles become shorter and the lamina less divided. They are of a dark green colour on the upper surface, paler beneath and glabrous, or nearly so.

The flower is zygomorphous; the calyx consists of five blue petaloid sepals (fig. 103, *b*), of which the upper (posterior) is shallow helmet-shaped; of the five petals, two only are easily found as hammer-shaped nectaries (fig. 103, *c*) concealed within the helmet-shaped upper sepal; the other three are small and inconspicuous (fig. 103, *d*); the stamens are numerous. The fruit consists of from three to five divergent follicles.

The plant has little odour, but produces slowly, when chewed, an unpleasant, acrid, burning taste.

The student should observe

(*a*) The characteristic *shape of the leaf*,

(*b*) The *shallow, helmet-shaped sepal*,

(*c*) The characteristic *taste*.

**Constituents.**—Aconite herb contains certain alkaloidal constituents, but of the exact nature of these and the proportion in which they exist there is no very definite information. The dried leaves have been stated to contain from 0·12 to 0·96 per cent. of total alkaloid, of which a part at least is undoubtedly the highly toxic crystalline alkaloid aconitine, but to what extent other alkaloids are associated with it, and what their nature may be, is not at present known. Probably the non-toxic alkaloids picraconitine and aconine, which have been isolated from the root, are also present in the herb. (Compare 'Aconite Root').

The herb also contains aconitic acid and tannin.

Aconitic acid, $C_3H_3(COOH)_3$, is a crystalline acid obtained by heating citric acid to 175°, or by heating it with water and sulphuric acid; it is widely distributed throughout the vegetable kingdom.

**Uses.**—Aconite administered internally produces a steady fall in the body temperature, and is therefore given to alleviate certain febrile conditions; it also relieves the pain of neuralgia, and may be used internally or externally. The tincture prepared from the dry root is much more generally used than the green extract or juice made from the fresh herb. (Compare also 'Aconite Root.')

The very poisonous nature of the herb renders care necessary in tasting it.

## BROOM TOPS

### (Cacumina Scoparii)

**Source, &c.**—The broom, *Cytisus Scoparius*, Link (N.O. *Leguminosæ*), is a woody shrub, attaining a height of 2 metres, indigenous to England, and distributed over temperate Europe. It was employed medicinally by the Anglo-Saxons and by the Welsh. The younger herbaceous tops are collected, and are official in the fresh state for the preparation of the juice, and dried for making the infusion.

**Description.**—The stem of the broom produces numerous long, slender, alternate, erect branches which are glabrous, tough, and flexible. The upper part of each branch is dark green, and bears five distinct wings, which, however, are thrown off as the stems increase in size. The leaves are small, alternate, and hairy whilst young; the lower are trifoliate and stalked, but those near the ends of the twigs are reduced to single sessile leaflets. They easily fall off from the stems when dried, and the dry drug therefore frequently contains but few.

The bright yellow, fragrant, papilionaceous flowers are borne on solitary, axillary peduncles, and succeeded by oblong, flattened legumes with hairy margins.

When fresh, the flowering plant has an agreeable odour which, however, disappears on drying. The taste is bitter and unpleasant.

**Constituents.**—The chief constituents of broom are the liquid volatile alkaloid, sparteine, and an indifferent, yellow, crystalline substance, scoparin; in addition to these a crystalline, volatile alkaloid, genisteine, and a non-volatile alkaloid, sarothamnine, have been isolated from it.

Sparteine, $C_{15}H_{26}N_2$, has been obtained as a colourless, viscid oil (b. pt. 325°) forming crystallisable salts of which the sulphate has been used medicinally. It is present in largest proportion in March (0·45 per cent.) and smallest in August (0·15 per cent.).

Scoparin belongs to the quercetin group.

Genisteine, $C_{16}H_{28}N_2$ has been isolated from commercial sparteine.

FIG. 104.—Broom, flowering branch. (Maisch.)

**Uses.**—Broom is largely used as a diuretic in dropsy, an action which is due to the scoparin it contains. Sparteine exhibits physiologically a close resemblance to coniine, weakening the heart and depressing nerve cells.

## HEMLOCK HERB

### (Herba Conii, Folia Conii)

**Source, &c.**—The common or spotted hemlock, *Conīum maculatum*, Linné (N.O. *Umbelliferæ*), is a biennial plant widely spread throughout temperate Europe, and generally distributed over Great Britain. It was in all probability the plant employed by the Greeks in the preparation of poisonous draughts, and was much used in Anglo-Saxon medicine, but latterly has lost much of its reputation owing to the uncertain action of preparations made from it. The herb is cultivated for medicinal use, but wild plants are also collected.

**Description.**—The plant produces, usually in its second year, an erect, cylindrical, glabrous, hollow stem, reaching a height of 1·5 to 2 metres, the lower part of which bears purplish spots which, however, usually disappear on drying.

The leaves are dark green on the upper, paler on the under surface, quite glabrous, and attached to the stem by amplexicaul petioles. The lower are large and decompound, attaining 50 cm. in length, the upper being less divided ; the ultimate segments are ovate or lanceolate, and acute, terminating in smooth, colourless, horny points (fig. 105, *a*).

The umbels are about twelve-rayed, and provided with both general and partial involucres, the latter consisting of three short, lanceolate bracts directed outwards. The fruits are broadly ovate and

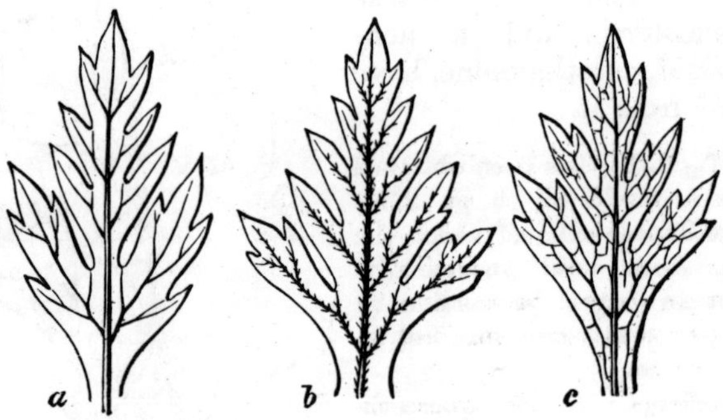



FIG. 105.—Portion of leaf of *a, Conium maculatum* ; *b, Anthriscus sylvestris* ; *c, Æthusa Cynapium.* Magnified 2 diam. (Vogl.)

characterised by the irregular, crenate ridges and grooved endosperm. (Compare 'Hemlock Fruits.')

The plant has a bitterish taste and unpleasant odour, especially when crushed ; the addition of solution of potash produces a strong, disagreeable odour of mice.

The student should observe

(*a*) The *glabrous, spotted hollow stem,*
(*b*) The *much divided leaves,* which are paler on the under surface, *quite glabrous,* the *ultimate divisions terminating in smooth colourless points,*
(*c*) The *general and partial involucres,*
(*d*) The *crenate ridges* and *grooved endosperm* of the *fruit.*

**Constituents.**—Hemlock herb contains coniine and conhydrine. These alkaloids are present in both stem and leaves in largest quantity when the plant is in full flower, the stem then containing 0·064 per cent., the leaves 0·187 per cent., and the flowers and flower-stalks

0·236 per cent. (Farr and Wright). The herb should therefore be collected when the fruit begins to form (compare p. 111).

**Uses.**—Fresh hemlock herb is used for the preparation of the green extract and juices. These are administered as sedatives in various spasmodic diseases, in asthma, whooping cough, and spasmodic affections of the larynx.

**Adulterations.**—Several indigenous Umbelliferous plants have been mistaken for hemlock, but the characters detailed above are sufficient to ensure the identification of the latter. The following may be particularly alluded to:

*Anthriscus sylvestris*, Hoffmann, Wild Chervil.—One of the commonest Umbelliferous plants; it has, in common with the other less frequent species of the genus, hairy leaves by which it is easily distinguished from hemlock (fig. 105, *b*). The involucels are not directed outwards, and the fruit is elongated.

*Chærophyllum temulum*, Linné, Rough Chervil, also has hairy leaves.

*Æthusa Cynapium*, Linné, Fool's Parsley.—The ultimate divisions of the leaves terminate in short brownish points. The under surface is dark green and glossy, and exhibits a more or less distinct network of veinlets (fig. 105, *c*). The umbel has no general involucre, and the bracts of the partial involucre are long and narrow. A transverse section of the leaf-stalk examined under the microscope shows on the upper surface a large central cell developed into a trichome.

*Oenanthe crocata*, Linné, has tuberculated roots; the juice of both stem and roots turns yellow when exposed to the air; very toxic.

*Cicuta virosa*, Linné, has narrow lanceolate, acute, serrated leaf-segments.

## COLTSFOOT

### (Herba Farfaræ)

**Source, &c.**—The coltsfoot, *Tussilago Farfara*, Linné (N.O. *Compositæ*), is a small herb with perennial creeping rhizome, abundant in Great Britain. The leaves appear much later than the flowering stems, and these two parts of the plant are therefore collected separately and usually sold separately. They have been long used as a domestic remedy for coughs.

**Description.**—The flowering stems, which appear in the early spring, are about 15 cm. high, and bear numerous small, narrow, alternate, reddish bracts; they are simple and hairy, many of the hairs terminating in dark, reddish brown glands. The flowerheads are terminal and solitary and surrounded by reddish involucral bracts; they possess numerous ray-florets with short, very narrow, bright yellow ligulate corollas, the disc-florets being less numerous and tubular. The receptacle is flat and naked, the fruits cylindrical, tapering towards the base and provided with an abundant pappus of white simple hairs.

The leaves appear much later than the flowers, and arise from separate shoots. They are radical and provided with long stalks; the lamina is cordate about 10 or 15 cm. in breadth, but occasionally much larger. The margin is sinuate-dentate, each tooth terminating in a hard brown point. The upper surface is dull greyish green and minutely wrinkled. Both surfaces are covered when young with loose, white, felted, woolly hairs, but those on the upper surface fall off as the leaf expands.

FIG. 106.—Coltsfoot leaf. About two-thirds natural size.

After the leaves have died down the shoot rests, and produces in the following spring a flowering stem, whilst other shoots develop leaves.

Neither leaves nor flowers have, when dried, any characteristic odour or taste.

The student should observe

(*a*) The *short, very narrow, ligulate corolla,*

(*b*) The *cylindrical fruit with pappus of white, simple bristles,*

(*c*) The *shape and margin of leaves and their hairy under surface,*

(*d*) The *characteristic glandular hairs on the peduncle.*

**Constituents.**—No active constituent is known. The drug contains a little tannin, and Bondurant (1887) found indications of a bitter glucoside which was not isolated.

**Uses.**—Coltsfoot is used as a domestic remedy for coughs.

## GRINDELIA

(Herba Grindeliæ)

**Source, &c.**—The drug consists of the dried leaves and flowering tops of *Grindelia camporum*, Greene (N.O. *Compositæ*), the common 'gum plant' of California, and is collected in quantity near San Francisco.

Part of the drug is said to be derived from *G. cuneifolia* and its variety *paludosa*; the leaves of this plant are cuneate and less coriaceous than those of *G. camporum.*

*G. robusta*, Nuttall, and *G. squarrosa* (Pursh), Dunal are closely allied species and were formerly regarded as yielding the commercial drug.

Before the flowerheads expand they secrete a white, sticky resin; in May and June the whole plant is resinous, and then the leaves and flowering tops are collected and dried.

**Description.**—The commercial drug consists of the slender upper part of the flowering stem, together with the flowerheads and a few leaves.

The stems, often 50 cm. in length, are rounded, yellow in colour, and smooth. They bear alternate, pale green leaves which, however, are easily broken off, and therefore frequently lie loose in the package.

The leaves are oblong or spathulate, 3 to 5 cm. long, with a serrate margin; they are rigid, brittle, smooth, sessile, and sometimes amplexicaul, and have a glabrous, minutely dotted surface.

The flowerheads are sub-conical, yellowish, hard and resinous, and are surrounded by several rows of lanceolate-acuminate, imbricated, recurved bracts. They contain numerous compressed fruits, each of which is crowned by a pappus consisting of two stiff, thick bristles, and when mature is bi-auriculate or more rarely uni-dentate at the summit.

The drug has a slight odour and a somewhat balsamic taste.

The student should observe

(*a*) The *stalky appearance of the drug*,
(*b*) The *resinous character, especially of the flowerheads*,
(*c*) The *pale green, rigid, brittle leaves*.

**Constituents.**—The chief constituents of grindelia are amorphous resins (up to 21 per cent.). These include a soft, greenish resin soluble in petroleum spirit and two dark coloured resins, one of which is soluble in ether. To these resins the activity of the drug appears to be due (Power and Tutin, 1905). The drug also contains a considerable quantity of lævoglucose, tannin (1·5 per cent.), and a trace of volatile oil. It leaves about 8 per cent. of ash.

Two glucosides resembling the glucosides of quillaja bark and senega root, an alkaloid and a crystallisable acid, robustic acid, are also said to be contained in the drug, but later researches have not confirmed these statements.

**Uses.**—The action of grindelia resembles that of atropine. The drug has the reputation of being almost a specific for certain forms of asthma, and has been recommended in cystitis and catarrh of the bladder. It is not much used in this country.

## LOBELIA

### (Herba Lobeliæ)

**Source, &c.**—Lobelia or Indian tobacco, *Lobelia inflata*, Linné (N.O. *Campanulaceæ*), an erect annual herb with acrid milky juice (latex), is distributed over the eastern States of North America, and cultivated for medicinal use in the States of New York and Massachusetts. The drug was a domestic medicine of the North American Indians, and was introduced into European practice about 1830. The plant is cut as soon as the lower capsules turn brown, and dried; it is sent into commerce either in bales or in oblong compressed packets.

**Description.**—The stem of the plant, which varies in colour from green to yellowish or often purplish, is hairy and winged in the upper part, but quadrangular, channelled, and nearly glabrous below. The leaves are alternate, broadly oval to ovate-lanceolate in outline, and vary in length from about 3 to 8 cm. They are sessile or shortly petiolate, and bear, especially on the veins of the under surface, scattered bristly hairs; the margin is irregularly crenate-dentate.

The flowers, which are arranged in a long leafy raceme, have a pale blue bilabiate corolla, a tubular calyx with five long spreading teeth and an inferior two-celled ovary. The latter develops into an ovoid, inflated, ten-ribbed fruit, about 7 or 8 mm. long, crowned with the remains of the calyx, and containing numerous minute, brown, oval-oblong seeds, which, under the lens, exhibit a beautifully reticulated surface.

In the commercial drug the green hairy and winged or purplish and channelled stems with alternate leaf-scars are easily found. The leaves are mostly in a fragmentary condition, but recognisable by their hairy under surface. The flowers are seldom to be found, but the characteristic inflated fruits containing the very minute seeds are always present.

The stems and leaves contain laticiferous vessels in the bast, a character that can be ascertained by microscopical examination only, but one that may afford valuable aid in identifying doubtful specimens.

The drug has a somewhat irritant odour, and, when chewed, an unpleasant acrid burning taste.

It is said that the plants are sometimes allowed to mature their seeds and are then thrashed, the seeds being sold separately, whilst the herb is pressed into packets. This would account for the frequent absence of the flowers from the drug, and for the presence of numerous capsules but comparatively few ripe seeds.

The student should observe

(*a*) The *hairy winged stem*,
(*b*) The *inflated fruits*,
(*c*) The *minute oblong reticulated seeds*.

**Constituents.**—That lobelia contains a toxic principle is beyond

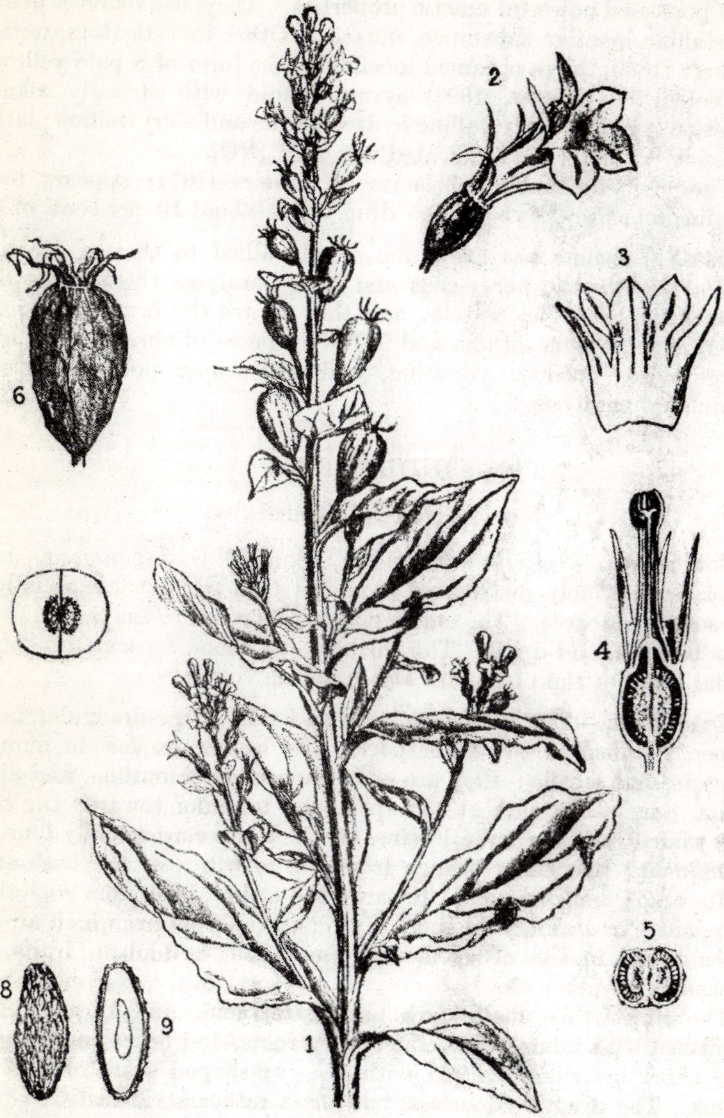


FIG. 107.—*Lobelia inflata*. 1, upper part of a plant ; 2, flower ; 3, corolla ; 4, flower, cut vertically ; 5, ovary, cut transversely ; 6, ripe fruit ; 7, the same, cut transversely ; 8, seed ; 9, the same cut longitudinally ; 1, slightly reduced ; 2–7, enlarged. 8, 9, highly magnified. (Bentley and Trimen.)

question, but of the nature of that principle there is some doubt. J. U. and C. J. Lloyd (1887) isolated a white, odourless, amorphous alkaloid, lobeline, which gave easily crystallisable salts with acids, and possessed powerful emetic properties. They also found a neutral crystalline inactive substance, inflatin. Other investigators, notably Siebert (1890) have obtained lobeline in the form of a pale yellowish or colourless (Dreser, 1890) syrupy liquid with strongly alkaline reaction, yielding a crystalline hydrochloride and a crystalline platinochloride. The formula indicated was $C_{18}H_{23}NO_2$.

The uncrystallisable lobelacrin of Enders (1871) appears to be lobeline lobelate (Lewis). The drug yields about 10 per cent. of ash.

**Uses.**—Lobeline has an action closely allied to that of nicotine; it first excites the nerve-cells and then paralyses them. The drug relaxes the bronchial muscles, and thus dilates the bronchioles; it is given in spasmodic asthma and in the dyspnœa of chronic bronchitis. Large doses produce vomiting, and may cause collapse through medullary paralysis.

## INDIAN PINK

### (Herba Spigeliæ)

**Source, &c.**—*Spigelia marilandica*, Linné (N.O. *Loganiaceæ*), is an erect herb, widely distributed over the United States, especially in the south and west. The entire plant is collected in the autumn, tied into bundles, and dried. The rhizome and roots, separated from the aerial parts of the plant, are also found in commerce.

**Description.**—The stem is smooth and simple, quadrangular in the upper, rounded in the lower part. The leaves are few in number, opposite and sessile; they are ovate-lanceolate in outline, and about 8 cm. long, acuminate at the apex, and tapering towards the base. The midrib, and the lateral veins, of which there are usually four, are prominent; the latter branch from the midrib near the base at an acute angle, curving round towards the apex. The stem sometimes terminates in a unilateral spike of brilliant red flowers, which are not often found in the drug, though the small two-lobed fruits, are occasionally met with.

The rhizome is small, dark brown, tortuous, and knotty. It is furnished with numerous, slender, wiry roots, and bears on the upper side short branches marked with the cup-shaped scars of previous stems. The drug is odourless, but has a rather acrid taste.

The student should observe

(*a*) The *opposite sessile leaves* with *prominent lateral veins*,
(*b*) The *small tortuous rhizome* with *cup-shaped scars*,
(*c*) The *smooth cylindrical stem*.

**Adulterations.**—Spigelia root has been frequently adulterated with the rhizome of *Ruellia ciliosa* (N.O. *Acanthaceæ*) which is longer, straighter, and thicker; the roots are less wiry; it contains sclerenchymatous cells and cystoliths in the bark.

**Constituents.**—The drug contains an acrid, bitter principle, soluble in water, and a liquid volatile alkaloid, spigeline (Dudley, 1881); the latter statement requires confirmation.

**Uses.**—Indian pink is used as a vermifuge. It possesses poisonous properties allied to those of gelsemium, depressing the action of the heart and respiration, and causing loss of muscular power when given in large doses.

## CHIRETTA

### (Chirata, Herba Chirettæ)

**Source, &c.**—Chiretta, *Swertia Chirata*, Hamilton (N.O. *Gentianeæ*), is an erect annual herb growing to a height of about a metre, and indigenous to the mountainous districts of Northern India. It has long been used by the Hindus, but was not introduced into European medicine till about 1830. The entire plant is collected when the flowering is well advanced, and made into bundles about a metre long, weighing nearly a kilogramme each, which are often compressed for exportation.

**Description.**—The stem, which attains about 6 mm. in thickness, is of a yellowish brown or purplish brown colour, glabrous, and slightly winged. The lower part is rounded, and exhibits, when cut longitudinally, a narrow wood enclosing a large, continuous, easily separable pith; the upper part of the stem produces in the axils of opposite leaves numerous slender, elongated, decussate branches which ramify further, bearing numerous fruits and occasional flowers.

The few leaves to be found are opposite and sessile, ovate or lanceolate in outline, acuminate, entire, and glabrous. The fruits are superior, ovoid and pointed; they are formed from two carpels, but are one-celled and contain numerous, minute, reticulated seeds. There are not many flowers present in the drug.

The tapering root attains about 10 cm. in length and 12 mm. in thickness at the crown, and is frequently oblique.

The drug has no marked odour, but all parts have an extremely bitter taste.

The student should observe

(*a*) The *purplish brown colour of the stem*,
(*b*) The *large continuous pith*,
(*c*) The *intensely bitter taste*,
(*d*) The *opposite leaves*,
(*e*) The *bicarpellary, unilocular fruits*.

The first three characters will suffice to distinguish the genuine drug from other species of *Swertia*, which sometimes are mixed with it or substituted for it, as well as from other substitutions that have been occasionally noticed. The last two are characteristic of the natural order *Gentianeæ* and are also useful in identifying the drug.

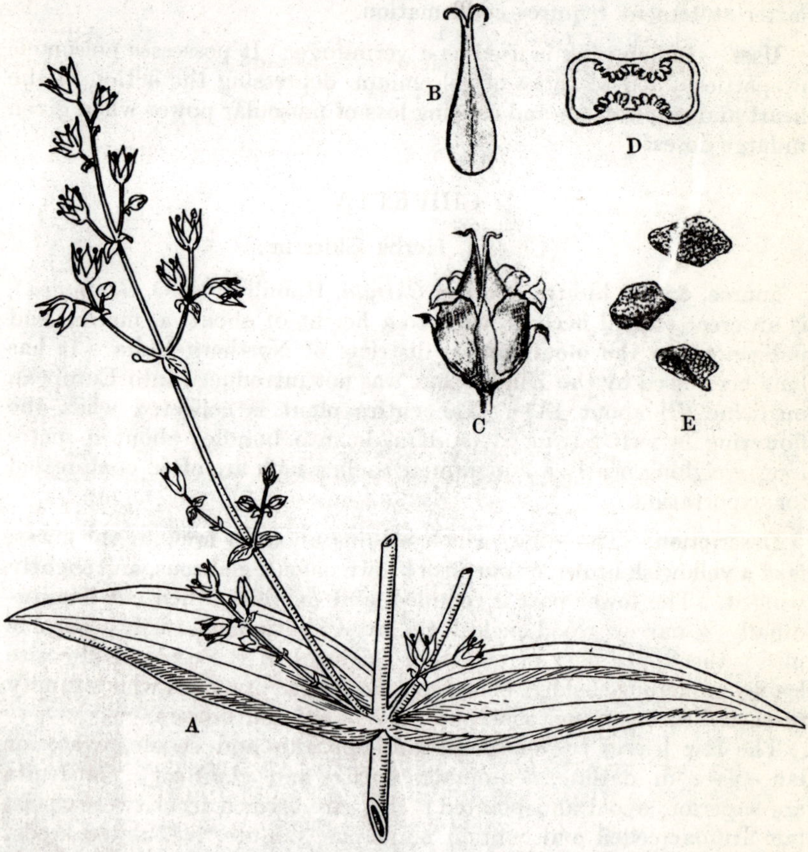


FIG. 108.—Chiretta. A, portion of flowering panicle; B, pistil; C, fruit with persistent calyx; D, transverse section of fruit; E, seeds. A, natural size. B, C, D, enlarged; E, highly magnified. (Bentley and Trimen.)

**Constituents.**—Chiretta contains two intensely bitter principles, ophelic acid and chiratin, both amorphous or indistinctly crystalline yellow substances.

Ophelic acid exists in the larger proportion; it has been obtained as a viscid yellow substance with a persistent bitter taste, soluble in water, alcohol, and ether. Chiratin is a yellow powder soluble in warm water, alcohol, and ether. Boiled with hydrochloric acid it yields ophelic acid and chiratogenin but no sugar, and is therefore not a glucoside.

**Uses.**—This drug, which has bitter and tonic properties, is highly esteemed in India, and much used as a tonic. In this country it is now but seldom prescribed, probably on account of the very disagreeable nature of its bitterness.

**Adulterations.**—The name chiretta being applied in India to a number of bitter plants, it is not surprising that other more or less similar bitter drugs are occasionally mixed with or substituted for the official chiretta. Sometimes, too, plants that resemble true chiretta in appearance but are much less bitter make their appearance under the name of chiretta. Among the substitutions and adulterations may be mentioned *Swertia angustifolia*, Buch-Hamilton ; *S. alata*, Royle ; *S. trichotoma*, Wallich, &c. ; *Andrographis paniculata*, Nees ; the root of *Rubia cordifolia*, Linné, &c. The large continuous pith, dark colour, and intensely bitter taste are sufficient to distinguish *S. Chirata* from other species of the same genus ; the opposite leaves and bicarpellary, unilocular fruits, from plants belonging to natural orders not possessing these characters.

*Japanese Chiretta*, derived from *S. chinensis*, Franchet, is a much smaller plant ; the stem varies from 10 to 35 cm. in length and 1 to 2 mm. in thickness, and is brown or purplish brown ; the root is straight or only slightly oblique ; it yields more alcoholic extract and is more bitter than *S. Chirata*.

## BITTER-SWEET

### (Stipites Dulcamaræ)

**Source, &c.**—The bitter-sweet or woody nightshade, *Solanum Dulcamara*, Linné (N.O. *Solanaceæ*), is a perennial shrubby plant with long climbing or straggling stems, common in England in hedges and thickets. It produces small purplish blue flowers and red berries. During the Middle Ages it was much used as a medicine ; now it is seldom employed.

The young stems are green and hairy, but as they grow older the hairs fall off, and they become quite smooth ; the stems are gathered when about two or three years old, cut into short pieces, and dried.

**Description.**—Bitter-sweet stems occur in commerce in short pieces about 6 mm. in diameter, of a light greenish or brownish yellow colour, bearing occasional alternate scars. They are nearly cylindrical, quite glabrous, and more or less longitudinally furrowed and wrinkled. The yellow, glossy, corky layer can easily be scraped off, and the green primary cortex disclosed. The wood is yellowish, and exhibits, in older pieces, distinct annual rings. The stems are usually hollow in the centre, the remains of the pith being attached to the inner surface of the ring of wood.

The drug has no marked odour ; the taste is at first bitter, but afterwards sweetish.

The student should observe

(*a*) The *alternate scars*,
(*b*) The *glabrous*, *glossy surface*, and *hollow centre*,
(*c*) The *bitter-sweet taste*.

Bitter-sweet has been reported to contain the amorphous glucoside dulcamarin to which the bitter-sweet taste is due, and also the glucosidal alkaloid solanine (which is also contained in the berries and young shoots of the potato, *S. tuberosum*, Linné) and other Solanaceous plants. According to Masson (1912) dulcamarin consists of two acid saponins, dulcamaretic acid which is non-glucosidal and dulcamaric acid which is glucosidal; the glucosidal alkaloid present resembles, but is not identical with, the solanine of potatoes, and is better termed solaceine.

**Uses.**—Bitter-sweet was formerly given in rheumatic and cutaneous affections, but is now seldom prescribed.

## WHITE HOREHOUND

### (Herba Marrubii)

**Source, &c.**—White horehound, *Marrubium vulgare*, Linné (N.O. *Labiatæ*), is an erect herbaceous plant with perennial root, widely distributed over Europe, but not very common in England. It is, however, cultivated as well as collected wild in various parts of the country for medicinal use, and is also imported from the south of France. The Romans esteemed horehound as one of the most valuable drugs, but at present it is used only as a domestic remedy and that not to any great extent. The plant is cut when in flower, and dried.

**Description.**—White horehound grows to a height of about 50 cm., and possesses a quadrangular, branching stem densely clothed with white woolly hairs. The leaves are from 3 to 5 cm. long, opposite and petiolate; the lower rounded-ovate, the upper ovate-acuminate, with dentate or dentate-crenate margin. They are much wrinkled, and both the upper and under surfaces, but especially the latter, are covered with white felted hairs.

The flowers are arranged in dense verticillasters in the axils of the upper leaves. The hairy calyx is provided with ten recurved, hooked teeth ; the whitish bilabiate corolla is characterised by its small, erect, cleft upper lip.

The drug has an agreeable though not powerful odour, and a somewhat aromatic but very bitter taste.

The student should observe

(*a*) The *dense felted hairs* on the stem and leaves,

(*b*) The *ten hooked teeth of the calyx.*

**Constituents.**—White horehound contains a bitter principle, marrubiin (0·25 per cent.), crystallising in colourless plates melting at 154°; it is almost insoluble in water, but soluble in alcohol and ether; a little volatile oil and tannin are also present.

**Uses.**—Horehound has tonic, stimulant, and expectorant properties; it is used as a domestic remedy for coughs and pulmonary complaints generally.

**Substitutes.**—The following plants have been substituted for *M. vulgare*:—

*M. peregrinum*, Linné; has ten straight calyx teeth.

*M. candidissimum*, Linné, has five nearly straight calyx teeth; is less aromatic and less bitter.

*Ballota nigra*, Linné (black horehound); is dark green; hairs rough, not woolly; five broadly ovate, pointed calyx teeth.

*B. hirsuta*, Bentham; calyx funnel-shaped, densely hairy, with ten broadly acute or mucronate teeth,

## INDIAN HEMP

### (Cannabis Indica)

**Source, &c.**—The drug known as Indian hemp, or Cannabis indica, is derived from *Cannăbis sativa*, Linné (N.O. *Urticaceæ*), cultivated in tropical districts of India.

The plant is an annual diœcious herb, indigenous to Central and Western Asia, but largely cultivated in temperate countries for its strong fibres (hemp) and its oily seed (hempseed), and in tropical countries also for the resinous secretion which it there produces. This secretion possesses powerful medicinal properties, but it is not produced by the plant when grown in temperate climates; on the other hand, the fibre of the plant under the latter condition is stronger than that of the tropical plant.

The cultivation of hemp for its seed and fibre dates from very remote periods. It was used as an intoxicant by the Persians and Arabians in the eleventh and twelfth centuries, and probably much earlier, but was not introduced into European medicine till comparatively recently (1838). It is largely grown in the districts to the north of Calcutta, and westward thence through Central India to Gujerat. Very good qualities of the drug are produced in Madras.

As the staminate plants produce but little resin they are eradicated and the pistillate plants pruned in order to produce flowering branches.

The tops of these are collected, allowed to wilt, and then pressed, by treading them under the feet, into more or less compact flattened masses. This forms the drug known on the London market as 'guaza'; it is exported from Bombay, formerly in rounded canvas bales called 'robbins,' but now more frequently in wooden cases.

In the Calcutta districts a somewhat different procedure is adopted. The tops that are collected are short, and, after being allowed to wilt, they are rolled under the feet of men, who support their weight by grasping a horizontal bamboo pole. In this way rounded, instead of flattened masses are produced. This variety is known as 'round' or Bengal ganjah (or guaza); it is more active than the flat Bombay drug, but is not seen in commerce in England, such quantities as are imported being immediately re-exported to the West Indies for the use of the coolies.

The larger leaves and young twigs collected and dried form the drug known as 'bhang'; it is chiefly consumed in India, made up into an electuary.

In addition to these drugs the resin secreted by the plant is also roughly collected by beating the plants on cloths to which the resin adheres; in this way a greenish brown soft mass is obtained, which may be freed from vegetable *débris* by warming and pressing through cloths. This resin is known as 'charas' or 'churrus,' and, like ganjah, is usually smoked. The Pharmacopœia guards against the use of the plants thus exhausted by limiting the drug to that 'from which the resin has not been removed.'

All these varieties of the drug are largely used in India for producing an agreeable form of intoxication.

**Description.**—In English commerce the drug usually occurs in flattened, compressed, rough masses of dull dusky green colour, harsh and resinous to the touch. Sometimes the tops are only 5 to 10 cm. in length, but more frequently they are 15 to 30 cm., the former being preferred. They consist of a straight stem with ascending branches, longitudinally furrowed, and bearing numerous, small, curved, appressed hairs and occasional glands. The hairs are characterised not only by their shape, but by being enlarged at the base and containing cystoliths.

The leaves on the drug (that is, the upper leaves of the plant) are alternate; the lower are digitate, and consist of three or five linear-lanceolate leaflets with distinctly serrated margins; the upper are simple. The upper part of the stem and branches bear numerous flowers or fruits.

The pistillate flowers are small and consist of a single ovary, surrounded by a perianth and supported by an ovate bract, beyond which two long brown stigmas, easily visible under a lens, protrude. The fruit is ovoid, slightly reticulated, and contains a single oily seed.

All parts of the plant, but particularly the bracts, stipules, and upper leaves, bear numerous hairs and large stalked glands, the latter secreting a quantity of viscid adhesive resin.

The drug has a powerful odour, but is almost devoid of taste.

The student should observe

(*a*) The *rough, dusky green masses,*
(*b*) The *curved appressed hairs,*
(*c*) The *linear-lanceolate leaflets.*

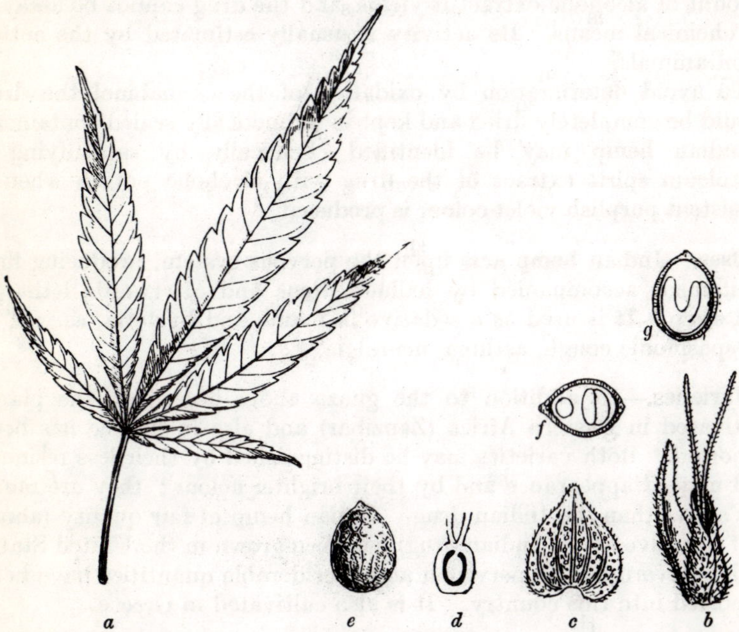


FIG. 109.—Indian Hemp. *a,* young leaf. *b,* pistillate flower enclosed by a bract and supported by a stipule. *c,* bract, showing glands. *d,* ovary, cut longitudinally. *e,* fruit (hemp-seed). *f* and *g,* fruit in transverse and longitudinal section (figures after Holmes).

It is advisable to free the drug from resin by macerating it in spirit, and then to soak it in water, when the leaves can easily be separated and the flowers dissected.

**Constituents.**—The resin secreted by the glands is undoubtedly the part of the plant that produces the narcotic effect of Indian hemp. This has been separated as a soft brown substance, cannabinone, from which, by distillation *in vacuo*, a viscous resin, cannabinol, which melts to an oily liquid when warmed, has been separated (Wood, Spivey, and Easterfield, 1896). Cannabinol produces a powerful narcotic action and is believed to be the active constituent of the drug.

14

On exposure to the air it resinifies rapidly and becomes less active. A similar change, due possibly to the presence of oxydase enzymes, takes place in the cannabinol in the drug. This rapid deterioration is well recognised in India, and persons addicted to the use of the drug refuse that which is more than a year old.

In addition to cannabinol and resin, the drug contains the alkaloid choline and traces of volatile oil; it yields from 10 to 18 (not less than 12·5, B.P. 1914) per cent. of alcoholic extract and about 15 per cent. of ash. The activity is not, however, accurately indicated by the amount of alcoholic extract it yields, and the drug cannot be assayed by chemical means. Its activity is usually estimated by the action upon animals.

To avoid deterioration by oxidation of the cannabinol the drug should be completely dried and kept in hermetically sealed containers.

Indian hemp may be identified chemically by saponifying a petroleum spirit extract of the drug with alcoholic potash when a persistent purplish violet colour is produced.

**Uses.**—Indian hemp acts upon the nervous system, producing first excitement accompanied by hallucinations and afterwards lethargy and sleep. It is used as a sedative in mania and hysteria, as well as for spasmodic cough, asthma, neuralgia, &c.

**Varieties.**—In addition to the guaza above described, the plant, cultivated in Eastern Africa (Zanzibar) and also in France has been imported. Both varieties may be distinguished by their less resinous and matted appearance and by their brighter colour; they are much less active than the Indian drug. Indian hemp of fair quality (about half as active as the Indian drug) has been grown in the United States under Government supervision and considerable quantities have been imported into this country. It is also cultivated in Greece.

## EUPHORBIA PILULIFERA

### (Herba Euphorbiæ)

**Source, &c.**—*Euphorbia pilulifera*, Linné (N.O. *Euphorbiaceæ*), is an annual herb indigenous to the hotter parts of India and occurring in all tropical countries. It has been used in India as a domestic remedy for ringworm, coughs, and other complaints.

The whole of the aerial part of the plant is collected whilst flowering and fruiting, and dried. It is exported chiefly from India.

**Description.**—The stem is slender, erect or ascending, 30 to 60 cm. high, cylindrical, and hispidly hairy. The leaves are from 2 to 4 cm. long, opposite, oblong-lanceolate, shortly petiolate and dentate or serrulate. They are dark green in colour, frequently blotched with red, and both upper and lower surfaces are hairy. The flowers are very minute, and crowded in dense axillary or terminal cymes about 1 cm. in diameter. The fruit is a minute, yellow, three-celled capsule, about 1 mm. in diameter, each of the three carpels being distinctly keeled, and containing a single, tetragonal, wrinkled seed.

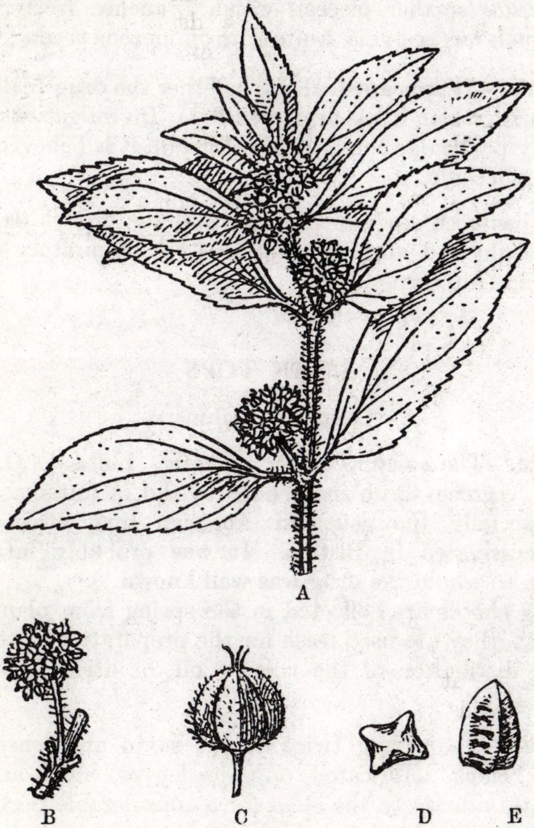


FIG. 110.—Euphorbia pilulifera. A, flowering top; B, inflorescence; C, fruit; D and E, end and side views of seed. C, D, E enlarged. (*Pharmaceutical Journal.*)

The most conspicuous part of the drug is frequently the slender rounded stems; the leaves are often much broken, but the fruits, which are very characteristic, can usually be found by examining the inflorescence with a lens. The drug is odourless, but has a bitter taste.

The student should observe

(*a*) The *slender cylindrical hairy stems*,

(*b*) The *minute three-celled fruits*, each carpel distinctly keeled;

and should compare the drug with

(i) *Indian pink*, which has larger leaves with prominent lateral veins and entire margins,
(ii) *Grindelia*, which has stouter yellowish stems bearing alternate leaf-scars,
(iii) *Chiretta* (smaller pieces), which branches freely, and bears much larger ovoid fruits with numerous seeds.

**Constituents.**—It appears doubtful whether the drug really possesses the efficacy with which it has been credited. Investigations have failed to isolate any particular constituent, although it is believed to contain a poisonous glucoside.

**Uses.**—It has been recommended for asthma, bronchitis, hay fever, whooping cough, and other affections of the respiratory organs, but has never come into general use.

## SAVIN TOPS

(Cacumina Sabinæ)

**Source, &c.**—The savin, *Junipĕrus Sabīna*, Linné (N.O. *Coniferæ*), is a small evergreen shrub indigenous to the mountains of southern Europe, especially the southern Austrian and Swiss Alps, and frequently cultivated in Britain. It was probably introduced by the Romans, to whom the drug was well known.

The young shoots are collected in the spring from plants grown in this country ; they are used fresh for the preparation of the ointment and for the distillation of the volatile oil, or dried for making the tincture.

**Description.**—The young twigs of the savin are densely covered with minute, thick, imbricated, opposite leaves, which are appressed and frequently adnate to the stem for a considerable portion of their length ; they are sessile, rhomboidal in shape, and bluntly pointed On the dorsal surface of each leaf a large oval depression is visible, corresponding to a large gland in the mesophyll of the leaf. Occasionally the small drooping baccate fruits may be found.

The lower leaves (and sometimes all) are more distant, linear, subulate, and more or less spreading.

The drug has a strong, characteristic odour and a bitter, acrid, unpleasant taste.

The student should observe

(*a*) The ***rhomboidal, appressed, bluntly pointed leaves,***
(*b*) The ***oil-glands on their dorsal surfaces,***
(*c*) The ***very characteristic odour.***

**Constituents.**—The principal constituent of savin is the volatile oil, of which it yields from 2 to 4 per cent. Tannin and resin are constituents of secondary importance.

The volatile oil (sp. gr. 0·910 to 0·930; O.R. + 40° to + 60°) consists chiefly of the alcohol sabinol and its acetic ester; it also contains cadinene, pinene, and probably other terpenes.

**Uses.**—Oil of savin is a powerful irritant, and is used externally to promote discharge from blisters; internally it acts as an emmenagogue, and is often used to procure abortion, frequently with fatal effect.

**Substitutes.**—The tops of *J. phœnicea*, Linné, have frequently been substituted in France for those of *J. Sabina.* They may be distinguished by the spiral arrangement of the leaves which also contain large sclerenchymatous cells not found in *J. Sabina.* The volatile oil which has a lower optical rotation (+ 4°), and contains less sabinol has been found in English commerce.

A B

FIG. 111.—Savin. A, natural size; B, magnified. (Moeller.)

## CARRAGEEN

### (Chondrus; Irish Moss)

**Source, &c.** — Irish moss, *Chondrus crispus*, Stackhouse (Class, *Algæ;* Subclass, *Rhodophyceæ;* Order, *Gigartinaceæ*), is not a moss, but a seaweed widely distributed on the northern shores of the Atlantic Ocean. It is collected for medicinal use on the north-western coast of Ireland, the coast of Brittany, and of Massachusetts in the United States. The plant, which is exceedingly variable in colour, in the amount of furcation, and in the width of the segments, grows just below low-water mark and is collected by raking. When fresh it varies in colour from green to dark purplish brown, but is bleached by exposing it to the sun and watering it, the colouring matter, which is soluble in water, being partly washed out and partly destroyed by the treatment. The bleaching is said to be completed by chemical means.

**Description.**—The drug, which consists of the entire plant, is usually yellowish white in colour, translucent, and cartilaginous or horny. The thallus is rounded near its attachment to the stones upon which it has grown, but becomes flattened in its upper part, repeatedly branching dichotomously, the branches being sometimes narrow, sometimes broadly wedge-shaped. It has a slight odour of seaweed

FIG. 112.—Irish Moss (*Chondrus crispus*). Three different forms of the plant, *a*, *b*, *c*; *a*, with reproductive organs, natural size. (Luerssen.)

and a mucilaginous saline taste. A decoction made with 20 times its weight of water solidifies on cooling to a jelly, which is not stained blue by iodine (distinction from Iceland moss jelly).

**Constituents.**—Stanford (1884) obtained from Irish moss 63 per cent. of a gelatinous substance which he termed carrageenin. This appears to be a complex mixture of carbohydrates; it yields by hydrolysis galactose, dextrose, and levulose (Sebor, 1900), and also hydroxymethylfurfuraldehyde (Müther and Tollens, 1904), and probably consists of galactan with glucosan and levulosan. The drug yields from 8 to 18 per cent. of ash, in which a little iodine can be detected, and about 7 per cent. of proteids.

**Uses.**—Irish moss possesses demulcent properties. It has been given in pulmonary complaints and for chronic diarrhœa, and has been used for the preparation of a nutritious jelly. It is also employed for various technical purposes, such as calico dressing, &c., as a cheap substitute for gum arabic.

**Varieties, &c.**—*Gigartina mamillosa*, J. G. Agardh, which is occasionally found mixed with Irish moss, may be distinguished by its stalked sporocarps, as may also *Gigartina pistillata*, Lamouroux, but the latter is a rare British seaweed and its presence would indicate that the drug had probably been collected in France; the sporocarps of *Chondrus crispus* are immersed in the thallus, the surface of which is slightly raised over them.

## BLADDERWRACK

### (Fucus vesiculosus)

**Source, &c.**—Bladderwrack, *Fucus vesiculosus*, Linné (Class, *Algæ;* Subclass, *Phæophyceæ;* Order, *Fucaceæ*), is a seaweed widely distributed on the shores of the Atlantic Ocean, and one of the commonest seaweeds on the coast of Great Britain, growing on the rocks and stones exposed at low tide. For medicinal use, the entire living plant should be collected and dried. That which is thrown up by the sea should be rejected, as by contact with the sea water it is liable to lose some of its constituents, which easily diffuse out from those cells the protoplasm of which has lost its vitality.

**Description.**—Bladderwrack, when fresh, is of a dark olive brown colour, changing, as the weed dries, to nearly black. The thallus of the plant is thin and flattened, about 2 cm. wide and up to a metre or more long. It has an entire margin, and branches dichotomously, bearing at intervals bladder-like swellings (air-vesicles) arranged mostly in pairs; some of the branches terminate in club-like enlargements, in which the organs of reproduction are situated. When slightly moist it is cartilaginous, but when quite dry, hard and brittle. It has a peculiar seaweedy odour and a disagreeable, mawkish, mucilaginous, and saline taste.

The student should observe

(*a*) The *entire margin*,

(*b*) The *vesicles usually in pairs.*

**Constituents.**—Bladderwrack yields to alkaline solutions a gelatine-like substance found in other seaweeds and termed algin. The plant yields by hydrolysis mannitol, fucose and arabinose.

It also contains mannite and a small proportion of iodine. The latter appears to exist in the form of an organic compound (Eschle, 1897), similar in its nature to the iodine compound found in thyroid glands. This is probably the most important medicinal constituent of the drug.

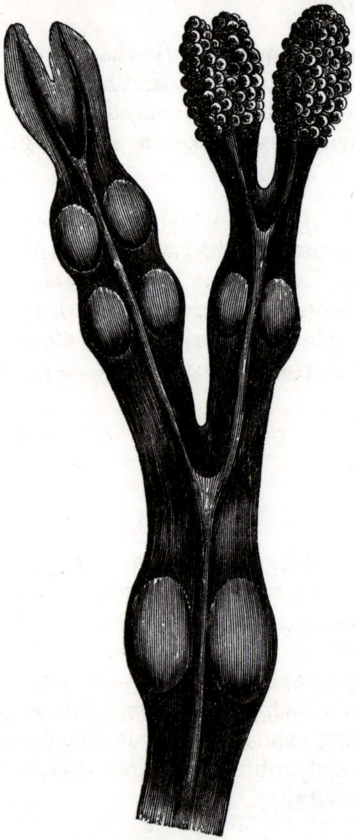

FIG. 113.—Bladderwrack. Branch with reproductive organs. Natural size. (Maisch.)

The dry plant yields about 0·0113 per cent. of iodine (v. Itallie, 1889), and 1·6 to 3·0 per cent. of ash, in which chlorides and bromides are present in addition to iodides.

Algin is prepared commercially by macerating seaweed in water and boiling the insoluble residue in a solution of sodium carbonate; the algin (or alginic acid) dissolves and is precipitated from the filtered solution by hydrochloric acid; the precipitate is redissolved in soda and the solution evaporated on glass plates; it is used as a calico dressing and for thickening the colours used in calico printing.

**Use.**—Preparations of bladderwrack have been used medicinally to reduce obesity.

**Varieties, &c.**—*Fucus serratus*, Linné, also a common seaweed occurring with *F. vesiculosus* on the rocky shores of Great Britain, has a serrated margin and no air-vesicles, whilst *F. nodosus*, Linné, has the vesicles usually single, not in pairs. The constituents of these seaweeds are probably similar to those of *F. vesiculosus*.

## AGAR-AGAR

### (Japanese Isinglass)

**Source, &c.**—The name agar-agar is now usually applied to a gelatinous substance prepared in Japan from various species of *Gelidium* viz. *G. elegans*, Kutz., *G. Amansii*, Kutz. (N.O. *Rhodophyceæ*). The name is also applied to various algæ, e.g. *Gracilaria lichenoides*, Greville (Ceylon agar-agar), *Eucheuma spinosum*, Agardh (Macassar agar-agar), &c.

Commercial agar-agar is made by boiling the carefully washed algæ with water, straining with pressure, cooling, cutting the jelly into strips and slowly drying.

**Description.**—Japanese agar-agar occurs in transparent strips about 6 dm. long and about as thick as a straw, or in yellowish white pieces about 3 dm. long, 2·5 to 25 mm. thick and more than 25 mm. wide. The decoction (1 to 200) solidifies on cooling to a jelly which melts at a much higher temperature than a similar jelly prepared from gelatin. In cold water it swells but does not dissolve.

**Constituents.**—Agar-agar consists chiefly of the carbohydrate gelose which is converted by boiling with dilute sulphuric acid into galactose. It always contains skeletons of diatoms (e.g. *Arachnoidiscus ornatus*) by which its presence in preserves can be readily detected.

**Uses.**—Agar-agar is largely used for the preparation of bacteriological culture media, the high melting-point (40°) of the jelly rendering it for this purpose particularly suitable. It passes through the intestinal canal almost unchanged, but absorbs water during its passage and promotes peristalsis.

## CETRARIA

### (Iceland Moss)

**Source, &c.**—Iceland moss, *Cetraria islandica*, Acharius (Class, *Fungi;* Subclass, *Ascomycetes;* Order, *Discomycetes*), is a foliaceous lichen indigenous to Great Britain and widely distributed over the northern hemisphere. It is collected chiefly in Sweden and Central Europe, growing usually amidst moss and grass on the lower mountain slopes.

**Description.**—The lichen consists of a very thin, erect, leafy thallus, branching fanlike into curled or flattened papery lobes about 6 mm. broad, fringed with minute projections each of which terminates in a spermagonium containing numerous spermatia capable of reproducing the plant. It is remarkably harsh and springy to the touch, tough when slightly moist, but brittle when quite dry. The upper surface is usually of a brownish or greenish brown colour; the under surface greyish and marked with numerous small, white, depressed spots. The apothecia are circular, of a dark reddish brown colour, and about 5 mm. in diameter; they are not often to be found on the plant. The drug is almost odourless, and has, when chewed, a mucilaginous, bitter taste. A decoction (1 to 20) yields on cooling a jelly which is stained blue by iodine

**Constituents.**—The principal constituent of Iceland moss is the carbohydrate lichenin, $C_6H_{10}O_5$, which is accompanied by isolichenin (dextrolichenin, Flückiger). Isolichenin is soluble in cold water, and behaves as a soluble modification of starch. Lichenin dissolves in boiling water, but the solution gelatinises on cooling. Both lichenin and isolichenin are converted by hydrolysis with dilute mineral acids into dextrose; as much as 72 per cent. of fermentable sugar can by this means be obtained from the drug.

FIG. 114.—Iceland Moss (*Cetraria islandica.*) Natural size. (Luerssen.)

Iceland moss also contains cetraric acid, $C_{20}H_{18}O_9$, a crystalline bitter substance, almost insoluble in water, but forming soluble acid salts with monovalent alkalies; hence the bitterness of the drug can be removed by soaking in dilute solution of sodium bicarbonate. Other constituents of the drug are bitter protocetraric acid and tasteless lichenostearic acid.

**Uses.**—The properties of Iceland moss are those of a bitter tonic and nutritive, but it is now seldom employed.

**Note.**—*Litmus* is a colouring matter obtained from various lichens, chiefly *Roccella tinctoria*, de Candolle (Cape Verde). *R. Montagnei*, Bel. (Madagascar), *Dendrographa leucophœa*, Darbish, &c. (N.O. *Discomycetes*). The method of preparation is guarded as a trade secret, but it appears to depend mainly upon the slow fermentation of the soaked and ground lichen in the presence of ammonium and potassium carbonates. A red colour is first produced which gradually changes to blue. The blue liquid is drawn off and evaporated, with the addition of chalk and gypsum; the mass is then cut into small rectangular cakes and dried. The chief constituents are erythrolitmin and azolitmin, together with erythrolein and spaniolitmin. The lichens themselves contain lecanoric acid, erythrin and orcin. By the action of alkalies, these yield orsellinic acid. Orsellinic acid by further change yields orsin, from which, by oxidation in the presence of ammonia the colouring matters are produced.

*Cudbear* (Persio) is a reddish colouring matter prepared by an analogous method from the same lichens.

## ERGOT

**Source, &c.**—Ergot is the sclerotium of *Claviceps purpūrea*, Tulasne, (N.O. *Pyrenomycetes*) originating in the ovary of the rye, *Secāle cereāle*, Linné (N.O. *Gramineæ*).

In the spring or early summer the spores of *Claviceps purpurea* are carried by the wind on to the flowers of various Graminaceous plants, in the case under consideration on to those of the rye. Here they germinate and produce colourless hyphæ, which envelop, with the exception of the apex, the very young ovary, and penetrate the outer part of the pericarp, covering it with a soft white felted mass, which gradually takes the place of the ovary, and is known as the *sphacelia*. During this period a saccharine secretion, 'honeydew,' is produced by the hyphæ, and at the same time numbers of conidia are formed, thus contributing to the further dissemination of the fungus by means of the insects attracted to the honeydew, a small weevil which feeds on the saccharine secretion being especially active. After the felted mass has reached its full development, the sclerotium is gradually produced at its base by the hyphæ forming a dense compact mass instead of a loose felt. The sclerotium grows and finally projects from the ear of rye (which has by this time ripened), bearing on its apex the remains of the felt.

FIG. 115.—Ear of Rye bearing two fully developed ergots. Natural size. (Luerssen.)

In this compact form the fungus is able to

resist the damp and cold of the autumn and winter. In the spring it produces small stalk-like projections (stromata, fig. 116, *b*) each of which is terminated by a globular head in which numberless spores (ascospores) are developed, and these, carried by the wind on to the flowers of the rye, complete the cycle.

Ergot is collected chiefly in Spain, Russia, Germany, and Austria. It is sometimes picked grain by grain by hand, or more usually separated from the rye after it has been thrashed by a machine specially designed for that purpose. Its exclusion from the grain is, in countries in which rye-bread forms the staple food of the people (as in Russia), a matter of the utmost importance, as the continued consumption of bread containing ergot has led to widespread disease (ergotism). The epidemics known as 'ignis sacer' and 'St. Anthony's Fire' were traced in 1747 to ergot. The chief centres for the collection of the drug in Russia are Tomsk, Omsk, and Viatka.

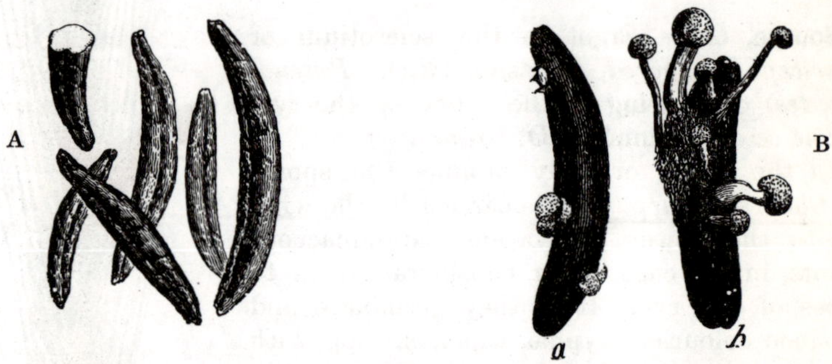


FIG. 116.—A, Ergot of Rye (Planchon and Collin). B, Ergot of Rye germinating (Luerssen.)

**Description.**—The grains of ergot are usually about 1·5 to 3·5 cm. in length, and of a very dark violet or nearly black colour. They are slender and curved, tapering towards both ends and rounded or obscurely triangular in section; to one extremity a small whitish appendage (remains of the sphacelia) is often attached. They are longitudinally furrowed, especially on the concave side, and often bear as well numerous small transverse fissures. Ergot breaks easily with a very short fracture, and is whitish or pinkish white within, but does not exhibit any definite structure when examined with a lens. It has a characteristic disagreeable odour and an unpleasant mawkish taste.

It should be thoroughly dried and kept in air-tight vessels to protect it from deterioration by damp and by the attack of insect pests.

**Constituents.**—The most important constituent of ergot is ergotoxine (Barger and Dale, 1906, hydroergotinine, Kraft, 1906), $C_{35}H_{41}N_5O_6$, an amorphous alkaloid yielding crystalline salts. This substance contracts the pregnant uterus, raises the blood pressure and darkens the cock'scomb, all characteristic effects of ergot, but does not produce convulsions. It is readily convertible into its lactone ergotinine (Kraft) which is crystalline but inactive. It is soluble in alcohol but insoluble in water.

A second active constituent is tyrosamine (Barger, 1909) also called ergotamine. It is parahydroxyphenylethylamine, a crystalline base contained in putrid meat (Barger and Walpole, 1909) and derivable from tyrosine, a constituent of muscular tissue by loss of $CO_2$. This base is soluble in water and like ergotoxine causes contraction of the uterus and rise in blood pressure. It is more stable than ergotoxine and is the chief active constituent in the official preparations of ergot.

A third active constituent is histamine (β-aminoethylglyoxaline) or β-iminazolylethylamine; also called ergotidine) a based erivable from histidine, also a constituent of muscle, by loss of CO, just as tyrosamine is derivable from tyrosine. Histamine is exceedingly potent, soluble in water, and causes contraction of the non-pregnant uterus but lowers the blood pressure.

To these three bases the activity of ergot is mainly to be ascribed. Other but much weaker constituents are isoamylamine (derivable from leucine) and agmatine (derivable from arginine).

Ergot also contains ergothioneine a base containing sulphur and of doubtful activity; fixed oil (about 30 per cent.); scleroxanthin and sclerocrystallin (crystalline yellow substances); sclererythrin (reddish violet colouring matter); ergotinic acid, an ill-defined organic acid which has no action on the uterus; sphacelinic acid, a resinous acid which contracts the blood vessels and also produces convulsions; a glucoside, clavicepsin.

Cornutine (Keller), secalintoxin and sphacelotoxin (Jacobj) are mixtures of ergotoxine with ergotinine. Chrysotoxin and ergochrysin (Jacobj) are also impure substances. Ergotinine (Tanret), sclerocrystallin (Podwissotzky), secalin (Jacobj), and picrosclerotin (Dragendorff) are impure ergotinine.

**Varieties.**—The two chief commercial varieties are Spanish and Russian, the former being distinguished by its larger size.

Ergot is also collected in Germany, Austria, in the Canary Islands (occasionally), and in this country (rarely).

Ergots will grow on many Graminaceous plants, e.g. *Triticum sativum*, Lamark, *Festuca elatior*, Linné, *F. arundinacea*, Vill. Prosp., *Lolium perenne*, Linné, *L. multiflorum*, Lamark, *Poa annua*, Linné. Ergots grown on *Festuca* and on *Lolium* appear to be specially active.

**Assay.**—As there is no known method of determining ergotoxine no chemical assay is at present available. It is, however, possible to obtain an insight into the relative activity of preparations of ergot by injecting them into fowls and observing the extent of gangrene produced in the combs and wattles. The determination of the inactive ergotinine which is frequently carried out is, of course, valueless.

**Action and Uses.**—The chief action of ergot is the stimulation of plain muscle, especially of the uterus and of the arterioles in peripheral parts of the body; it is employed to excite or increase uterine contraction and to control uterine hæmorrhage.

## HERBS IN LESS FREQUENT USE

**Pulsatilla** (Pasque Flower).—The leaves, stems and flowers of *Anemone Pulsatilla*, Linné (N.O. *Ranunculaceæ*). The plant produces a rosette of stalked leaves and an erect scape bearing a whorl of three bracteoles which form an involucre below the large, single, terminal flower. Leaves tripinnate, ultimate lobes linear, pointed, covered with silky hairs; petioles often purplish; stem hairy; bracteoles much divided; sepals six, light purple, silky externally; odourless; very acrid when fresh, becoming less so on keeping. Contains a crystalline vesicant substance which decomposes into acrid, crystalline anemonin and tasteless crystalline isoanemonic acid. Irritant: used in dysmenorrhœa and amenorrhœa.

**Pilewort** (Lesser Celandine).—The leaves and flowers of *Ranunculus Ficaria*, Linné (N.O. *Ranunculaceæ*), indigenous, very common. Leaves stalked, broadly ovate or reniform, glabrous, crenate, cordate at base. Flowers on long peduncles, three sepals, eight to twelve bright yellow petals, each with a nectary at the base. Several of the roots enlarge to tubercles. Taste, when fresh, acrid. Used in the fresh state as a remedy for hæmorrhoids. It probably contains traces of anemonin.

**Purging Flax.**—The leaves, stems, and flowers of *Linum catharticum*, Linné (N.O. *Lineæ*), indigenous. Stems very slender, 10 to 20 cm. long; leaves small, opposite, obovate or oblong, entire; flowers very small, white, on long slender pedicels; sepals pointed; petals obovate, minute. Taste acrid and bitter. Herb (and also seeds) used as a purgative. Active constituent unknown, probably glucosidal, yielding by hydrolysis colourless, crystalline, inactive, linin.

**Wormwood.**—The leaves and flowering tops of *Artemisia Absinthium*, Linné (N.O. *Compositæ*), Northern Asia and Europe. Stems 30 to 90 cm. high; leaves alternate, rounded-oval, bi- or tri-pinnate, segments lanceolate, ash grey on the upper, greyish green on the under surface, very hairy and glandular; hairs consisting of a long spindleshaped cell supported horizontally at its centre on a short 3-celled pedicel; flowerheads small, florets pale yellow, tubular. Odour aromatic; taste aromatic and bitter. Contains volatile oil (1 per cent.) and a crystalline bitter glucoside, anabsinthin (absinthin). Used in the preparation of the French liqueur absinthe.

**Lettuce.**—The fresh, flowering herb, *Lactuca virosa*, Linné (N.O. *Compositæ*) indigenous and cultivated. Stem 1 metre or more high, prickly near the base, producing leafy panicles of small, pale yellow flowers. Leaves lanceolate to broadly oblong, coarsely toothed or lobed with prickles on the undersurface of midrib and lateral veins; leaves and stem exude a white latex when incised; contains traces of hyoscyamine (see also 'Lactucarium').

**Centaury.**—The flowering herb of *Erythræa Centaurium*, Persoon (N.O. *Gentianeæ*) indigenous. Stem 20 to 30 cm. high; leaves opposite, glabrous, sessile, oblong, with 3 to 5 distinct veins; flowers in cymes; tube of corolla twice as long as that of the calyx, 5-partite, rose-pink, taste bitter. Contains crystalline erythro-centaurin. Used as a tonic.

# SECTION VI

# WOODS

Under this heading are grouped together a number of drugs consisting of the wood of the trunk, branch, or root of trees. Medicinal woods are derived from dicotyledonous plants, and the following structural details which may be of assistance in identifying woods apply to those of dicotyledonous origin.

1. *Medullary Rays.*—These are best examined on a smooth transverse and tangential section. On a transverse section they appear as fine, continuous lines radiating from the centre to the periphery. They vary in width in different woods and also frequently in the same wood; the wide ones are usually visible to the naked eye, but they are best examined under the lens. The distance between the rays also varies with the different woods, a variation which may be indicated by stating the number of rays that occur in 5 mm. of the wood.

2. *Vessels.*—The size and distribution of the vessels also afford valuable information. The vessels in the spring wood are usually larger and more numerous than those in the autumn wood, and hence more or less well-defined concentric lines are produced; in some woods, however, the distribution is irregularly radial. They are often large enough to be visible to the naked eye (over 0·1 mm.), frequently, however, scarcely perceptible even under the lens (under 0·02 mm.); they may occur singly or be arranged in radial or tangential groups of two or more.

3. *Parenchyma.*—The parenchymatous tissue of the wood may also exhibit a characteristic distribution usually assuming the form of more or less distinct concentric circles (false annual rings).

The woods dealt with in the succeeding section are easily distinguished from one another by their colour, odour, or taste, but the student should examine them carefully with the lens, as above described, in order to distinguish them from other similar woods that might be substituted for them.

## BERBERIS

### (Berberis)

**Source, &c.**—Berberis is the dried stem of *Berběris aristata*, de Candolle (N.O. *Berberidĕæ*) a shrub indigenous to India and Ceylon.

**Description.**—The drug occurs in slightly undulating pieces, from 2·5 to 5 cm. in diameter, yellowish brown and striated externally, bright yellow internally. The transverse section exhibits a narrow brown cork, a broader yellowish-brown bast, a large yellow wood and small pith. The bast is traversed by conspicuous, yellow medullary rays, and is laminated. The wood exhibits numerous distinct medullary rays and small vessels. It has a bitter taste.

**Constituents.**—The chief constituents of berberis are the alkaloid berberine (see p. 235) and other substances of alkaloidal nature not yet fully investigated, tannin, and resin.

**Uses.**—It is used in India as a bitter tonic in intermittent fevers.

## GUAIACUM WOOD

### (Lignum Guaiaci, Lignum Vitæ)

**Source, &c.**—The lignum vitæ or guaiacum wood of commerce is derived from *Guaiacum officinale*, Linné, and *G. sanctum*, Linné (N.O. *Zygophyllĕæ*), both evergreen trees, the former a native of the West Indian Islands and the north coast of South America, the latter of southern Florida and the Bahamas. Both occur in Cuba and Hayti, whence the wood is largely exported. The Spaniards became acquainted with the drug when they conquered San Domingo ; it was soon brought to Europe, where it acquired an immense reputation in the sixteenth century as a cure for syphilis and certain other diseases, the resin extracted from the trunk being introduced subsequently.

The trees are felled, the bark stripped off, and the wood exported in logs varying commonly from 1 to 2 metres in length and from 10 to 50 cm. in thickness.

**Description.**—The logs are exceedingly hard, heavy, and compact, and consist of a dark greenish brown heartwood surrounded by a yellowish sapwood. The exterior is yellowish brown in colour and either smooth or furrowed, the furrows being oblique and varying in direction (corresponding to the arrangement of the wood fibres in the wood).

The medullary rays, which can be seen under a lens, are narrow, straight, and closely approximated ; the vessels are distinct, usually single, and arranged in concentric zones.

The wood splits very irregularly, owing to the oblique and varying course taken by the wood fibres. Both these and the vessels—in fact, all the elements of the heartwood—are filled with a dark resin, which is sometimes also found in cavities in the trunk. It exhales, when warmed, a faint aromatic odour, recalling benzoin, and has, when chewed, an acrid taste. Its toughness and hardness render it valuable for many technical purposes, it being used in making blocks, pulleys, &c. The chips or turnings, in which state it is usually employed in pharmacy, should consist of the dark-coloured heart wood alone, but they are frequently mixed with the pale yellow sapwood from which they may be separated by a 30 per cent. solution of sodium chloride in which the splinters of heartwood sink. The sapwood contains about 3 per cent. of resin, not identical with the resin of the heartwood.

The student should observe

(a) The *pale colour of the sapwood*,
(b) The *dark greenish brown heartwood*,
(c) The *distribution of the vessels*.

**Microscopical Characters.**—The vessels are large and isolated, often extending from one medullary ray to the next. The medullary rays are one cell wide and three to six cells high. The wood-fibres are abundant and have very thick walls. The wood parenchyma occurs in narrow bands and some of the cells contain prismatic crystals of calcium oxalate.

**Constituents.**—The heartwood of guaiacum contains between 20 and 25 per cent. of resin, which has been found to consist of $\alpha$- and $\beta$-guaiaconic acids, guaiaretic acid, and guaiacic acid. Guaiaconic acid is converted by oxidising agents into guaiac blue and accordingly tincture of guaiacum wood strikes a deep blue colour with dilute solution of ferric chloride, a reaction which is useful in identifying the wood. (Compare ' Guaiacum Resin.')

Guaiacum wood also contains guaiacsaponic acid and guaiac saponin, two non-toxic bodies belonging to the class of saponins ; they are present in larger quantity in the sapwood than in the heartwood. Guaiaguttin, which resembles gutta-percha, is also present.

Guaiaconic acid, though a characteristic constituent of the wood, has been found in other woods (e.g. species of *Bulnesia* and *Porlieria*), and its presence therefore is not an infallible diagnostic character of guaiacum wood.

**Uses.**—Guaiacum has a local stimulant action which is sometimes useful in sore throat. The resin is used in chronic gout and rheumatism, whilst the wood is an ingredient in the compound concentrated solution of sarsaparilla, which is used as an alterative in syphilis.

**Substitutes, &c.**—Commercial guaiacum wood turnings frequently contain the sapwood as well as chips of other woods. These may be detected by their floating in brine and by the amount of alcoholic extract which should not be less than 22 per cent.

## QUASSIA WOOD

### (Lignum Quassiæ)

**Source, &c.**—The official quassiawood is derived from *Picræna excelsa*, Lindley (N.O. *Simarubĕæ*), a tree of moderate size, common on the plains and lower mountains of Jamaica.

Quassia wood was introduced into medicine about the middle of the eighteenth century, but was then obtained from *Quassia amara*, Linné, a smaller tree than *Picræna excelsa*, and indigenous to the north of South America, whence its usual, distinctive name of Surinam quassia. The wood of *P. excelsa* was found to possess the same properties, and has been substituted for it in England, but Surinam quassia remains official on the Continent.

The trunks and larger branches with the bark attached are exported in logs and billets about 1½ to 2 metres in length and 20 to 30 cm. in diameter.

**Description.**—The logs of Jamaica quassia wood are commonly covered with a thin, dark grey or nearly black bark. The wood is pale yellow in colour, light, rather dense, and easily split. When the smoothed transverse section is moistened and examined with a lens numerous, narrow medullary rays can be seen traversing somewhat irregular concentric rings. The latter are not annual rings, but are produced by the distribution, in more or less concentric zones, of bands of parenchyma (false annual rings). The vessels are usually in groups of two or three, and frequently extend from one medullary ray to the next.

Not unfrequently dark grey patches are visible in the wood; they are caused by a fungus, the dark hyphæ of which penetrate the wood through the cells of the medullary rays and wood parenchyma.

The wood has no odour, but the taste is purely and intensely bitter. For use in pharmacy it is usually cut across the grain by large knives like chisels and the chips kiln-dried to prevent them becoming mouldy, as the wood often contains much moisture.

The student should observe

(*a*) The *pale colour* and *intensely bitter taste* of the drug,
(*b*) The *distribution of the vessels*.

**Constituents.**—According to Massute (1890), Jamaica quassia wood contains two closely allied crystalline bitter principles, *a-picrasmin*, $C_{35}H_{46}O_{10}$ (m.pt. 204°), and *β-picrasmin*, $C_{36}H_{48}O_{10}$

(m.pt. 209° to 212°), which are to be regarded as the active constituents; it contains, further, a very small amount of a crystalline bitter principle melting at 234°, as well as a minute quantity of a yellow crystalline substance which exhibits in acidified alcohol a magnificent blue fluorescence. It contains no tannin.

**Use.**—Quassia is used as a pure bitter tonic. It is also largely used for the preparation of horticultural insecticides and non-poisonous fly papers. Cups turned from the wood were formerly in use under the name of bitter cups.

**Substitutes.**—*Surinam quassia* is usually in smaller billets than Jamaica quassia; it is best distinguished by its microscopical characters, most of the medullary rays being one cell wide, whereas in Jamaica quassia they are two or three cells wide; the wood parenchyma is free from crystals of calcium oxalate, which are contained in Jamaica quassia. The width of the medullary rays and the presence of calcium oxalate are therefore specified in the Pharmacopœia in order to exclude Surinam quassia; for although the physiological action is presumably the same, it is desirable for the sake of uniformity to use one variety of the wood only. The bitter principles contained in Surinam quassia are distinct from those of Jamaica quassia, and have been called 'quassiins' (Massute, 1890).

Quassia chips that have been exhausted and re-dried yield less aqueous extract (2·7 as compared with 6·3 to 8·6 from genuine wood) and are less bitter.

## LOGWOOD

### (Lignum Hæmatoxyli)

**Source, &c.**—The logwood tree, *Hæmatoxylon campechianum*, Linné (N.O. *Leguminosæ*), is a tree of moderate size, indigenous to Central America, but naturalised in the West Indian Islands. The use of the wood as a dye was probably known to the Mexicans, for its introduction into Europe followed closely on the conquest of Mexico by Cortés; in 1746 it was introduced into the London Pharmacopœia as a mild astringent, but is now not much used medicinally.

The wood is exported in the form of billets and logs from which both bark and sapwood have been separated; the heartwood alone contains the colouring and astringent principles.

**Description.**—The logwood of commerce consists of the heartwood of the tree, and is imported in logs and billets from 6 to 15 cm. in diameter and 1 to 2 or more metres in length. Externally these are of a dull dark orange or purplish red colour, internally they are reddish brown ; they are hard and heavy, but easily split. The transverse section exhibits under the lens very narrow and closely approximated medullary rays and narrow concentric dark zones alternating with paler ones, a difference due to the colouring matter secreted in the former. The freshly felled trunk wood is pale in colour ; the gradual darkening is due to oxidation of the hæmatoxylin to hæmatein.

The odour of the chips is faint but pleasant, recalling that of violets ; the taste is sweetish and astringent. It imparts a violet colour to dilute aqueous solutions of caustic alkalies.

Logwood is cut by suitable machinery into chips or turnings of a reddish brown colour, and these are usually subjected to a process of fermentation ('ageing'). They are well moistened, heaped together, and exposed to the air for a period of from four to six weeks, the heaps being frequently turned over ; they are then dried. By this process the chips darken in colour and exhibit patches of a dark beetle-green lustre. The unfermented chips are alone official. Fermentation of logwood is not much practised, as it has been found that the oxidation of the hæmatoxylin to hæmatein that is necessary in the dyeing process can be effected by using an oxidising mordant (potassium dichromate).

The student should observe

(a) The *reddish brown colour* of the chips,

(b) The *violet colour they produce in contact with alkaline solutions.*

**Constituents.**—The principal constituent of unfermented logwood is hæmatoxylin, $C_{16}H_{14}O_6,3H_2O$, of which it contains about 10 per cent. This, when pure, forms colourless crystals which acquire a reddish colour on exposure to the air. It is sparingly soluble in water, but dissolves readily with purple coloration in solutions of the caustic or carbonated alkalies and ammonia. The latter solution absorbs oxygen from the air, forming hæmatein-ammonia, from which acids separate hæmatein, $C_{10}H_{12}O_6$, a dark violet crystalline body with green metallic lustre. This change takes place during the fermentation of logwood, the hæmatoxylin being partially converted into hæmatein. Hæmatoxylin is hydroxybrasilin (see below) and haematein hydroxybrasilein.

Logwood contains, further, tannin, resin, quercetin, and a trace of volatile oil. The sweetish taste is produced by the hæmatoxylin, the astringency by the tannin.

**Varieties.**—Several commercial varieties of the wood are recognised, that from Yucatan (Campeachy) being considered the best, while British Honduras and San Domingo also furnish wood of good quality ; Jamaica logwood is less esteemed as it is inferior in colouring power. Bastard logwood is the name given to a variety of the wood of paler colour and much less colouring power than the genuine ; it appears to be derived from a variety of *H. campechianum.*

**Uses.**—Logwood is largely used as a dye and in the manufacture of inks. Much of it is converted for these purposes into an aqueous extract containing about 50 per cent. of hæmatoxylin and from 10 to 30 per cent. of hæmatein. Medicinally, it is employed occasionally as a mild astringent.

## SAPPAN

### (Sappan)

**Source, &c.**—Sappan is the heartwood of *Caesalpinia Sappan*, Linné (N.O. *Leguminosæ*), a tree indigenous to India and the Malay Archipelago.

**Description.**—The wood occurs in hard, heavy billets of varying size, usually 2 to 6 inches or more in diameter, consisting of the orange red heartwood to which a little of the whitish sapwood still adheres. Sometimes the wood is reduced to orange red chips. The transverse section exhibits well-marked concentric rings, numerous narrow medullary rays and large vessels. It has no odour, but a slightly astringent taste. The tincture and decoction have a reddish colour which changes to carmine on the addition of solution of sodium hydroxide (compare Logwood).

**Constituents.**—Sappan wood contains brasilin (see below).

**Note.**—*Brazil wood* is the heartwood of *Caesalpinia brasiliensis*, Linné, Brazil, Guiana, West Indies, &c., and other species. Outer surface (after exposure to the air) dark reddish or nearly black, the freshly cut surface reddish brown; vessels smaller than in sappan; decoction coloured carmine by alkalies; contains brasilin, $C_{16}H_{14}O_5$, yellowish, which in alkaline solution readily oxidises to brasilein, $C_{16}H_{12}O_6,H_2O$, deep red.

*Peachwood* is the heartwood of *C. echinata*, Lamark and also contains brasilin.

*Fustic* from *Chlorophora tinctoria*, Gaudich, South America, contains a yellow colouring matter morin or moric acid, $C_{15}H_{10}O_7$, $2H_2O$, together with maclurin; with alum mordants it dyes intense yellow.

*Hungarian fustic* is obtained from *Rhus Cotinus*, Linné (N.O. *Anacardiaceæ*) contains a glucoside fustin which yields by hydrolysis the yellow colouring substance, fisetin, $C_{15}H_{10}O_6,4H_2O$, a flavonol derivative closely allied to morin.

## RED SANDERS WOOD

### (Red Sandal Wood, Lignum Pterocarpi)

**Source, &c.**—Red sanders wood is the heartwood of *Pterocarpus santalinus*, Linné filius (N.O. *Leguminosæ*), a small tree indigenous to Southern India and the Philippine Islands. During the Middle Ages it was classed as a spice and used for culinary purposes, and at one time it was supposed to possess medicinal properties; it is now employed solely as a colouring agent.

The tree is regularly cultivated in districts situated to the west and north-west of Madras. The felling of the trees is controlled by the inspectors of forests, and the revenue thus obtained is considerable. The dark red heartwood alone is exported.

**Description**—Red sanders wood is imported in irregular often branching billets, commonly 7 to 15 cm. thick and about a metre long, deprived of both the rugged bark and the pale sapwood. It is of a deep blood-red colour both internally and externally, the transverse section exhibiting alternating darker and lighter zones. The medullary rays are just visible under the lens ; the vessels are large, mostly isolated, and connected by fine, bright red lines (wood parenchyma). The wood is very hard, but can be easily split. It is inodorous, has but a very slight astringent taste.

The red resinous colouring matter is produced in the form of droplets in all the elements of the heartwood (wood fibres, wood parenchyma, and vessels). It is readily soluble in alcohol, but only very sparingly in water, and in this respect it differs essentially from logwood, which freely yields its colouring matter to water.

In pharmacy the wood is usually employed in the form of small, hard, splintery raspings of a dull purplish red colour.

The student should observe

(*a*) The *dull purplish red colour*,
(*b*) The *hardness*,
(*c*) The *slight solubility of the colouring matter in water*.

**Constituents.**—The chief constituent of red sanders wood is santalin (santalic acid), $C_{15}H_{14}O_5$, which has been obtained in blood-red microscopic needles melting at 226°. It is insoluble in water, but yields a blood-red solution with alcohol, yellow with ether, and violet with caustic alkalies. Santal (Weidel, 1869), pterocarpin, and homopterocarpin (Cazeneuve and Hugouneng, 1887, 1890) are colourless crystalline substances also contained in red sanders wood.

**Use.**—Red sanders wood is used solely as a colouring agent, and even in this respect it has only a limited application, as the colouring matter requires alcohol to dissolve it, and is precipitated when the alcoholic solution is diluted with water.

## SANDAL WOOD

### (Yellow Sandal Wood, Lignum Santali)

**Source, &c.**—The yellow sandal wood tree, *Santălum album*, Linné (N.O. *Santalăceæ*), is a small tree distributed over India and the Malay Archipelago. It is found especially in Southern India, from Mysore to Madras, and is regularly cultivated there for the sake of its wood, which has long been used in India in religious ceremonies. It was known in Europe in the eleventh century, and has been used medicinally since the middle of the fifteenth century.

The yellow sandal wood tree is a Government monopoly in Mysore, whence nearly all the sandal wood of commerce is obtained. The tree is plentiful, but must be grown slowly in arid situations upon poor and stony soil to yield the largest proportion of oil. The tree is uprooted and roughly deprived of its bark and part of the sapwood. It is then taken to certain depôts (of which there are nine in Mysore) where the trunks are sawn into lengths of about a metre and trimmed, and the roots are freed from bark. The logs are sent either direct to London or by native craft to Bombay or other ports on the west coast of India, whence they are exported to London. Periodical auctions of the wood are held in Mysore, where also a certain amount of oil is distilled.

**Description.**— Sandal wood logs are about a metre in length and up to 15 or 20 cm. in diameter, consisting of the heartwood only of the tree. This is yellowish or pale reddish in colour, hard, heavy, and dense, but easily split. The transverse section shows alternating lighter and darker zones ; the medullary rays, which are very fine and close together, are visible under a lens ; the vessels are mostly solitary, being only occasionally arranged in small radial groups.

The wood has a slightly bitter taste, but strong, very fragrant odour.

In the yellow sandal wood the formation of heartwood is due to the production of a volatile oil, not, as in the case of guaiacum or red sandal wood, of a resin, or, as in the case of logwood, of a mixture of resin, tannin, and colouring matter. The volatile oil is found in all the elements of the wood ; it is not secreted by or contained in any particular cells or glands.

The student should observe

(*a*) The *pale colour*,
(*b*) The *characteristic odour and taste*,
(*c*) The *vessels usually single*.

**Constituents.**—The only important constituent of the wood is the volatile oil (sp. gr. 0·973 to 0·985 ; O.R.˚— 13° to — 21°), of which it yields from 2 to 5 per cent. The chief constituent of the oil is the alcohol santalol, $C_{15}H_{26}O$, (probably a mixture of $\alpha$- and $\beta$-santalols), of which it contains over 90 per cent.

**Uses.**—The wood is used as a source of the volatile oil, and technically for the manufacture of various articles. Sandal wood oil is a stimulant and disinfectant of the whole genito-urinary tract.

**Varieties.**—*West Indian Sandal wood*, *Amyris balsamifera*, Linné (N.O. *Burseraceæ*) ; vessels in radial groups ; sp. gr. of volatile oil

0·960 to 0·967; O.R. + 24° to + 29°; contains 50 per cent. of amyrol).

*West Australian Sandal wood, Fusanus spicatus*, Robert Brown (N.O. *Santalaceæ*); vessels in radial groups; sp. gr. of volatile oil 0·953 to 0·965; O.R. + 5°; contains 75 per cent. of santalol.

*South Australian Sandal wood, Santalum Preissianum*, Miquel (=*Fusanus acuminatus*, Robert Brown); sp. gr. of volatile oil 1·022; rose-like odour.

*Fiji Sandal wood, Santalum Freycinetianum*, Gaudich (= S. *Yasi*, Seem.); sp. gr. of volatile oil 0·9768; O.R. – 25·5°; odour slight.

Several other fragrant woods are known, e.g. those of *Osyris tenuifolia*, (N.O. *Santalaceæ*) from East Africa; *Olearia Traversii*, F. Mueller (N.O. *Compositæ*) New Zealand; *Brachyleana Hutchinsii*, Hutchinson (N.O. *Compositæ*), Nairobi.

## STEMS IN LESS FREQUENT USE

**Tinospora** (*Gulancha*). The dried stem of *Tinospora cordifolia*, Miers. (N.O. *Menispermaceæ*), a climbing shrub indigenous to tropical India. Straight or twisted, light, cylindrical pieces, or in slices, about 2 cm. or less in diameter; cork thin, brown, papery, with raised scars of numerous lenticels; cortex greenish; section shows a yellowish wood with radially arranged, wedge-shaped wood bundles, containing large vessels, separated by narrower medullary rays; taste bitter. Contains berberine and also a glucosidal bitter principle.

## SECTION VII

## BARKS

The term 'bark' is commonly employed by pharmacognosists to denote all the tissues of the stem and root of trees and shrubs exterior to the cambium. It comprises, therefore, botanically, such tegumentary tissue as may have been formed, primary and secondary cortex, together with primary and secondary bast.

In examining a bark attention should first be directed to the shape of the pieces in which it occurs.

Barks are said to be in *flat* pieces when they are quite flat; in *curved* pieces when they present a curved but not deeply concave transverse section; in *recurved* pieces when the concave surface is the outer portion of the bark; in *channelled* pieces when the transverse section is deeply concave. Should it be so deeply concave that the edges nearly or quite overlap, a *quill* is produced, and should both edges be inrolled a *double quill* is formed. Single or double quills packed inside one another form *compound* quills.

Next the physical characters should be noted, such as colour odour, taste, nature of the outer and inner surface, &c.

The appearance of the edges when a piece is broken ('fracture' of a bark) often affords useful information; the fracture may be short, granular, splintery, or fibrous—terms which sufficiently explain themselves.

As the bark of any particular tree may be, and frequently is, collected from axes varying in size from a twig to a large trunk, the drug thus produced will exhibit not only corresponding variations in size, but also considerable variation in external appearance caused by the changes in the tegumentary tissue due to the growth of the axis. The internal structure, however, retains its important characteristics unchanged, and the student should therefore carefully examine a smooth transverse section which he will find an excellent and often indispensable guide to the identification of a bark. Before passing to this part of the work he should study the anatomy of the bark in his text-book of botany; the following brief notes may, however, be useful.

*Tegumentary Tissue.*—Few barks are so young when collected as to retain their epidermis; this tissue has usually been replaced by a cork of varying thickness. Sometimes the development of successive layers of phellogen deeper and deeper in the cortex, and possibly in the secondary bast, has resulted in the formation of corresponding layers of cork, the tissues thus cut off losing their vitality. To the mass of protective tissue (the 'bark' of botanists) thus formed the name of 'outer bark' is commonly assigned by pharmacognosists.

The characters of the cork, influenced in some cases by the nature of the cell contents (cascarilla, alder buckthorn), often afford a valuable means of identifying a bark.

*Phelloderm* is seldom formed in quantity sufficient to be diagnostically important. Canella bark alone contains a sclerenchymatous phelloderm easily visible under a lens.

*Cortex.*—This tissue, often termed 'primary,' or 'middle' bark by pharmacognosists, is the tissue extending from the epidermis, cork, or phelloderm, as the case may be, up to and including the endodermis. Beyond the presence or absence of groups of sclerenchymatous cells, of oil-cells, oil-glands, or calcium oxalate, it seldom affords any well-marked characters.

*Bast.*—The primary bast being seldom distinguishable even under a microscope, this tissue is practically composed of secondary bast. It comprises the tissue extending from the endodermis to the cambium, and corresponds to the inner bark of many pharmacognosists. Its structure should be carefully examined, as it is frequently of great diagnostic importance.

It occasionally happens that the cells of the pericycle thicken and lignify, and thus form a band of sclerenchymatous tissue, readily visible under a lens, in the mature bark (sassy bark, nux vomica bark); otherwise the extent of the secondary bast is not readily fixed; it continues at least as far as the medullary rays extend.

The chief feature of importance in the bast, and visible under a lens, are the presence or absence of: (i) sclerenchymatous cells; (ii) bast fibres; (iii) oil-cells or glands; (iv) calcium oxalate; (v) mucilage.

Sclerenchymatous cells commonly occur in rounded, less often in tangentially elongated, groups exhibiting a uniform semi-translucent appearance. They impart hardness to the bark and render the fracture short and granular (sassy bark).

Bast fibres may occur scattered or in groups, which are then often tangentially elongated and arranged in tangential lines. They impart to the bark a tough and fibrous character; from the fractured surface of such a bark the fibres usually project (oak, elm, &c.).

If the bast fibres approach sclerenchymatous cells in nature the the fracture becomes splintery (quillaja).

Oil-glands or oil-cells usually appear as minute yellowish, reddish or brownish translucent points; their presence is often very characteristic (canella bark).

Calcium oxalate is generally visible as colourless points or lines (cusparia bark); if present in large quantities the tissue itself may appear white (canella).

## BARBERRY BARK

### (Cortex Berberidis)

**Source, &c.**—The common barberry, *Berbĕris vulgaris*, Linné (N.O. *Berberidĕæ*), is a shrub about 2 metres high, with three-lobed thorns at the bases of the tufts of leaves. It occurs scattered over Great Britain, and over the greater part of Europe and temperate Asia. The yellow flowers, in elegant drooping racemes, are succeeded by oblong scarlet fruits.

The stem bark is collected by shaving, and dried.

**Description.**—Small, thin, nearly flat pieces not often exceeding 5 cm. in length or 1 cm. in breadth, dark in colour, with a decided yellow or yellowish grey tinge. The outer layer (cork) is dark yellowish grey, and marked with shallow, longitudinal furrows or deeper fissures, becoming ultimately scaly; it frequently bears the minute black apothecia of small inconspicuous lichens. The inner surface is dark yellowish brown, longitudinally striated and fibrous, and occasionally has fragments of yellow wood adhering to it. The fracture is short in the outer portion (cork and cortex), but fibrous and strongly laminated in the inner (bast). The transverse section exhibits under the lens a narrow cork and a dark brown bast traversed by paler, yellow medullary rays. The bast rays contain narrow, tangentially elongated bundles of bast fibres, which easily separate from the bast parenchyma in strands and produce the laminated appearance and fracture. The bark has a bitter taste, and colours the saliva yellow.

**Constituents.**—Barberry bark contains several alkaloids, of which berberine is the most important.

Oxyacanthine and berbamine are colourless alkaloids that have been obtained from the bark of barberry root, and doubtless exist also in that of the stem.

Berberine, $C_{20}H_{19}NO_5$, crystallises in brilliant yellow needles with four or six molecules of water, and melts when anhydrous at 145°; it forms well crystallised compounds with acetone and chloroform, and is coloured in aqueous solution blood-red by chlorine water. Berberine is one of the few alkaloids that occur in plants belonging to several different natural orders (*Berberideæ*, *Ranunculaceæ*, *Rutaceæ*, *Leguminosæ*).

**Uses.**—Barberry bark is sometimes employed as a remedy for fevers, and as a tonic.

## CANELLA BARK

### (Wild Cinnamon Bark, Cortex Canellæ)

**Source, &c.**—Canella bark is obtained from *Canella alba*, Murray (N.O. *Canellaceæ*), a small tree distributed over the West Indian Islands and found also in Florida. With this tree the Spaniards became acquainted when they discovered America, and thinking, from its aromatic bark, that it was a kind of cinnamon (which was known to them as a valuable Asiatic spice) they brought it to Europe, where it received the name of white cinnamon or 'canella alba.' It is now used by the negroes as a condiment.

FIG. 117.—Canella bark. Natural size.

The bark of the tree is covered with a thick layer of ash-grey cork ; by gentle beating, this layer of cork is detached, and the remainder of the bark, which has been at the same time loosened, can then be stripped off and dried. It is exported chiefly from the Bahamas.

**Description.**—Canella bark occurs in channelled pieces and single quills of very varying size, evidently obtained from small trunks and from large and small branches. The quills vary in diameter from 5 to 25 mm. or more ; channelled pieces may be as much as 50 mm. broad and 5 mm. thick. Much of the bark shows evidence, in the shape of irregular longitudinal fractures, of the beating to which it has been subjected. The outer surface is of a bright pale reddish or yellowish buff colour, very hard and granular, and usually marked at somewhat distant intervals with circular, crateriform scars or with whitish spots, as well as with numerous shallow, transverse or longitudinal depressions. The inner surface is paler, and finely striated longitudinally.

The fracture is very short and granular. The smoothed transverse section exhibits under the lens a narrow, irregular, translucent, brown outer layer (phelloderm of sclerenchymatous cells), a paler cortex in which numerous brown oil-cells can be seen, and, in the bast, white wavy medullary rays.

The odour is agreeably spicy, recalling cinnamon ; the taste pungent and bitter.

The student should observe

(*a*) The *hard buff outer surface* with its characteristic *scars or spots*.

(*b*) The *oil-cells* and *medullary rays*,
(*c*) The characteristic *odour and taste*.

**Constituents.**—Canella bark contains about 1 per cent. of volatile oil which has a pungent aromatic taste and contains eugenol, cineol, and terpenes. The bitter principle has not yet been isolated, and it is doubtful whether the pungency is due entirely to the volatile oil. The bark contains no tannin, a negative character by which it may be distinguished from that of *Cinnamodendron corticosum*, Miers.

**Uses.**—Canella has aromatic stimulant and tonic properties.

## WINTER'S BARK

### (True Winter's Bark, Cortex Winteranus)

**Source, &c.**—True Winter's bark is derived from *Drīmys Wintĕri*, Forster (N.O. *Magnoliaceæ*), a tree distributed throughout the whole of South America. It was first brought to Europe in 1579 by Captain Winter from the Straits of Magellan. It is now imported from Colombia.

**Description.**—True Winter's bark occurs in channelled pieces or strongly inrolled quills from 3 to 8 mm. in thickness, the quills being often small in diameter in comparison with their great thickness. Both surfaces are usually of a rusty-brown colour, the outer being occasionally whitish or silvery; the thinner pieces frequently bear lichens. The inner surface is well characterised by deep and close striations, the projecting ridges being the inner margins of the bast rays, which are strongly lignified, and therefore do not contract on drying. A transverse section shows radiating lines of white bast, the intervening parenchyma as well as that of the cortex being dark brown. The bark has a terebinthinate odour and an extremely pungent taste, which, however, disappears on keeping.

**Constituents.**—Winter's bark contains 0·6 per cent. of a volatile oil consisting chiefly of the terpene, winterene. It also contains resin, tannin, and mucilage. A cold aqueous infusion is coloured dark violet by solution of potassium hydroxide, a reaction that is sometimes useful in identifying the bark.

**Uses.**—Winter's bark has stimulant, tonic, and antiscorbutic properties. It is much used in Brazil as an astringent and stimulant.

**Substitutes.**—*Cinnamodendron Bark*, *C. corticosum*, Miers (N.O. *Canellaceæ*); for many years substituted for true Winter's bark; dingy brown, rarely pale yellow, the freshly cut surface has a distinct odour of coriander.

*Malambo Bark*, *Croton Malambo*, Karsten (N.O. *Euphorbiaceæ*);

hard curved or channelled pieces with silvery white cork and reddish brown inner surface; fracture short, slightly aromatic; taste very bitter.

## COTTON ROOT BARK

### (Gossypii Radicis Cortex)

**Source, &c.**—Many species and varieties of *Gossypium* are cultivated in the United States, India, Egypt, &c. for the sake of the hairs that cover the seeds (see 'Cotton'); the most important are *G. herbaceum*, Linné, *G. barbadense*, Linné, and *G. arboreum*, Linné. The bark stripped from the root and dried constitutes the drug.

**Description.**—The bark occurs in thin, tough, fibrous strips to which long, thin, tapering rootlets are attached at intervals. The outer layer is a rough, cinnamon-brown cork which easily separates, exhibiting a paler cortex; the inner surface is whitish, silky and finely striated. The bast is laminated, the outer laminæ exhibiting, when separated, surfaces bearing minute, brownish spots. No odour; taste somewhat acrid and astringent. The bark may be distinguished from the bark of *D. Mezereum* (see 'Mezereon Bark') by its rougher, darker surface and absence of leaf-scars or buds.

**Constituents.**—Cotton root bark yields a small amount of volatile oil containing probably furfuraldehyde and acetovanillone. It yields also about 10·6 per cent. of deep purplish resin. From an alcoholic extract of the bark, dihydroxybenzoic acid, salicylic acid, and two substances of phenolic nature, have been separated, together with betaine, a phytosterol, ceryl alcohol and fatty acids. No tannin is present (Power and Browning).

**Uses.**—It has been used as a substitute for ergot in labour and as an emmenagogue, but is of doubtful value.

## CUSPARIA BARK

### (Angostura Bark, Cortex Cuspariæ)

**Source, &c.**—Cusparia bark is obtained from *Galipēa officinalis*, Hancock (N.O. *Rutaceæ*), a tree indigenous to the mountains of Venezuela. It was introduced into European medicine about the end of the eighteenth century, being brought from Angostura (on the Orinoco) to Trinidad, whence its name 'Angostura' bark. It was long considered to be derived from *Cusparia febrifuga*, de Candolle, a tree closely allied to *G. officinalis* botanically, but characterised by the presence of calcium oxalate in rosettes, while in *Galipea* this takes the form of acicular crystals.

**Description.**—Cusparia bark occurs in somewhat thin curved or channelled pieces or single quills, usually about 8 or 10 cm. long and 1·5 to 3 mm. thick, but sometimes much longer, one of the longitudinal margins being frequently obliquely cut. The outer layer (cork) is sometimes buff-coloured, friable, and easily removed by the finger-nail, sometimes dark grey, thin and firmly adherent, a difference due to the alternate production of layers of thin and thick-walled cork cells.

Below the cork is a hard, dark brown middle layer (cortex), whilst the inner surface of the bark (bast) is of a cinnamon or chocolate brown colour and finely striated. This portion frequently exhibits a laminated structure and bears numerous minute short white lines, longitudinally arranged, which are usually easily visible under a lens, especially after the inner surface has been smoothed with a knife. They are caused by axially elongated cells filled with acicular crystals of calcium oxalate. Similar white masses of calcium oxalate may be seen on the smoothed radial and transverse sections.

The fracture is short and resinous, the transverse section exhibiting

FIG. 118.—Cusparia bark, showing buff spongy cork and obliquely cut edges. Natural size.

a whitish cork, a yellowish brown cortex, and, in the bast, yellowish oblique or wavy medullary rays alternating with darker bast-rays. Throughout both cortex and bast numerous cells filled with white crystals of calcium oxalate, as well as minute dark ones filled with oil, may be distinguished under the lens.

**Microscopical Characters.**—A transverse section examined under the microscope shows a cork varying in thickness and in the nature of the cells, layers with thin-walled cells alternating with thick-walled. The primary cortex usually contains but few sclerenchymatous cells; when present they are generally near the cork and probably belong to the phelloderm. The secondary bast contains occasional tangentially elongated groups of bast fibres and slender axially elongated prisms of calcium oxalate. All the tissues except the cork contain small oil-cells with droplets of yellowish oil, and tangentially elongated cells filled with acicular crystals of calcium oxalate.

The bark has an unpleasant musty odour and a bitter taste.

The student should observe

(*a*) The *cork, which is often spongy*,
(*b*) The *laminated inner portion*,

(*c*) The *calcium oxalate*,
(*d*) The *characteristic odour*.

**Constituents.**—The bitterness of cusparia bark is due chiefly to angosturin, $C_9H_{12}O_5$, a substance that has been obtained in colourless pulverulent crystals melting at 58°, easily soluble in water, alcohol, and ether. The bark also contains a number of alkaloids (2·4 per cent.), —viz. galipine, cusparine, galipoidine, and a fourth which has not yet been closely examined. A glucoside yielding by hydrolysis a fluorescent substance is also present. By distillation the drug yields about 1·5 per cent. of an aromatic volatile oil, the most important constituent of which is the aromatic sesquiterpene alcohol, galipol. Cusparidine and galipidine, alkaloids reported to be present, are probably mixtures of galipine and cusparine.

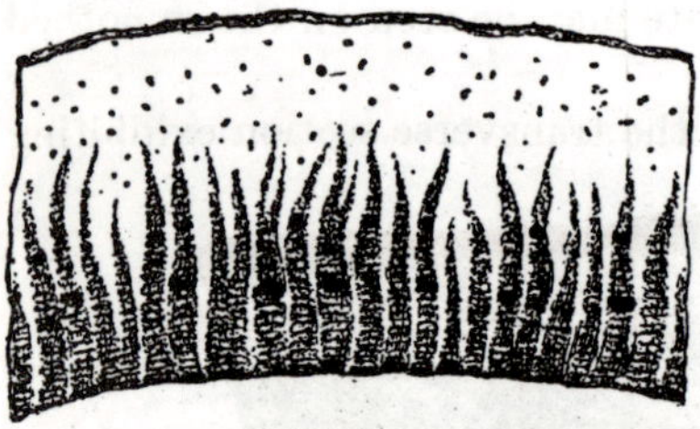

FIG. 119.—Cusparia bark. Transverse section, showing oil-cells and groups of bast-fibres. Magnified. (Planchon and Collin.)

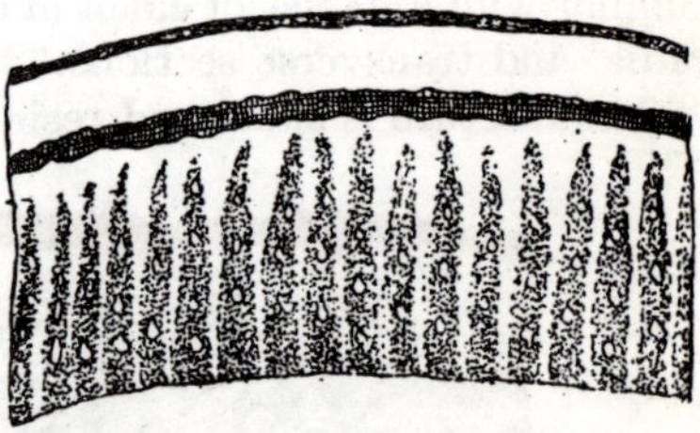

FIG. 120.—Nux Vomica bark. Transverse section, showing band of sclerenchymatous cells. Magnified. (Planchon and Collin.)

*Cusparine*, $C_{20}H_{19}NO_3$, forms colourless needles or compact crystalline masses, readily soluble in alcohol, ether, and chloroform; its salts are sparingly soluble in water.

*Galipine*, $C_{20}H_{21}NO_3$, crystallises in colourless needles, but yields deep yellow salts.

**Uses.**—Cusparia belongs to the group of aromatic bitters. It formerly had a high reputation as a febrifuge and tonic, being preferred to cinchona bark on account of its freedom from astringency.

**Substitutes.**—The accidental substitution in Hamburg in 1804 of the bark of *Strychnos nux-vomica*, Linné, for cusparia bark led to several cases of poisoning. Such a substitution or admixture is not likely to occur again, and would moreover be easily detected, as the barks do not bear much resemblance to one another. Nux vomica bark is harder and thicker; it occurs in small, often recurved, pieces, with dark greyish, yellowish, or rusty-red cork, usually bearing numerous greyish warts. The dark transverse section exhibits under the lens a distinct paler line of sclerenchymatous cells separating the cortex from the bast; this line of sclerenchymatous cells is never found in cusparia bark.

*Brazilian Angostura bark*; *Esenbeckia febrifuga*, A. Jussieu (N.O. *Rutaceæ*); greyish brown with reddish brown patches; internally brown; taste bitter.

## SIMARUBA BARK

(Cortex Simarubæ)

**Source, &c.**—Simaruba bark is obtained from various species of *Simaruba* (*S. officinalis*, de Candolle, *S. amara*, Aublet, *S. glauca*, de Candolle), (N.O. *Simarubeæ*), tall trees with long horizontal roots, natives of Guiana, northern Brazil, the West Indian Islands, and Florida. It was brought from Guiana to Paris in 1713 as the bark of a tree called by the natives simaruba and used by them with great success in dysentery. In Europe it soon gained renown, and was imported in considerable quantity. The bark is stripped from the root, probably after a preliminary beating to loosen it, freed from the outer layer (cork), and dried.

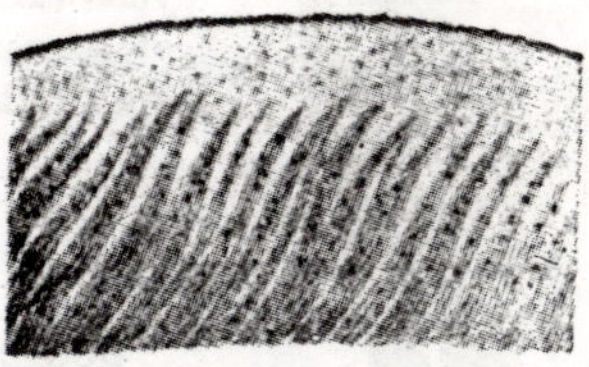

FIG. 121.—Simaruba bark. Transverse section. Magnified. (Berg.)

There appear to be three varieties of simaruba bark in commerce, viz. Orinoco, Surinam and Maracaibo. The first two are derived from *Simaruba officinalis*, de Candolle; the last named from a species of *Simaba* closely allied to *S. suffruticosa*, Engler.

**Description.**—True simaruba bark occurs in long, very fibrous strips, sometimes as much as a metre in length, a decimetre wide, and between 3 and 6 mm. thick. These pieces are more or less fissured and rent longitudinally, probably the result of beating. Externally they are of a buff or yellowish brown colour, and rough as though they had been deprived of the outer cork layer by rasping. They are frequently marked with brownish raised corky warts or the depressions left after their removal. The inner surface is yellowish, longitudinally striated, and fibrous.

The transverse section exhibits numerous, narrow medullary rays traversing the bark from the inner almost or quite to the outer margin, and showing therefore that the drug consists almost entirely of bast tissue, the cork and part or all of the cortex having been removed.

The drug has no odour, but a very bitter taste.

The student should observe

(*a*) The *extremely fibrous* nature of the bark,

(*b*) Its *yellowish colour* and *bitter* taste,

(*c*) The characters of the *transverse section*.

**Constituents.**—Simaruba bark contains 0·05 to 0·1 per cent. of a colourless, crystalline bitter principle possessing neither alkaloidal nor glucosidal properties and quite distinct from other crystalline bitter principles occurring in plants of the same natural order (*e.g.* samaderin, quassiin). It also contains a tasteless crystalline substance, fixed oil, a yellow resin, and traces of a fluorescent substance (Gilling, 1908).

FIG. 122.—Euonymus root-bark. Natural size.

**Use.**—Simaruba bark was formerly used for dysentery ; it is now seldom employed in European medicine.

**Note.**—The simaruba bark at present (1920) is the Maracaibo variety. It is much harder and less fibrous than the true ; the outer surface is hard and finely marbled yellowish and white ; the section shows abundant groups of sclerenchymatous cells in the cortex and bast ; taste very bitter.

## EUONYMUS BARK

### (Cortex Euonymi)

**Source, &c.**—Euonymus bark is the dried root-bark of the wahoo tree, *Euonymus atropurpūrĕus*, Jacquin (N.O. *Celastrineæ*), a tall erect shrub with small dark purple flowers succeeded by crimson fruits ; it is common in the eastern United States, extending westward to Wisconsin and southward to Florida. The root-bark is alone official, but the stem-bark also is collected.

**Description.**—Euonymus root-bark occurs in small, more or less irregular quilled or curved pieces, not usually exceeding 8 cm. in length or 12 mm. in width. The outer layer is a soft, spongy, finely fissured cork, of a light ash-grey colour marked with darker lines or patches (due to adhering particles of earth) and occasional small transverse scars. The inner surface is of a pale tawny yellow or buff colour and nearly smooth ; occasionally a thin shaving of pale yellow, dense wood adheres to it, indicating that the bark is separated with difficulty from the root.

FIG. 123.—Euonymus root-bark. Transverse section. Magnified. (Planchon and Collin.)

It breaks with a very short fracture, and if the two pieces be separated very gently from one another delicate silky threads will be seen connecting them ; these threads consist of a substance resembling caoutchouc or guttapercha which is secreted in laticiferous cells in the bast, and is found in all species of *Euonymus*.

The section, which is of a greyish white colour, exhibits, when moistened, a narrow whitish cork, a pale cortex, and darker bast.

The bark has a faint but characteristic odour recalling that of liquorice root, and a disagreeable and persistent, bitter, acrid taste.

**Microscopical Characters.**—The transverse section exhibits a thick layer of narrow cork cells, a cortex containing abundant cluster crystals of calcium oxalate and a wide layer of secondary bast in which small cells filled with a granular substance are scattered. The bark contains no bast fibres or sclerenchymatous cells. The medullary rays are one cell wide.

The student should observe

(*a*) The *spongy grey cork*,
(*b*) The characteristic *silky threads from the fractured surface*,
(*c*) The *bitter taste*.

**Constituents.**—A bitter, crystalline alcohol, euonymol ($C_{21}H_{30}O_4$), has been isolated from it. From the resin (3·2 per cent.) contained in it the crystalline alcohols euonysterol, homoeuonysterol, atropurol and citrullol have been separated.

The bark also contains considerable quantities of dulcitol (dulcite), $C_6H_8(OH)_6$, a hexatomic alcohol readily crystallising in large plates melting at 188·5°; it has a sweetish taste and has been found in a considerable number of plants. The presence of the crystalline glucoside euonymin has not been confirmed.

**Uses.**—Euonymin (the powdered extract) is an hepatic stimulant, direct cholagogue, and mild cathartic. It is used in constipation and in hepatic derangements.

**Substitutes.**—The stem-bark, which, although not official, is a commercial article, occurs in long thin narrow strips; it has a dark greenish grey cork, green cortex, and fibrous bast; it is therefore easily distinguished from the root-bark.

*Wafer Ash bark*, *Ptelea trifoliata*, Linné (N.O. *Rutaceæ*) has been frequently mixed with or substituted for euonymus bark; it is thicker, bears long, transverse, whitish scars; medullary rays three cells wide; below the cork a layer of yellow sclerenchymatous cells; in the secondary bast large oleo-resin cells.

## ALDER BUCKTHORN BARK

### (Cortex Rhamni Frangulæ)

**Source, &c.**—The alder buckthorn, *Rhamnus Frangŭla*, Linné (N.O. *Rhamnĕæ*), is distributed generally over Europe. It is distinguished from the common buckthorn (*R. catharticus*, Linné), the only other indigenous species, by the entire leaves, hermaphrodite flowers with five stamens, absence of thorns, and tree-like habit.

The bark is stripped from the stem and branches, the wood of which was formerly valued for making charcoal for gunpowder. When fresh it has an unpleasant odour and taste, and acts as an emetic, but these properties are lost when the bark is dried and kept : alder buckthorn bark should not therefore be employed medicinally until it has been kept for at least a year.

FIG. 124.—Alder Buckthorn bark. A, young, B, older bark. Natural size.

**Description.**—The commercial drug occurs in single or double quills varying from 0·5 to 4 cm. in diameter and commonly 15 cm. or more in length. Young bark is usually extremely thin and has a smooth, glossy, dark purplish exterior marked with small, circular or transversely elongated, whitish lenticels. The cork frequently exfoliates, or at least easily separates, disclosing a yellowish brown cortex, but the inner part of the cork is of a dark crimson colour easily seen by gently scraping off the outer cork cells. The inner surface is dark cinnamon-brown in colour and nearly smooth, exhibiting under the lens fine longitudinal striations. The fracture is short in the outer, but rather fibrous in the inner part, groups of bast fibres projecting a short distance beyond the fractured surface.

The traverse section shows under the lens a narrow dark purple cork and yellowish brown cortex and bast.

Older bark is commonly much rougher. It has usually a dull dark purple colour, and is marked with transversely elongated lenticels and shallow longitudinal fissures, the cork exhibiting less disposition to exfoliate. The transverse section shows a similar purple cork; the bast is thicker and allows of very thin medullary rays being distinguished. The absence of sclerenchymatous cells is a character occasionally of value in excluding substitutes.

The bark has no odour and a taste that is scarcely bitter.

The student should observe

(*a*) The *dark purple cork*, showing, when scraped, a *deep crimson* colour,

(*b*) The *light-coloured lenticels*,

(*c*) The *shortly fibrous fracture* of the bast.

FIG. 125.—Alder Buckthorn bark. Transverse section. Magnified. (Berg.)

**Microscopical Characters.**—A transverse section exhibits a cork consisting of narrow cells many of which contain a bright purplish crimson colouring matter. The cortex contains small starch grains, cluster-crystals of calcium oxalate and elongated mucilage cells, but no sclerenchymatous cells. The secondary bast contains numerous tangentially elongated groups of thick-walled bast fibres; the medullary rays are mostly two cells wide. The cells of the medullary rays and bast parenchyma contain a yellowish amorphous substance dissolving in solution of potassium hydroxide with production of a bright purple colour.

**Constituents.**—The active constituents of alder buckthorn bark are but imperfectly known. It contains a glucoside, frangulin, $C_{21}H_{20}O_9$, which crystallises in lemon-yellow needles melting at 228° to 230°, is slowly volatile at ordinary temperature and stains the paper in which the drug is kept; it is soluble in caustic alkalies with purple coloration. Boiled with alcoholic hydrochloric acid it is converted into rhamnose and frangula-emodin. Rhamnose (isodulcite) is a pentatomic alcohol (pentose, pentaglucose), and is produced by the hydrolysis of quercitrin and a number of other glucosides, which are therefore termed rhamnosides. Frangula-emodin, $C_{15}H_{10}O_5$, occurs in reddish yellow crystals melting at 254°. It is a derivative of anthraquinone, viz. trioxymethylanthraquinone, $C_{14}H_4(CH_3)(OH)_3O_2$, and appears to be identical with rheum-emodin, but different from aloe-emodin and senna-emodin.

Frangula-emodin has been known under various names, viz. frangulin (Casselmann, 1857), frangulinic acid (Faust, 1869, Keussler 1873), avornic acid (Kubly, 1866), rhamnoxanthin (Binswanger, 1850).

Frangulin is said not to be present in the fresh bark, but to be produced from some unknown constituent during the maturing of the bark, the change in the physiological action of the bark occurring simultaneously. The bark also contains free frangula-emodin, chrysophanic acid and an iso-emodin.

The total quantity of emodin present in the bark either free or in the form of a rhamnoside has been estimated at 1 per cent. in old bark, 2 per cent. in bark of medium age, and 3·8 per cent. in very young bark. This proportion is insufficient to account for the laxative action of the drug, and the assumption is that either other, at present unknown, purgative bodies are present or that the frangulin exists in an unknown more active form.

The presence of emodin in the bark is readily demonstrated by moistening 0·1 gm. of the powdered bark with 10 drops of alcohol and boiling for a few moments with 10 c.c. of water. The cooled decoction is shaken with 10 c.c. of ether, the yellowish ethereal solution separated and shaken with ammonia, which will acquire a reddish colour.

**Uses.**—Alder buckthorn bark has been used as an agreeable laxative, preferable to cascara sagrada on account of its less disagreeable taste.

**Substitutes.**—The bark of *R. carniolicus*, Kerner, is not unfrequently substituted for that of *R. Frangula*; the cork contains a dull red, not crimson, colouring matter; medullary rays 4 to 7 instead of 2 to 3 cells wide; outer bark, when formed, contains groups of sclerenchymatous cells.

The bark of *Alnus glutinosa*, Gaertner (N.O. *Betulaceæ*) exhibits in transverse section a ring of sclerenchymatous cells. The bark of *R. catharticus*, Linné, is glossy reddish brown and has very distant lenticels.

## CASCARA SAGRADA

(Cortex Rhamni Purshiani, Sacred Bark, Chittem Bark)

**Source, &c.**—Cascara sagrada is the bark of *Rhamnus Purshianus*, de Candolle (N.O. *Rhamneæ*), a shrub or small tree abundant in North California and in the States of Washington and Oregon, the bark being collected in these States and exported from San Francisco. The bark of *R. californicus*, Eschscholz, was known to the Spanish settlers in California as cascara sagrada, a name which has since been applied to the bark of the closely allied *R. Purshianus*. The latter has been much used as a laxative since 1883.

The bark is collected in the spring and early summer, and dried in the shade; if left till later in the year it adheres so firmly to the wood that it has to be cut off, and then brings shavings of wood with it.

**Description.**—Cascara sagrada occurs in straight, stiff, single quills or in channelled pieces. The quills vary from 5 to 25 mm. or more in diameter, whilst the channelled or sometimes flattish pieces may be as much as 10 cm. wide; commonly the drug is seen in pieces about 10 to 20 cm. long, 2 cm. wide, and from 1·5 to 4 mm. thick, the thinnest being most esteemed.

The outer layer is a smooth, dark purplish brown cork marked with transversely elongated whitish lenticels. The bark, however, is usually more or less completely covered with silvery grey patches of lichens which conceal the purple cork and the lenticels and give to the drug its pervading greyish white colour. The inner surface is of a dark, or even very dark, reddish brown colour and longitudinally striated, with faint transverse corrugations.

Fig. 126.—Cascara Sagrada bark. Natural size.

The fracture is short, that of the bast being shortly fibrous. The section exhibits under the lens a narrow purplish cork, a yellowish grey cortex in which darker translucent points (groups of sclerenchymatous cells) can be distinguished, and a brownish yellow bast in which wavy, somewhat distant medullary rays may sometimes be discerned.

The bark has a characteristic though not strong odour, and a persistent, nauseously bitter taste.

Like alder buckthorn bark, this drug should be kept for at least a year before it is used medicinally; the action is then milder and less emetic. It also shares with alder buckthorn bark the property of imparting a yellow colour to the paper in which it is kept.

**Microscopical Characters.**—The structure of cascara sagrada closely resembles that of alder buckthorn bark, the chief differences being in the contents of the cork cells which are reddish brown, and the presence of groups of sclerenchymatous cells in the cortex and secondary bast.

The student should observe

(*a*) The *purplish cork* and the *grey lichens* covering it,
(*b*) The *groups of sclerenchymatous cells* in the cortex,
(*c*) The *characteristic odour and taste*;

and should compare the bark with

*Alder Buckthorn bark*, which has deep crimson inner cork layers and no sclerenchymatous cells.

**Constituents.**—Our knowledge of the constituents of cascara sagrada is very deficient. The presence of emodin and an allied substance, possibly frangula-emodin, has been definitely proved, but the latest research (Jowett, 1904) failed to afford any evidence of chrysophanic acid, chrysarobin, or any glucoside yielding by hydrolysis emodin, chrysophanic acid, or rhamnetin. The total amount of emodin and frangula-emodin (1·4 to 2·0 per cent.) present in the bark either normally or after boiling with dilute sulphuric acid is quite insufficient to account for the purgative action of the drug, and the real laxative principle remains therefore still unknown.

The presence of emodins may be demonstrated by the test given under Alder Buckthorn bark; the ammonia acquires a yellowish red colour.

The bitter taste of the bark appears to be due to a lactone which is converted into less bitter salts by treatment with alkalies or alkaline earths, but this change is accompanied by simultaneous loss of activity.

The bark also contains fat (about 2 per cent.), glucose, and a hydrolytic enzyme. No difference could be detected in the chemical constituents of the fresh (one year old) bark and the matured (three years old) bark. Purshianin (Dohme) and cascarin (Leprince) appear to be impure substances.

Cascara bark yields about 27 per cent. of aqueous extract and 5 per cent. of ash.

**Uses.**—Cascara sagrada is tonic and stomachic in small doses, aperient in large doses, and cathartic if freely given. It is said to be more active and more certain than alder buckthorn.

**Substitutes.**—The bark of *R. californicus*, Eschscholz, is said to have been substituted for that of *R. Purshianus*. The shrub occurs sparingly in North California, but abundantly in the south and east of the State, as well as in Arizona, New Mexico, and Texas. The bark is said to be distinguished by its dull grey, slightly reddish cork, its fewer lenticels and uniform coat of lichens; the inner surface is said to be paler, and the medullary rays commonly 3 to 4 cells wide, those of *R. Purshianus* being only 2 to 3 cells wide. The two plants are, however, so closely allied that some botanists refer them to the same species.

The bark of *Rhamnus catharticus*, Linné, also contains frangula-emodin together with chrysophanol, rhamnosterin and a fluorescent body, rhamnofluorin.

## ACACIA BARK

### (Acacia Bark)

**Source, &c.**—The dried bark of *Acacia arabica*, Willdenow, or of *A. decurrens*, Willdenow (N.O. *Leguminosæ*). Both of these species are large trees, the former indigenous to India, Arabia, and Africa, the latter to Australia. Both are largely cultivated for their bark which is much used in the tanning industries. The bark should be obtained from trees at least seven years old and be kept for one year before use.

**Description.**—*Acacia arabica* (babul bark) occurs in rather large pieces, hard and woody, rusty brown, readily dividing into several layers. Periderm thick, blackish, rugged, fissured longitudinally and transversely. Inner surface red, longitudinally striated and fibrous; taste astringent and mucilaginous.

*Acacia decurrens* occurs usually in curved or channelled pieces 1·5 to 3·0 mm. thick, externally greyish brown, becoming darker when kept, often with irregular longitudinal ridges and transverse cracks; inner surface reddish brown, longitudinally striated; fracture coarsely fibrous; freshly fractured surface pale.

**Constituents.**—Both barks contain tannin (about 22 per cent.) and gallic acid.

**Uses.**—As an astringent.

## SASSY BARK

### (Red Water Bark, Cortex Erythrophlei)

**Source, &c.**—Sassy bark is derived from *Erythrophleum guineense*, G. Don (N.O. *Leguminosæ*), a large tree widely distributed on the west coast of Africa (Upper Guinea and Senegambia). Possibly other species than *E. guineense* yield the sassy bark of commerce, as certain differences have been observed in the alkaloid obtained from the bark imported at different times.

It has poisonous, emetic, and purgative properties, and is used by the West African negroes as an ordeal poison.

The bark is collected from the trunk as well as from the larger and smaller branches.

**Description.**—Sassy bark varies exceedingly in size and thickness according to the age of the stem or branch from which it has been collected. Most common are hard heavy curved or flat pieces about 8 or 10 cm. long, 4 to 8 cm. wide, and about 5 to 10 cm. thick; but small quills, not 1·5 cm, in diameter, may sometimes be found.

In pieces of medium thickness the outer layer (cork) is usually of a dull grey colour, but sometimes it is so dark as to be nearly black; it is interrupted by reddish warts or circular spots that eventually fuse together into longitudinal bands. Older and therefore thicker barks are rugged, and have a more uniform reddish brown colour. They often exhibit conchoidal depressions of considerable size (1 to 2 cm. long), the whole surface, elevations as well as depressions, being ruggedly ridged. In these barks most of the primary cortex has

A B

FIG. 127.—Sassy bark. A, portion of bark of medium thickness, showing fissured surface; B, portion of very thick bark showing rugged surface, with depressions produced by exfoliation of the outer portion.

been exfoliated by the formation of cork. Very young bark is comparatively smooth and dark in colour; it bears occasional small reddish warts and exhibits longitudinal reddish bands.

The inner surface exhibits shallow longitudinal striations or elevations, and is of a dark reddish brown or, more commonly, dull black colour.

The bark is extremely hard, and breaks with a very short granular fracture. The transverse section examined under the lens is most characteristic. The cork appears as a narrow brownish line; the cortex is narrow and darker in colour, and separated from the bast by a pale complete or interrupted line of sclerenchymatous cells. The bast, which constitutes the major part of the larger and thicker pieces, exhibits numerous large closely approximated groups of sclerenchymatous cells embedded in reddish brown (parenchymatous) tissue. The structure of this portion of the bark is discernible with the naked eye, and is identical in both young and old bark. The bark has no odour, and only a slightly bitter and astringent taste.

The student should observe

(a) The *reddish brown warts* (*or outer surface*) and the *nearly black inner surface*,
(b) The *hard granular fracture*,
(c) The *sclerenchymatous cells* in the bast.

**Constituents.**—Sassy bark, which is used by the West African negroes as an ordeal poison, contains a toxic alkaloid erythrophlœine; other constituents are resin (13·5 per cent.) and tannin together with traces of ipuranol, luteolin, small quantities of fatty acids, &c.

*Erythrophlœine* is amorphous and yields amorphous salts; it has not yet been sufficiently investigated.

*Luteolin* is a yellow colouring matter, first separated from *Reseda luteola.* Linné; it is identical with digitoflavone; quercetin is hydroxyluteolin.

**Uses.**—Erythrophlœine has been found useful in certain forms of heart disease. The hydrochloride has local anæsthetic properties and has been used in dental operations.

## WILD CHERRY BARK

### (Virginian Prune Bark, Cortex Pruni Virginianæ)

**Source, &c.**—The bark that is commonly known as 'wild cherry bark,' is obtained from *Prunus serōtina*, Ehrhart (N.O. *Rosaceæ*), the black cherry, a tree widely distributed over North America, especially throughout the northern and central States. The wood is highly valued for cabinet work, whilst the bark is employed medicinally for which purpose it is collected from the branches in the autumn, experiments proving that such bark is the most active.

**Description.**—Wild cherry bark varies considerably in appearance. It occurs usually in flattened, curved, or recurved pieces, attaining 12 cm. in length and 5 cm. in width, but generally smaller, and about 2 mm. in thickness. Young bark is frequently covered with a thin, smooth, often glossy, reddish brown cork, much interrupted by whitish lenticels which are strongly tangentially elongated; it can easily be peeled off in thin, membranous tangential strips, disclosing a smooth greenish brown cortex. Old bark is darker and rougher. Much of the commercial drug has been deprived of its cork, and then the smooth greenish brown cortex, bearing scars corresponding to the lenticels, constitutes the outer layer. Sometimes even this has been removed, and the exposed part is then the outer layer of bast, which has a rough or rasped appearance and is of a uniform, dark cinnamon-brown colour ; examined under the lens such bark exhibits pale longitudinal strands (sclerenchymatous cells) alternating with darker parenchymatous tissue (medullary rays).

The inner surface of the bark is of a cinnamon-brown colour and is finely longitudinally striated or rough, with reticulately anastomosing pale strands, the interstices of which are only partially filled with the brown (parenchymatous) tissue of the medullary rays.

FIG. 128.—Virginian Prune bark, showing transverse lenticels, and papery cork peeling off. Natural size.

The fracture is short and granular ; the fractured surface has a reddish grey colour, and usually exhibits numerous tortuous, pale red medullary rays alternating with bast rays containing much sclerenchymatous tissue, the latter projecting beyond the medullary rays, owing to their contracting less on drying.

The bark has a slight odour of bitter almonds, which becomes much more apparent when it is moistened ; the taste is astringent, aromatic, and bitter, resembling that of bitter almonds.

The students should observe

(*a*) The *reddish brown cork with numerous lenticels*,
(*b*) The *irregularly reticulated or fissured inner surface*,
(*c*) The *granular fracture*,
(*d*) The *taste of bitter almonds* when chewed.

**Constituents.**—Wild cherry bark yields, when moistened with water, hydrocyanic acid and benzaldehyde. This reaction has been shown by Power and Moore (1909) to be due to lævo-mandelonitrile glucoside, $C_{14}H_{17}O_6N$, which is isomeric, but not identical with prulaurasin and sambunigrin (compare p. 37) ; it is hydrolysed by an enzyme also contained in the bark, yielding hydrocyanic acid, benzaldehyde and dextrose.

Experiments have shown that thin green bark collected in the autumn from trees of moderate size yields most hydrocyanic acid (0·12 to 0·16 per cent.) ; the bark of the root is said to be more active than that from the stem or branches (Stevens, 1896).

The bark contains also a brown resin, trimethylgallic acid, para-cumaric acid and traces of benzoic acid and volatile oil; a green resin yielding by acid hydrolysis $\beta$-methylæsculetin is also present.

The commercial drug afforded Power and Moore 0·075 per cent. of hydrocyanic acid.

**Uses.**—The bark has mild tonic and sedative properties; it is frequently given for coughs and chest complaints.

**Substitutes.**—The bark of other North American species of *Prunus* is occasionally substituted for the official. The latter is well characterised by its short granular fracture and distinctive taste, as well as by the presence of abundant sclerenchymatous cells and absence of bast fibres. Spurious barks may be fibrous or more astringent or almost devoid of taste. Old (trunk) bark is characterised by the numerous depressions on the outer surface and the absence of lenticels.

## QUILLAJA BARK

### (Panama Wood, Soap Bark, Cortex Quillaiæ)

**Source, &c.**—Quillaja bark, or, as it is often called, soap bark, is obtained from *Quillāja saponāria*, Molina (N.O. *Rosaceæ*), a large tree indigenous to Chili and Peru. The bark, which is called 'cullay' by the natives and has apparently been long used by them for washing silk and wool, was known to Europeans in the early part of the eighteenth century, but was not regularly imported until about 1857, when it was sent to France under the name of 'Bois de Panama,' the name indicating the route by which it was sent. It is evidently stripped from the trees, freed from the outer dark brownish portion (bark), and dried. The tree has been introduced into India, where it yields a bark identical with the commercial drug.

**Description.**—Soap bark comes into commerce in large flat pieces about a metre in length and 10 or 15 cm. in breadth; it is usually about 6 mm. thick, and evidently the produce of trunks of considerable size.

The outer surface is usually of a pale brownish or yellowish white colour longitudinally striated and streaked with reddish brown where the outer portion (bark) has been imperfectly removed. Sometimes, from insufficient trimming, the bark is of a uniform dark dull red colour and bears patches of the outer bark still adhering to it. The inner surface is smooth and white or yellowish white.

The bark breaks with a splintery fracture, and the fractured surface exhibits a disposition to separate into thin plates or laminæ. Here and there, especially on the freshly fractured laminated surfaces, but also on the smooth inner surface of the bark, minute, glittering, prismatic crystals (calcium oxalate) can be seen with the naked eye, or better with a lens; sometimes these are present in considerable numbers.

The transverse section is seen under a lens to be traversed by parallel tangential and radial lines, which give it a chequered appearance ; the tangential lines are tangentially arranged bands of bast parenchyma, the radial are medullary rays, the darker portions between being groups of bast fibres.

The bark is almost odourless, but the powder is extremely irritating to the nostrils and fauces and gives rise to prolonged fits of sneezing ; the taste is acrid and unpleasant.

FIG. 129.—Quillaja bark, showing splintery fracture. Natural size.

The student should observe

(*a*) The *splintery, laminated fracture,*
(*b*) The *glittering crystals of calcium oxalate,*
(*c*) The *very smooth inner surface,*
(*b*) The *appearance of the transverse section* ;

and should compare this bark with

(i) *Elm bark*, which is fibrous and has a roughish, not smooth, inner surface,
(ii) *Slippery elm bark*, which has a decided odour of fenugreek and is very fibrous.

**Constituents.**—The principal constituents of quillaja bark are two colourless, amorphous, toxic glucosides, quillajic acid and quillajasapotoxin. Both of these substances impart to water the property of frothing, and possess other characters common to the class of substances known as ' saponins ' (see below). The drug also contains sucrose.

The saponins are nitrogen-free glucosides of acid or neutral reaction ; most of them are soluble in water and in hot dilute alcohol, less soluble in strong alcohol, insoluble in ether, and all of them are capable of imparting to water the property of frothing freely like soap solution. They are widely distributed, having been found in 200 plants belonging to 50 natural orders, but are most strongly represented in the orders *Caryophyllaceæ* and *Sapindaceæ*. They are all glucosidal, and yield by hydrolysis a non-toxic sapogenin together with sugar, a pentose and a hexose being often simultaneously produced.

All saponins are toxic when introduced directly into the circulation, but the degree of toxicity varies greatly ; they are protoplasmic poisons, irritating and killing protoplasm and dissolving red blood-corpuscles. Taken by the mouth some can be borne in considerable quantity. Almost all of them stupefy fish even when present in small proportion (1 : 200,000) only. With concentrated sulphuric acid they produce a characteristic red coloration; with a mixture of alcohol and sulphuric acid and ferric chloride a green colour is obtained.

The saponins have been classified in two series, viz. acid saponins (quillajic acid, polygalic acid, guaiacsaponic acid, &c.) and neutral saponins (quillaja-sapotoxin, senegin, parillin, guaiacsaponin, &c.). The composition of many of them may be expressed by the general formula $C_nH_{2n-8}O_{10}$ (Kobert); for neutral saponins of the formula $C_{17}H_{26}O_{10}$ the name of sapotoxin is reserved.

Commercial saponin is usually obtained from quillaja bark, and is a mixture of quillajic acid, quillaja-sapotoxin, and frequently also a non-toxic modification of quillajic acid produced during the preparation; as the term saponin has become a generic one, a prefix to indicate the source would be desirable.

Quillajic acid, $C_{19}H_{30}O_{10}$, has been obtained as a colourless amorphous mass, the powder of which is strongly sternutatory. The aqueous solution is acid and has an acrid taste. Boiled with a mineral acid it yields quillaja-sapogenin, galactose, and another sugar, which is non-fermentable and dextrorotatory.

Quillaja-sapotoxin, $C_{17}H_{26}O_{10}$, is also white, amorphous, sternutatory, and acrid.

**Uses.**—Quillaja bark has been recommended as a stimulant and expectorant, but has not met with much favour. A tincture is largely used as a means of emulsifying tars, &c.

**Substitutes.**—A quillaja bark differing from the foregoing in being thinner and in having a more or uniform, distinctly reticulated outer layer has been imported; its botanical origin is unknown (possibly *Q. Poeppigii*, Walp.). Another variety in quills 7 to 15 cm. long, 1 to 2·5 cm. wide and about 3 mm. thick, softer and not laminated, has been referred to *Q. smegmadermos*, de Candolle.

## WITCH-HAZEL BARK

### (Cortex Hamamelidis)

**Source, &c.**—The witch hazel, *Hamamelis virginiana*, Linné (N.O. *Hamamelideæ*), is a common shrub in the United States and Canada. It attains a height of about 3 metres and resembles the common hazel both in its leaves and its fruit, which is edible. The bark should be collected in the spring.

**Description.**—Witch-hazel bark occurs in thin channelled pieces of a characteristic, pale reddish pink colour, occasionally as much as 15 or 20 cm. long and 2·5 cm. wide, but usually much smaller. They are sometimes covered with an ash-grey smooth cork which in older pieces becomes darker in colour, fissured, and scaly. Frequently the cork has been removed and the cortex forms the outer layer, which is then pale reddish brown in colour and nearly smooth, exhibiting under the lens slight transverse striations.

The inner surface is pale reddish pink in colour and finely striated longitudinally; small portions of white wood, which is seen in transverse section to be dense and traversed by numerous thin medullary rays, are frequently found adhering to it.

The outer portion of the bark (cork and cortex) breaks with a short fracture, but the inner portion (bast) is coarsely fibrous and disposed to separate into laminæ, due to its containing numerous tangentially elongated groups of bast fibres. The smoothed transverse section exhibits under the lens a dark narrow cortex and a pale tangential line (sclerenchymatous cells) separating this from the bast. In many of the pieces the cork and much of the cortex have been removed, and the ring of sclerenchymatous cells may then form the outer layer.

FIG. 130.—Witch-hazel bark, showing scaly cork. Natural size.

The bark has no odour, and an astringent, slightly bitter taste.

The student should observe

(*a*) The *reddish pink colour*,
(*b*) The *pale grey but not glossy cork*,
(*c*) The *line of sclerenchymatous cells between the cortex and bast*;

and should compare this bark with

(i) *Oak bark*, which has a glossy, silvery cork and usually a brownish colour,
(ii) *Willow bark*, which has a dull greenish brown cork, is usually longitudinally striated on the outer surface, and does not exhibit a line of sclerenchymatous cells.

**Constituents.**—The bark contains about 6 per cent. of tannin, part of which, hamamelitannin, is said to be crystalline and part amorphous; gallic acid is also present (Gruttner, 1898).

**Uses.**—It is astringent and hæmostatic, and is useful in hæmorrhages from the nose, lungs, rectum, or uterus.

## POMEGRANATE BARK

### (Cortex Granati)

**Source, &c.**—The pomegranate tree, *Punīca Granātum*, Linné (N.O. *Lythrarieæ*), is a shrub or small tree indigenous to North-western India, but cultivated generally in the warmer parts of the temperate regions, especially in the countries bordering on the

Mediterranean. More than one variety is known, but that producing crimson flowers is the commonest.

The vermifuge action of the pomegranate was well known to the ancients, who employed the juice of the fruit mixed with wine, as well as the small roots and a decoction of the root-bark. The drug appears to have been subsequently forgotten, but at the beginning of the present century its use among the Hindus attracted the attention of English physicians, and it began to be successfully employed. The bark of the root is said to be the most efficacious, especially when administered in the fresh state, but analyses have shown that the stem-bark is only slightly inferior in the proportion of alkaloid it contains (Ewers, 1899). The commercial drug usually consists of both.

**Description.**—Pomegranate bark occurs in irregular, curved or channelled pieces, varying usually from 5 to 10 cm. in length and from 1 to 3 cm. in width ; it seldom forms quills.

Fig. 131.—Pomegranate root-bark. Transverse section. Magnified. (Berg.)

The root-bark has a rough outer surface of an earthy yellow colour with darker patches and is marked with conchoidal depressions, due to exfoliation of the outer portion. The inner surface is smooth and yellow in colour, with irregular, darker, brown blotches. It breaks with a very short fracture, the fractured surface being nearly white and exhibiting under the lens numerous fine tangential and still finer radial lines.

The stem-bark differs from the root-bark in being smoother. It exhibits no conchoidal depressions, the formation of cork being less abundant, but it presents occasional, shallow, longitudinal furrows and bands of pale cork. Very frequently the minute apothecia of lichens can be detected on it ; these are not to be found on the root-bark. The latter, too, is usually in more irregular, curved, flattish or even recurved fragments, whilst the stem-bark is in straighter, channelled pieces and sometimes in quills.

The bark of both stem and root is odourless, but has an astringent, slightly bitter taste.

The student should observe

(*a*) The *short fracture* and *pale interior*,

(*b*) The *yellow colour of the inner surface with brown patches*,

(*c*) The comparatively *smooth surface of the stem-bark* and the frequent *presence of lichens*,

(*d*) The *rough surface, conchoidal depressions*, and *absence of lichens* characteristic of *root-bark*.

**Constituents.**—The most important constituents of pomegranate bark are the alkaloids it contains. Four of these have been isolated (Tanret, 1878–1880); three, viz. pelletierine, isopelletierine, and methylpelletierine, are liquid, but pseudo-pelletierine is crystalline. These alkaloids exist to the extent of about 0·5 per cent. in the stem-bark, and 0·6 to 0·7 per cent. in the root-bark (Ewers, 1899). The average of commercial bark appears to be about 0·35 per cent. Carr and Reynolds (1908) found only 0·12 to 0·29 per cent.; other authorities give 0·5 to 0·7 per cent. The freshly dried bark has been found (in Java) to yield as much as 3·0 per cent., and indications are not wanting that the percentage of alkaloid diminishes on keeping.

Pomegranate bark contains further about 22 per cent. of gallotannic acid.

*Pelletierine* ($C_8H_{15}NO$) (Tanret, 1878) or punicine (Bender, 1885) is a colourless liquid boiling at 195°, but rapidly assuming a brown colour. *Isopelletierine* closely resembles it; *methylpelletierine* boils at 215°. *Pseudopelletierine* has also been called *n*-methylgranatonine (Cimician and Silber, 1893).

**Uses.**—Pomegranate bark has an anthelmintic and slightly irritant action, but is somewhat astringent unless taken freely. It is used in the treatment of tapeworm, which is expelled (not actually killed) by the decoction, or by the sulphate of pelletierine.

## BLACK HAW BARK

### (Cortex Viburni Prunifolii)

**Source, &c.**—Black haw bark is the root-bark of *Viburnum prunifolium*, Linné (N.O. *Caprifoliaceæ*), a shrub indigenous to the eastern and central United States.

**Description.**—The drug occurs in short, quilled, channelled or curved pieces from 1 to 4 mm. (usually about 2 or 3 mm.) in thickness, and of a dull brown or reddish brown colour. In the youngest pieces the outer surface is slightly longitudinally wrinkled, older pieces exhibit small, rounded, or oval lenticels, while in old bark the surface is irregularly wrinkled, fissured, and scaly. Inner surface longitudinally striated, rough and reddish brown. Fracture short and granular, the fractured surface exhibiting a dark brown cork and a whitish or reddish cortex and bast, in which numerous pale yellowish groups of sclerenchymatous cells are distinguishable. The odour is slight, the taste bitter and astringent. Portions of the smaller roots often occur in the drug.

The student should observe

(*a*) The *dull outer surface*,
(*b*) The *granular fracture*,
(*c*) The *bitter taste*;

and should compare the bark with

(i) *Wild Cherry bark*, which has a glossy outer surface,

(ii) *Pale Cinchona Bark*, which has a more fibrous fracture and (usually) rougher surface.

**Constituents.**—Black haw bark contains viburnin (a water-soluble bitter glucoside), tannin, and a little resin.

**Uses.**—The drug has been used for dysmenorrhœa and asthma it is supposed to prevent threatened abortion and to check hæmorrhage.

## CINCHONA BARK

### (Cortex Cinchonæ)

**Source, &c.**—The genus *Cinchona* embraces, according to Baillon, about twenty species, all of which are indigenous to South America and restricted in that country to the chain of the Andes extending from western Venezuela, through New Granada, Ecuador, and Peru, to Bolivia. On the spurs of this mountain range, at an elevation of about 5,000 to 7,000 feet, where the climate is warm and moist, the cinchona trees occur usually singly, not forming forests and seldom groups. They are evergreen shrubs or trees, frequently of handsome appearance and considerable size, attaining upwards of 100 feet in height.

The natives of Peru and Bolivia appear to have been only imperfectly acquainted with the febrifuge properties of the bark of these trees; at least they seldom employed it. After the conquest of Peru the bark found its way into Spain, probably by the aid of the Jesuits, who were frequently instrumental in introducing new drugs into Europe; it was known as Countess bark, Jesuit's bark, or Peruvian bark, and early in the eighteenth century the trade in it at Loxa in Ecuador had assumed considerable proportions.

About this time (1736) the expedition sent by the Paris Academy of Sciences to measure a degree of the earth's surface at the equator, which was accompanied by the botanist Jussieu, found trees, hitherto unknown, that yielded valuable cinchona bark. The botany of these trees was subsequently specially investigated by Mutis (1760), Ruiz and Pavon (1778–1788), Weddell (1845–1848), and others.

The method of collecting the bark—viz. by felling the tree and stripping the bark from it—very soon aroused fears that the trees would eventually be exterminated. Attempts were made by the Jesuits to induce the bark collectors to plant young trees to replace those that they destroyed, and suggestions and attempts to cultivate the trees were not wanting. These eventually culminated in Markham's expedition to Peru and Bolivia (1859), which was successful in introducing *Cinchona succirubra*, *C. officinalis*, and other species into

British India, the cultivation being commenced at positions in the Himalayas and Neilgherries that had already been recommended by Royle.

A few years previously the German botanist Hasskarl was commissioned by the Dutch on a similar expedition, and succeeded in bringing plants and seeds to Java.

From small beginnings the cultivation of cinchona trees has rapidly assumed such enormous dimensions that the world is practically independent of South America for its supply of cinchona bark, and consequently of quinine. At present about nine-tenths of the bark required is derived from *C. Ledgeriana*, cultivated in Java. The cultivation of this species is now being pushed forward in the British Government plantations in India.

The chief species of cinchonas that yield commercially valuable barks are:

1. *C. Ledgeriana*, Moens (Southern Peru and Bolivia).
2. *Cinchona Calisaya*, Weddell (Southern Peru and Bolivia).
3. *C. officinalis*, Linné (Ecuador and Peru).
4. *C. succirubra*, Pavon (Ecuador).

The first three of these species—viz. *C. Ledgeriana*, *C. Calisaya*, and *C. officinalis*—yield barks rich in quinine, and it is to the cultivation of these three species, especially *C. Ledgeriana*, that attention is now being directed. *C. succirubra* has been largely grown in India, but is now being replaced by *C. Ledgeriana*. *C. lancifolia*, Mutis, and certain other species yield barks that are poor in quinine; they are therefore not cultivated, but nevertheless these barks are occasionally imported from South America and used to some extent as sources of the cinchona alkaloids.

**Collection.**—The following methods have been adopted in collecting cinchona bark:

1. *Felling.*—In South America the bark is collected by felling the tree, stripping the bark from it, and drying it either in the sun or over a gentle fire in huts. Large thick pieces from the trunk and large branches are pressed under weights, and often freed from the dead outer portions (bark), and occur therefore in commerce in the form of thick, flat, heavy pieces (*flat bark*); the bark from the smaller branches curls as it dries into quills (*quilled bark*). Hence the bark from the same tree may occur in two forms differing very considerably in appearance. Cultivated trees are not allowed to attain the age or size of the South American forest trees, and yield therefore no flat bark.

2. *Mossing* and *Renewing.*—MacIvor found in 1863 that, if the bark were removed in longitudinal strips and the trunk were afterwards protected by covering it with moss, the cambium rapidly replaced the bark that had been removed by a fresh growth, and this fresh

growth proved to be richer in alkaloid than the natural bark of the tree. This method of collecting the bark was largely adopted in India, and was practised as follows : When the trees had reached a sufficient age longitudinal incisions were made and alternate strips of bark about 4 or 5 cm. wide removed, leaving the intervening strips untouched. The bark thus collected, which had not been subjected to any artificial treatment, was known as 'natural bark.' The trunk was then covered with moss, paper, straw, or other protecting material, and left to replace the bark that had been stripped off. After a sufficient time had elapsed the covering was removed, and the strips that had been left untouched on the first occasion were then collected ; these had spent part of their life under a protecting covering of moss, and were called 'mossed' bark. The tree was then again covered, and having been deprived by the first two strippings of all its original bark, yielded, when visited for the third time, bark that had been entirely developed under the covering of moss to replace that which had been removed ; this was 'renewed' bark.

The process is interesting but laborious and expensive.

3. *Shaving*.—By this modification of renewing, only a portion of the bark is removed by shaving ; the remainder is left on as a protection to the tree, which therefore does not require any covering of moss.

Both these methods have been abandoned in favour of the following :

4. *Uprooting*.—The tree is allowed to grow until it has attained the age at which it yields the maximum proportion of alkaloid. It is then uprooted and the bark stripped from the root as well as from the stem. By this means the valuable root-bark is secured, and the land can then be re-planted.

5. *Coppicing*.—The tree is cut down to form stools, from which adventitious shoots arise. These shoots yield handsome quills of bark, and the method is specially suited for the production of the quilled bark sold to the pharmacist.

The bark is usually collected in the rainy season, when it separates easily from the stem ; the colour of the inner surface of the fresh bark is always pale, but by the action of the air a change in the tannin rapidly takes place, and the bark assumes a brown or red colour. The drying is usually effected in the sun, or frequently by artificial heat in a specially constructed drying machine. For exportation, the bark is usually pressed by hydraulic pressure into firm bales (as in Ceylon) or stamped into sacks (as in Java) ; fine quills are carefully packed in cases.

**Description.**—The commercial varieties of cinchona bark yielded by the following species of *Cinchona* may be briefly described.

1. *C. Calisaya*.—The bark yielded by this species is sometimes called 'yellow' bark, but as this term is also applied to all cinchona barks exhibiting a distinctly yellowish brown colour it is better to specify

this variety of yellow bark as calisaya bark. It was formerly imported in two distinct varieties, viz. (a) *flat* and (b) *quill calisaya*, but the former is now seldom seen.

(a) *Flat calisaya* was formerly imported in thick, flattish, heavy pieces 20 cm. or more in length, from 5 to 10 cm. in width, and varying frequently from 6 to 12 mm. in thickness. The inner surface was tawny yellowish brown in colour and showed a close fibrous structure,

A B

FIG. 132.—Calisaya bark. A, showing the longitudinal fissures and transverse cracks, natural size; B, showing in addition exfoliating cork, natural size.

the undulating course of the fibres often communicating a wavy appearance to the bark. The outer surface was darker and marked with broad, shallow, longitudinal depressions (digital furrows). These were caused by the formation in the bark of concave lines of cork, by which shallow curved pieces of the bark had been cut off. This outer bark was commonly removed from the drug before exportation.

(b) *Quill calisaya.*—This, which at present is the variety of calisaya bark commonly seen, is principally obtained from plantations of *C. Calisaya* in Bolivia and Java. It occurs in quills varying usually from 12 to 25 mm. in diameter and 30 cm. or more in length, fine specimens attaining 60 cm. in length and 5 to 8 cm. in diameter. The outer surface is of a dull dark grey or dull brownish colour marked with lighter, whitish patches. The outer layer is rugged, and exhibits shallow, rather broad longitudinal fissures that are frequently of a brownish colour, and hence, even if not deep, are easily seen. Transverse cracks mark the bark at distances of 6 to 12 mm. This layer shows in many pieces a decided disposition to exfoliate, the inner portion, which is of a dull yellowish brown colour, bearing impressions corresponding to the cracks of the cork.

The bark breaks with a shortly fibrous fracture; the section exhibits a narrow dark brown outer layer (cork) and brown inner portion.

The taste is distinctly bitter and astringent, the former quality predominating.

2. *C. Ledgeriana*.—This species, which is considered by some botanists to be a variety of *C. Calisaya*, yields a bark that very closely resembles quill calisaya in general appearance and is remarkable for its richness in quinine. On this account the tree is being largely cultivated in Java and also in India in preference to other species; most of the 'yellow' bark of commerce is Ledger bark imported from Java.

Commercial Ledger bark occurs usually in single, sometimes in double, quills about the same size as quill calisaya. The colour internally is a dull tawny brown, often with a reddish tinge; externally the quills mostly exhibit large light grey or whitish patches of lichen.

They are marked with longitudinal furrows and transverse cracks, which, however, are usually more numerous and less conspicuous than in quill calisaya; hence the bark is rougher than quill calisaya. Some pieces bear distinct longitudinal ridges and scattered reddish warts that recall typical red bark (see below), but from this bark they are easily distinguished by their colour and by their taste, which is bitter but not markedly astringent. Sometimes the cork shows a distinct disposition to exfoliate as it does in quill calisaya, but in this feature the bark varies considerably.

As already observed, Ledger bark bears a close resemblance to quill calisaya; it is best distinguished by its more numerous and less conspicuous longitudinal fissures and transverse cracks.

3. *C. officinalis*.—The bark of this species, commercially known as *pale cinchona* or *crown bark*, is obtained chiefly from India, although some is imported from South America. It occurs in quills only, and these are much narrower than those of quill calisaya, seldom exceeding 12 mm. in diameter, the bark itself being usually less than 1·5 mm. thick. They are frequently inrolled on both margins, so as to form double quills. The outer surface is of a dull brown colour, and often has foliaceous lichens adhering to it. Typical pieces are marked with numerous transverse cracks often less than 6 mm. apart, in addition to which there are numerous less prominent longitudinal cracks, all of which, but especially the transverse, tend to impart to the bark a roughness to the touch that is characteristic.

The colour of the inner surface is usually yellowish brown, and the taste resembles that of calisaya bark in being bitter as well as astringent.

4. *C. succirubra.*—This species yields a drug that is characterised by a more or less distinct reddish colour, and is therefore usually known as 'red bark'; it occurs in two forms, viz. 'flat' and 'quill.' The former is occasionally imported from South America; the latter is obtained from cultivated trees, and is imported principally from Java.

**Fig. 133.—Pale bark (*C. officinalis*), showing very numerous longitudinal and transverse cracks. Large specimen, natural size.**

(a) *Flat red bark* occurs in flattish pieces, often of considerable size, and attaining 20 mm. in thickness, though usually thinner; in these respects it resembles flat calisaya, but it differs from that bark essentially in having the outer bark attached; the latter is rugged, of a dusky, ferruginous red colour, and marked with longitudinal ridges of cork as well as brighter red warts. The inner surface has also a distinctly red colour, and does not exhibit the wavy fibrous structure characteristic of flat calisaya.

The bark has a bitter and markedly astringent taste.

(b) *Quill red bark.*—The only official cinchona bark is that obtained from the stem and branches of cultivated plants of *C. succirubra.* The quills vary in size, but are often about 25 mm. in diameter. The outer surface is of a dull brownish grey or reddish brown colour, and often bears numerous greyish lichens attached to it. It is always more or less strongly wrinkled longitudinally, and marked with warts which are sometimes small and numerous or sometimes larger and scattered; in the latter case they are usually either reddish in colour or exhibit a reddish colour when broken. Some varieties of the bark bear small transverse cracks and reddish warts, the longitudinal wrinkles being less pronounced. In thickness the bark varies from 2 to 4 mm. The colour of the inner surface is in typical specimens reddish brown, but that of the interior of the bark is yellowish brown. This colour is due to a change in the tannin of the bark, by which a reddish phlobaphene is produced. The bark has a distinctly astringent and bitter taste. The Pharmacopœia requires that it shall yield between 5 and 6 per cent. of total alkaloid, of which not less than half should consist of quinine and cinchonidine.

*Branch bark* may generally be distinguished from trunk bark by being thinner, by bearing more numerous wrinkles and small warts, and by being more strongly inrolled.

Both varieties are characterised by their taste, which is more markedly astringent than that of pale or yellow bark.

5. *C. lancifolia*.—The cinchona barks yielded by *C. lancifolia* (Columbian, Cartagena barks) occur both in single quills and in flattish pieces; they are usually more or less spongy in texture and reddish brown in colour. They are characterised and easily dis-

Fig. 134.—Red Cinchona bark, showing longitudinal wrinkles (A), reddish warts and small transverse cracks (B). Slightly reduced.

tinguished by the presence of smaller or larger patches of silvery cork which are to be found on almost every piece. The bark has an astringent bitterish taste.

**Microscopical Characters.**—The transverse section of quilled red cinchona of moderate thickness exhibits an abundant cork layer, a cortex containing small starch grains, cells filled with sandy crystals of calcium oxalate, and, on its inner margin, large isolated oval laticiferous cells. The secondary bast contains large, striated bast fibres either singly or in small radially elongated groups; these bast fibres are spindle-shaped and have conspicuous more or less funnel-shaped pits. The walls of all the parenchymatous cells are dark reddish brown in colour.

The student should carefully compare the principal varieties of cinchona bark and note the following prominent characters:

1. Quilled barks:
   (*a*) *C. succirubra*; longitudinal wrinkles and reddish warts; a more or less spongy bark.
   (*b*) *C. Calisaya*; longitudinal furrows, transverse cracks, cork often exfoliating; a firm hard bark.
   (*c*) *C. officinalis*; very numerous small transverse and longitudinal cracks; quills usually very small and rough to the touch; a firm bark.
   (*d*) *C. lancifolia*; more or less uniformly smooth surface with patches of silvery-grey cork.

FIG. 135.—Bark of *Cinchona lancifolia*, showing patches of silvery cork. Natural size.

   (*e*) *C. Ledgeriana*; cracks more numerous than in *C. Calisaya*; bark rougher.

2. Flat barks:
   (*a*) *C. succirubra*; ferruginous red colour, reddish warts, and raised ridges; usually in thick pieces.
   (*b*) *C. lancifolia*; exhibits the characteristic silvery cork.
   (*c*) *C. Calisaya*; thick pieces with a wavy fibrous surface exhibiting depressions.

**Constituents.**—The chief constituents of cinchona barks are the alkaloids they contain, of which a number have been isolated. Some of these are well-defined crystalline substances; some have been obtained only in an amorphous condition; of others it is questionable whether they are not produced from other pre-existing alkaloids during the process of isolation.

The principal alkaloids in cinchona are quinine, cinchonidine, cinchonine, and quinidine; next in importance are hydroquinine, hydrocinchonidine, quinamine, and homocinchonidine. (For a complete list see Henry, 'Plant Alkaloids,' 1913, p. 130.)

In addition to the alkaloids, many cinchona barks also contain a very bitter amorphous glucoside, quinovin; a crystalline organic acid, quinic or kinic acid, also found in coffee, in the whortleberry (*Vaccinium Myrtillus*, Linné), and other plants; a particular tannin, cinchotannic acid, which by oxidation rapidly yields a dark-coloured phlobaphene, cinchona red; starch, calcium oxalate, &c.

The amount of total alkaloid present in cinchona barks is subject to great variation. Quilled red bark from India averages about 6·5 per cent.; that from Java is richer, yielding about 8·25 per cent. Calisaya quills afford about 6 or 7 per cent. and pale bark about 6 per cent. Ledger bark from Java is the richest of all, yielding from 5 to 10 or even more per cent. of total alkaloid. Root-bark is the richest, and stem-bark is better than branch-bark. The relative value of a bark is, however, determined by the proportion of quinine it contains.

Quinine was first isolated by Pelletier and Caventou in 1820, after Gomez in 1811 had produced from cinchona a crystalline combination of quinine and cinchonine. It occurs in the largest proportion in Ledger bark, the highest recorded yield being 14·5 per cent. This, however, is quite exceptional, the quinine in commercial Ledger bark averaging from 3·0 to 8·0 per cent. Bolivian cultivated calisaya contains from 3·0 to 4·0, and cultivated *C. officinalis* about 3·0 per cent. Indian red bark contains about 1·5 per cent. of quinine, but that from Java is richer, attaining as much as 5 per cent.

Cinchonine is found in small quantity in most of the cinchona barks, especially *C. lancifolia* and some specimens of *C. succirubra*, this alkaloid is frequently more abundant in the root-bark than in the stem-bark.

Quinidine, discovered by Henry and Delondre in 1833, seldom exceeds 0·5 per cent. in any bark; it occurs chiefly in certain varieties of *C. Calisaya*.

Cinchonidine, isolated in 1847 by Winckler, is found more generally distributed and in much larger proportion than quinidine. The cultivated *C. succirubra* at present exported from India contains as a rule more cinchonidine than quinine, from 3 to 4 per cent. being frequently present.

All these alkaloids exist, according to de Vrij, combined with cinchotannic acid in the parenchymatous tissue of the bark.

**Uses.**—The cinchona barks are far too bulky for use as antiperiodics and antipyretics if quinine can be obtained. They are therefore given only as bitter stomachics and tonics. The amount of tannin contained in them indicates that they may be used when an astringent effect also is desired.

**Allied Drugs.**—*Cuprea Bark*; the bark of *Remijia pedunculata*, Triana, and *R. purdieana*, Triana (N.O. *Rubiaceæ*, Columbia); coppery red, dense, very hard and breaks with a short granular fracture; contains quinine, cupreine, quinidine, cinchonine, and cinchonamine; not now in commerce.

Various species of *Cascarilla*, *Exostemma* and *Stenostoma*; these barks contain none of the cinchona alkaloids.

## ALSTONIA

### (Alstonia, Dita Bark)

**Source, &c.**—The official alstonia bark may be obtained either from *Alstonia scholaris*, Robert Brown (N.O. *Apocynaceæ*), a tree indigenous to India and the Philippine Islands or from *A. constricta*, P. Mueller, a native of Australia; the latter bark is also known as Australian fever bark. The former is the variety usually found in commerce.

**Description.**—*Alstonia scholaris*. This bark, also known as dita bark, occurs in single quills or in irregular curved pieces of varying size and appearance. That from older stems or branches is commonly in small, curved or channelled pieces, or quills, and is of a light, yellowish brown colour, rough, irregularly fissured, and spongy externally; internally it is darker. Such pieces break with a short fracture, the section exhibiting a narrow inner portion (cortex and secondary bast) traversed by numerous, fine medullary rays and a spongy outer portion (cork) of varying thickness. The bark from young branches bears scattered, pale lenticels and is very fibrous. Both kinds contain numerous, pitted, sclerenchymatous cells, laticiferous vessels, and prismatic crystals of calcium oxalate, but the young bark contains abundant pericyclic fibres which are seldom visible in old bark. Taste bitter, no odour.

*Alstonia constricta*. The bark is in quills or channelled pieces, often of considerable size. Outer surface brown or yellowish brown in colour and deeply fissured both longitudinally and transversely; inner surface cinnamon brown in colour and deeply striated. The transverse section exhibits an abundant, dark brown cork, within which is a yellowish brown secondary bast. The latter exhibits under the microscope abundant bast fibres in tangentially arranged groups. The taste is very bitter. The aqueous infusion is yellowish and shows a well marked blue fluorescence.

**Constituents.**—*Alstonia scholaris*. The chief constituents are the alkaloids ditamine (bitter, crystalline powder), echitenine (amorphous, bitter powder) and echitamine or ditaine. The bark also contains echicerin (crystalline, non-nitrogenous), echicaoutchin (resembles caoutchouc), echitin and echitein (both crystalline), and echiretin (amorphous); all these constituents appear to be devoid of therapeutic value.

*Alstonia constricta.* This bark contains the alkaloids alstonine (chlorogenine) and porphyrine; also alstonidine and porphyrosine concerning which little definite is known. Porphyrine in acid solution exhibits a blue fluorescence.

**Uses.**—The bark is used in India for malaria but is inferior in value to cinchona bark; it has also been used as a tonic, anthelmintic, and in chronic diarrhœa.

## QUEBRACHO

### (White Quebracho, Quebracha, Aspidosperma)

**Source, &c.**—The dried bark of *Aspidosperma quebracho-blanco*, Schlechtendal (N.O. *Apocynaceæ*), a large tree indigenous to the Argentine.

**Description.**—Nearly flat pieces, 1 to 3 cm. thick; outer surface, yellowish grey or brownish, deeply fissured, inner surface yellowish brown or reddish brown, distinctly striated. Fracture granular in the outer, lighter portion, splintery in inner darker portion. Transverse section shows numerous whitish dots and striæ arranged tangentially. Taste very bitter and slightly aromatic; no odour.

**Constituents.**—Quebracho bark contains the alkaloids aspidospermine, aspidospermatine, aspidosamine, quebrachine, hypoquebrachine, and quebrachamine; further starch, tannin, and the sugars inosite and quebrachite. Commercial aspidospermine is usually a mixture of alkaloids.

**Uses.**—The bark has been used as a tonic and also in dyspnœa; large doses may cause vomiting.

**Note.**—This bark must not be confused with quebracho colorado, the bark of *Loxopterygium Lourenzii*, Grisebach (N.O. *Terebinthaceæ*, Argentine); an extract of the wood is largely used in the tanning industries.

## CINNAMON BARK

### (Cortex Cinnamomi)

**Source, &c.**—The cinnamon tree, *Cinnamomum zeylanicum*, Breyn (N.O. *Laurineæ*), is a small evergreen tree indigenous to and cultivated in Ceylon.

Cinnamon bark appears to have been collected from wild plants and exported towards the end of the thirteenth century. After the occupation of the island by the Portuguese in 1536, the exportation became more regular. In 1770 the cultivation of the tree was successfully carried out by the Dutch, who, as in other cases, made strenuous efforts to retain the cinnamon trade in their own hands, controlling the supply and the price. Soon afterwards the English obtained possession of the island, and the trade in cinnamon was diverted from Amsterdam to London.

Cinnamon is now almost entirely obtained from cultivated plants. These are cut down to form stools, from which adventitious shoots arise. When these are from 1·5 to 2 metres long and about a year and a half or two years old, they are cut down during the rainy season. The shoots are trimmed from the leaves, &c., ringed at the nodes with a brass or copper knife (to avoid the discoloration that steel would cause), and the bark removed in strips, which are allowed to remain exposed in heaps for about twenty-four hours. Each strip is then stretched upon a stick and the epidermis and cortex scraped off, great care being taken that neither too much nor too little is removed. They are then packed inside one another so as to form sticks, which are dried, cut to a definite length, sorted, and made into bundles. In the London docks each bundle is opened, carefully sorted and again made up; the bundles are then classified as 'firsts,' 'seconds,' 'thirds,' or 'fourths.' The trimmings are either exported as such (cinnamon chips), or are used in the island for the distillation of the volatile oil. The leaves and petioles also yield an oil, which, however, is less valuable than that from the bark.

**Description.**—Cinnamon occurs in long, slender, flexible sticks about 1 metre in length and 6 mm. in width, each consisting of numerous (about forty) channelled pieces or single quills, about 1 to 2 dm. in length, skilfully packed into one another, the largest on the outside, so as to form a long stick of compound double quills; such a stick may easily be separated into its component parts after it has been soaked in water.

Each of the pieces of bark of which the stick is composed is of papery thickness and of a dull, pale brown colour. The outer surface is marked with paler, glossy, undulating, longitudinal lines (bundles of bast fibres), and show here and there scars or holes, indicating the insertion of leaves or lateral shoots, but is quite free of epidermis or cork. The inner surface is rather darker than the outer, and finely striated longitudinally. The drug consists almost entirely of secondary bast, the epidermis (or cork) and cortex having been removed by scraping.

The fracture is short and rather splintery; the transverse section shows an outer pale layer (sclerenchymatous cells) and an inner dark layer (bast).

Cinnamon has a fragrant odour and a warm, sweet, aromatic taste.

The student should observe

(*a*) The *uniform colour* and *absence of cork*,
(*b*) The *narrow wavy longitudinal lines*,
(*c*) The *odour and taste*.

**Constituents.**—The principal constituent of cinnamon bark is the volatile oil, of which it yields 0·5 to 1·0 per cent.; the bark contains also tannin and mucilage. Inferior qualities are generally more mucilaginous and contain a volatile oil of inferior fragrance. The drug yields about 4 per cent. of ash.

The volatile oil (sp. gr. 1·000 to 1·030; O.R. – 0·5° to – 1°) contains from 55 to 65 per cent. of cinnamic aldehyde together with eugenol (4 to 8 per cent.), terpenes and small quantities of numerous other bodies. The amount of cinnamic aldehyde present may be determined by the official process of assay.

**Uses.**—Cinnamon is used chiefly as a flavouring agent in astringent powders and tinctures. It has aromatic and mildly astringent properties.

**Varieties.**—As already mentioned four grades of cinnamon bark are recognised on the London market. In addition the following may be noted.

*Jungle Cinnamon.* Obtained from wild plants; the bark is darker, coarser, less carefully trimmed and less aromatic; it closely resembles cassia bark (see below).

*Saigon Cinnamon.* Official in the U.S. Pharmacopœia; obtained from an undetermined species of *Cinnamomum*; quillsa bout 15 cm. long, 10 to 15 mm. wide, and 2 to 3 mm. thick; greyish or greyish brown with lighter patches; warty and ridged; taste sweeter, odour stronger than Ceylon cinnamon.

*Java Cinnamon.* From *C. Burmanni*, Blume; odour less delicate than that of cinnamon; volatile oil contains about 75 per cent. of cinnamic aldehyde; used in Holland; the cells of the medullary rays contain small tabular crystals of calcium oxalate, whereas in cinnamon they contain minute needles.

## CASSIA BARK

### (Cassia Lignea, Chinese Cassia)

**Source, &c.**—Cassia bark is obtained from *Cinnamomum Cassia*, Blume (N.O. *Laurineæ*), a medium-sized tree, probably a native of Cochin China, but cultivated now in the south-eastern provinces of the Chinese Empire (Kwang-si and Kwang-Tung). This tree yields the bark known in English commerce as cassia bark or Chinese cassia lignea. Other species of *Cinnamomum* growing in the country between these provinces of China and north-eastern India yield barks to which the name of Cassia vera (or sometimes also Cassia lignea) is given; they are exported from Calcutta and Saigon. The cassia vera of the London market is a firm, rather thick bark with a very mucilaginous taste; it is said to be the bark of *C. Burmanni*, de Candolle.

The bark is collected entirely from cultivated trees. When about six years old the branches are cut and all the small twigs and leaves are stripped off; two longitudinal slits are then made, and three or four transverse incisions are cut round the circumference through the bark at intervals of about 40 cm. The bark is then removed in pieces about 40 cm. long and half the circumference of the branch. These are next laid with the concave surface downwards and a small plane passed over them, by which the cork and part of the cortex are more or less completely removed. The bark is then tied up into bundles and exported in boxes resembling tea chests, which are sometimes wrapped in bast-mats.

**Description**—Cassia bark is imported in bundles about 30 or 40 cm. long and weighing about 500 grammes. The pieces of which the bundle is composed vary from 5 or 6 to 40 cm. in length and average from 1 to 2 cm. in width and 3 to 5 mm. in thickness. They are either channelled pieces or single (but not double) quills, of a dark, earthy brown colour and smooth, but with patches of the thin greyish cork still adhering to the outer surface, indicating a want of care in trimming them.

The fracture is short, the section of the thicker pieces showing a faint white line (sclerenchymatous cells) sometimes near the centre, sometimes near the outer margin and parallel to it. In odour and taste cassia bark resembles cinnamon, but it is less delicate in aroma and more mucilaginous and astringent.

Cassia bark occurs in much larger and thicker pieces than cinnamon, seldom in double quills, and never packed into sticks. It is darker in colour, and frequently exhibits patches of cork on the outer surface.

The student should observe

(*a*) The *thickness of the bark* and its *dark colour*,
(*b*) The *patches of cork*,
(*c*) The *characteristic odour* and *taste*.

**Constituents.**—The constituents of cassia bark are similar to those of cinnamon. It yields from 1 to 2 per cent. of *volatile oil*, resembling that of cinnamon but having a higher specific gravity (1·050 to 1·070) and containing more cinnamic aldehyde (75 per cent.) but no eugenol.

## OLIVERI CORTEX

### (Oliver Bark, Black Sassafras)

**Source, &c.**—Oliver bark is the dried bark of *Cinnamomum Oliveri*, Bailey (N.O. *Laurineæ*), a tree indigenous to New South Wales and Queensland.

**Description.**—Flat strips about 20 cm. long, 4 cm. wide and 1 cm. thick; outer surface brownish with patches of whitish cork, very coarsely granular or warty; inner surface umber brown, finely striated, satiny. Fracture short, somewhat fibrous. Section exhibits a somewhat thick periderm, often separated from the inner part of the secondary bast by a paler line of cork cells. Odour aromatic, recalling that of sassafras; taste aromatic bitter and camphoraceous.

**Constituents.**—The chief constituent is a yellow, volatile oil containing safrol, eugenol, cineol and cinnamic aldehyde; the bark also contains tannin.

**Uses.**—As a substitute for cinnamon.

FIG. 136.—Greenheart bark, showing short fracture. Natural size.

## BEBEERU BARK

### (Bibiru Bark, Greenheart Bark, Cortex Nectandræ)

**Source, &c.**—The greenheart tree, *Nectandra Rodiæi*, Hooker (N.O. *Laurineæ*), is a large forest tree growing abundantly on the hills in *British Guiana*. Its tall, straight stem yields a hard and resistant wood that is highly valued for shipbuilding; the bark was recommended early in the present century as a substitute for cinchona bark, and the alkaloid obtained from it in 1835 by Rodie as a substitute for quinine. It aroused some interest at first, but now neither the bark nor the alkaloid obtained from it is much used.

**Description.**—Bebeeru bark occurs in flat, heavy pieces, frequently 10 to 15 cm. long, 5 to 8 cm. wide, and 3 to 10 mm. thick. It is of a more or less uniform greyish brown colour, and frequently marked on the outer surface with broad shallow depressions left by the exfoliation of the outer portions by the formation of bands of cork; these exfoliating portions are occasionally but not often found adhering to the bark. The outer layer is usually a very thin greyish brown, often warty cork, which can easily be scraped off, disclosing a darker inner portion. The inner surface is walnut-brown, bears shallow, rather broad longitudinal depressions, and is coarsely longitudinally striated by strands of sclerenchymatous tissue, which, under the lens, can be seen slightly projecting beyond the remaining parenchymatous tissue.

The bark is extremely hard, and breaks with a short granular fracture. The transverse section, smoothed and moistened, exhibits under the lens a very narrow, pale grey cork; the remainder of the bark is completely traversed by closely approximated, yellowish, wavy medullary rays, showing that in the majority of cases the drug consists (with the exception of the cork) of bast tissue. Between the medullary rays numerous minute groups of sclerenchymatous cells can be distinguished, arranged in radial lines.

The bark has no odour, but a bitter taste.

The student should observe

(*a*) The *flat heavy pieces*,
(*b*) The *thin grey cork*,
(*c*) The *structure of the transverse section*;

and should compare this bark with

(i) *Sassy bark*, which is seldom flat, is often nearly black on the inner surface, and exhibits in transverse section large conspicuous groups of sclerenchymatous cells;
(ii) *Coto bark*, which has a reddish brown colour, is usually much thicker, and has a characteristic odour and taste;
(iii) *Elm bark*, which is much paler in colour and fibrous.

**Constituents.**—The chief constituents of bebeeru bark are the alkaloids beberine and siperine and probably others that have not yet been isolated.

Beberine, $C_{19}H_{21}NO_3$, crystallises in colourless prisms melting at 214°. It has a bitter taste, and forms colourless crystalline salts. It is identical with pelosine (see 'Pareira brava'), but distinct from buxine, a similar alkaloid occurring in box.

*Siperine* has been obtained in the form of dark scales of very doubtful purity.

The commercial sulphate of beberine in dark brown scales, which was official in the British Pharmacopœia of 1885, is not a pure sulphate of the alkaloid. It contains about 30 per cent. of beberine associated with siperine and probably other alkaloids, as well as with much colouring matter.

**Uses.**—Bebeeru bark is a bitter stomachic and tonic; the alkaloid is, to a small extent, antipyretic, but these effects being insignificant, its use in fever and ague has now been abandoned.

## COTO BARK

### (True Coto and Paracoto Bark)

**Source, &c.**—Coto bark was first sent to Europe about 1873 from Bolivia under the name of coto-cinchona. Its botanical source is unknown, but it appears to be derived from a tree belonging to the natural order *Laurineæ*, and probably closely allied to the genus *Cryptocarya* (Hartwich, 1899). The bark was employed in the form of powder or alcoholic tincture for diarrhœa and for neuralgia, and rapidly gained a reputation in Europe as an astringent. Owing, however, partly, to the difficulty of obtaining the genuine bark, it fell into disuse, and is now seldom prescribed.

This, the true coto bark, was in commerce but a very short time. In 1876 it was replaced by a very similar bark, also imported from Bolivia. This later bark was found on examination to yield constituents resembling, but not identical with, those of true coto; one of these was called 'paracotoin,' and the bark that yielded it 'paracoto bark.'

Paracoto bark, as it is properly termed, is at present the current variety of coto; it is commonly sold as coto bark, and to it the following description refers.

FIG. 137.—Paracoto bark, showing a fissured cork and splintery-fibrous fracture. Natural size.

**Description.**—Paracoto bark occurs in flat or curved pieces commonly up to 30 cm. long, 8 or 10 cm. wide, and 10 to 15 cm. thick. They are hard, heavy, and of a cinnamon-brown colour; some of the pieces bear on their outer surfaces patches of a thin whitish, others of a brownish cork; they are frequently nearly smooth, or bear transverse wrinkles, or are irregularly chequered by longitudinal fissures and transverse cracks. The inner surface is brown and distinctly and coarsely striated, this being due to projecting strands of sclerenchymatous tissue.

The bark breaks with a fibrous, splintery fracture, coarse stiff strands of sclerenchymatous tissue projecting from the fractured surface. The inner part easily separates into coarse longitudinal strips, on the surface of which, especially after they have been exposed to the air for some time, glistening prismatic crystals can be observed even with the naked eye.

The section is very characteristic. The outer layer, a thin brownish cork, encloses a narrow brown cortex separated by a distinct paler line (of sclerenchymatous cells) from the very thick bast which in older barks constitutes nine-tenths of the drug. This bast contains an abundance of sclerenchymatous tissue arranged in rounded or tangentially elongated groups. The section resembles that of sassy bark, but the groups of sclerenchymatous cells are much smaller, and the colour is reddish brown.

The drug has a very characteristic odour and a pungent taste.

The student should observe

(*a*) The *brownish cork* and *brown striated inner surface*,
(*b*) The *characteristic transverse section*, and
(*c*) The *distinctive odour* ;

and should compare this bark with

(i) *Sassy bark*, which has a nearly black inner surface, darker outer surface, and fewer and larger groups of sclerenchymatous cells ;
(ii) *Bebeeru bark*, which is smoother and exhibits distinct medullary rays, but very inconspicuous masses of sclerenchyma ;
(iii) *Flat red cinchona bark*, which is characterised by its rusty red colour, reddish warts, and bitter astringent taste.

**Constituents.**—Paracoto bark contains several crystalline bodies, the chief of which, paracotoin, $C_{12}H_8O_4$, occurs in pale yellow scales melting at 152°, and give a yellowish brown colour with nitric acid ; it is dioxymethylenephenylcumalin. Other crystalline bodies present in the bark are phenylcumalin (leucotin), hydrocotoin (dimethyl ether of benzoylphloroglucinol), methylhydrocotoin, protocotoin (dioxymethylene-phloroglucinol), and methylprotocotoin (oxyleucotin).

The bark also contains piperonylic acid, resin, tannin, and a little volatile oil.

**Uses.**—The drug has been employed with success for diarrhœa, but is not much used now.

**Varieties.**—True coto bark closely resembles paracoto ; it is said to be distinguished by its odour, which recalls cinnamon, that of paracoto recalling nutmeg, and by its taste, which is more pungent. The most definite distinctive characters lie in the constituents. True coto contains cotoin and phenylcumalin, the substance formerly known as dicotoin being probably a mixture of these two substances. Cotoin, $C_4H_{12}O_4$, is the monomethyl ether of benzoylphloroglucinol ;

it crystallises in pale yellow prisms (melting point 130°) which colour nitric acid blood red.

The following simple test is the only definite means of distinguishing these two varieties of coto bark: Shake 10 grammes of the powdered bark with 100 grammes of ether frequently during an hour. Pour off the ethereal solution, add 50 grammes of water and distil off the ether. Cool, shake with 30 grammes of petroleum spirit, transfer to a separator (disregarding the resin which adheres to the sides of the flask), filter the aqueous solution into a porcelain dish, evaporate to dryness, dissolve the residue in a little glacial acetic acid, and add a drop of fuming nitric acid. True coto thus tested will afford a blood red coloration (Cæsar and Loretz).

Resaldol, the ethyl ester of resorcinol-benzoyl carbonic acid is a synthetic product allied to cotoin and having a similar action.

## MEZEREON BARK

### (Cortex Mezerei)

**Source, &c.**—Mezereon bark may, according to the British Pharmacopœia 1898, be obtained from either of the following species of *Daphne* (N.O. *Thymelæacæ*):

1. *Daphne Mezēreum*, Linné, *Mezereon*, a small shrub attaining a metre in height, growing in moist woods in hilly parts of Europe, and found also in the southern counties of England. It bears in early spring, before the leaves appear, purple, sweet-scented flowers on the preceding year's shoots. The bark is collected chiefly in Thuringia.

2. *Daphne Laurēola*, Linné, *Spurge Laurel*, a small indigenous evergreen shrub, not uncommon in woods; it bears inconspicuous green odourless flowers in the axils of the leaves, the latter being crowded towards the summit of the stem. The bark is seldom collected.

3. *Daphne Gnidium*, Linné, a small shrub with numerous straight slender branches and small white flowers; it is a native of the south of France and the north coast of Africa, and is found generally on the shores of the Mediterranean. It is largely collected in Algeria and the south of France, and has been long known and used as an irritant.

The bark of all these plants is collected in the winter or early spring, when it separates readily from both stem and root in long flexible strips. These are dried, and either sold loose or made into small bundles or flat disc-like rolls.

**Description.**—The bark of *Daphne Mezereum* sometimes occurs in quills of varying length, but more usually in long, thin, more or less flattened fibrous strips that are remarkable for their extreme toughness and flexibility; they can easily be torn lengthwise, but it is almost impossible to break them.

The outer surface (cork) is of a yellowish or olive brown colour, very thin, and transversely wrinkled, that of the stem-bark being marked with scattered rounded scars of leaves and buds, and often bearing the minute black apothecia of small lichens. It easily separates in papery fragments from the cortex, which is either green (stem-bark) or yellowish (root-bark) in colour. The inner portion of the bark (bast) has a pale yellowish or nearly white and silky inner surface; it is extremely tough and fibrous from the presence of numerous strands of tough bast fibres.

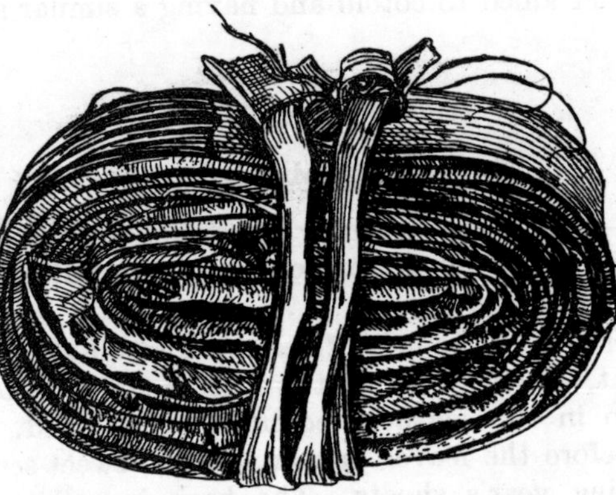

FIG. 138.—Mezereon bark. Slightly reduced.

The dry bark has little or no odour, but a persistent burning acrid taste. So marked is the acrid nature of the bark that, moistened and applied to the skin, it produces inflammation and even vesication.

*Spurge laurel* bark is very similar to the foregoing; it may be distinguished by the purplish grey colour of the cork and by elongated, pointed oval leaf- and bud-scars which are crowded at intervals.

The bark of *D. Gnidium* has a dark purplish brown cork; the leaf-scars resemble those of *D. Mezereum.*

The student should observe

(*a*) The *thin, easily separable cork,*
(*b*) The *silky inner surface,*
(*c*) The *extreme toughness and flexibility.*

**Constituents.**—Mezereon bark contains a greenish brown amorphous resin, mezerein, possessing extremely acrid and sternutatory properties. It easily changes into an acid bitter resin, mezereic acid, which is present in the ethereal and alcoholic extracts of the bark.

A crystalline bitter glucoside, daphnin, has also been isolated, as well as a fixed oil, and a substance resembling euphorbone, neither of which is acrid.

**Uses.**—Mezereon is a powerful local irritant capable of producing vesication. Internally it is stimulant, and in large doses an irritant poison ; it is, however, seldom administered internally, and although frequently employed in the south of France as an irritant, finds but little use in this country.

## CASCARILLA BARK

### (Cascarilla, Cortex Cascarillæ)

**Source, &c.**—Cascarilla bark is obtained from *Croton Eleutĕria*, J. J. Bennett (N.O. *Euphorbiaceæ*), a shrub or small tree indigenous

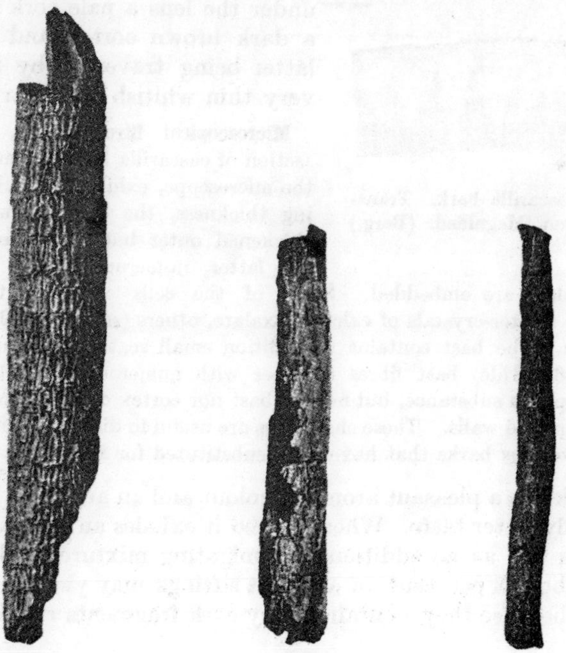

A B C

FIG. 139.—Cascarilla bark. A, large ; B, medium ; C, small. All natural size.

o the Bahama Islands, the name cascarilla having its origin probably n the resemblance this bark bears to small quilled cinchona bark, vhich was formerly called cascarilla (the diminutive of the Spanish *áscara*, bark).

**Description.**—Cascarilla bark is usually imported in single quills or channelled pieces, commonly varying from 5 to 10 cm. in length and from 4 to 6 mm. in width, rarely exceeding the latter limit. The outer layer of the bark is a white or greyish white cork which owes its characteristic chalky appearance to the presence in the cells of numerous crystals of calcium oxalate. It is longitudinally wrinkled, and often, at more distant intervals, both longitudinally and transversely furrowed, thus assuming a chequered appearance. It frequently bears the minute black apothecia of a small lichen, and easily exfoliates, disclosing a brown or dark grey inner layer (cortex) marked with furrows corresponding to those in the cork. On some portions of the bark the white cork is so thin that the brown cortex shows through and imparts to the outer surface a brown or dark grey colour.

The inner surface of the bark is dark in colour and longitudinally striated. The fracture is short and resinous. The section exhibits under the lens a pale cork layer and a dark brown cortex and bast, the latter being traversed by numerous very thin whitish medullary rays.

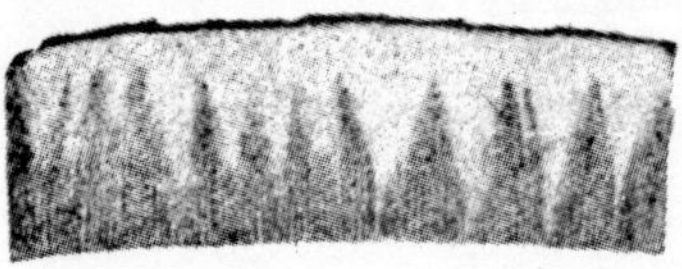

FIG. 140.—Cascarilla bark. Transverse section. Magnified. (Berg.)

**Microscopical Structure.**—A transverse section of cascarilla bark, examined under the microscope, exhibits a cork of varying thickness, the cells of which have thickened outer but thin inner walls; in the latter, numerous minute crystals of calcium oxalate are embedded. Some of the cells of the cortex contain prismatic or cluster-crystals of calcium oxalate, others (secretion cells) droplets of oleo-resin. The bast contains in addition small scattered groups of small (not over 26$\mu$ wide) bast fibres together with numerous cells filled with a brown amorphous substance, but neither bast nor cortex contains any cells with thickened lignified walls. These characters are useful in distinguishing cascarilla bark from various barks that have been substituted for it.

The bark has a pleasant aromatic colour and an aromatic but rather disagreeably bitter taste. When burned it exhales an agreeable odour, whence its use as an addition to fumigating mixtures, tobacco, &c. It yields about 8 per cent. of ash, but siftings may yield much more, probably because they contain many cork fragments rich in calcium oxalate.

The student should observe

(a) The *chalky cork*,
(b) The *longitudinal wrinkles* and *longitudinal and transverse fissures*,
(c) The *short resinous fracture* and the appearance of the *section* under the lens,
(d) The *aromatic odour* and *bitter taste*.

**Constituents.**—Cascarilla contains about 1 per cent. of volatile oil, a crystalline bitter principle, cascarillin, which is neither alkaloidal nor glucosidal, and at least two alkaloids, viz. betaine and cascarilline, the latter crystallising in prismatic plates; it yields from 7·5 to 10·5 per cent. of ash, siftings commonly affording a higher figure.

**Use.**—Cascarilla is used as an aromatic, bitter stomachic.

**Substitutes.**—*Copalchi bark*, from *Croton niveus*, Jacquin (West Indies, Venezuela), occurs in long quills, much larger than those of cascarilla; it has a thin, greyish cork, below which is the cortex marked with minute transverse furrows.

The bark of *Croton lucidus*, Linné (?): the cork is greyish and firmly adherent; the inner surface is striated and pinkish brown, and the transverse section shows numerous groups of sclerenchymatous cells.

The barks of several other species of *Croton* have been found in the commercial drug; genuine cascarilla may be identified by the characters given above.

## SLIPPERY ELM BARK

### (Cortex Ulmi Fulvæ)

**Source, &c.**—The slippery elm, *Ulmus fulva*, Michaux (N.O. *Urticaceæ*), is a small tree indigenous to the central and northern United States. The bark is collected in the spring from the trunk and large branches, deprived of its outer dead portions, and dried; the tree is thereby destroyed, and as the wood has no commercial value no effort is made to replace it. Large quantities of the bark are collected in the lower peninsula of Michigan.

**Description.**—The slippery elm bark of commerce consists entirely of secondary bast. It is commonly imported in large flat strips half to one metre long, but only 3 mm. or less in thickness. The outer surface is reddish yellow in colour, with patches of the reddish brown outer portion (bark), and is distinctly striated longitudinally; the inner surface is tawny yellow and also longitudinally striated.

It is extremely tough and fibrous. The section, examined under the lens, is seen to be completely traversed by medullary rays, between which small tangential bands (bast fibres and bast parenchyma) are arranged, giving the section a chequered appearance. If the transverse section is moistened and allowed to remain for a minute or two and again examined, numerous cells full of transparent swollen mucilage can be detected. No trace of the cortex can be found, but portions of the dark outer layer (bark) are frequently present.

The bark has a strong odour resembling fenugreek, and a very mucilaginous taste.

The student should observe

(*a*) The *striated inner* and *outer surface*,

(*b*) The *fibrous fracture*,

(*c*) The *odour of fenugreek* and the *mucilaginous taste* ;

and should compare the bark with

*Quillaja bark*, which has a splintery fracture, a smooth inner surface, and is odourless.

**Constituents.**—The principal constituent is mucilage, which appears to swell but not dissolve in water ; it is contained in large mucilage cells in the bast, and is present in such proportions that 1 gm. of the powdered bark will convert 50 c.c. of water into a thick jelly.

**Uses.**—The bark has demulcent and emollient properties. It is chiefly used as an external application in the form of a poultice.

## OAK BARK

### (Cortex Quercus)

**Source. &c.**—The British oak, *Quercus robur*, Linné (N.O. *Cupuliferæ*), is widely diffused over Europe and largely cultivated for its wood and especially for its bark, which, from the large proportion of tannin it contains, is valued highly for tanning. Although the astringent properties of the bark have long been known, and although it has been used from time immemorial in preparing leather, it never appears to have been much used in medicine.

In the collection of oak bark the trees are usually felled when they have reached an age of twelve to thirty-five years, and in the early spring when the buds are opening. Longitudinal incisions are made through the bark, which can then be removed in strips, and after drying is ready for the market.

Young bark is preferable to old, because as the trees increase in age the outer portions are cut off by the production of layers of cork in the bast (formation of outer bark), and the tannin in the portions thus cut off undergoes certain changes. Bark from older stems is also collected and freed from its dead outer portions, but such bark is not so valuable for tanning or fit for medicinal use.

From the stools that are formed when the trees are felled adventitious shoots arise, and these, when they have attained a sufficient age, are cut and peeled.

While the bark is young (up to about twenty years old) it

possesses a smooth, glossy, silvery cork; such bark is to be preferred and should alone be used medicinally.

**Description.**—Oak bark usually occurs in channelled pieces 10 to 20 cm. in length and 2 to 3 cm. in breadth. The outer layer is a thin, smooth, shining, silvery-grey, firmly adherent cork, which in young barks is marked with darker transverse lenticels and in older barks is frequently longitudinally fissured and bears darker spots and patches; beneath the cork is a reddish brown cortex.

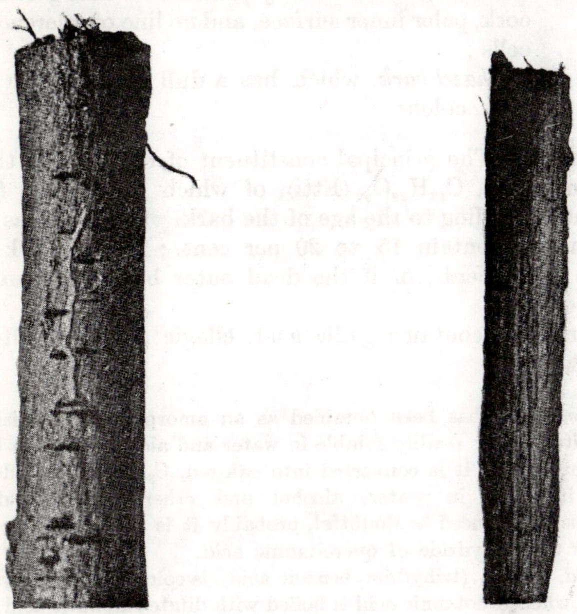

A B

FIG. 141.—Oak bark. A, outer surface, showing the smooth, glossy cork and transverse lenticels. B, inner surface. Natural size.

The inner surface is strongly striated longitudinally and fibrous, and varies in colour from yellowish to reddish brown.

The bark breaks with a short fracture in the outer part (cork and cortex), but is coarsely fibrous in the inner part (bast). Under a lens the section exhibits a thin cork, a narrow yellowish cortex, occupying about one fourth of the total width, separated by a pale line (sclerenchymatous cells) from the reddish brown bast, which is chequered by tangentially arranged groups of bast fibres. Touched with dilute solution of ferric chloride the section assumes a black colour.

The bark has a scarcely perceptible odour, but a strongly astringent taste.

The student should observe

(*a*) The *glossy silvery cork*,
(*b*) The *line of sclerenchymatous cells*,
(*c*) The *striated fibrous inner surface* ;

and should compare the bark with

(i) *Willow bark*, which usually possesses a dull greenish brown cork, paler inner surface, and no line of sclerenchymatous cells ;
(ii) *Witch-hazel bark*, which has a dull grey cork and reddish pink colour.

**Constituents.**—The principal constituent of oak bark is the tannin, quercitannic acid, $C_{17}H_{16}O_9$ (Etti), of which it contains from 5 to 20 per cent. according to the age of the bark. Oak bark, as described above, should contain 15 to 20 per cent. ; trunk bark contains only 5 to 8 per cent., or if the dead outer bark has been removed, 8 to 10 per cent.

The bark also contains gallic acid, ellagic acid, quercite, phloroglucinol, &c.

*Quercitannic acid* has been obtained as an amorphous, yellowish brown or reddish white powder readily soluble in water and alcohol. When boiled with dilute sulphuric acid it is converted into oak-red, $C_{38}H_{26}O_{17}$, a reddish brown substance insoluble in water, alcohol and ether. Whether dextrose is simultaneously produced is doubtful, probably it is not, and oak-red may be regarded as the anhydride of quercitannic acid.

*Gallic acid*, $C_7H_6O_5$, (trihydroxybenzoic acid), is colourless and crystalline ; it is produced when gallotannic acid is boiled with dilute sulphuric acid and occurs in a number of plants.

*Ellagic acid*, $C_{14}H_6O_8$, is pale yellow and crystalline ; it forms a deep yellow solution with caustic alkalies and dark blue with ferric salts.

*Quercite* $C_6H_{12}O_5$, is a sweet, crystalline pentahydric alcohol also found in acorns.

*Phloroglucinol* (phloroglucin), $C_6H_3(OH)_3$, occurs in several plants and may be obtained from many resins, tannins, &c., by fusing with caustic potash.

**Use.**—Oak bark has been used medicinally as an astringent, but is not much prescribed at present.

**Note.**—*Quercus Suber*, Linné, the cork oak (Mediterranean, Algeria) yields commercial cork. The cork first formed is unsuitable for technical purposes and is sold as 'virgin' cork. That which is subsequently produced from the phellogen is stripped from the tree every 8 or 10 years, steamed, scraped and pressed into flat strips.

### *Other Substances used for Tanning*

The following are the chief substances used in the tanning industry together with the approximate percentages of tannin contained in them:

| | Per cent. |
|---|---|
| Oak bark, bark of *Quercus robur*, Linné | 8 to 13 |
| Valonia, acorn cups of *Quercus Ægilops*, Linné | 25 to 35 |
| Hemlock, bark of *Tsuga canadensis*, Carr. | 10 to 14 |
| Mangrove, bark of *Rhizophora mangle*, Linné | 9 to 33 |
| Wattle, bark of *Acacia dealbata*, Link; and of *A. decurrens*, Willdenow | 24 to 30 |
| Catechu, extract from *Acacia Catechu* | 45 to 55 |
| Gambier, extract from *Uncaria Gambier* | 36 to 40 |
| Sumac, leaves of *Rhus Cotinus*, Linné; and of *R. coriaria*, Linné | 15 to 33 |
| Quebracho, bark of *Aspidospermum Quebrachocolorado*, Schlechtendal | 14 to 33 |
| ,, Extract | 55 to 60 |
| Myrobalans, fruit of *Terminalia Chebula*, Retz. | 20 to 40 |
| Divi-divi, fruit of *Cæsalpinia coriaria*, Willdenow | 30 to 50 |
| Galls | 60 to 70 |
| Tamarisk Galls, *Tamarix gallica*, Linné | 50 to 54 |

## WILLOW BARK

### (Cortex Salicis)

**Source, &c.**—The willow bark of commerce is generally referred to the white or common willow, *Salix alba*, Linné (N.O. *Salicinĕæ*), a common tree on river-banks and marshy ground in England and throughout central and southern Europe. Its astringent properties were well known to Dioscorides, but the bark appears to have fallen into disuse until the latter end of the last century, when it was recommended as a remedy for ague. For this purpose, however, salicin, the principal active constituent of willow bark, has entirely taken the place of the bark, which is now seldom used medicinally. Salicin is usually obtained from the bark of other species of *Salix*, which contain it in greater abundance than *S. alba*. Very frequently, too, the structure of commercial willow bark shows that it is not derived from *S. alba*.

**Description.**—The willow bark of commerce usually occurs in channelled pieces several inches in length and 10 to 20 cm. in width; they are usually of a greenish or greyish brown colour.

The outer surface is in young bark a smooth, sometimes glossy cork; in older barks it is dull, slightly longitudinally wrinkled, and usually of a greenish brown colour. The inner surface has a pale reddish colour, and, in the younger pieces at least, appears smooth to the naked eye, but is seen under the lens to be finely striated longitudinally. Older pieces are more coarsely striated. The fracture is short in the outer but fibrous in the inner portion.

The section examined under a lens shows numerous minute tangentially arranged groups of bast fibres. Under the microscope the cork is seen to consist of but few rows of cells, the outer wall of which is strongly thickened and bulges outwards; one such row of cork cells is formed each year. The bast contains no groups of sclerenchymatous cells. These characters sharply distinguish willow barks from poplar and other barks.

FIG. 142.—Willow bark, showing, A, outer surface; B, inner surface. Natural size.

Willow bark often has a slight agreeable odour, and an astringent, slightly bitter and aromatic taste.

**Constituents.**—The principal constituent of commercial willow bark is tannin together with a little salicin.

Salicin, $C_{13}H_{18}O_7$, is a crystalline bitter glusoside present in the bark of various species of *Salix* and *Populus*. Emulsin hydrolyses it in aqueous solution to saligenin, $C_6H_4<^{OH}_{CH_2OH}$ (orthohydroxy-benzyl alcohol), and dextrose. Boiled with dilute sulphuric acid it yields saliretin and dextrose. Cautiously heated with dilute sulphuric acid and potassium bichromate, salicylic aldehyde with an odour of meadow sweet is evolved.

The student should observe

(*a*) The *greenish* or *greyish brown colour* ;
(*b*) The *smooth* or *wrinkled outer surface* ;
(*c*) The *pale reddish inner surface* ;

and should compare this bark with

(i) *Oak bark*, which has a silvery cork and dark brown inner surface ;
(ii) *Witch-hazel bark*, which has a pale greyish cork and reddish pink inner surface.

**Uses.**—Willow bark is astringent and has been used in rheumatism and ague ; it has, however, been completely replaced by salicin.

**Allied Species.**—*Salix discolor*, Muehlenberg, and *S. nigra*, Marsh, yield black willow bark which occurs in long, thin, tough, fibrous strips ; brownish or greenish brown externally ; taste bitter, astringent, somewhat aromatic ; contains tannin (about 4 per cent.) and salinigrin (about 1 per cent.) a colourless, crystalline glucoside.

Different species of the genus *Salix* are known as willows, osiers, and sallows. The name willow is generally applied to those species that form trees, osiers to those that form long slender shoots with few lateral branches, and sallows to those that have a shrubby growth.

The chief willows are *S. alba*, the wood of which is used for making cricket-bats, chip-boxes, &c., *S. triandră*, which is often pollarded close to the ground to give long shoots (also called osiers) for white basket work, and *S. fragilis*, Linné, the shoots from which have a reddish colour. *S. fragilis* is largely grown in Belgium, and the bark peeled from the shoots constitutes the 'rood scorce' (*i.e.* red bark) from which salicin is made. Rood scorce is largely imported into this country for that purpose, and contains about 3 per cent. of salicin.

The chief osiers are *S. viminālis*, which is used for hampers, *S. vitellīna*, which is employed for binding, and *S. purpurĕa*, which is used for fine basket work. The bark of the last-named species is especially rich in salicin, containing 6 or 7 per cent. of salicin. The bulk of the salicin employed medicinally appears, however, to be made from *S. fragilis*.

## BARKS IN LESS FREQUENT USE

**Indian Azadirach** (*Neem Bark ; Margosa Bark*).—The dried stem-bark of *Melia Azadirachta*, Linné (N.O. *Meliaceæ*), a tree indigenous to India and Malay Archipelago. Channelled, tough, fibrous pieces up to 10 mm. thick. Externally brownish grey, rough and scaly or fissured ; internally yellowish, conspicuously laminated and coarsely fibrous. Transverse section minutely chequered, pale narrow medullary rays and tangential bands of parenchyma alternating with darker groups of bast fibres. Under the microscope the latter are seen to be surrounded by lignified parenchyma. Contains a bitter, amorphous resin, a crystalline, bitter alkaloid (margosine), margosia acid, a crystalline substance and tannin. Used in India as a bitter tonic.

NOTE.—The seeds yield margosa oil; the pulp of the fruit is used in India as an ingredient in curries.

**Piscidia** (*Jamaica Dogwood*). The root-bark of *Piscidia Erythrina*, Linné (N.O. *Leguminosæ*), a West Indian and South American shrub. Quills or curved pieces, 5 to 15 cm. long, 4 to 6 mm. thick. Externally orange brown to dark reddish brown, with thin longitudinal and transverse ridges, somewhat fissured. Inner surface brown, smooth or fibrous. Fracture tough, fibrous, showing greenish patches. Odour characteristic, taste bitter, acrid. Contains crystalline, piscidin, bitter glucoside, resin. Used as a sedative in neuraglia, irritant coughs; also for dysmenorrhœa and nervous debility. Used in Jamaica as a fish poison.

**Calotropis** (*Mudar Bark*).—The dried root-bark of *Calotropis procera*, Robert Brown (N.O. *Asclepiadeæ*), a tree indigenous to India and Ceylon. Short quilled pieces, 2 to 5 mm. thick, 2 to 4 cm. wide, occasionally with rootlets attached. Cork soft, pale buff, longitudinally furrowed, wrinkled; inner surface pale yellow, granular. Section exhibits a thick buff coloured cork and white inner portion. Cortex and bast contain abundant laticiferous vessels. Starch very characteristic; simple grains $3\mu$ to $10\mu$ long, with distinct hilum and conspicuous striations; compound grains with two component grains. Taste bitter and acrid. Contains a yellow bitter resin; black acid resin; crystalline colourless madaralbin; yellow madarfluavil. Used as a diaphoretic and expectorant; as a substitute for ipecacuanha.

**Condurango.**—The bark of *Gonolobus Condurango*, Triana (N.O. *Asclepiadeæ*), a climbing plant indigenous to Ecuador. Quilled or curved pieces, 5 to 10 cm. long, 0·5 to 2 cm. wide and 2 to 6 mm. thick. Cork thin, greyish brown, often warty, sometimes scaly; inner surface paler, coarsely striated. Section pale, exhibiting scattered groups of sclerenchymatous cells. Almost odourless; taste bitter, rather acrid. Contains numerous laticiferous vessels, groups of sclerenchymatous cells and bast fibres, and abundant cluster-crystals of calcium oxalate. Constituents imperfectly known; one or more toxic glucosides and a toxic resin. Condurangin is the name applied to a mixture of glucosides; soluble in cold water, precipitated by boiling, re-dissolved on cooling. Used as a cure for cancer, but is useless.

**Elm Bark.**—The inner branch and trunk bark of *Ulmus campestris*, Linné (N.O. *Urticaceæ*). Flattened pieces, 10 to 12 cm. long, 3 to 5 cm. wide and about 5 mm. thick. Outer surface yellowish or brownish, often with dark brown patches of the outer portion; fracture short, somewhat fibrous; section shows numerous dark, usually oblique, medullary rays and tangentially elongated, whitish groups of bast fibres. Inodorous; tastes lightly astringent, mucilaginous. Contains tannin, mucilage and starch. Formerly used as an astringent.

**Larch Bark.**—The bark of *Larix europæa*, de Candolle (N.O. *Coniferæ*), central and southern Europe. Flat, curved, or channelled pieces; seldom in quills. Outer portion dark brownish red, often several centimetres thick, laminated; tissue between the laminæ often rose pink; inner portion nearly white. Odour terebinthinate; taste astringent and terebinthinate. Contains tannin and crystalline larixin, $C_{10}H_{10}O_5$, allied to pyrogallol. Formerly used as an astringent, stimulant and expectorant; now seldom employed.

## SECTION VIII

# SUBTERRANEAN ORGANS

## (Rhizomes, Corms, Bulbs, and Roots)

The subterranean organs of perennial plants serve as storage-rooms for the reserve material destined for the subsequent use of the plant. These organs often contain accumulated in them alkaloids, glucosides, and other medicinally valuable constituents, in addition to such carbohydrates as starch, sugar, inulin, &c.; hence the utilisation of such subterranean organs for medicinal purposes. They are generally collected in the autumn after the summer foliage has filled them with reserve material and before the development of the stem in the spring has partially exhausted them. Whether, however, they are in all cases richest in active constituents at this particular period has not yet been satisfactorily demonstrated.

In many herbaceous plants, such as dandelion, pellitory, &c., the aerial stem dies down in the autumn, leaving the lower modified portion attached to the root into which it passes more or less imperceptibly. This lower portion of the stem bears the buds destined to develop into new stems; being a subterranean stem it is to be regarded as an erect rhizome (often called rootstock), and is collected together with the root. Such drugs as dandelion and pellitory root consist therefore of both root and rhizome.

In other cases the transition from root to rhizome is more abrupt, as it is in valerian, serpentary, &c., but here also both are collected and dried together. In other cases, again, the rhizomes are separated, as they are in ginger and turmeric, the roots being rejected; or both roots and rhizome occur simultaneously in the drug, as with liquorice, gelsemium, &c. Comparatively few drugs consist either of root or of rhizome (corm or bulb) alone. This being the case it is evidently impracticable in classifying the drugs satisfactorily to separate roots from rhizomes.

Rhizomes may be defined as stout or slender, prostrate, oblique, or erect, hypogæic or epigæic stems. They may be distinguished from roots, which in many cases they closely resemble, by the following characters:

(i) They bear more or less evident cataphyllary leaves in the axils of which buds are present.
(ii) In many instances they exhibit a more or less evident pith, though frequently this is not easily discernible in organs as old as these are when collected.
(iii) The transverse section exhibits under the microscope leaf-traces in the cortex.

Bulbs may be regarded as subterranean buds, modified by the enlargement of the leaves to receive reserve material.

Corms are also modified buds, the basal stem portion of which serves as a reserve organ.

Roots may be distinguished from rhizomes

(i) By the absence of cataphyllary leaves (or their scars) and of buds.
(ii) By the absence of pith.
(iii) By the absence of leaf traces in the cortex.

The treatment that these organs receive after their removal from the ground varies with their nature. Small and slender rhizomes and roots are usually dried entire; larger ones are often peeled, cut, or sliced in various ways to facilitate rapid drying. In some cases slow drying is preferred in order to induce certain changes resulting in the production of aromatic or other bodies (e.g. valerian, gentian, orris).

## HYDRASTIS RHIZOME.

### (Golden Seal Rhizome, Rhizoma Hydrastis)

**Source, &c.**—Golden seal, *Hydrastis canadensis*, Linné (N.O. *Ranunculaceæ*), is a small herbaceous plant with perennial rhizome, widely distributed in woods in Canada and the eastern United States, being collected in Ohio, Minnesota, West Ontario, Georgia, and Missouri. It has been successfully cultivated in Europe (Holland, Switzerland). The plant produces but a single leaf, or two leaves and a single flower. The rhizomes are collected in the autumn after the leaves have withered; as the stems persist for some time to ripen the fruits, the remains of them are frequently found attached to the drug. The introduction of the drug into American and European medicine has been of very recent date, although its yellow colour had long attracted attention.

**Description**—The drug consists of small yellowish brown rhizomes varying from 1 to 4 cm. in length and averaging about 5 mm. in thickness; although they appear to be horizontal and creeping they are often oblique or even erect, as is indicated by the direction of the buds with which some of the branches terminate, as well as by the presence of rootlets on the (apparently) upper as well as under surface. They are knotty, tortuous, and rough, frequently giving off short upright branches terminated by cup-shaped scars left by the aerial stems of previous years. These branches usually bear distinct encircling scars of cataphyllary leaves, and similar scars are borne by the rhizome also, although there they are less distinct.

Thin, shrivelled, wiry, brittle roots of the same colour as the rhizome proceed from all parts of it, but especially from the lateral and ventral surfaces. Many of them break off, leaving small protuberances on the rhizome, which is often therefore rough and comparatively free from roots.

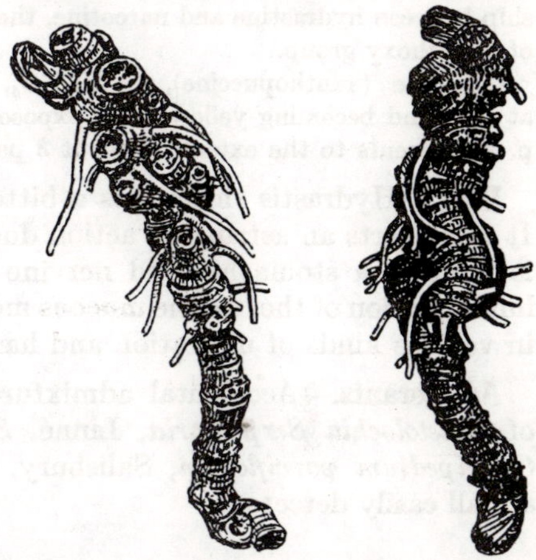

FIG. 143.—Hydrastis rhizome. Natural size. (*Pharmaceutical Journal.*)

The rhizome is hard, and breaks with a short resinous fracture. The transverse section varies in colour from dark yellow to very dark yellowish brown, and exhibits a comparatively thick bark[1] and a ring of bright yellow, somewhat distant narrow wood-bundles surrounding a large pith. The root also exhibits a dark bark and small, bright yellow wood.

The drug has a faint but characteristic odour; when it is chewed a bitter taste is developed and the saliva is coloured yellow.

The student should observe

(*a*) The *yellow colour*,

(*b*) The *structure visible in the transverse section* when examined with a lens.

(*c*) The *characteristic odour*;

[1] I have employed the term 'bark' to designate the collection of tissues exterior to the cambium of the rhizome and root as well as of the stem (compare p. 233). The Pharmacopœia has adopted the word 'cortex,' but this is liable to misinterpretation owing to the restricted sense in which the term is employed by modern botanists. For the tissue exterior to and including the endodermis (in monocotyledonous rhizomes) I have used the term 'cortex,' which is correct both in its ordinary and its botanical sense.

and should compare the drug with

*Bloodroot*, which usually has a dark reddish brown colour, and exhibits in transverse section a more or less prominent red colour without evident wood-bundles.

**Constituents.**—The principal constituents of golden seal are the alkaloids hydrastine, berberine, and canadine. The drug contains in addition resin, starch, and a trace of volatile oil. It leaves from 5 to 8 per cent. of ash on incineration.

Hydrastine, $C_{21}H_{21}NO_6$ (1·5 to 3·2 per cent.), crystallises in colourless bitter prisms melting at 132°. It yields by oxidation hydrastinine, $C_{11}H_{13}NO_3$, and opianic acid, $C_{10}H_{10}O_5$, the latter body being also obtained when narcotine is boiled with solution of potassium hydroxide, thus indicating a close relationship between hydrastine and narcotine, the latter differing only by the presence of a methoxy group.

Canadine (xanthopuccine), $C_{20}H_{21}NO_4$, forms colourless crystals melting at 132° and becoming yellow when exposed to the light. Berberine (compare p. 235) occurs to the extent of about 3 per cent.

**Uses.**—Hydrastis rhizome is a bitter tonic resembling nux vomica. It also exerts an astringent action due to the alkaloid hydrastine. It is used as a stomachic and nervine stimulant, in menorrhagia and inflammation of the uterine mucous membrane, and is employed locally in various kinds of ulceration and hæmorrhage.

**Adulterants.**—Accidental admixture of other rhizomes such as that of *Aristolochia Serpentaria*, Linné, *Stylophorum diphyllum*, Nuttall, *Cypripedium parviflorum*, Salisbury, have been observed, but they are all easily detected.

## BLACK HELLEBORE RHIZOME

### (Rhizoma (Radix) Hellebori Nigri)

**Source, &c.**—Black hellebore rhizome, or, as it is often, but less correctly, termed, root, is obtained from the Christmas rose, *Hellebŏrus niger*, Linné (N.O. *Ranunculaceæ*), a low herb with a perennial rhizome, abundant on the lower Alps of southern Europe, especially in Austria, and much cultivated in this country for its white flowers, which, as the name of the plant indicates, appear in midwinter. Our supplies of the drug come chiefly from Germany. The rhizome enjoyed a considerable reputation in the later Middle Ages as a stimulant, purgative, and digestive, but it is now seldom employed. It should be collected in the autumn.

**Description.**—The Christmas rose produces a horizontal creeping or frequently oblique or even upright rhizome, which is usually, when dried, about 3 to 5 cm. long, 6 mm. thick, and nearly black in colour. It is generally very irregular, tortuous and branched, the older pieces often forming small knotty masses. The branches are short and erect, marked with encircling leaf-scars, and usually terminated by the scar of the aerial stem or occasionally by the remains of a stem or bud. On

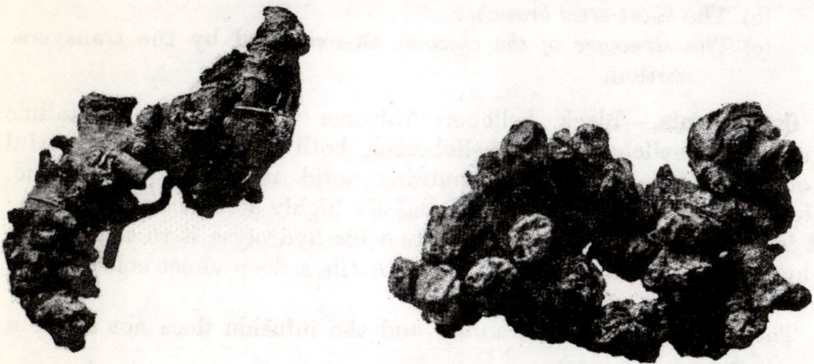

A B

FIG. 144.—Black Hellebore rhizome. A, young rhizome, natural size. B, old knotty rhizome, natural size.

the under surface the scars or short portions of numerous roots may be seen ; in the fresh plants these are long, rather stout and straight, but they are commonly removed from the drug.

The rhizome breaks easily with a short fracture ; the section is yellowish and exhibits a thick bark, within which is a ring of small

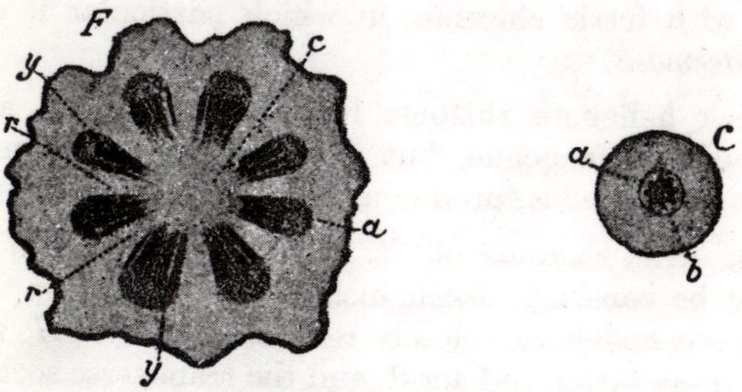


FIG. 145.—Black Hellebore rhizome. *F*, transverse section of rhizome : *a*, bark ; *y*, wood-bundles ; *c*, pith ; *r*, medullary rays ; magnified 3 diam. *C*, transverse section of root : *a*, cortex ; *b*, stele ; magnified 3 diam. (Berg.)

wood-bundles, all or some of which are narrow and radially elongated, enclosing a large pith. In the root the bark is thick and the wood tends to assume a stellate form, which, however, even in the older roots in which it is most marked, is never so conspicuous as it is in the root of *Cimicifuga racemosa*.

The odour of the drug is slight, the taste somewhat bitter and acrid. The dry powder, when inhaled, produces violent sneezing.

The student should observe

(*a*) The *dark colour* and tortuous appearance,

(*b*) The *short erect branches*,

(*c*) The *structure of the rhizome* as exhibited by the transverse section.

**Constituents.**—Black hellebore rhizome contains two crystalline glucosides, helleborin and helleborein, both of which are powerful poisons. Helleborin has a burning acrid taste and is narcotic. Helleborein has a sweetish taste and is a highly active cardiac poison; it is a chromogenic saponin; dilute acids hydrolyse it to acetic acid, glucose, an acid and a neutral helleboretin, a deep violet colour being developed.

The drug is free from tannin, and the infusion does not strike a

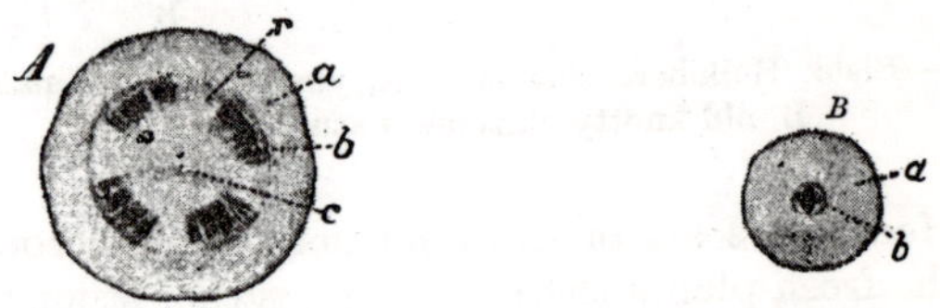


FIG. 146.—Green Hellebore rhizome. Transverse section of rhizome (*A*) and root (*B*): *a*, cortex; *b*, wood; *c*, pith. Magnified 3 diam. (Berg.)

dark colour with ferric chloride, in which particular it differs from *Cimicifuga racemosa*.

**Uses.**—Black hellebore rhizome has been employed as a drastic purgative and emmenagogue, but is now seldom administered. In large doses it is poisonous, producing violent inflammation.

**Substitutes.**—The rhizome of *H. viridis*, Linné, green hellebore—which should be carefully distinguished from *Veratrum viride*, often called also green hellebore—closely resembles that of *H. niger*; it is, however, far more bitter and acrid, and the transverse section exhibits wood-bundles that are broader and shorter than those of *H. niger*.

In the rhizome and root of *H. fœtidus*, Linné, the wood is more strongly developed and radiate in appearance, there being little or no visible pith.

## ACONITE ROOT

### (Radix Aconiti)

**Source, &c.**—The aconite, monkshood, or wolfsbane, *Aconitum Napellus*, Linné (N.O. *Ranunculaceæ*), is a perennial herb growing abundantly on the lower mountain slopes of central Europe. It is cultivated in England as a garden plant as well as for medicinal use, and is found apparently wild in some localities, but in these cases has probably escaped from cultivation. The drug has only recently been introduced into medicine, although the poisonous properties of the plant have long been known. Both the fresh leaves and flowering tops, as well as the dried roots, have been used; the latter are, however, now alone official.

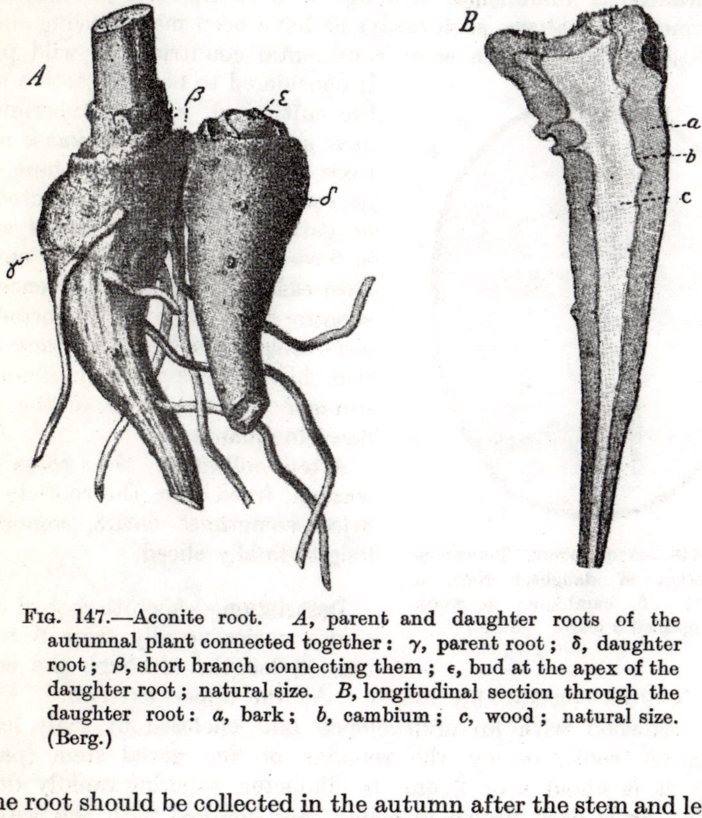


FIG. 147.—Aconite root. *A*, parent and daughter roots of the autumnal plant connected together: $\gamma$, parent root; $\delta$, daughter root; $\beta$, short branch connecting them; $\epsilon$, bud at the apex of the daughter root; natural size. *B*, longitudinal section through the daughter root: *a*, bark; *b*, cambium; *c*, wood; natural size. (Berg.)

The root should be collected in the autumn after the stem and leaves have died down, but before the bud destined to produce the stem for the following year has begun to develop. This bud is protected by cataphyllary leaves, in the axils of which lateral buds are situated. As the terminal bud grows and forms a flowering stem, certain (usually from one to three) of these lateral buds develop into short lateral shoots, each of which produces a long, slender, descending, adventitious root crowned with a bud. These roots rapidly enlarge and fill with reserve material produced by the parent plant, the root of which shrivels and perishes in proportion as the daughter roots increase in size. Towards the autumn the parent plant dies down, and the daughter roots, which have then attained their maximum development, are plump and full of starch. If allowed to remain in the soil the buds that crown the daughter roots begin to grow in the late winter or very early spring, and this growth is effected at the expense of the starch in the root, which for that purpose is converted into sugar. The root therefore becomes gradually exhausted of its reserve material, and it is generally assumed that the proportion of alkaloid it contains simultaneously diminishes, although this assumption has not been confirmed by analysis, such assays as have been made having shown but little difference. In some continental countries the wild plant is considered to be more active than the cultivated, and as experiments have shown that *A. Napellus* is more toxic than *A. variegatum*, Linné, and its varieties, the root is directed to be gathered from wild plants when in flower, at which time they can be most easily distinguished. Hence the aconite root that is imported in large quantities from Germany consists chiefly of the partially exhausted summer (parent) roots of the wild flowering plants.

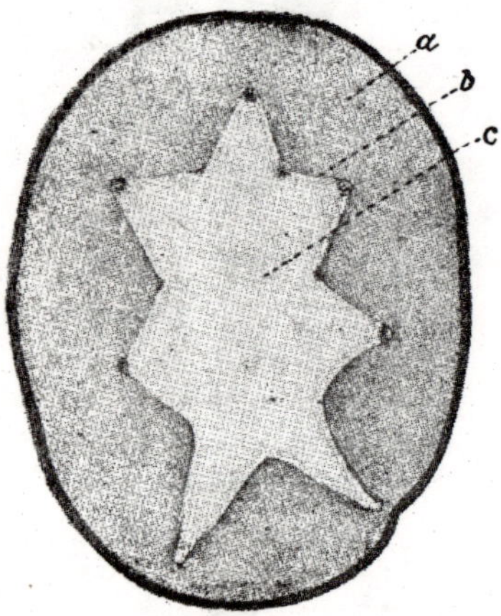

FIG. 148.—Aconite root. Transverse section of daughter root. *a*, bark; *b*, cambium; *c*, wood. Magnified 3 diam. (Berg.)

After collection the roots are washed, freed from the rootlets and dried, sometimes entire, sometimes longitudinally sliced.

**Description.**—Aconite root of commerce varies usually from 5 to 10 cm. in length, although the entire root is often considerably longer. At the upper extremity, where it is crowned with an undeveloped bud enclosed by scaly leaves (daughter root), or by the remains of the aerial stem (parent root), it is about 1 or 2 cm. in diameter, tapering rapidly downwards. It is dark brown in colour, and marked with the scars of rootlets. The surface is usually longitudinally wrinkled, especially if it has been dried entire. The root breaks with a short fracture, and should be whitish and starchy internally. The smoothed transverse section exhibits a thick bark, separated from the inner portion by a well-marked, darker, stellate line (cambium), with a group of vessels distinctly visible at each of the projecting angles, usually five to seven in number. Commercial aconite root is, however, often yellowish or brownish internally, and does not readily show the stellate cambium; this is probably due to its having been collected too early. The taste is at first slight, but is followed by a persistent sensation of tingling and numbness in the mouth. Although the daughter roots are usually more active than the parent root, the British Pharmacopœia admits the use of either, uniformity being ensured by the assay.

The student should observe

(a) The *elongated conical shape*,
(b) The *starchy interior* and *stellate cambium*,
(c) The *characteristic tingling taste.*

**Constituents.**—Aconite root contains three closely allied alkaloids, viz. aconitine, picraconitine (benzoylaconine), and aconine.

In addition to these alkaloids the root contains starch and aconitic acid, $C_3H_3(COOH)_3$, an acid fairly widely distributed in plants. The drug yields about 5 per cent. of ash.

Aconitine, $C_{34}H_{47}NO_{11}$ (m.pt. 198° ), is the only one of these three that is crystalline and extremely toxic, the lethal dose being about 0·00013 gm. per kilo weight. It produces even in minute quantity the persistent tingling on the tongue that characterises aconite root. It is very readily hydrolysed by acids as well as by alkalies, yielding first acetic acid and picraconitine; the latter by further hydrolysis yields benzoic acid and aconine. Aconitine is, therefore, acetylbenzoylaconine, and picraconitine is benzoylaconine. Aconitine may be distinguished from benzoylaconine and aconine by its taste, by its melting point, by its solubility in ether, and by its being readily crystallisable.

Picraconitine (benzoylaconine), $C_{32}H_{45}NO_{10}$, is amorphous, but yields crystalline salts; it has a bitter taste, is much less toxic than aconitine, and is insoluble in ether.

Aconine, $C_{25}H_{41}NO_9$, is also amorphous, bitter, and insoluble in ether; it is practically non-toxic.

The relation in which these alkaloids stand to one another may be made clear by the following equations:

$$\underset{\text{aconitine}}{C_{34}H_{47}NO_{11}} + H_2O = \underset{\text{acetic acid}}{CH_3.COOH} + \underset{\text{benzoylaconine}}{C_{32}H_{45}NO_{10}}$$

$$\underset{\text{benzoylaconine}}{C_{32}H_{45}NO_{10}} + H_2O = \underset{\text{benzoic acid}}{C_6H_5.COOH} + \underset{\text{aconine}}{C_{25}H_{41}NO_9}$$

The amount of total alkaloid present in the root appears to be subject to great variation, from 0·2 to 1·5 per cent. having been reported. No process has yet been devised by which the aconitine can be accurately determined in the presence of the other alkaloids. That portion of the total alkaloid that is dissolved by ether is, however, mostly crystalline, and consists almost entirely of aconitine. The ether-soluble alkaloid in aconite root of good quality ranges usually from about 0·3 to 0·5 per cent.

**Uses.**—Preparations of aconite and its principal alkaloid, aconitine, when applied to the skin, produce tingling followed by numbness; they are used in certain forms of neuralgia and rheumatism. Administered internally, aconite produces a steady fall of temperature, moistening of the skin, increase in the amount of urine, and lowering of the sensibility; it is given in cases of fever and pain, usually in the form of small doses of the tincture frequently repeated.

**Varieties and Substitutes.**—*English Aconite Root*, as described above, may generally be recognised by the bud which crowns most of the roots, indicating that the root is a daughter root and not a parent root.

*Foreign Aconite Root.*—Much aconite root is imported from Germany. It is collected from wild plants when in flower, and consists largely of the parent roots, as shown by their being crowned with the lower portions of the stems. Comparative assays of German and English aconite root tend to show that the German is scarcely inferior to the English in the proportion of ether-soluble alkaloid it contains.

*Japanese Aconite Root* (*A. uncinatum*, Linné, var. *japonicum*, Regel), is regularly imported in considerable quantities. It tapers gradually, and is either dark grey and nearly smooth (daughter root) or brownish, and marked with not very prominent, paler, longitudinal ridges (parent root). Both kinds of root are smaller than the English; they are, however, very starchy, less wrinkled than the English root, and exhibit, in transverse section, the groups of vessels arranged in a more regular circle. It contains japaconitine, which closely resembles, but is not identical with, aconitine, being acetylbenzoyl-japaconine.

*Indian Aconite Root* (*A. laciniatum*, Stapf) is much larger than the English, measuring frequently 15 cm. in length and 4 cm. in thickness near the crown; it is crowned with the remains of a bud and is coarsely wrinkled; internally it may be either starchy or yellowish and horny; the horny character is due to the starch having been completely gelatinised by the prolonged application of heat, the roots being, according to some accounts, boiled in cow's urine to preserve them. The drug contains pseudaconitine (acetylveratroyl-pseudaconine), which is about twice as toxic as aconitine.

*Atis Root* (*A. heterophyllum*, Wallich), small, ovoid, greyish roots with a bitter but not numbing taste, contain the alkaloid atisine.

The root of *A. chasmanthum*, Stapf, contains indaconitine which resembles aconitine in action; that of *A. paniculatum*, Lamark, is much less toxic than that of *A. Napellus* and is believed to contain benzoylaconine only. The rhizome of A. *lycoctonum*, Linné, contains lycaconitine and myoctonine: its action resembles that of atis root.

## CIMICIFUGA

### (Rhizoma Cimicifugæ, Radix Actææ Racemosæ, Black Snakeroot)

**Source, &c.**—*Cimicifuga racemosa*, Elliot (*Actæa racemosa*, Linné), is a tall herbaceous plant belonging to the natural order *Ranunculaceæ*.

It grows freely in shaded woods in Canada and the United States, extending southwards as far as Florida. In the autumn after the plant has fruited and the leaves have died down, the stout, perennial rhizome is collected, cut into pieces, and dried. It was introduced into medical practice in the United States about 1823, and in this country about 1860.

**Description.**—The drug consists of a thick, hard, and knotty horizontal rhizome with numerous stout ascending branches. The rhizome averages about 5 to 8 cm. in length and 1 cm. in diameter, but may attain as much as twice these dimensions ; the commercial drug consists, however, usually of pieces 2 to 5 cm. long. The branches, which are so close together as almost to conceal the rhizome, are about 1 cm. thick and 3 cm. or more long ; they curve upwards, and

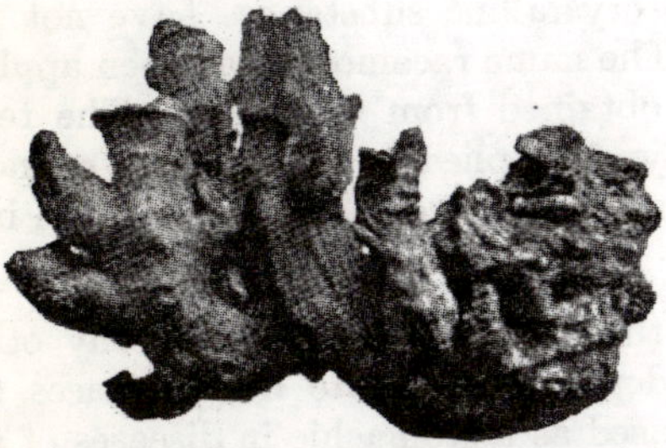

FIG. 149.—Black Snakeroot, showing branches curving upwards and marked with leaf-scars. Natural size.

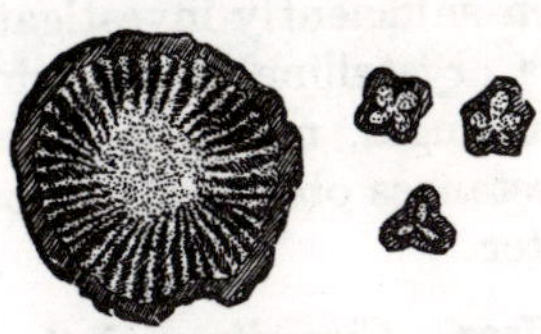

FIG. 150.—Black Snake-root. Transverse section through a branch of the rhizome, and through roots. Natural size. (Maisch.)

terminate either in the remains of a bud or more frequently in a circular cup-shaped scar exhibiting a distinctly radiate structure. Both the rhizome and its branches bear encircling scars of cataphyllary leaves, those on the branches being more conspicuous.

From the under surface of the rhizome numerous straight stout roots are given off. These are dark brown in colour and obscurely quadrangular or longitudinally furrowed ; they are rather brittle, and are usually broken off near the rhizome. The interior of the drug is hard and horny. The transverse section of the rhizome exhibits a thin dark horny bark surrounding a ring of numerous paler narrow wedges of wood alternating with wide dark medullary rays ; the centre is occupied by a large dark pith. The branches have a similar structure.

The transverse section of the root frequently exhibits a thick bark and four (or sometimes five or six) distinct wedges of porous whitish wood, arranged in the form of a Maltese cross, the medullary rays being broad and dark ; this characteristic structure is usually best seen in the stoutest roots.

The drug has no marked odour, but a bitter acrid taste.

The student should observe

(*a*) The numerous *stout branches curving upwards*,

(*b*) The *structure of the rhizome*, and of the *root*, exhibited by the transverse section;

and should compare the drug with

*Black hellebore rhizome*, which is tortuous and provided with irregular branches not exhibiting a prominent curve upwards; the section of the rhizome exhibits a thicker bark and few wood-bundles; in that of the roots t .e wood is much less distinctly cruciate.

**Constituents.**—The drug contains three crystalline subst .nces, iso-ferulic (hesperetic) acid, sugar, tannin, phytosterol, etc.; it yields about 7 per cent. of ash. The crystalline substances have not yet been sufficiently investigated. The name racemosin has been applied to a crystalline bitter principle obtained from the drug. The term cimicifugin, macrotin, or macrotyn is applied to a mixture of resinous substances obtained by pouring a concentrated alcoholic tincture into water

**Uses.**—Cimicifuga influences the gastric secretion, like any other bitter, and, to a slight extent, depresses the rate but increases the force of the pulse. It has been used as a stomachic in diseases of the heart, but is more frequently employed as a remedy for rheumatism, neuralgia, dysmenorrhœa, dyspepsia, &c. Its value is questionable.

## PAREIRA BRAVA

### (Pareira Root, Radix Pareiræ)

**Source, &c.**—The drug formerly official in the British Pharmacopœia under the name Pareira Root is derived from *Chondrodendron tomentosum*, Ruiz and Pavon (N.O. *Menispermaceæ*), a climbing plant with a stout woody stem, growing to a considerable height. It is a native of Peru and Brazil, and is regarded by the Brazilians as a valuable medicine. Its botanical origin remained long in an obscurity that was increased by Linnæus, who founded a species of *Cissampelos* (*C. Pareira*) and quoted it as the source of Pareira brava.

**Description.**—True Pareira brava occurs in long, woody, nearly cylindrical pieces averaging 2 to 4 cm. in diameter, but attaining 5 cm. or even more. It is nearly black in colour, tortuous and knotty, and marked externally with longitudinal furrows and transverse ridges and fissures. It is hard and heavy, breaking with a coarsely fibrous fracture; internally it is yellowish or brownish grey. The fractured root, when cut with a knife, exhibits a glossy, waxy, rather than woody surface. The transverse section shows four or five crenate, concentric or more or less eccentric zones, separated from each other by lighter lines of parenchymatous tissue. Each zone consists of a varying number of wedge-shaped wood-bundles with large pores alternating with wide medullary rays. A similar abnormal structure is found in many Menispermaceous stems and roots. The innermost zone is usually from 6 to 12 mm. in diameter.

The drug has no odour, but a decidedly bitter taste.

The student should observe

(*a*) The *nearly black outer surface*,
(*b*) The *crenate outline of the zones*,
(*c*) The *waxy cut*,
(*d*) The *bitter taste*.

**Constituents.**—Pareira brava contains about 2·5 per cent. of beberine (compare bebeeru bark, p. 273), to which its bitter taste is due; to this alkaloid the name pelosine (derived from *Cissampelos*) was given until its identity with beberine was proved. An unusually large quantity (about 9 per cent). of fatty acids, chiefly stearic, is said to be present, but this statement requires confirmation. Other constituents are an amorphous alkaloid (chondrodine), tannin and starch. The genuine drug yields about 12 per cent. of (cold) aqueous extract and about 4 per cent. of ash. Spurious Pareira brava often yields much less aqueous extract (see below). According to Fattis, in addition to beberine, which is soluble in benzene, two other alkaloids insoluble in benzene, one crystalline, the other amorphous, are present.

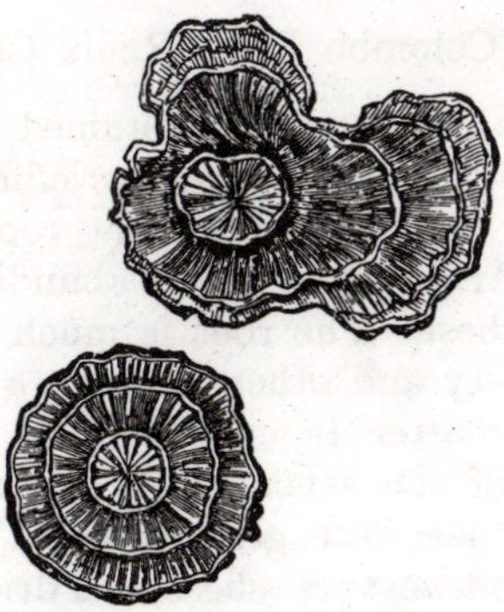

FIG. 151.—True Pareira Brava. Portion of a root, and transverse sections. (Bentley, after Hanbury.)

**Uses.**—Pareira brava is used in inflammatory affections of the urinary tract; it is considered to relieve pain and promote healing and cessation of muco-purulent discharge. It is not much used now, owing possibly to the substitution of other roots for the genuine drug.

**Varieties and Substitutes.**—Genuine Pareira brava appears at uncertain intervals on the London market. Its place has for some years been frequently taken by the root of a Menispermaceous plant of unknown botanical origin. This, the common substitute, may be distinguished by its distinctly brownish colour, the larger number of narrower zones which are not distinctly crenate, the larger vessels in the wood, and the less bitter taste. The drug is, further, not so heavy as true Pareira brava, and does not exhibit, when cut transversely, the same waxy nature. In addition to the above differences this false Pareira contains scarcely any fat, yields less ash (about 1·3 per cent.), and less aqueous extract (about 6 per cent.).

Occasionally the stems of the plant are mixed with true Pareira brava ; these are paler in colour, frequently bear the minute apothecia of lichens, and exhibit a small pith.

Other substitutes have occurred, and are likely to occur ; it is essential therefore, in this as in all cases, that the student should make himself familiar with the characters of the true drug, so as to be able to distinguish it from any substitute that may occur at any time.

## CALUMBA ROOT

### (Colombo Root, Radix Calumbæ)

**Source, &c.**—Calumba root is obtained from *Jateorhīza Columba*, Miers (N.O. *Menispermaceæ*), a lofty climbing plant with annual herbaceous stems and swollen fleshy roots. It is indigenous to Portuguese East Africa, growing in abundance in the forests in the region of the Zambesi. The root is much used by the natives as a remedy for dysentery and other diseases, and also on account of the yellow colouring matter it contains. It was brought to Europe towards the end of the seventeenth century, and after being long neglected came at last into general use. The roots are dug in the dry season, cut into transverse slices, and dried. The drug as imported is of a dingy brown colour and has a quantity of brown powder (soil) adhering to it ('natural' calumba root). It is cleaned by washing and brushing, and is then graded for sale ('washed' calumba root). The following description applies to the washed root, which is the official variety.

**Description.**—Calumba root occurs in commerce in irregular, elliptical or nearly circular slices averaging about 4 cm. in diameter and 6 mm. in thickness, but often much larger and thicker. They are usually depressed in the centre on both sides, the tissue there being apparently less loaded with starch and less woody than the outer portions, and consequently contracting more on drying. The exterior of the root is covered with a thin, dark, brownish, wrinkled cork, which readily separates, disclosing the yellowish brown bark beneath. The transverse surface is of a dull greyish colour passing to greenish yellow towards the outer margin of the wood and bark. When smoothed with a knife the interior is seen to be much brighter and the section exhibits a thick bark marked with radiating lines (sieve tissue), separated by a dark line (cambium) from the large central portion (wood), in which the vessels are arranged in narrow, rather distant, radially elongated groups. The parenchyma of the wood, like that of the bark, is loaded with starch-grains which under the microscope are seen to be large ($20\mu$ to $70\mu$) and mostly simple with

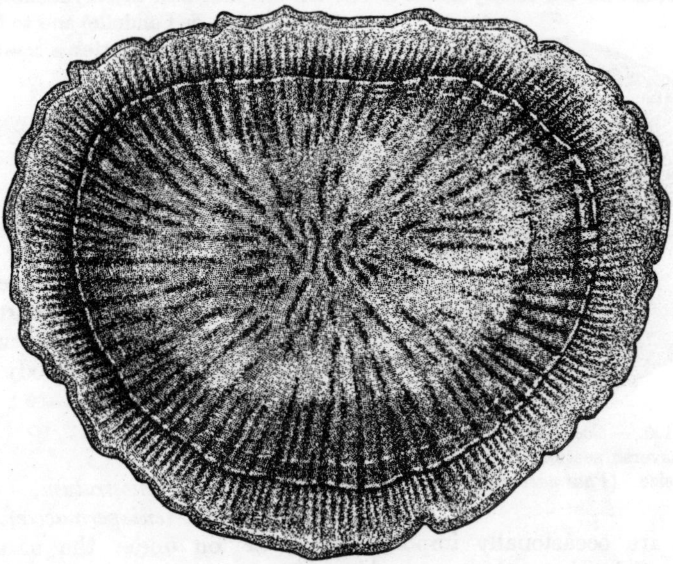

FIG. 152.—Calumba root. Transverse section, magnified. (Moeller.)

eccentric hilum; near the cork characteristic sclerenchymatous cells with yellow, irregularly thickened walls and containing prismatic crystals of calcium oxalate are to be found.

The drug breaks with a short starchy fracture; it has a slight musty odour and a marked bitter taste.

The student should observe

(a) The *yellow colour* and *depressed centre* of each slice,
(b) The *thick bark* and largely developed *parenchymatous tissue*,
(c) The *short fracture* and *abundance of starch*.

**Constituents.**—Calumba root contains three yellow crystalline alkaloids, viz. jateorhizine, columbamine, and palmatine, the latter in small quantity only.

The drug also contains two colourless, crystalline, non-alkaloidal bitter principles, one only of which, columbin, has been investigated; this appears to be a lactone, and yields, when treated with acid or alkali, yellow amorphous columbic acid, previously believed to be a constituent of the root. Traces of a fluorescent substance, also obtainable from columbin, are present in the drug.

In addition to these principles, mucilage and abundance of starch are present, but no tannin. The drug yields from 4 to 7 per cent. of ash.

Jateorhizine, $C_{20}H_{20}NO_5,OH$, columbamine, $C_{21}H_{22}NO_5,OH$, and palmatine, $C_{21}H_{22}NO_6,OH$, are closely allied to one another and also to corydaline (*Corydalis tuberosa*, de Candolle) and to berberine, forming, like the latter, a crystalline compound with acetone.

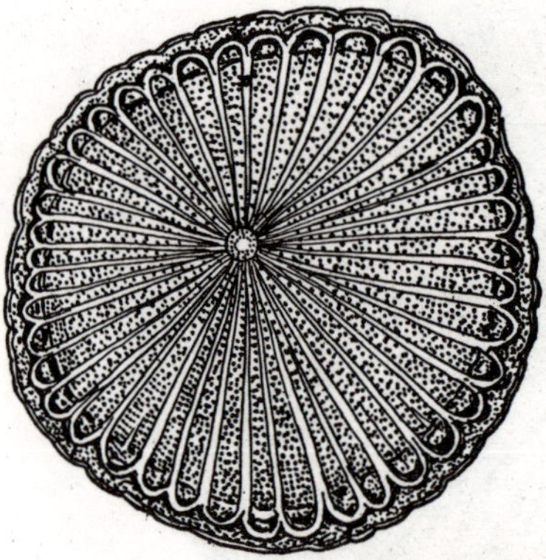

FIG. 153.—*Coscinium fenestratum.* Transverse section of stem. Natural size. (*Pharmaceutical Journal.*)

**Uses.**—Calumba is employed as a stomachic and bitter tonic.

**Substitutes, &c.**—*Calumba Rhizome.*—This is not unfrequently present in small quantity and occasionally to a considerable amount; the pieces are narrower (often about 2 to 3 cm. wide), deeper in colour, more woody, and more conspicuously radiate; they also yield more ash (12 to 17 per cent.).

*Coscinium fenestratum*, Colebrooke (N.O. *Menispermaceæ*). The stems are occasionally imported from Ceylon under the name of Ceylon Calumba; these may be cut into slices about the size of calumba, but are readily distinguished by their dark yellow colour, flat surface, not depressed in the centre, and hard, woody (not starchy) nature. They contain berberine (3·5 per cent.).

*Frasera caroliniensis*, Walter (N.O. *Gentianeæ*).—Slices of the root of this plant have been found substituted for calumba, but the occurrence is rare; the slices are smaller, thicker, and free from starch, but contain tannin.

## PODOPHYLLUM RHIZOME

### (May-apple Root, Rhizoma Podophylli)

**Source, &c.**—The May-apple, *Podophyllum peltatum*, Linné (N.O. *Berberideæ*), is a small herb with a long perennial creeping rhizome. It is a native of and common in the eastern United States and Canada, producing, when it flowers, but two leaves and a single flower. The drug was well known to the North American Indians as an emetic and vermifuge; it was introduced into the British Pharmacopœia in 1864.

The rhizome, which grows to a length of several feet, is collected in the late summer and dried; it is usually cut into pieces about 10 cm. in length.

**Description.**—Podophyllum root, or more correctly rhizome, is seen in commerce usually in nearly cylindrical pieces 10 or more cm. in length and about 5 mm. in thickness. These are of a dark reddish brown colour, and are nearly smooth or slightly longitudinally

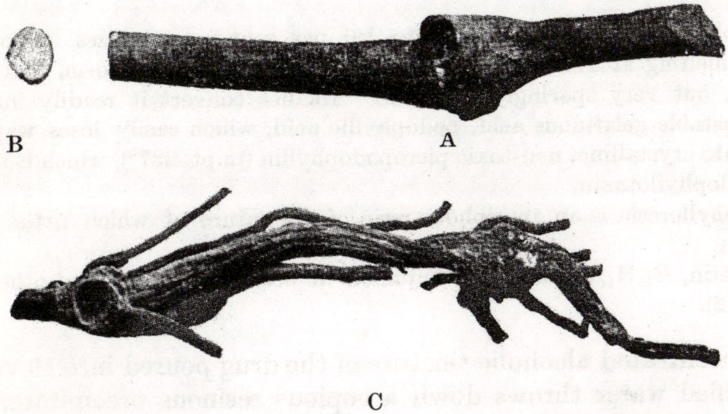


FIG. 154.—Podophyllum rhizome. A, plump autumnal rhizome without roots showing the scar left by the aerial stem. B, transverse section of the same. C, shrivelled summer rhizome with roots attached. All natural size.

wrinkled. Summer rhizomes contain less starch and are flatter and more deeply wrinkled. At intervals of about 5 cm. the rhizome is enlarged and bears on the upper surface a concave scar surrounded by several circular leaf-scars; these have been left by an aerial flowering stem and its cataphyllary leaves. Below the stem-scar, on the under surface of the rhizome, are the scars of several stout roots, which are occasionally left attached to the rhizome, but which more commonly have been removed. The rhizome occasionally forks, but produces very few lateral branches. When a flowering stem is produced the growth of the main axis is terminated; a bud in the axil of one of the cataphyllary leaves then develops, forming a sympodial system and continuing the growth of the plant.

The rhizome breaks with a short fracture. The transverse section is usually white and starchy, but if the heat employed in drying the rhizome has been sufficient to gelatinise the starch it is yellowish and horny; it exhibits a very thin cork and a circle of small, oval, distant, fibro-vascular bundles.

The drug has a characteristic but not strong odour, and a bitter acrid taste.

The student should observe

(*a*) The *straight cylindrical rhizome* with occasional stem, root, and leaf scars,

(*b*) The *structure shown by the transverse section.*

**Constituents.**—The most important constituents of podophyllum rhizome are a neutral crystalline substance, podophyllotoxin, and a resin, podophylloresin, both of which are purgative. In addition the drug contains picropodophyllin, quercetin, and starch. It yields about 3 per cent. of ash on incineration.

Podophyllotoxin, $C_{15}H_{14}O_6$ (0·2 to 1·0 per cent.) crystallises in colourless needles melting at 117°, and easily soluble in alcohol, chloroform, and boiling benzene, but very sparingly in water. Alkalies convert it readily into salts of an unstable gelatinous acid, podophyllic acid, which easily loses water and passes into crystalline, non-toxic picropodophyllin (m.pt. 227°), which is isomeric with podophyllotoxin.

Podophylloresin is an amorphous resin of the nature of which little definite is known.

Quercetin, $C_{15}H_{10}O_7,2H_2O$, is contained in bearberry leaves and other drugs (see p. 42).

A concentrated alcoholic tincture of the drug poured into 10 volumes of acidified water throws down a copious resinous precipitate, podophyllin (yield from 3 to 7·7 or even 16 per cent.; roots as rich as rhizome). This podophyllin contains as its chief constituents podophyllotoxin (about 20 per cent.), podophylloresin, and picropodophyllin, together with quercetin and other bodies. The yield of podophyllin is very variable, but whether this is due to the time at which the rhizome is collected or to other conditions is not at present definitely known. The rhizome is said to be most active in the spring when beginning to shoot.

**Uses.**—Podophyllum rhizome, or at least the podophyllin obtained from it, is a gastro-intestinal irritant. In large doses it produces inflammation of the stomach and intestines which has proved fatal. In moderate doses it is a drastic purgative with some cholagoguic action, and is much used in cases of constipation from hepatic trouble.

## INDIAN PODOPHYLLUM RHIZOME

### (Rhizoma Podophylli Indici)

**Source, &c.**—Indian podophyllum rhizome is obtained from *Podophyllum Emōdi*, Wallich (N.O. *Berberideæ*), a plant growing freely on the lower slopes of the Himalayas.

**Description.**—The drug consists of the rhizome together with numerous stout roots which frequently break off. The rhizome is cylindrical or flattened, about 10 mm. in diameter and contorted. The upper surface bears the short remains of aerial stems, with cup-shaped scars; to the under surface numerous stout roots are attached. The transverse section is pale brown, starchy or horny, with a ring of radially elongated bundles. The colour, odour and taste resemble those of podophyllum rhizome.

**Constituents.**—The constituents also are similar to those of the American drug but the yield of resin is usually higher (10 to 12 per cent.) and the proportion of podophyllotoxin greater (1 to 4 per cent.), the resin containing approximately twice as much podophyllotoxin as the resin from the American drug, and being about twice as active. According to Dott the two resins may be distinguished by the following test: Mix 0·5 gm. with 5 cc. of 10 per cent. solution of ammonia and 5 cc. of water; stir well for twenty minutes, filter, wash and dry; the insoluble residue from American podophyllin should not exceed 0·13 gm.; from Indian about twice as much will be obtained.

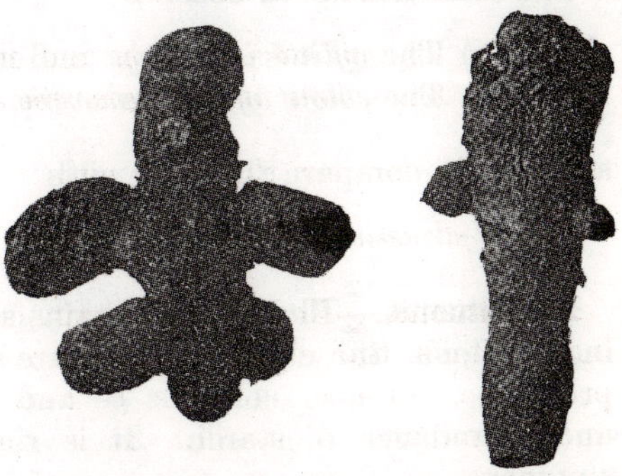

Fig. 155.—Bloodroot. Natural size.

## BLOODROOT
### (Rhizoma Sanguinariæ)

**Source, &c.**—Bloodroot is the rhizome of *Sanguinaria canadensis*, Linné (N.O. *Papaveraceæ*), a herb with a perennial rhizome, widely distributed throughout Canada and the United States, growing freely in shady places on rich soils; the plant produces in early spring a single flower succeeded by one or two leaves 10 to 15 cm. high, and exudes, when cut, an orange-red juice. The rhizome is collected in the autumn and dried.

**Description.**—Bloodroot varies usually in length from 2 to 5 cm., and in thickness from 5 to 10 mm. It has a dark grey or dark reddish brown colour, and is generally plump, straight, or somewhat curved, and nearly cylindrical, but it is sometimes much shrivelled and shrunken, and then contains less starch, due probably to its having been gathered before the end of the period of active growth. It is bluntly conical at the apex, and shows only traces of a bud or aerial stem. From the lower surface numerous dark, thread-like, brittle, wiry roots spring, commonly more or less interlaced, but easily breaking off, leaving inconspicuous raised scars; in the commercial drug the roots are mostly detached. The rhizome bears also more or less distinct encircling leaf-scars, and sometimes short knob-like branches at right angles to it.

The rhizome breaks with a short fracture; the fractured surface is sometimes whitish and starchy, with numerous minute deep red dots (cells containing secretion), sometimes of a more or less uniform deep blood-red or nearly black colour, and then hard and resinous instead of starchy, a difference due to the escape of the deep red secretion from the cells in which it was originally contained into the surrounding tissue. The appearance of the section varies not only in different rhizomes, but even in different parts of the same rhizome. The fibro-vascular bundles are distributed in a circle near the bark, but they are so inconspicuous as to be with difficulty discernible even under a lens. The drug has little odour, but an unpleasant bitter and acrid taste.

The student should observe

(*a*) The *cylindrical shape* and comparative *absence of roots*,
(*b*) The *colour of the transverse section*;

and should compare the drug with

*Hydrastis rhizome* (see above).

**Constituents.**—Bloodroot contains several remarkable and interesting alkaloids, the chief of which are sanguinarine, chelerythrine, and protopine. It also contains β- and γ-homochelidonine, a red resin and abundance of starch. It is richest in alkaloid in the early summer.

*Sanguinarine*, $C_{20}H_{15}NO_4,\frac{1}{2}H_2O$ (m.pt. 213°), forms colourless crystals, but yields with acids deep red crystalline salts. It is also found in small quantity in the root of the greater celandine (*Chelidonium majus*, Linné), of *Glaucium luteum*, Scopoli, &c.; it causes tetanus and excitement. This alkaloid should not be confused with the mixture of resinous substances known as sanguinarin, which is obtained in the same way as cimicifugin, podophyllin, and other so-called 'eclectic' remedies.

*Chelerythrine*, $C_{21}H_{17}NO_4$ (m.pt. 203°), a colourless bitter alkaloid yielding bright yellow salts; it is a toxic alkaloid and is also found in *Chelidonium majus*, *Glaucium luteum*, &c.

*Protopine*, $C_{20}H_{19}NO_5$ (m.pt. 207°), also found in opium, celandine, &c.

β- and *γ-homochelidonine*, two colourless closely related alkaloids.

**Uses.**—Bloodroot in full doses depresses the action of the heart, and produces nausea and vomiting; in smaller doses it increases the appetite and improves digestion. It has been used in atonic dyspepsia, croup, bronchitis, and asthma. The powdered rhizome is a powerful irritant to the respiratory passages.

## HORSERADISH ROOT

### (Radix Armoraciæ)

**Source, &c.**—The horseradish, *Cochlearia Armoracia*, Linné (N.O. *Cruciferæ*), is indigenous to eastern Europe, but is naturalised in several parts of Britain, and is cultivated in this country as well as in many others. It possesses a large perennial root, and produces stout, erect, flowering stems about 1 metre high.

**Description.**—The root, which is used in the fresh state only, attains a length of over half a metre and a thickness of 3 cm. or more. It is nearly cylindrical except at the crown, where it often divides into a few short branches, each of which is enlarged in its upper part and marked with closely approximated, semi-amplexicaul leaf-scars. It gives off but few slender, lateral roots, is pale yellowish or brownish white in colour, and fleshy in consistence. The transverse section exhibits a thick bark and a distinct cambium, within which is a wood consisting principally of parenchymatous tissue, groups of vessels being visible as minute points, especially near the cambium.

The root is odourless until broken, bruised, or scraped, when a pungent mustard-like odour is evolved ; the taste is also pungent.

The student should observe

(*a*) The *pale yellowish colour* and *cylindrical shape*,
(*b*) The *pungent odour* (*when crushed*) and *taste*.

**Constituents.**—Horseradish root contains sinigrin and myrosin, and yields, when crushed and distilled with water, about 0·05 per cent. of allyl isothiocyanate (compare p. 147).

**Uses.**—Horseradish root has properties similar to those of black mustard seed, viz. stimulant and rubefacient.

## SENEGA ROOT

### (Radix Senegæ)

**Source, &c.**—Senega root is obtained from *Polygala Senega*, Linné (N.O. *Polygalĕæ*), a small plant producing a perennial knotty rootstock, from which numerous slender stems 15 to 30 cm. high arise. It is widely distributed over the United States and the southern parts of British America, the root being collected largely in Minnesota and Manitoba (western senega) and in the north-western of the United States (northern senega). It was formerly collected in the more southern States, but is now nearly exterminated there. The root was used by the Seneca Indians as a remedy for snake-bite, and was introduced into medicine about the middle of the eighteenth century.

**Description.**—Senega root consists of a slender, greyish or brownish yellow root surmounted by a knotty crown, to which are attached the remains of numerous slender aerial stems and small shoots beset with the scars or remains of purplish scaly leaves. The root is usually about 3 to 6 mm. thick at its upper extremity, but it soon divides into two or three spreading branches. It is frequently curved and contorted, and is longitudinally and sometimes, especially near the crown, transversely wrinkled. Very frequently, but not always, it exhibits a prominent keel resembling a contracted sinew and following a gently spiral course; this keel may generally be found on the concave surface of the curves of the tap-root, and often extends a considerable distance.

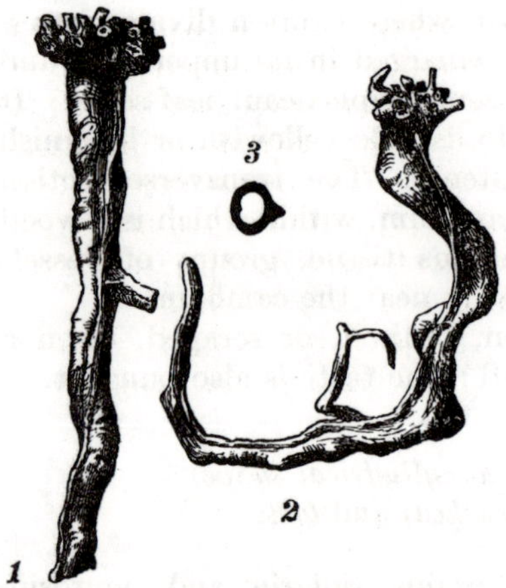


FIG. 156.—Senega root. 1 and 2, roots, showing the keel; 3, transverse section. Natural size. (Vogl.)

The root breaks with a short fracture, the fractured surface exhibiting a whitish wood and yellowish translucent bark. The former frequently presents an abnormal appearance. Instead of a complete circle of wood a wedge-shaped portion, or sometimes two, of varying extent, is replaced by parenchymatous tissue, and the wood therefore appears to have had a segment cut out of it. This appearance varies in different parts of the same root, the segment that is missing being sometimes narrow and wedge-shaped or occasionally increasing to nearly a semicircle, thus reducing the wood to one-half its normal amount. If a keeled root is soaked in water and the bark stripped from it, the wood will be seen to have transverse cracks or a longitudinal fissure on the convex surface, the latter usually extending for some distance and widening from a narrow crack into a broad fissure. These cracks and fissures are filled with easily removable parenchymatous tissue.

The concave sides of the curved roots bear the keels, and these are seen in the transverse sections to be due to a largely developed bast; the keels do not arise from any abnormal development of the wood.

The root has a distinct odour, recalling wintergreen; the taste is at first somewhat sweet, but soon becomes sour and acrid. The powdered root is very irritating to the throat and nostrils when inhaled, and imparts to water the property of frothing.

The student should observe

(*a*) The *prominent keel*,

(*b*) The *irregular wood in transverse sections*,

and should strip the bark from the root and examine the wood.

**Constituents.**—Senega root contains as principal constituents senegin and polygalic acid. These substances are both glucosides, and resemble, but are not identical with, quillaja-sapotoxin and quillajic

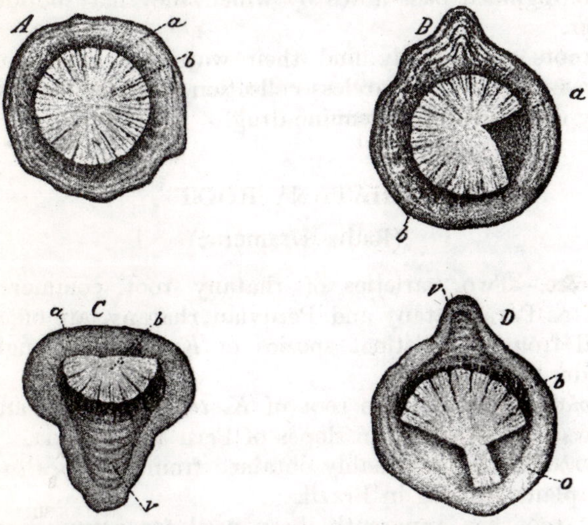


FIG. 156A.—Senega root. *A*, transverse section of normally developed root; *B, C, D*, of abnormally developed roots. *a*, bark; *b*, wood; *v*, bast. Magnified. (Berg.)

acid, constituents of quillaja bark, their action being qualitatively the same, but quantitatively different (Atlass, 1890). Polygalic acid is sternutatory, and imparts to water the property of frothing. Senegin is decidedly toxic. Both these bodies require, however, further investigation.

The drug often contains a small percentage of methyl salicylate (oil of wintergreen), probably produced by the gradual decomposition of an unknown glucoside; its presence has been utilised as a test of the identity of the drug, but it is unreliable. Senega root also contains about 5 per cent. of fixed oil, but is free from starch; it yields about 4 per cent. of ash.

**Use.**—Senega is used as a stimulant expectorant in bronchitis.

**Varieties.**—*Northern senega*, collected in the north-western States, is considerably larger than the usual variety (western senega), and darker in colour; it is less contorted and shows the keel less distinctly, but it has a very acrid taste, and is undoubtedly a good senega. It is said to be derived from *Polygala Senega*, var. *latifolia*.

*White senega*, from *Polygala alba*, Nuttall, is collected to some extent in the southern States. The root is more slender than western senega, has descending branches rather than spreading, and is lighter in colour. It has no keel and a normal wood. The taste is much less acrid than that of western senega, and it is presumably less active.

*Senega stems*; these are about 2 mm. thick, rounded, hollow; they contain non-lignified bast fibres by which they may be identified in the powder.

Other roots occasionally find their way into parcels of senega, frequently as a result of careless collection, but the characters given sufficiently distinguish the genuine drug.

## RHATANY ROOT

### (Radix Krameriæ)

**Source, &c.**—Two varieties of rhatany root, commercially distinguished as Pará rhatany and Peruvian rhatany, are official; they are derived from two distinct species of *Krameria* belonging to the natural order *Polygaleæ*.

1. *Peruvian rhatany* is the root of *K. triandra*, Ruiz and Pavon, which grows on the mountain slopes of Peru and Bolivia.

2. *Pará rhatany* is presumably obtained from *K. argentĕa*, Martius, a shrubby plant growing in Brazil.

Rhatany root has apparently been used for many years by the Peruvians for cleaning and preserving the teeth. The Spaniards became acquainted with it in Lima and introduced it into Europe.

**Description.**—The root of Peruvian rhatany is large and knotty at its upper extremity, but divides near the crown into several long stout branches, from which again smaller branches proceed. The larger pieces have a dark reddish brown colour and a rugged, scaly bark; the smaller are usually rather brighter and smoother, not exhibiting any conspicuous transverse fissures. The bark, which can easily be separated from the smaller roots, is rather fibrous, but the wood breaks with a splintery fracture. The transverse section exhibits a pale reddish or yellowish wood and a comparatively narrow reddish brown bark occupying about one-fourth of the radius of a root of medium size. The wood is dense, and marked with numerous narrow medullary rays and minute vessels. The drug is quite odourless; the bark has a strongly astringent taste, but the wood is practically tasteless.

*Pará rhatany* is usually imported in long, nearly straight, cylindrical pieces, not often exceeding 15 mm. in thickness. They are well characterised by their dark purplish brown colour, and by the presence of deep transverse cracks at more or less regular intervals in the bark, which in this variety is never rough or scaly, although transversely fissured and longitudinally wrinkled. The transverse section shows a reddish wood and a dark reddish brown bark, which, in this case

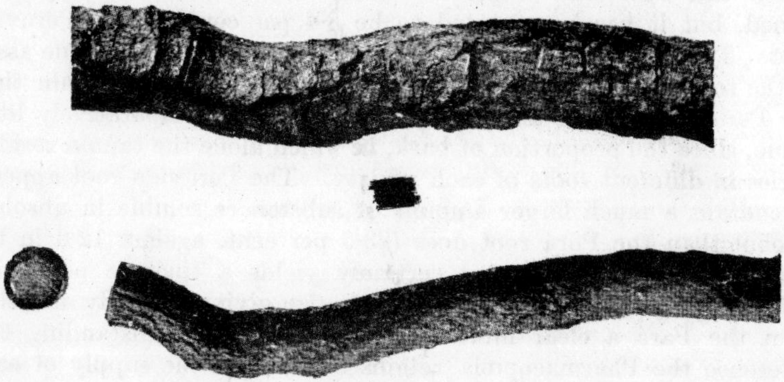

B

FIG. 157.—Peruvian Rhatany root. A, portion of an older root, showing scaly cork; natural size. B, portion of a younger root, natural size.

occupies about one-half the radius of a root of medium size. In respect to odour and taste Pará rhatany resembles Peruvian.

The student should carefully compare these two roots, and observe in Peruvian rhatany

(a) The *reddish brown bark*, which is *scaly in large pieces*, *smoother in the smaller*, and *free from deep transverse cracks*,

(b) The *proportion of bark to wood* as exhibited in the transverse section;

FIG. 158.—Parà Rhatany. Natural size.

in Pará rhatany

(a) The *dark purplish brown colour of the bark*, which is *not scaly*, but *exhibits deep transverse cracks*,

(b) The *proportion of the bark to wood* in the transverse section, which is greater than in Peruvian rhatany.

Large pieces of Indian sarsaparilla (see p. 340) occasionally bear a considerable resemblance to small pieces of Pará rhatany, both in colour and in the presence of transverse cracks; they may be distinguished by their agreeable odour and by the difference in the transverse section.

**Constituents.**—The principal constituent of rhatany root is the tannin (krameria-tannic acid) that it contains. The proportion in which this substance is present has not been satisfactorily determined, but it has been stated to be 8·4 per cent. in the Peruvian root. The Pará variety contains about the same quantity, one assay of the root showing rather more, and another rather less tannin than the Peruvian. These analyses, are, however, of comparatively little value, since the proportion of bark, in which alone the tannin resides, varies in different roots of each variety. The Peruvian root appears to contain a much larger amount of substances soluble in absolute alcohol than the Pará root does (23·0 per cent. against 12·6 in the Pará—Dunwody, 1890), and certainly yields a tincture possessing different properties, that from the Peruvian giving a cloudy and that from the Pará a clear mixture with water. Notwithstanding this difference the Pharmacopœia retains both, since the supply of each is subject to considerable irregularity.

The root contains in addition a dark red phlobaphene, krameria-red, produced by decomposition of the tannin, as well as starch, and about 2 per cent. of inorganic matter.

**Uses.**—Rhatany root is employed solely as an astringent.

**Varieties.**—Several other species of *Krameria* are known to yield astringent roots, but none are of commercial importance.

## MARSHMALLOW ROOT

### (Radix Althææ)

**Source, &c.**—The marshmallow, *Althœa officinalis*, Linné (N.O. *Malvaceæ*), is not uncommon in moist or marshy places in southern England, and is widely distributed in similar situations in central and southern Europe. Our supplies are derived from plants cultivated in Germany, France, and Belgium.

The roots are collected in the autumn from plants about two years old, trimmed, deprived of their cork by scraping, and dried. Sometimes the root is also split longitudinally, or cut into transverse slices.

**Description.**—The roots as met with in commerce are generally in straight, tapering, whitish pieces about 15 or 20 cm. in length, and attaining 2 cm. or more in diameter at their upper extremity. They are usually obscurely quadrangular or rounded, with a few broad and deep longitudinal wrinkles. The surface is softly fibrous from the presence of bast fibres that have been liberated by the scraping, and bears brownish scars of lateral roots. The bark, which can readily be removed in long strips, is tough and fibrous, but the wood breaks with a short granular fracture; internally the root is whitish and starchy. It can easily be cut, and the transverse section exhibits a bark of moderate thickness, separated by a yellow sinuate cambium line from the wood. Both bark and wood possess a radiate structure that is more distinct when the surface of the section is moistened; numerous cells containing a translucent mucilage then also become visible.

The drug has a faint but characteristic odour, and a mawkish, mucilaginous taste. To improve its appearance it is sometimes limed, but this sophistication can easily be detected by rinsing the drug with 1 per cent. hydrochloric acid, filtering, and adding excess of sodium carbonate, when the liquid should remain clear.

The student should observe

(a) The *fibrous bark*,
(b) The *yellow cambium line*,
(c) The *radiate structure of the wood*,
(d) The *presence of mucilage*;

and should compare the root with

*Belladonna root*, which closely resembles *unpeeled* marsh mallow root, but may be distinguished by the non-fibrous nature of the bark, by the absence of mucilage, and by the scarcely radiate structure of the wood.

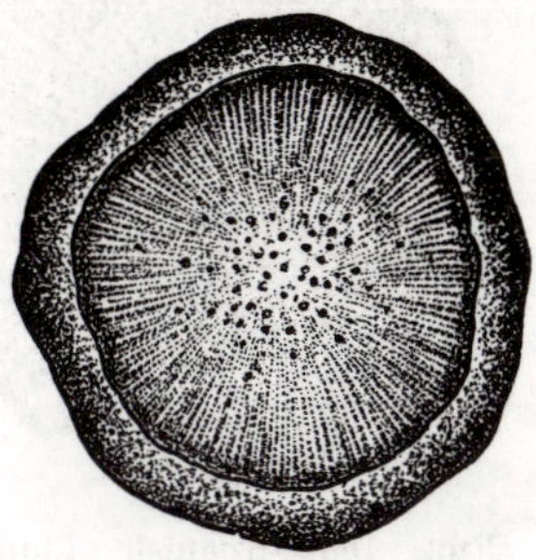

FIG. 159. — Marshmallow root. Transverse section. Magnified. (Moeller.)

**Constituents.**—The principal constituent is mucilage, of which the root is said to contain as much as 25 to 35 per cent., but these figures require confirmation. The drug contains also an abundance of starch together with asparagin and a substance allied to lecithin. Asparagin, $C_4H_8N_2O_3$, is the amide of aspartic (amidosuccinic) acid, and is found in many plants, especially in marshmallow root and in etiolated plants.

**Uses.**—Marshmallow root is used as an emollient and demulcent.

## LIQUORICE ROOT

### (Radix Glycyrrhizæ)

**Source, &c.**—The chief plants that yield the liquorice root of commerce are *Glycyrrhiza glabra*, Linné, *G. glandulifera*, Waldstein and Kitaibel, and *G. glabra*, var. *β-violacea*. The British Pharmacopœia permits the collection from any species of *Glycyrrhiza*, but the product must comply with the official description.

*Glycyrrhiza glabra*, Linné (N.O. *Leguminosæ*), is widely distributed over southern Europe, extending to Central Asia; it is cultivated to a limited extent in England (Yorkshire), but our supplies are derived chiefly from Sicily and Spain.

The plant produces a tall, erect, herbaceous stem, and a stout perennial root, dividing, a few inches below the surface, into several long, straight, descending branches. Near the surface it also throws out long horizontal runners provided with scaly cataphyllary leaves with buds in their axils. In England the plant is dug up in the late autumn, and either sold in the fresh state or cut transversely and dried. The drug consists therefore of both runners and roots, the former constituting the major part.

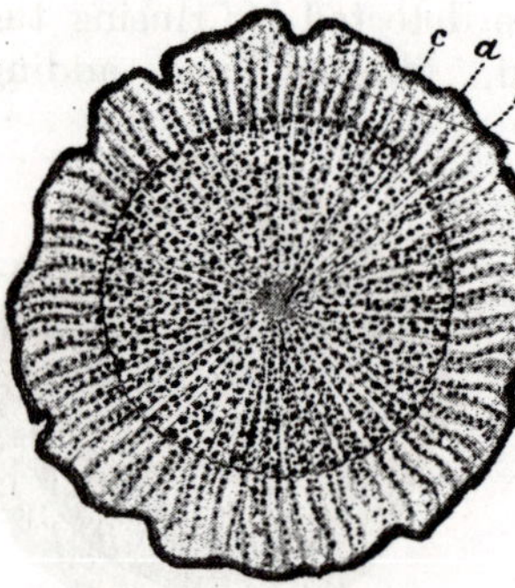


FIG. 160.—Spanish Liquorice root. Transverse section of rhizome (runner). *a*, bark; *b*, wood; *c*, pith. Magnified 3 diam. (Berg.)

Spain (Murcia, Aragon and Toledo, valleys of the Ebro, Guadalquivir, &c.), Sicily and the south of France furnish considerable quantities of carefully dried liquorice root. In southern Italy large quantities of liquorice root are grown, but it is chiefly converted into extract.

**Description.** Spanish liquorice is exported in large bales or in bundles of long, straight, cylindrical pieces from 1 to 2 cm. thick. These have a rather dark reddish brown colour, and are usually longitudinally wrinkled. On the surface they also bear small scars of roots, and, on the majority of pieces (the runners), here and there minute dark buds may be seen.

The drug breaks with a fracture that is fibrous in the bark, splintery in the wood; the section exhibits a yellow wood surrounded by a moderately thick yellowish grey bark, the pieces of runner being distinguished by the presence of a small pith. Under the lens the wood is seen to consist of very numerous medullary rays, between which are very narrow, porous wedges of vascular tissue; opposite to these in the bark are radial rows of dark points (groups of bast fibres).

Peeled root (including runner), which alone is official, has a pale yellow, slightly fibrous exterior and exhibits no trace of the small buds; otherwise it resembles the unpeeled.

The drug has a characteristic but not powerful odour, and a very sweet taste without perceptible bitterness or acridity.

The student should observe

(*a*) The *yellow colour of the section*, and *fibrous bark*,
(*b*) The *minute buds on the unpeeled root*,
(*c*) The *pith in most of the pieces*,
(*d*) The *characteristic sweet taste, free from acridity*;

and should carefully compare this (the Spanish) variety with the Russian.

**Microscopical Characters.**—In transverse section the cork-cells are brown and flattened. The cells of the cortex contain small (1·5$\mu$ to 20$\mu$) starch grains

FIG. 161.—Bales of decorticated Liquorice root.

and prismatic crystals of calcium oxalate. The bast-ring is characterised by the presence of numerous bundles of strongly thickened, striated, yellowish bast fibres accompanied by calcium oxalate cells. The wood contains vessels of varying size (often 100$\mu$) accompanied by wood fibres closely resembling the bast fibres.

The powder is characterised by the very thick-walled, grouped bast fibres with calcium oxalate cells frequently attached; by the portions of large, thick-walled vessels with closely packed, bordered pits; by the small starch grains and by the prevailing yellowish colour which is changed to orange yellow by sulphuric acid. It should be free from polygonal, brownish cork cells (unpeeled root) and from sclerenchymatous cells (olive-stones, almond shells, &c.).

**Constituents.**—The principal constituent of liquorice root is the sweet principle, glycyrrhizin, which, when quite pure, is a white, crystalline, intensely sweet powder, soluble in water. This has been shown to consist of the potassium and calcium salts of glycyrrhizinic acid, a colourless, crystalline acid, slightly soluble in water, but imparting a sweet taste to it in a dilution of 1 in 20,000; it is not a glucoside as was formerly supposed, since it yields on hydrolysis glycyrrhetinic acid and glycuronic acid, but no sugar. Liquorice root contains about 5 to 7 per cent. of it. The drug also contains glucose (1·4 per cent.), sucrose (2·5 per cent.), starch (29 per cent.), an acrid bitter principle, proteids, asparagin, fat, and resin. It yields from 3 to 4 per cent. of ash, and from 15 to 27 per cent. of aqueous extract dried at 100° (official minimum 20 per cent.).

Many other plants are remarkable for their sweet taste, but the presence of glycyrrhizin has been definitely proved only in *Periandra dulcis*, Martius (N.O. *Leguminosæ*, Brazil), *Pradosia latescens*, Radlkofer (Monesia bark, N.O. *Sapotaceæ*, Brazil). The root of *Abrus precatorius*, Linné (Indian liquorice, N.O. *Leguminosæ*), *Trifolium alpinum*, Linné (mountain liquorice, N.O. *Leguminosæ*, Europe), *Astragalus Glycyphyllos*, Linné (N.O. *Leguminosæ*), *Polypodium vulgare*, Linné (N.O. *Polypodiaceæ*), *Myrrhis odorata*, Scopoli (N.O. *Umbelliferæ*), *Ononis spinosa*, Linné (N.O. *Leguminosæ*), &c., contain sweet principles, possibly glycyrrhizin. The leaves of *Eupatorium Rebaudianum* (N.O. *Compositæ*), contain eupatorin, and rebaudin said to be 180 times as sweet as sugar.

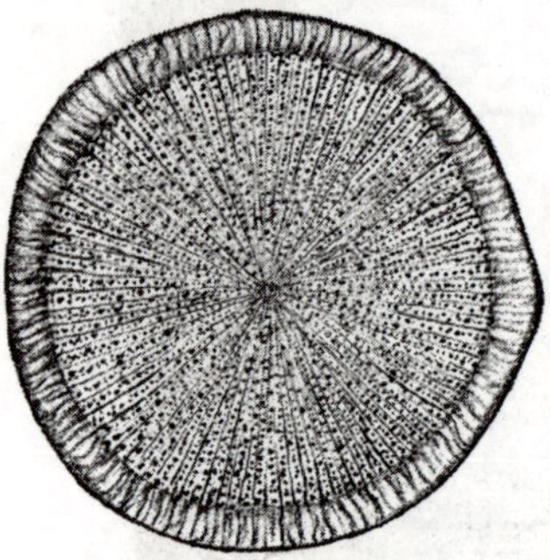

FIG. 162.—Russian Liquorice root. Transverse section. Magnified 1½ diam. (Berg.)

According to Kobert eupatorin is a neutral and rebaudin an acid saponin (compare p. 254). Glycyrrhizin also belongs to the saponins; it has no hæmolytic action, but its decomposition product, glycyrrhetinic acid, has; Kobert considers that this explains the value of liquorice as a remedy for coughs.

**Uses.**—Liquorice root is a demulcent and expectorant; the liquid extract is frequently employed to mask the taste of nauseous medicines.

**Varieties.**—*Russian Liquorice Root.* Very large quantities of liquorice root are collected in southern Russia, where the plant grows wild on the banks of the Volga and other rivers, needing no cultivation. The plant that yields this root is *G. glandulifera*, Waldstein and Kitaibel. Instead of producing numerous runners this plant forms a large rootstock, from which long perennial roots are given off. These are usually freed from the purplish brown cork by scraping.

Nearly all the Russian liquorice root that reaches this country has been peeled, and presents therefore a smooth yellow exterior, to which loose fibres are attached, the larger pieces being often longitudinally split. It attains a much larger size than the Spanish, the crown of the root, which shows the remains of several stems, being occasionally as much as 10 cm. in diameter. The texture is commonly looser and more fibrous than that of the Spanish drug, and the taste, although sweet, is accompanied by a more or less perceptible but not strong bitterness or acridity.

Unpeeled Russian liquorice shows a close resemblance to the Spanish. It consists, however, chiefly of roots (instead of runners)

FIG. 163.—Compressed bales of natural (unpeeled) Liquorice root.

which are destitute of pith and exhibit no traces of buds; their colour is purplish rather than brown, and the cork is often scaly. The official description is framed so as to include peeled Russian (and other) liquorice root provided that it complies with the official characters ('taste sweet and almost free from bitterness').

*Persian Liquorice Root.*—Liquorice root is also largely collected in the valleys of the Tigris and Euphrates from *G. glabra*, var. *β. violacea*, and exported in bales from Bussorah. It is usually unpeeled and in rather large coarse pieces closely resembling the unpeeled Russian root; it is largely consumed in America.

Anatolian and Syrian liquorice are exported from Smyrna and

Alexandretta respectively to the United States; they are probably derived from *G. glabra.*

*Stick Liquorice.*—The manufacture of stick or block liquorice is carried on chiefly in southern Italy, but also to some extent in Spain, Anatolia, &c. The runners and roots of both wild and cultivated plants are collected, crushed, boiled with water, and pressed. The decoction thus obtained is allowed to clear by standing, and is then run off into large pans, where it is concentrated by boiling until it has acquired a suitable consistence, when it is formed into sticks, which are stamped with the name of the manufacturer (e.g. Solazzi), or blocks (largely Anatolian) and dried.

Stick liquorice contains approximately 10 to 13 per cent. of glycyrrhizin, 13 per cent. of sugars, 23 per cent. of starch and gum, and 22 per cent. insoluble in water.

## BRYONY ROOT

### (Radix Bryoniæ)

**Source, &c.**—The common bryony, *Bryonia dioica*, Jacquin (N.O. *Cucurbitaceæ*) is a climbing and trailing plant, with rough, hairy leaves, common in hedges and thickets in southern England. It must not be confounded with black bryony (*Tamus communis*, Linné), a very different plant with entire shining leaves. The root is collected in the autumn and used in the fresh state.

**Description.**—The plant produces in the spring aerial stems attaining a great length and arising from a large, tuberous rhizome which is continuous with a thick, fleshy root. This subterranean part of the plant is often of very considerable size and weight, measuring occasionally at the upper extremity 15 cm. or more in diameter, and reaching a length of half a metre, the whole weighing several kilograms. It tapers more or less gradually towards the tip, and is usually simple. When fresh it is of a greyish yellow colour externally, and marked at close intervals with prominent transverse corky ridges often extending half round the root. Internally the root is whitish and fleshy, exuding when cut a juice that is milky from the presence of numerous, minute starch grains. The transverse section exhibits a fine line separating a narrow bark from a large, fleshy wood; the latter contains, more or less uniformly distributed over it, small groups of vessels, radially arranged and extending from the centre to the bark.

The fresh root has an unpleasant odour and a nauseously bitter and acrid taste.

It is sometimes cut into transverse slices and dried; the slices average about 5 cm. in diameter and have a thin yellowish grey cork, a whitish wood marked with concentric rings and radially arranged vessels; they somewhat resemble calumba, but may be distinguished by their yellowish grey cork and whitish wood.

The student should observe

(*a*) The *large size* and *gradually tapering shape*,
(*b*) The *transverse corky ridges*,
(*c*) The *nauseously bitter taste*;

and should compare small specimens with horseradish root, which is cylindrical and smooth, and has a pungent taste.

**Constituents.**—An intensely bitter amorphous alkaloid and a dark brown resin, both of which are purgative; an alcohol, bryonol, and various fatty acids. The autumnal root contains, further, large quantities of starch, the grains of which are exceedingly small and impart a milky appearance to the juice that exudes from a freshly cut surface.

**Uses.**—The root when taken internally in full doses acts as a cathartic and diuretic; applied to the skin it is irritant and may cause vesication. It has been recommended for pleurisy, whooping cough, and bronchitis, and has been given in cases of dropsy.

**Varieties.**—*Bryonia alba*, Linné, a Continental species, is distinguished by its monœcious flowers and black berries, the common bryony having diœcious flowers and scarlet berries. It produces a similar root containing probably the same constituents.

## SUMBUL RHIZOME

### (Sumbul Root, Musk Root, Radix Sumbul)

**Source, &c.**—The botanical origin of the drug at present known in commerce and official in the British Pharmacopœia under the name of Sumbul Root is not definitely known. It is generally referred to *Ferula Sumbul*, Hooker filius (N.O. *Umbelliferæ*), a plant of considerable size, growing in Turkestan. This plant, however, according to Holmes, produces a large napiform root that could scarcely furnish the cylindrical pieces, 3 to 6 cm. long, that form much of the commercial drug, although it was probably the source of the drug of twenty-five years ago, which was apparently obtained from a large fragrant root. *F. suaveolens*, Aitchison and Hemsley, has been suggested as yielding the commercial drug.

**Description.**—Sumbul root occurs in short, more or less cylindrical pieces that are remarkable for their extreme lightness; they vary considerably in size, but are usually from 3 to 6 cm. in width and about the same in length, often dividing in the upper part into two, three, or more branches. The latter are occasionally not more than

1 cm. in diameter, and some of them bear a depressed scar left by the aerial stem after it has perished. Most of the pieces bear numerous short bristly fibres arranged in encircling lines, or at least show the scars of such ; these fibres are the remains of the fibro-vascular bundles of leaves that have perished, and prove that most of the commercial drug consists of rhizome, not root. All the pieces exhibit regular transverse wrinkles and are covered with a thin but tough cork, which often shows a disposition to exfoliate, or at least can easily be stripped off.

Internally the drug is whitish or yellowish, spongy and irregularly fibrous, exhibiting numerous fissures which have possibly originated

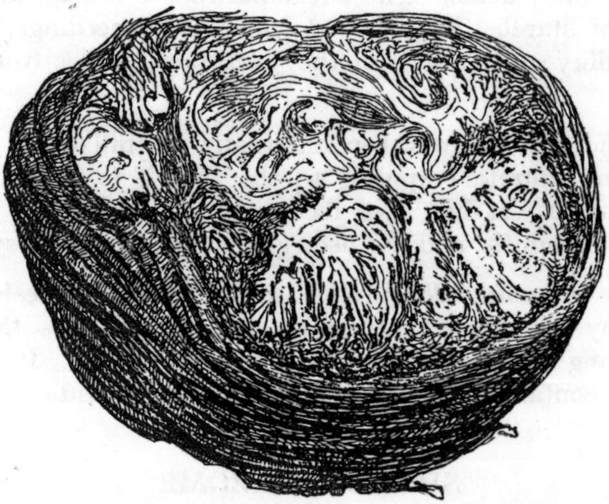

FIG. 164.—Sumbul root. Slightly reduced. (*Pharmaceutical Journal.*)

during the process of drying, and more or less abundant soft resin to which dust, &c., rapidly adheres. The transverse section of a small rhizome shows a pale bark, within which is a ring of narrow, finely porous yellow wood-bundles ; the central portion is parenchymatous tissue through which vascular bundles pass in varying directions, a peculiarity that is exhibited by many rhizomes, and to which in this case the fibrous nature of the drug is due. In the larger pieces the structure is usually less distinct. The drug has an agreeable musky odour and a bitter, slightly aromatic taste.

The student should observe

(*a*) The *transverse wrinkles*, from which *short fibres* proceed,
(*b*) The *whitish, fibrous and spongy interior*,
(*c*) The *musky odour*.

**Constituents.**—Of the constituents of the sumbul root of commerce very little is known. Hahn (1896) found it to yield to petroleum benzin 17·25 per cent. of a yellow viscid oil, from which crystals of a substance not further investigated were obtained. The drug yielded 8 per cent. of ash.

Utech (1893) obtained 6·1 per cent. of an aromatic amber-coloured resin having a bitter taste and possessing the aromatic odour of the root.

Heyl and Hart (1916) found neutral and glucosidal resins yielding umbelliferone by hydrolysis, sucrose, levulose, betaine, &c.

The drug also contains free umbelliferone and traces of volatile oil.

**Uses.**—Sumbul is considered to possess stimulant and antispasmodic properties resembling those of valerian. It has been given in hysteria and certain nervous disorders, but is now not much prescribed.

## IPECACUANHA ROOT

### (Radix Ipecacuanhæ)

**Source, &c.**—Ipecacuanha root is obtained from *Psychōtria* (*Cephalëis*) *Ipecacuanha*, Stokes (N.O. *Rubiaceæ*), a small plant about 30 cm. in height with a stem that is at first prostrate or ascending and afterwards becomes erect. It is found in most parts of Brazil, but especially in the province of Matto Grosso, in the interior, whence much of the drug is obtained; it is also cultivated in the province of Minas Geraes. It is conveyed down the Paraguay and Parana rivers to Montevideo whence it is exported.

Endeavours have not been wanting to cultivate the ipecacuanha plant in other countries, and they have met with some success. From the state of Selangor (near Singapore) ipecacuanha root of unusually fine appearance and rich in alkaloid is exported in considerable quantity. In Java, Ceylon, and China the cultivation has not been successful.

From the slender, prostrate stem fibrous roots are given off at intervals; some of these in the course of their growth develop an abnormally thick bark in which abundance of starch is deposited, whilst the wood remains comparatively small. These thickened roots should constitute the commercial drug; they are collected by raising the plants, which usually form clumps, from the earth, removing the roots, drying them, and sifting them from adherent sand and earth. The drug is exported in compact canvas bales. Very frequently much of the prostrate or ascending stem finds its way into the drug.

Ipecacuanha appears to have long been used in Brazil for dysentery; it was introduced into Europe about 1672.

**Description.**—The official ipecacuanha root, which is distinguished as Brazilian or Rio ipecacuanha, is slender and rather tortuous, seldom exceeding, in the commercial drug, 15 cm. in length or 5 mm. in thickness. The colour varies from dark brick-red to dark brown, the former being due, partly at least, to adhering particles of earth. Very characteristic of the root is the annulated appearance that it presents, the bark of typical pieces being constricted at short intervals so as to resemble a number of discs somewhat irregularly strung

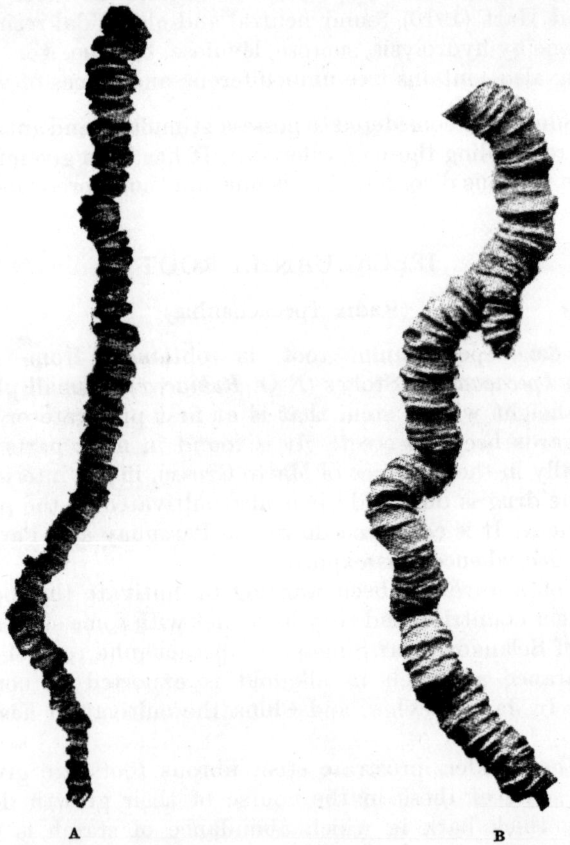

FIG. 165.—A, Brazilian Ipecacuanha (*P. Ipecacuanha*) cultivated in Selangor. B, Cartagena Ipecacuanha. Large specimens. Natural size.

together; the constrictions are sometimes quite shallow, but sometimes they penetrate nearly to the wood. These annulations seldom, however, take the form of distinct, rather distant, narrow, raised ridges (compare Cartagena ipecacuanha).

The root is hard, and breaks with a very short fracture. The transverse section exhibits a thick, dark grey bark which is usually horny, but sometimes starchy, and a small wood in which no distinct pores or pith can be discerned when examined with a lens. The bark, when examined under the microscope, is found to contain abundance of starch grains that are mostly compound and, in addition, acicular crystals of calcium oxalate; the wood is free from vessels. These characters are useful in distinguishing ipecacuanha from certain substitutes that appear from time to time, and are referred to below.

The drug has a slight odour which to many persons is particularly unpleasant; the taste is slightly bitter. The powder is often very irritating to the throat and nostrils, producing violent coughing and sneezing.

The student should observe

(*a*) The *closely approximated disc-like annulations*,
(*b*) The *thick, starchy bark*,
(*c*) The *small, dense wood*;

and should compare this variety with Cartagena ipecacuanha, which is usually rather thicker, and in which the annulations assume the form of distinct, somewhat distant, narrow, raised ridges.

**Microscopical Characters.**—Below the narrow cork is a largely developed cortex consisting of thin-walled parenchymatous cells most of which are filled with starch, but a few contain acicular calcium oxalate. The bast ring is narrow, and both this and the cortex are free from sclerenchymatous cells and fibres and from cells containing colouring matter. The wood is composed of tracheids, wood fibres, and parenchyma and is free from typical vessels. The tracheids, when isolated, exhibit moderately thick walls, and often near the pointed extremities a large perforation.

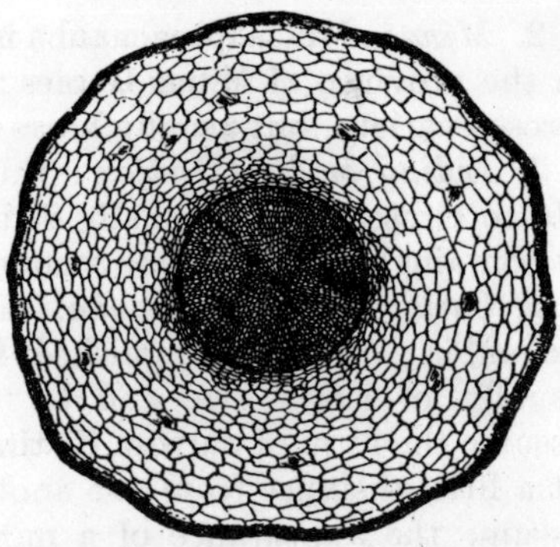

FIG. 166.—True Ipecacuanha (*Psychotria Ipecacuanha*). Transverse section, showing dense wood. Magnified. (Planchon and Collin.)

The powder is characterised by the starch grains which are either oval or rounded (not over 15$\mu$) or compound with from 2 to 5 constituent grains, by the acicular calcium oxalate, by the characteristic tracheids and by the absence of sclerenchymatous cells, spiral vessels (ipecacuanha stem), cells containing colouring water, calcium oxalate in other than acicular crystals, and typical sclerenchymatous fibres or vessels.

**Constituents.**—Ipecacuanha root contains three alkaloids; two of which, emetine and cephaëline, have been more closely examined, whilst the third, psychotrine, which occurs in much smaller quantity, awaits further investigation. These alkaloids exist in good root to the extent of from 2 to 3 per cent., and are contained chiefly in the bark, the wood yielding only about 1 per cent. The alkaloids ipecamine and hydroipecamine have also been reported as present.

Ipecacuanha also contains about 0·4 per cent. of a crystalline glucoside, ipecacuanhin, which is sparingly soluble in cold water, more freely in hot, insoluble in ether; it is apparently devoid of marked physiological action.

Other constituents are a glucosidal acid, ipecacuanhic acid, allied to quillajic acid, calcium oxalate and an abundance of starch.

*Emetine*, $C_{15}H_{22}NO_2$, is a white amorphous alkaloid which darkens on exposure to light, gradually assuming a yellow colour; it yields crystalline salts and has been shown to be methyl-cephaëline, but is less powerfully emetic and about half as toxic as cephaëline. It constitutes about 72 per cent. of the total alkaloid.

*Cephaëline*, $C_{14}H_{20}NO_2$, crystallises in white, silky needles, but yields uncrystallisable salts; though a more powerful emetic, it is a less efficacious expectorant than emetine. It constitutes about 26 per cent. of the total alkaloid.

*Psychotrine* exists in small quantity only (about 2 per cent. of the total alkaloid).

**Varieties.**—1. *Matto Grosso*: this is the variety above described.

2. *Minas*: Minas ipecacuanha is the produce of the plant cultivated in the province of Minas Geraes; it very closely resembles the Matto Grosso variety, but usually bears evidence of careful cleaning.

3. *Indian or East Indian*: this variety also closely resembles the Matto Grosso, but is usually distinctly larger and brighter in colour. It contains about the same amount of total alkaloid.

4. *Cartagena Ipecacuanha.*—This variety is imported from Colombia; it is distinguished from the Brazilian drug by being usually of larger size and by the annulations which in this case assume the form of narrow, distinct, transverse, raised ridges arranged at a little distance from one another, and the root does not therefore assume the appearance of a number of discs strung together. The starch grains are also rather larger than those of Brazilian ipecacuanha. The drug has been attributed to *Psychotria acuminata*, Karsten, but definite evidence on this point has not yet been adduced. Cartagena ipecacuanha contains the same alkaloids as Brazilian, and about the same amount. The chief difference in this respect lies in the proportion in which each of the alkaloids emetine and cephaëline is present rather than in the amount of total alkaloid contained in the root. In Cartagena ipecacuanha the emetine amounts to only 40 per cent. of the total alkaloid, whilst the cephaëline reaches as much as 57 per cent.; in other words, it contains twice as much of the latter alkaloid as the Brazilian root, and its physiological action is therefore much more emetic and less expectorant. For this reason the Cartagena variety has been excluded from the Pharmacopœia.

**Uses.**—Ipecacuanha is largely used as an expectorant and emetic; it also possesses diaphoretic and cholagogue properties. One of its most important uses is as a remedy for dysentery, for which purpose large doses are given (30 to 90 grains). In such cases a remarkable tolerance of the drug is often readily established, or the drug deprived of its alkaloids (de-emetinised ipecacuanha) is administered.

**Substitutes.**—1. *Ipecacuanha Stem.*—Much of the Brazilian drug is imported mixed with the stems of the plant. These are slender and longitudinally striated, exhibiting no annulations. The transverse section shows a thin bark surrounding a ring of wood within which is a distinct pith. They contain less alkaloid than the root, various

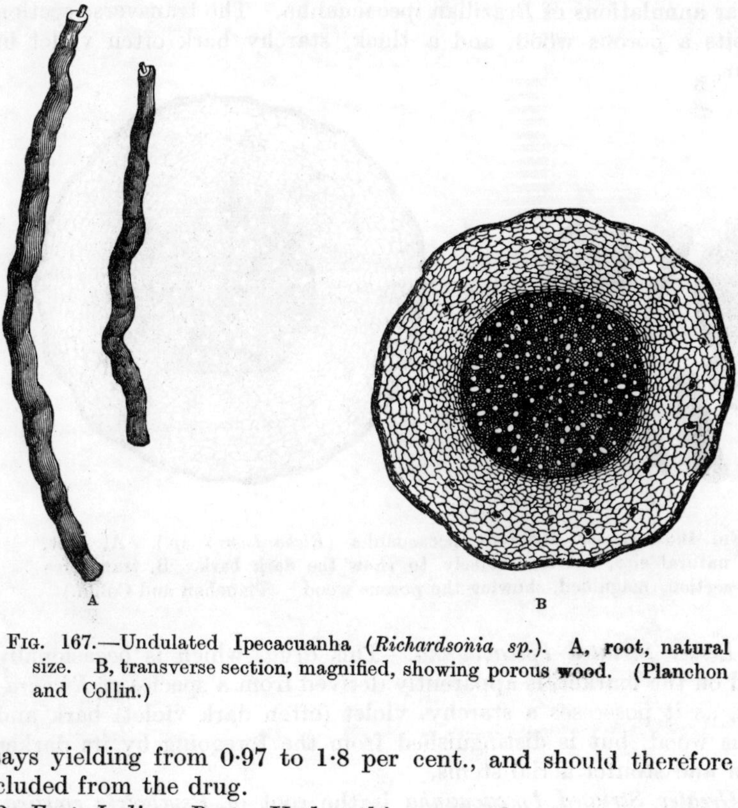

FIG. 167.—Undulated Ipecacuanha (*Richardsonia sp.*). A, root, natural size. B, transverse section, magnified, showing porous wood. (Planchon and Collin.)

assays yielding from 0·97 to 1·8 per cent., and should therefore be excluded from the drug.

2. *East Indian Root.*—Under this name the rhizome of a small monocotyledonous plant, probably *Cryptocoryne spirālis*, Fischer (N.O. *Aroideæ*), has been imported from the south of India. It occurs in short, nearly cylindrical pieces, 1·5 to 5 cm. long, often tapering at one end and there bearing the remains of leaves; they are slightly enlarged and constricted at regular intervals, which gives them an annulated appearance somewhat resembling ipecacuanha, and are marked with encircling leaf scars. The transverse section exhibits a typically monocotyledonous structure, and distinguishes this drug at once from genuine ipecacuanha; for instead of the central column

of dense wood there is a parenchymatous stele, through which scattered bundles run, surrounded by a white starchy cortex. This substitute for ipecacuanha contains no emetine or cephaëline.

3. *Undulated Ipecacuanha*, derived from *Richardsonia scabra*, Saint Hilaire (N.O. *Rubiaceæ*, Brazil), occurs in tortuous pieces, the upper part being cylindrical and bearing at the crown the remains of numerous very slender aerial stems. On one side of the root the bark is occasionally fissured nearly to the wood, and this gives the drug a somewhat annulated appearance, which, however, is quite different from the regular annulations of Brazilian ipecacuanha. The transverse section exhibits a porous wood, and a thick, starchy bark often violet in colour.

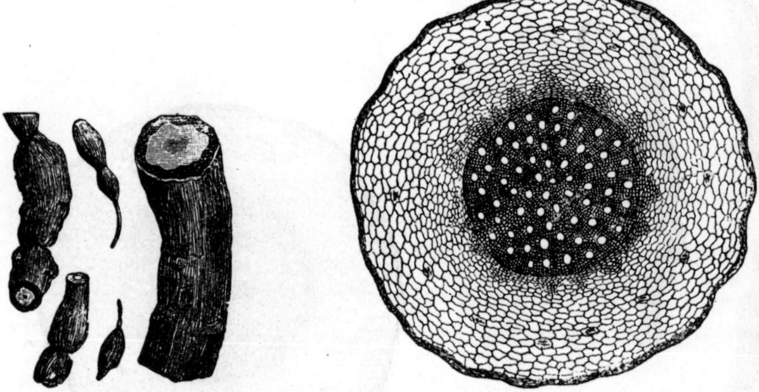

A B

FIG. 168.—Lesser Striated Ipecacuanha (*Richardsonia sp.*). A, root, natural size, cut transversely to show the dark bark. B, transverse section, magnified, showing the porous wood. (Planchon and Collin.)

4. *Lesser Striated Ipecacuanha.*—This drug, which is occasionally found on the market, is apparently derived from a species of *Richardsonia*, as it possesses a starchy, violet (often dark violet) bark and porous wood, but is distinguished from the foregoing by its darker colour and stouter aerial stems.

5. *Greater Striated Ipecacuanha* is the root of *Psychotria emetica*, Linné. It is about the size of Cartagena ipecacuanha and closely resembles the foregoing, being irregularly constricted, dark in colour, and exhibiting in transverse section a dark, violet-coloured bark; it may be distinguished by its dense wood, and by the presence of sugar in the bark, starch being completely absent. The colour of the bark and the absence of starch easily distinguish this root from either Brazilian or (which it closely resembles in appearance) Cartagena ipecacuanha.

6. *White Ipecacuanha* is the root of *Ionidium Ipecacuanha*, Ventena

(N.O. *Violaceæ*, Brazil). It is greyish white or yellowish in colour, 1·5 to 3 mm. thick, and frequently branches. The section exhibits a large, porous, yellowish wood and narrow, darker bark.

7. *Trinidad Ipecacuanha.*—Under this name the rhizome and root of *Asclēpias curassavīca*, Linné (N.O. *Asclepiadeæ*), have been offered for sale. It is 2 to 4 mm. in diameter, yellowish brown externally and whitish internally; it bears wiry rootlets and has an unpleasant bitter taste.

Other roots are occasionally found in the market mixed with or substituted for ipecacuanha; the enumeration of these would be beyond the scope of the present work.

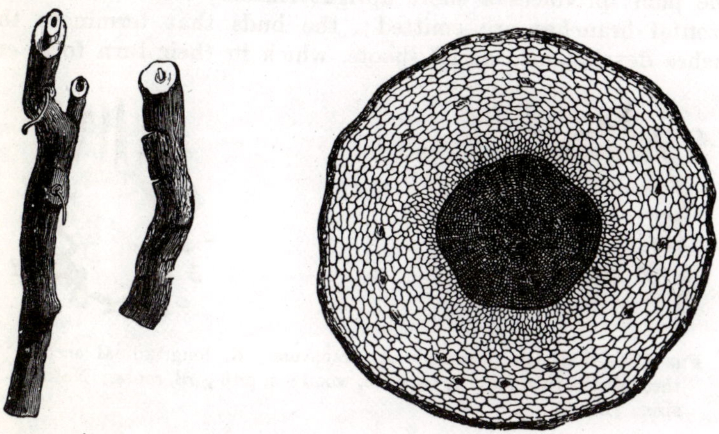

A B

FIG. 169.—Greater Striated Ipecacuanha (*Psychotria emetica*). A, root, natural size. B, transverse section, magnified, showing the dense wood. (Planchon and Collin.)

With the exception of ipecacuanha stem, none of these substitutes, all of which reappear from time to time on the London market, contains any emetine or cephaëline.

The following test for emetine is useful in distinguishing roots containing that alkaloid from numerous substitutes which do not:—0·5 gramme of the finely powdered root is mixed with 20 c.c. of strong hydrochloric acid and 5 c.c. of water; to 2 c.c. of the filtrate 0·01 gramme of potassium chlorate is added; if emetine is present the liquid assumes a yellow colour, changing in the course of an hour to red.

## VALERIAN RHIZOME

### (Rhizoma Valerianæ)

**Source, &c.**—The common valerian, *Valeriana officinalis*, Linné (N.O. *Valerianeæ*), is an herbaceous plant widely diffused over Europe and Northern Asia. It is common in England in moist situations, attaining a height of 1 to 1·5 metres, and producing terminal corymbs of small white or pinkish flowers, each with a gamopetalous corolla and three stamens. It is cultivated in England (Derbyshire), and also in Holland and Germany, our supplies being derived partly from English and partly from foreign sources, but chiefly from cultivated plants. The drug, or at least a species of valerian, was well known to the Greeks and Romans; whilst the common valerian, or 'all-heal' as it is sometimes called, was a domestic medicine of the Anglo-Saxons. In the Middle Ages the root was used not only as a medicine but also as a spice, and, curiously enough, even as a perfume.

The plant produces a short upright rhizome from which slender horizontal branches are emitted; the buds that terminate these branches develop into aerial shoots, which in their turn form erect

FIG. 170.—Valerian rhizome. *A*, transverse, *B*, longitudinal section through the rhizome: *a*, bark; *b*, wood; *c*, pith; *rd*, roots. Natural size. (Berg.)

rhizomes, and continue the life of the plant, the mother-rhizome perishing after flowering and fruiting. The drug is collected in the autumn, the lateral shoots are cut off for the propagation of the plant, and the rhizomes usually, especially if of large size, sliced longitudinally and dried.

In Derbyshire the wild plants are collected and planted out; the stems of the larger which would otherwise produce flowers are topped to prevent this from happening. In September or October all the tops above ground are cut off with a scythe, the rhizomes dug up, washed and dried, the larger being generally sliced.

**Description.**—The commercial drug consists of short erect conical rhizomes not more than 12 mm. thick and 25 mm. long, of a dull yellowish brown colour, usually halved, or quartered if large, but entire if small, and sometimes crowned with the remains of the stem and leaves. Internally the rhizome is firm and horny and of a whitish or yellowish colour; sometimes it becomes hollow, portions of the tissue remaining as transverse septa.

From the rhizome numerous brittle roots of the same colour and about 1·5 mm. thick are given off, as well as occasionally short, slender, lateral branches. In the commercial drug the roots seldom exceed 10 cm. in length and 2 mm. in diameter; they are plump or longitudinally striated, but usually not much shrivelled.

Sometimes the drug consists of small rhizomes about 5 mm. long, which are crowned with the remains of several leaves and bear rather short slender roots. These are formed from the lateral branches (runners) produced by the principal rhizome, which bear cataphyllary leaves; the buds in the axils of these develop leaves, roots are developed below, and independent plants are produced. Such plants may occasionally be found mingling their roots with those of the principal rhizome. Had they been allowed to continue their growth, they would ultimately have developed rhizomes as large as those of the parent plant.

The transverse section of the rhizome is irregular in outline and exhibits a comparatively narrow bark separated by a dark line (cambium) from an irregular circle of wood-bundles of varying size. The section of a root shows a thick bark and small wood.

Valerian rhizome has a powerful characteristic disagreeable odour and a camphoraceous, slightly bitter taste. The odour of the fresh rhizome, though disagreeable, is not strong, but it develops, during the drying, into the penetrating unpleasant odour of the drug. This change is due to the action of an enzyme, and may be prevented by boiling the root previous to drying it.

The student should observe

(*a*) The *colour* and *odour* of the drug,
(*b*) The *short, erect rhizome* surrounded by numerous roots,
(*c*) The characters of the *section of the rhizome* as given above.

**Constituents.**—The principal constituent of valerian root is the volatile oil which is contained in the sub-epidermal layer of cells in the root, not in isolated oil-cells or glands. Good valerian root yields about 1 per cent. of volatile oil, one of the constituents of which is bornyl isovalerianate. This constituent is gradually decomposed by an enzyme present, yielding free isovalerianic acid, an oily liquid possessing an unpleasant odour of valerian; to this body the unpleasant odour of valerian root is to be ascribed, and its gradual production from bornyl isovalerianate explains the development of the odour as the root dries.

Valerian root also contains two alkaloids, chatinine and valerianine, which require further investigation.

Chevalier (1907) found in the fresh rhizome an alkaloid, a glucoside, and a resin, all of which were physiologically active, and considered that the fresh root was preferable from a medicinal point of view to the dry, the reason being that the bornyl isovalerianate was present as such.

**Uses.**—The drug is used as a powerful carminative, stimulant, and antispasmodic; it is given chiefly in hysteria, palpitation of the heart, &c.

**Allied Drugs.**—*Japanese Valerian, Kesso*; the rhizome and roots of *Valeriana officinalis*, var. *latifolia*, Miquel; rhizome small, erect, crowned with scars or with the remains of aerial stems; brownish; odour resembling that of valerian but much stronger and somewhat aromatic; yields up to 8 per cent. of volatile oil containing *l*-bornyl isovalerianate and acetate, kessyl acetate, and terpenes.

*Radix Valerianæ Majoris* is the root of *V. Phu*, Linné, it has very little odour.

*Mexican Valerian* is the rhizome of *V. officinalis* growing in Mexico; it yields traces of volatile oil and a little free valerianic acid; the odour is feeble.

*Nardus root* is the rhizome and roots of *Nardostachys Jatamansi*, de Candolle (N.O. *Valerianeæ*), Alpine Himalayas; rhizome short, thick, dark grey, crowned by a bundle of fibres; odour resembles valerian; yield of oil about 1 per cent.

## INDIAN VALERIAN

### (Rhizoma Valerianæ Indicæ)

**Source, &c.**—Indian valerian is the rhizome and roots of *Valeriana Wallichii*, de Candolle (N.O. *Valerianeæ*), which is widely distributed in the temperate Himalayas.

**Description.**—The rhizome is nearly cylindrical, about 5 cm. long and 5 to 10 mm. thick, usually curved or bent, dull brown in colour and marked with raised, encircling leaf-scars and numerous prominent root scars; roots thick and few. The rhizome is hard and tough, breaking with a short fracture and exhibiting a dark section with a large pith and diffuse ring of small wood-bundles. The odour resembles that of valerian.

**Constituents.**—Indian valerian yields about 1 per cent. of volatile oil of similar nature to that of valerian.

## ELECAMPANE ROOT

### (Radix Inulæ)

**Source, &c.**—Elecampane, *Inŭla Helenium*, Linné (N.O. *Compositæ*), is a tall, herbaceous plant, with perennial rhizome, distributed over temperate Europe and Asia. For medicinal use the plant is cultivated in Holland, Thuringia, and Switzerland.

The plant produces a short, thick, fleshy rhizome and large, fleshy roots. Both rhizome and roots are collected when the plants are about three years old and dried, the larger being often sliced longitudinally.

**Description.**—The drug consists of both rhizome and roots. The former is usually cut longitudinally, and occurs in thin, irregularly

FIG. 171.—Elecampane root. A, longitudinally cut ; B, entire, with portion of the rhizome attached. (Planchon and Collin.)

rounded slices about 4 or 5 cm. in diameter. The roots vary much in size, the smaller being the thickness of a pencil or even less, the larger sometimes exceeding 3 cm. in diameter. They are nearly cylindrical (if entire), tapering very gradually towards the tip, but are seldom quite straight, usually curling irregularly as they dry, especially if they have been sliced.

Both rhizome and root are hard and horny, or, if slightly moist, tough, and are of a dark brownish grey colour externally, whitish or pale brownish internally. They break with a short fracture. The transverse section is more or less uniform in colour and exhibits a number of shining, brown oil-glands scattered over the whole of the surface, both in the bark and in the central portion (wood) which is separated from it by a dark and often very indistinct cambium line. In the wood small, radially elongated groups of vessels occur, but are difficult to discern with a lens. The root has an agreeable aromatic odour and an aromatic, slightly bitter taste.

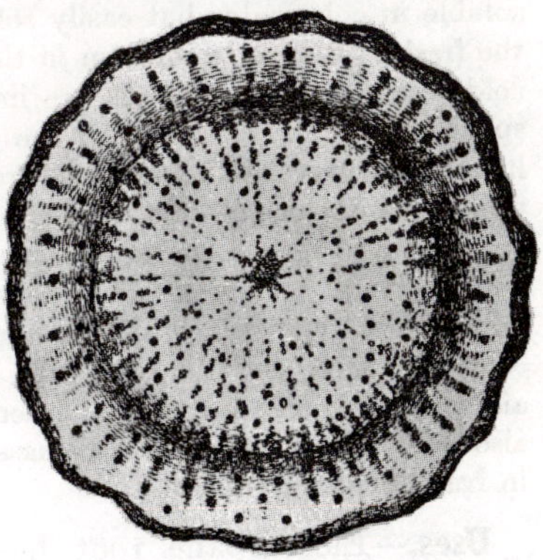

FIG. 172.—Elecampane root. Transverse section, showing the distribution of the oil-glands. Magnified. (Planchon and Collin.)

The student should observe

(*a*) The *horny* (*not starchy*) *nature of the drug*,

(*b*) The *presence of oil-glands*,

(*c*) The *absence of distinct radiate structure in the wood*;

and should compare the root with

(i) *Belladonna root*, which has no oil-glands, and which should possess a starchy fracture,

(ii) *Dandelion root*, which has no oil-glands, a small yellow wood, and thick ringed bark,

(iii) *Marshmallow root*, which has no oil-glands, a radiate structure, and a fibrous, easily separated bark,

(iv) *Pellitory root*, which has oil-glands, but is distinguished by its yellow, radiate wood, distinctive odour and taste.

**Constituents.**—By distillation with water elecampane root yields from 1 to 2 per cent. of a crystalline mass associated with a little volatile oil. The crystalline mass consists of alantolactone, isoalantolactone (helenin), and alantolic acid, all of which are nearly colourless and crystalline and have but slight odour or taste. Alantol is an oily liquid with a peppermint-like odour found in the distillate. Elecampane root contains also in large quantity, as reserve material, inulin. The roots gathered in the autumn, when they are richest in inulin, contain as much as 45 per cent. of this substance, in the spring about 19 per cent.

Inulin $C_6H_{10}O_5$, can be obtained as a white crystalline powder, slightly soluble in cold water but easily soluble in hot water without gelatinising. In the fresh root it is in solution in the cell sap, although only slightly soluble in cold water; when the roots are immersed in alcohol it is slowly deposited in sphero-crystalline masses; in the dry root it occurs in transparent, irregular lumps. It may be distinguished from starch by these characters as well as by the cooled decoction not assuming a blue colour with iodine. It takes the place of starch as a reserve material in many plants belonging to the order *Compositæ*, and in several belonging to *Campanulaceæ*, &c. Hydrolysis by means of a dilute mineral acid converts it into levulose.

In the dried root the oil-glands contain crystals of alantolactone which are also to be found in the powder when mounted in dilute nitric acid; the powder also contains irregular black masses of phytomelan which is chiefly found in fruits of *Compositæ*.

**Uses.**—Elecampane root has been used in bronchitis and tuberculosis; helenin is an effective antiseptic particularly for the tubercle bacillus.

## PELLITORY ROOT

### (Radix Pyrethri)

**Source, &c.**—Pellitory root is obtained from *Anacyclus Pyrethrum*, de Candolle (N.O. *Compositæ*), a small plant with perennial root, indigenous to Algeria. In habit and general appearance the plant resembles the chamomile. The root is collected in the autumn and dried.

**Description.**—Pellitory root varies from 7 to 15 cm. in length and is usually simple. It is nearly cylindrical in shape, tapering slightly towards the tip, and often towards the crown as well, where there is frequently to be found a tuft of grey hairs on the remains of leaves. The outer surface is brown and deeply and irregularly wrinkled.

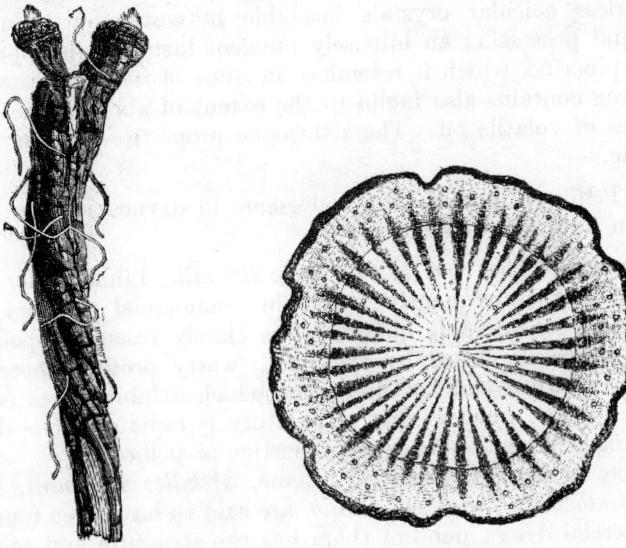

A B

FIG. 173.—Pellitory root. A, root, natural size. B, transverse section, showing radiate structure of the wood and the distribution of the oleo-resin glands, magnified. (Planchon and Collin.)

The root is tough and breaks with a short fracture. The bark is closely adherent to the wood, which exhibits a prominently radiate structure, narrow yellowish wedges of vascular tissue alternating with whitish medullary rays of equal or even greater breadth. In the latter, as well as in the bark, yellow or brown oleo-resin glands are scattered. The drug when cut is of a horny, not starchy, or fibrous nature. It has a characteristic though not strong odour, and, when chewed, a pungent taste, exciting a copious flow of saliva.

The student should observe

(*a*) The *distinctly radiate structure of the wood*,
(*b*) The *wide medullary rays*,
(*c*) The *presence of oleo-resin glands* ;

and should compare the root with

(i) *Belladonna root*, which has no oil-glands and should be starchy,
(ii) *Dandelion root*, which has a small yellow wood and thick, ringed bark,
(iii) *Liquorice root*, which has no oil glands, and much narrower medullary rays.

**Constituents.**—Pellitory contains, as principal constituent, an alkaloid, pyrethrine (also called pellitorine), which can be obtained in colourless acicular crystals insoluble in water but soluble in alcohol and possessing an intensely pungent taste. It is apparently allied to piperine, which it resembles in some of its properties.

The drug contains also inulin to the extent of about 50 per cent., and traces of volatile oil. The sialogogue properties are due to the pyrethrine.

**Uses.**—Pellitory is used as a sialogogue in dryness of the throat and as an addition to dentifrices.

**Adulteration.**—The root of *Corrigiōla littoralis*, Linné (N.O. *Illecebrācĕæ*), is not unfrequently found in commercial pellitory root, sometimes in considerable quantity. It closely resembles pellitory, but may be distinguished by the small, warty protuberances with which it is crowned and by the section which exhibits three or four concentric circles, whereas that of pellitory is radiate. It is devoid of the pungency and sialogogue properties of pellitory.

The roots of *Sonchus oleraceus*, Linné, *Achillea Ptarmica*, Linné and *Chrysanthemum frutescens*, Linné, are said to have been found in the commercial drug ; none of them has the structure and taste of pellitory root.

German pellitory root, *Anacyclus officinalis*, Hayne, cultivated in Germany, is thinner, has fibrous rootlets, is often crowned with the remains of leaves, and has a similar pungent taste.

Pará Cress or Paraguay-Roux (Brazil, E. and W. Indies), the herb *Spilanthes oleracea*, Jacquin (N.O. *Compositæ*) has cordate, sinuate-dentate, opposite leaves and yellow capitula without ray florets ; taste similar to pellitory root.

## ARNICA RHIZOME

### (Arnica Root, Rhizoma Arnicæ)

**Source, &c.**—Arnica rhizome, or arnica root as the drug is commonly termed, consists of the rhizome and roots of *Arnica montana*, Linné (N.O. *Compositæ*), a small plant with a creeping perennial rhizome, indigenous to central Europe and common in the meadows on the lower mountain spurs. It should be collected in the autumn, after the aerial parts have died down, and dried.

**Description.**—The rhizome, which is horizontal or oblique, is slender, nearly cylindrical in shape, and usually curved. It averages about 5 cm. in length and about 5 mm. in thickness, and varies in colour from yellowish brown to nearly black. From the flanks and under surface numerous dark, brittle, wiry curved and twisted roots about 8 cm. in length are given off, and the scars that are left, when these break off or perish, together with the encircling scars of cataphyllary leaves, render the surface of the rhizome, which is in addition longitudinally shrivelled, distinctly rough. Usually it is simple and often terminated by the hairy remains of leaves; sometimes, after the plant has flowered, the growth of the rhizome, which has thus been terminated, is continued in the same direction by the development of a bud on its under surface; the rhizome thus formed will in due course itself flower, and the further growth will again be ensured by the development of one of the lower buds. Thus the rhizome occasionally assumes a jointed appearance.

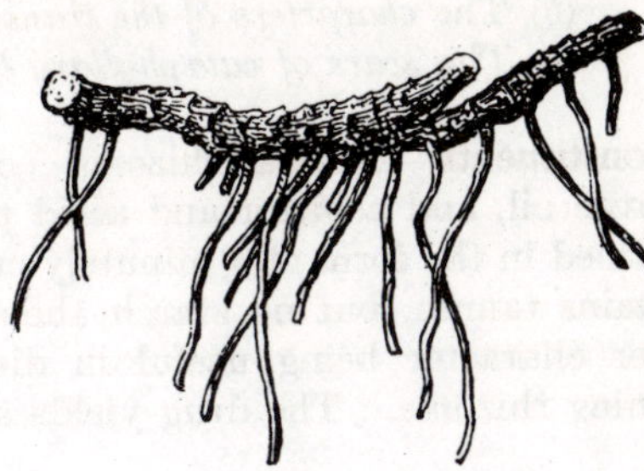

FIG. 174.—Arnica rhizome. Natural size. (Holmes.)

The rhizome breaks with a short fracture, and is often discoloured in the interior. The transverse section of a favourable specimen shows a comparatively thick whitish or yellowish bark, in which near the wood is a circle of darker oleo-resin ducts. The wood consists of yellow wedge-shaped wood-bundles and broad medullary rays enclosing a large whitish pith which in longitudinal sections is seen to be continuous. The root also shows a thick white bark with a ring of oleo-resin ducts and a yellowish wood.

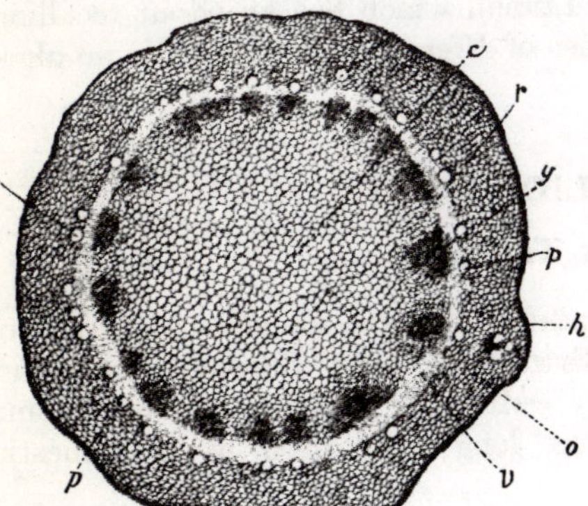


FIG. 175.—Arnica rhizome. Transverse section of rhizome. *v*, bast; *o*, primary cortex; *y*, wood-bundles; *r*, medullary rays; *c*, pith; *p*, oleo-resin ducts. Magnified 10 diam. (Berg.)

The drug has a faint but rather agreeable apple-like odour and a bitter acrid taste.

The student should observe

(*a*) The *curved rhizome* with *wiry roots*,
(*b*) The *characters of the transverse section*,
(*c*) The *scars of cataphyllary leaves.*

**Constituents.**—Arnica rhizome contains about 0·5 per cent. of volatile oil, and a bitter and acrid principle, arnicin, which has been obtained in the form of a minutely crystalline yellow powder. It also contains tannin, but no starch, the reserve material being inulin, the latter character being useful in distinguishing it from any starch-forming rhizome. The drug yields about 8 per cent. of ash.

**Uses.**—The tincture diluted with water is a popular application for bruises, preventing swelling and hastening the absorption of effused blood. It appears to increase the activity of the circulation in the skin, but should be used with caution, as it is liable to produce extensive dermatitis.

Arnica is not often given internally; it has a depressing or, in over doses, irritant action, causing vomiting, pain, and purging.

**Substitutes, &c.**—Foreign roots and rhizomes are occasionally present in the drug as imported, and have to be separated by hand-picking. Amongst these may be mentioned the erect rhizome of the common avens (*Geum urbanum*, Linné), which has an odour recalling cloves, and the rhizomes of species of *Hieracium*, which have no oleo-resin ducts.

## DANDELION ROOT

### (Radix Taraxaci)

**Source, &c.**—The common dandelion, *Taraxacum officinale*, Wiggers (N.O. *Compositæ*), is widely distributed over Europe, Asia, and North America, and is in many cultivated districts a troublesome weed. In this country it has long been used as a domestic medicine.

The fresh root alone is official and is directed to be gathered in the autumn, when, judging from analogy, it may be expected to be most active. This, however, has been denied, the assertion being made that the root is most active in the spring or summer. A method for the quantitative determination of the active constituent not being at present known, this point cannot be definitely settled, but roots collected in the spring or summer contain much sugar, which is more objectionable than inulin, so that, unless positive evidence can be adduced to the contrary, the autumn must be regarded as the proper season in which to collect dandelion root.

The roots are collected from wild plants, washed, and, if necessary, dried. Much of the drug is imported from Germany.

**Description.**—Dandelion root consists of a simple straight root, which, towards the upper part, passes imperceptibly into an erect rhizome; the latter sometimes remains simple, but often divides into several erect branches. It attains a length of 30 cm. or more, and a thickness varying from 15 to 25 mm. Whilst fresh it is yellowish brown externally, whitish and fleshy within. From the freshly cut surface, as indeed from all parts of the plant, an abundance of a very bitter, milky juice exudes, which on careful examination may be observed to rise from concentric rings of tissue. In the centre of the root is a small yellow wood.

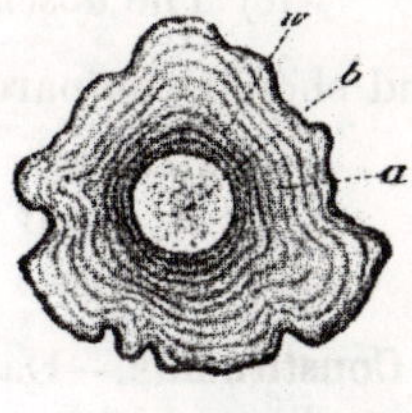


FIG. 176.—Dandelion root. Transverse section. *a*, bark · *b*, wood; *w*, cambium Magnified 4 diam. (Berg.)

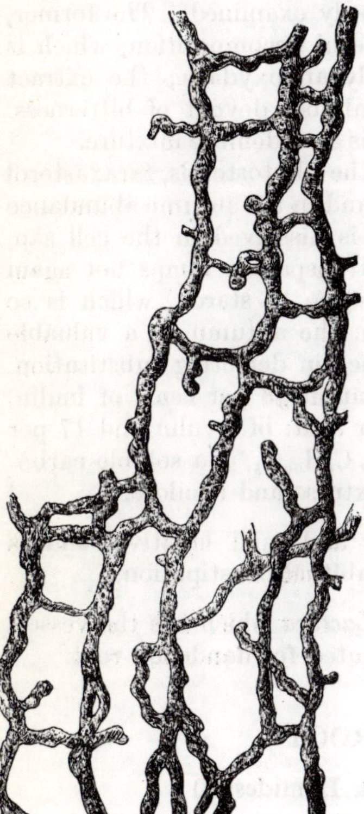

FIG. 177.—Anastomosing laticiferous vessels of Dandelion root. Magnified 140 diam. (Vogl.)

The dried root has a dark brown colour and is much shrivelled and wrinkled longitudinally; it tapers but little below, and often divides in the upper part (rhizome) into several erect branches, the rhizome being distinguishable from the root only by its slightly varying structure. These branches (or the rhizome itself) are crowned with the short remains of the leaves which bear brownish hairs near the point of insertion.

It breaks when dry with a short fracture, the section exhibiting a very small, yellow, porous, central wood surrounded by an abnormally thick, whitish bark in which numerous brownish concentric rings (of laticiferous tissue) are visible. The root, which is rather hygroscopic, becomes tough when slightly moist. It has no odour, but a bitter taste, which, however, is often not nearly so pronounced as it is in the milky juice that exudes from the fresh root.

The student should observe

(*a*) The *small, yellow wood,*

(*b*) The *thick, whitish bark* marked with *distinct, darker, concentric rings,*

(*c*) The *absence of starch* (see below);

and should compare the root with

(i) *Pellitory root,* which has oil glands and a large radiate wood with conspicuous medullary rays,

(ii) *Liquorice root,* which also has a large radiate wood.

**Constituents.**—Dandelion root contains a small quantity of a crystalline, bitter substance taraxacin, and also an acrid principle, neither of which, however, has been closely examined. The former, taraxacin, appears to be very susceptible of decomposition, which is probably induced by an enzyme, possibly an oxydase; the extract prepared from the fresh root is often almost devoid of bitterness. According to recent researches taraxacin is an indefinite mixture.

The drug also contains choline, resin, the phytosterols, taraxasterol and homotaraxasterol, various fat acids, and in the autumn abundance of inulin. In the fresh root the inulin is dissolved in the cell sap, but in the dry root forms amorphous, transparent lumps not again readily soluble in cold water. The absence of starch, which is so commonly present in roots, especially in the autumn, is a valuable negative character, and is often of service in detecting substitution. The autumn root has been found to contain 25 per cent. of inulin, whereas the spring root contained 18 per cent. of levulin and 17 per cent. of uncrystallisable sugar. Levulin, $C_6H_{10}O_5$, is a soluble carbohydrate converted by hydrolysis into dextrose and levulose.

**Uses.**—Dandelion is a simple, bitter, and mild laxative, and is given in atonic dyspepsia attended by habitual constipation.

**Substitutes.**—The root of a species of *Lactuca* which has the vessels arranged in radial rows has been substituted for dandelion root.

## HEMIDESMUS ROOT

### (Indian Sarsaparilla, Radix Hemidesmi)

**Source, &c.**—Indian sarsaparilla or hemidesmus root is obtained from *Hemidesmus indicus*, Robert Brown (N.O. *Asclepiadeæ*), a climbing shrub indigenous to India and Ceylon.

**Description.**—Hemidesmus root usually occurs in long, slender, rigid pieces, often more than 30 cm. in length, but seldom exceeding 6 mm. in thickness, which branch occasionally and bear a few fibrous rootlets. Portions of the slender aerial stems with opposite leaf-scars are attached to the upper extremity. The root varies in colour from brick-red to dark brown or nearly black ; it is rather tortuous, nearly cylindrical in shape, and marked with longitudinal wrinkles, and, at rather distant intervals, with conspicuous transverse cracks. The thin cork with which the root is covered shows a disposition to separate from the cortex, and on one side may often be observed to be distinctly raised above it. The transverse section exhibits a large, porous, but not distinctly radiate yellowish wood surrounded by a thin greyish or sometimes dark grey bark. The drug has an agreeable odour, distinctly recalling tonco beans ; the taste is slightly aromatic and sweetish, but not otherwise characteristic.

FIG. 178.—Hemidesmus root. Reduced.

The student should observe

(*a*) The *rigid, tortuous character* of the root,
(*b*) The *transverse cracks*,
(*c*) The *easily separable cork*,
(*d*) The *large yellowish wood* ;

and should compare the root with sarsaparilla, which is destitute of transverse cracks, has a firmly adherent cork, and a small wood.

**Constituents.**—The agreeable odour is due to a crystalline odorous substance resembling, but not identical with, coumarin. Of other constituents little is known.

**Uses.**—It has been used as an alterative, but is now practically obsolete.

## GELSEMIUM RHIZOME

### (Yellow Jasmine Root, Radix Gelsemii)

**Source, &c.**—The yellow jasmine, *Gelsemium nitidum*, Michaux (*G. sempervirens*, Aiton) (N.O. *Loganiaceæ*), is an elegant climbing plant indigenous to the southern United States; it ascends lofty trees and forms festoons, scenting the atmosphere with its fragrant yellow flowers. It has long been known, but its medicinal use is of recent date. The rhizome and roots should be collected in the autumn.

This plant should not be confounded with *Jasminum nudiflorum*, Lindley, a yellow-flowering jasmine commonly cultivated in this country.

**Description.**—The drug consists of the rhizomes, to which portions of both slender aerial stems and small and large roots are sometimes attached; usually the rhizomes and larger roots, cut into pieces about 15 cm. in length, constitute the commercial drug.

The rhizomes are generally in nearly straight cylindrical pieces varying from 5 to about 20 mm. in thickness, of a dark purplish brown colour, or at least marked with a more or less distinct network of purplish lines, the intervening spaces being yellowish brown. This difference in colour is due to the fact that the outer cork-cells are filled with a dark reddish brown substance, the inner with a yellowish deposit; by the growth of the rhizome the outer dark layer is fissured, disclosing the paler inner layer. The rhizomes are hard and woody, and break with an irregular splintery fracture, frequently exhibiting in the bast silky fibres, which, however, are much more conspicuous in the aerial stems.

The roots are, on the average, rather smaller than the rhizomes; they may be distinguished by their uniform yellowish brown colour and finely wrinkled surface, and by their rather more sinuous course.

The aerial stems are usually slender, but may attain 15 mm. in thickness. They are of a dark purplish brown colour, longitudinally wrinkled or nearly smooth, internally whitish and hollow in the centre. The fractured bark exhibits projecting strands of bast fibres.

Both rhizome and root exhibit in transverse section a comparatively narrow bark enclosing a large yellowish white wood; the latter consists of narrow wood-bundles with small vessels alternating with distinct, straight, whitish medullary rays, the section thus assuming an elegant radiate appearance. The section of the rhizome is distinguished from that of the root by the presence of a small pith, which, however, is more evident in the smaller (younger) than in the larger (older) pieces; it differs also from that of the aerial stem in the arrangement of the fibres in the bark; in the stem these are grouped into bundles, whilst in the rhizome they form an interrupted ring of isolated fibres or groups of two or three (Sayre, 1897).

The drug has a bitter taste, especially conspicuous in the bark, and a very slight aromatic odour.

The student should observe

(*a*) The *splintery fracture*,
(*b*) The *radiate structure* of the transverse section,
(*c*) The *purplish colour* and small pith of the rhizome,
(*d*) The *yellowish colour* and slightly *tortuous appearance* of the root.

**Constituents.**—Gelsemium rhizome contains three alkaloids, one only of which, gelsemine, has been obtained crystalline, the others (gelseminine and gelsemoidine) being amorphous. Other constituents are β-methylæsculetin, emodin monomethyl ether, phytosterol, resin and fixed oil.

Gelsemine, $C_{20}H_{22}N_2O_2$, has been reported to owe any toxicity it may possess to the presence of accompanying gelseminine but this has been denied ; both of the amorphous alkaloids are strongly toxic.

β-methylæsculetin, $C_9H_5(CH_3)O_4$ (gelsemic or gelseminic acid, scopoletin, chrysatropic acid), is a product of the hydrolysis of methylæsculin, æsculin being a fluorescent substance found in horse-chestnut bark. β-methylæsculetin is present in belladonna root and in scopola rhizome ; in alkaline solution it exhibits an intense bluish green fluorescence, and its presence in the drug may easily be demonstrated by shaking a little of the powder with lime water or, better, by shaking with chloroform, filtering and shaking the filtrate with very dilute solution of ammonia.

Gelsemine must be carefully distinguished from gelsemin, which is a powdered alcoholic extract.

**Uses.**—Gelsemium resembles hemlock in action but is more strongly depressant. It has been much used for, and appears to relieve, certain forms of neuralgia and sick headache as well as rheumatic and ovarian pains.

## PICRORHIZA

(Picrorhiza)

**Source, &c.**—Picrorhiza is the dried rhizome of *Picrorhiza Kurroa*, Royle (N.O. *Scrophularineæ*), a small plant indigenous to the Alpine Himalayas.

**Description.**—The rhizomes are greyish brown, more or less cylindrical, about 2·5 to 5 cm. long and 4 to 8 mm. thick. They are often deeply wrinkled longitudinally and covered with a thin cork. The surface exhibits the transverse scars of cataphyllary leaves, numerous, small, black buds but only an occasional root or root-scar; near the apex are the black remains of closely approximated scaly leaves. They are light in weight, break readily with a short fracture disclosing a very dark, lacunous interior. The section exhibits a thin, pale grey cork and a narrow ring of tangentially extended wood bundles. Taste very bitter.

**Constituents.**—The chief constituent is the bitter, crystalline glucoside, picrorhizin.

**Uses.**—Tonic and antiperiodic.

## GENTIAN ROOT

### (Radix Gentianæ)

**Source, &c.**—Gentian root is the dried root and rhizome of the yellow gentian, *Gentiāna lutĕa*, Linné (N.O. *Gentianeæ*), a stately herb with large, opposite, broadly ovate leaves and yellow flowers. It is indigenous to central Europe, growing abundantly on the lower slopes of the Jura and Vosges mountains, in the Black Forest, in the Pyrenees and European Turkey. Large quantities are imported from Spain. The root has long been used as a medicine.

The plant produces an erect rhizome, from which large fleshy roots are given off; the latter take a more or less horizontal course a short distance below the surface of the earth, and attain a considerable length. Both rhizome and roots are collected in the autumn and dried; they are occasionally, but not often, sliced longitudinally. When fresh they are whitish internally and almost odourless, but during the slow drying to which they are subjected the colour changes to a yellowish brown, and a distinctive odour is developed. So much are the particular colour and odour required, that in some districts the practice prevails of heaping the fresh root and allowing it to heat and ferment before drying it; by this treatment it darkens in colour and acquires the desired odour. Gentian root that has been longitudinally sliced or strung upon strings and quickly dried is paler and more bitter, and has an odour less strong than that which has not been so treated.

**Description.**—Gentian root is usually seen in nearly cylindrical pieces of varying length and thickness, but seldom more than 15 mm. in diameter. The roots, which differ in appearance from the rhizomes, are of a yellowish brown colour, and much shrivelled and wrinkled longitudinally. They are tough and flexible when slightly moist, but when quite dry they are brittle. They break with a very short fracture, and internally are usually of a reddish brown colour and spongy. The transverse section exhibits a dark line (cambium) separating a rather thick bark from a large central portion (wood) in which vascular tissue can with difficulty be discerned, as, like the bark, it consists almost entirely of parenchyma; neither bark nor wood exhibits any radiate structure.

The rhizome is easily distinguished from the root by the absence of conspicuous longitudinal wrinkles, and the presence of close,

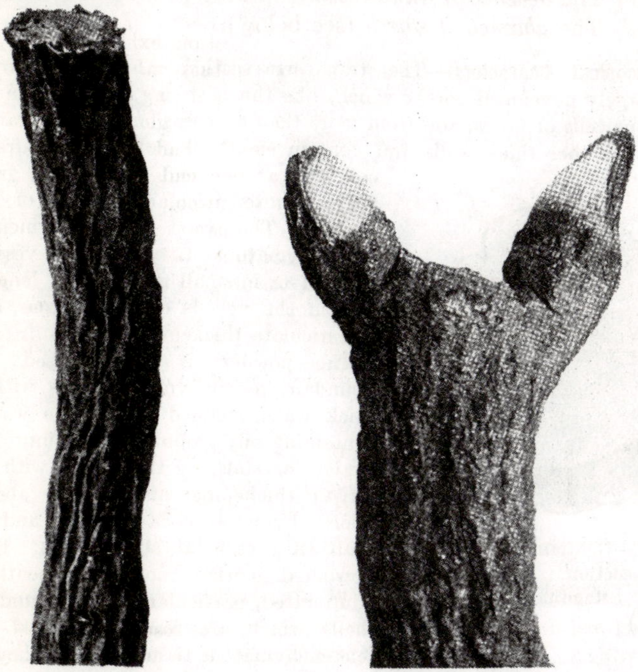

A B

Fig. 179.—Gentian root. A, root, showing surface with longitudinal wrinkles. B, rhizome, showing fine transverse lines (leaf scars), and dividing into two branches, each terminating in a bud. Natural size.

transverse annulations. It often exceeds the root in diameter, and may attain a length of 30 cm. or more. Each transverse annulation represents a year's growth, and bears the scars of fallen leaves, the minute fibro-vascular bundles of which are evident under a lens. The rhizome, which occasionally branches, is usually crowned with a large bud protected by dry, scaly leaves.

Roots that have been longitudinally sliced before they were dried frequently exhibit transverse instead of longitudinal wrinkles, and are paler internally; otherwise they correspond to the description given.

Both roots and rhizome have the same distinct and characteristic odour, and a sweetish, afterwards agreeably bitter, taste.

The student should observe

(*a*) The *uniform yellowish brown colour*,
(*b*) The *longitudinal wrinkles of the root*, the *transverse annulations and leaf scars on the rhizomes*,
(*c*) The *absence of any evident structure in the transverse section*,
(*d*) The *absence of starch* (see below).

**Microscopical Characters.**—The transverse section exhibits a narrow cork and a largely developed cortex which, like the bast-ring, is free from sclerenchymatous cells or fibres, and from more than an occasional starch grain. The cells have rather thick walls and contain small oil globules, and, frequently collected at one end of the cell, groups of very minute acicular crystals of calcium oxalate. The wood consists principally of parenchymatous tissue; the vessels are isolated or in small groups. In longitudinal section the vessels exhibit porous, annular, or reticulate thickening.

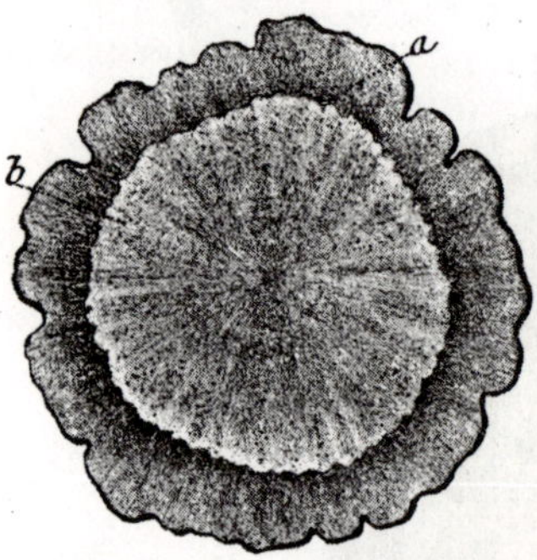

FIG. 180.—Gentian root. Transverse section. *a*, bark; *b*, wood. Magnified 2 diam. (Berg.)

The powder is characterised by the abundant parenchymatous cells with rather thick walls, inclined to swell in water, and containing oily globules and minute calcium oxalate crystals, by the vessels with characteristic thickening, and by the absence of sclerenchymatous cells and fibres and of more than an occasional starch grain. Imported powdered gentian root is frequently much adulterated, particularly with ground almond shells which are readily detected by the sclerenchymatous tissue of which they consist.

**Constituents.**—Fresh gentian root contains two bitter principles, viz. gentiopicrin and gentiin. During the drying of the root, however, the gentiopicrin disappears and the dry root contains gentiin and gentiamarin, the latter not being present in the fresh root; this decomposition is induced by the fermentative processes which take place during the drying; carefully dried root retains its gentiopicrin intact, but still contains enzymes which act upon the gentiopicrin when the tincture is made.

Gentiopicrin is a pale yellow crystalline bitter substance, and is hydrolysed by emulsin and also by dilute mineral acids to gentiogenin and dextrose; it is present in *G. punctata*, *G. asclepiadea* and certain other species as well as in *Chlora perfoliata*, Linné. Gentiin is also a crystalline glucoside, but gentiamarin is amorphous.

Gentian root contains a yellow crystalline acid, gentisic (gentianic) acid, sucrose, and (in the fresh state) a sugar gentianose, together with pectin and oily globules, probably of a cholesterol compound;

it is free from starch, or contains at most but minute traces of the latter. It yields from 3 to 4 per cent. of ash.

*Gentianose* is a hexatriose yielding by partial hydrolysis gentiobiose and levulose; further hydrolysis converts the gentiobiose into two molecules of dextrose. During the drying of the root the sucrose completely disappears and the gentianose is converted into dextrose, levulose, and gentiobiose. If the root is unduly exposed to fermentation these sugars are further converted into carbon dioxide and alcohol, so that highly fermented root may yield as little as 13 per cent. of extract to cold water, whereas good gentian root should yield 40 per cent. or even more. The British Pharmacopœia requires not less than 33 per cent. The darkening that takes place on the addition of ferric chloride to an infusion of gentian root is probably due to gentin.

Fresh gentian root is largely used in Germany and Switzerland for the production of an alcoholic beverage. The roots are cut, macerated with water, fermented and distilled; the distillate contains alcohol and a trace of volatile oil which imparts to it a characteristic odour and taste.

**Uses.**—Gentian is a favourite bitter tonic.

**Varieties, Substitutes, &c.**—The roots of other species of gentian are often collected and dried; *Gentiana purpūrea*, Linné (Switzerland &c.), *G. pannonica*, Scopoli (Austria), *G. punctata*, Linné (Austria), all yield gentian roots. They are, however, all smaller than those of *G. lutea*. The roots of *G. purpurea*, which approach nearest to the official gentian, attain about half the size and are crowned with several (eight to ten) aerial stems clothed below with many scaly remains of leaves. The top of the root has thus a peculiar branched appearance never found in the root of *G. lutea* (Pharmacographia). All these species appear to possess similar properties.

The rhizomes of *Rumex alpinus*, Linné, have also been found; they are reddish brown, astringent and bitter, and give a deep red colour with caustic alkalies. *Laserpitium latifolium*, Linné, yields the white gentian of Continental commerce. White gentian of English commerce is said to be derived from *G. Burseri*, Lapeyr.

## ALKANET ROOT

(Alkanna Root, Radix Anchusæ)

**Source, &c.**—Alkanet root is obtained from *Alkanna tinctoria*, Tauscher (N.O. *Boraginĕæ*), a small herbaceous plant with perennial root growing in sandy soil in southern Europe, Hungary, and Asiatic Turkey, our supplies being derived chiefly from Hungary.

**Description.**—The drug consists of deep reddish purple, simple, tapering roots averaging about 15 cm. in length, and 12 mm. in thickness near the crown, to which slender branches and the remains of leaves are attached; the latter have a purplish colour, and bear numerous, bristly hairs.

The root has a very remarkable appearance. The outer layers, which are deeply coloured, easily exfoliate, separating from the wood in large, papery flakes or strips; these are readily removed, and the exterior of the root is then the inner layers of cortical tissue. It is not smooth but more or less deeply furrowed longitudinally, the furrows being often so deep as to divide the wood into separate woody strands. Cut transversely in the upper part the root exhibits a dark violet spot in the centre, and a woody ring containing several radiating strands of dense woody tissue separated by large wedge-shaped masses of parenchyma; the bark is yellowish near the wood, but becomes coloured and laminated towards the exterior. From near the crown of the root downwards the violet spot in the centre becomes larger and the colouring matter in the bark penetrates deeper until by their union the separation of the strands of wood is complete.

The root when handled stains the fingers red. It has no perceptible odour or taste.

The students should observe

(a) The *dark reddish purple colour of the parenchymatous tissue* and *whitish colour of the wood*,

(b) The *hairy remains of the leaves*.

**Constituents.**—Alkanet root is said to contain two red colouring substances, viz. anchusic and alkannic acids, the former turning green with alkalies and the latter blue. Alkannin is the name given to an oleo-resinous mixture of deep red colour obtained by extracting the root with petroleum spirit or benzene and evaporating. These substances are produced first in the epidermal cells and hairs of the young roots. After the cortical parenchyma has been thrown off some of the parenchymatous cells within the pericycle rupture, and in the intact cells next to the ruptured ones the formation of the colouring matter proceeds rapidly; cork cells form a little deeper in the tissue, resulting in the formation and rejection of a succession of deep red scales which extends to the wedges of parenchyma that divide the wood into strands; these perish, and thus the remarkable appearance of the commercial drug is explained. (Ericksson, 1910.)

**Uses.**—Alkanet is used for colouring oils, ointments, &c.; the tincture, diluted with an equal volume of water, is used for the microscopical detection of fat and oil, which it colours red.

**Varieties.**—Many other plants furnish roots containing a similar red colouring matter, e.g. *Onosma echioides*, Linné (S. Europe), *Macrotāmia cephalōtes*, de Candolle (Armenia and Syria), &c.; they are occasionally substituted for the root of *Alkanna tinctoria*, the Syrian drug being apparently more than half as rich again in colouring matter as the Hungarian. Both of these substitutes are larger than alkanet root.

## COMFREY ROOT

(Symphyti Radix; Consolidæ Radix)

**Source, &c.**—Comfrey root is the dried rhizome and root of *Symphytum officinale*, Linné (N.O. *Boragineæ*), an indigenous and widely distributed perennial herb. The plant produces a large erect, branching rhizome and stout, fleshy roots.

**Description.**—The commercial drug consists chiefly of root and occurs in pieces of varying length (often 5 to 20 mm.) and about 5 to 15 mm. thick. Externally it is deeply longitudinally wrinkled and blackish or greyish black; internally it is pale grey, brownish, or even brown in colour. It may be hard, horny or tough in consistence and exhibits a waxy surface when cut. The structure is often indistinct, but favourable pieces show a dark cork and a grey cortex separated by a greyish cambium line from an indistinctly radiate wood within which, in pieces of the rhizome, there is a circular pith. It has a slight odour and a mucilaginous taste.

**Constituents.**—Comfrey root contains about 0·8 per cent. of allantoin together with tannin, resin, gum, and sugar.

Allantoin, $C_4H_6O_3N_4$, is a diureide of glyoxylic acid. It occurs in colourless, glistening prisms melting at 227°. Oxidising agents convert it into allantoic acid, $C_3H_4N_4O_2$. It is soluble in solution of sodium hydroxide, the solution evolving ammonia when heated, with simultaneous production of urea.

**Uses.**—The drug has been used as an application to wounds, sores and ulcers, the healing action being attributed to the allantoin contained in it.

## SCAMMONY ROOT

(Radix Scammoniæ)

**Source, &c.**—Scammony root is obtained from *Convolvulus Scammonia* Linné (N.O. *Convolvulaceæ*), a twining plant indigenous to the eastern Mediterranean countries, and resembling the common bindweed but much larger. It produces a root, often of very considerable size, from which a number of slender aerial stems spring. The root, which yields by incision the gum-resin scammony, is collected and dried. It is exported from Odessa, Batoum, &c.

**Description.**—Scammony root is usually simple and often of large size. Small roots measure about 12 mm. in diameter and 15 to 30 cm.

in length, but larger specimens attain as much as 8 or 10 cm. in thickness and a metre in length; they are nearly cylindrical in shape, tapering gradually towards the tip but slightly enlarged at the crown, and there

FIG. 181.—Portion of a small Scammony root. Natural size.

rugged from the remains of aerial stems. The drug is of a greyish brown colour, hard, heavy, and woody, rough and furrowed externally, and often spirally twisted. Lateral rootlets occasionally spring from it and commonly split into fibrous strands. It breaks with an irregular fracture, from which short fibres (strands of wood) project. Internally the root is whitish, and the transverse section, when smoothed, exhibits an abnormal and characteristic structure. There is no distinct central wood, but the root exhibits instead several not very well-defined circles, in each of which a central portion containing groups of vessels can be distinguished from a surrounding parenchymatous portion; in the latter numerous dark points (resin cells) can be discerned with a lens. The root is invested with a thin bark, which however is not very conspicuous.[1] The whole of the parenchymatous tissue contains starch.

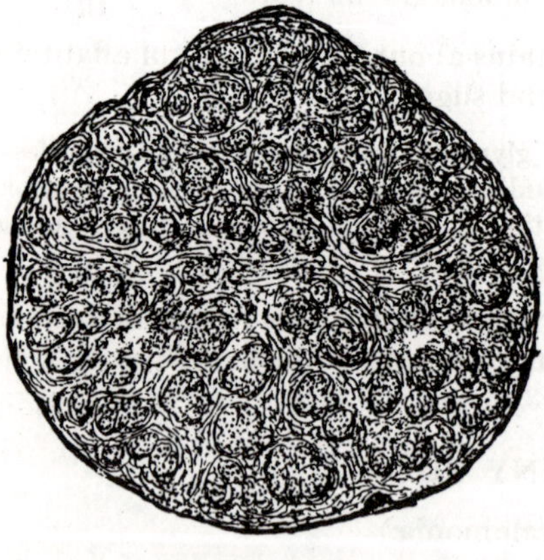

FIG. 182.—Scammony root. Transverse section. (*Pharmaceutical Journal.*)

The root has a characteristic odour distinctly recalling that of jalap, and especially noticeable when the surface is freshy cut; the taste is sweetish and slightly acrid.

The student should observe

(*a*) The *large size and greyish brown colour*,

(*b*) The *characteristic jalap-like odour*,

[1] For further anatomical details see Planchon and Collin, *Les Drogues simples*, i. 623.

(*c*) The remarkable *structure of the transverse section*,

(*d*) The *presence of resin cells* ;

and should compare the root with belladonna, which is usually smaller, darker externally, odourless, and has an entirely different structure.

**Constituents.**—Scammony root contains from 3 to 13 (average about 9) per cent. of a glucosidal resin which can be isolated by mixing a strong tincture with water, washing and drying the precipitated resin. The latter should be entirely, or almost entirely (97 per cent.) soluble in ether, and in this respect differs from the resin obtained in a similar manner from jalap, which is only partially (not more than 10 per cent.) soluble in that menstruum. The resin is extremely complex in composition and consists to a large extent of the glucosides and methylpentosides of jalapinolic acid and its methyl ester. The methylpentose obtained by the hydrolysis of the resin appears to be rhamnose and not identical with that from the resin of Orizaba root (see p. 355). The resin is hydrolysed by potassium hydroxide or barium hydroxide with production of methylbutyric, tiglic, formic, valeric, jalapinolic and other acids together with rhamnose. Jalapinolic acid has also been termed scammonolic acid. The resin also contains a small percentage of fat (palmitin, stearin, olein, linolin, &c.)

**Note.**—The British Pharmacopœia (1914) requires not less than 75 per cent. of the resin to be soluble in ether ; the difference between this figure and that given above is due to the fact that the Pharmacopœia permits the use of Orizaba root in the production of the resin, and that Orizaba resin is less soluble in ether (65 to 70 per cent.). When scammony resin is boiled with solution of potassium hydroxide, and the solution acidified, turbidity only, due to the fat in the resin, is produced ; most other resins produce a distinct precipitate.

Scammony root contains also a little sucrose and reducing sugar, starch, β-methylæsculetin (see p. 343), dihydroxycinnamic acid, a phytosterol, the alcohol ipuranol and a trace of volatile oil. The starch occurs in grains of characteristic appearance, an important fact, as it is sometimes necessary in examining scammony to distinguish the starch of the scammony root from other starches that may possibly have been fraudulently added.

**Uses.**—Scammony root is employed only as a source of scammony resin, which can be more economically prepared from the root than from the expensive gum-resin. Much of the commercial resin is, however, at present obtained from Orizaba root (male or woody jalap, Mexican scammony ; compare p. 355).

## JALAP
### (Radix Jalapæ)

**Source, &c.**—Jalap is obtained from *Ipomœa Purga*, Hayne (N.O. *Convolvulacæ*), a plant with climbing, twining stems indigenous to the eastern slopes of the Mexican Andes. It sends out slender runners which are provided at intervals with scaly, cataphyllary leaves in the axils of which are buds ; below the buds roots are produced, some of which thicken rapidly and form fusiform or napiform organs, often of considerable size. These tuberous roots (tubercules) are collected and dried in nets over fires, the smaller entire, but the larger longitudinally incised to allow of the free escape of moisture. When fresh they are fleshy and white internally, but by drying, especially in the manner indicated, they darken in colour. Jalap is imported in sacks from the east coast of Mexico, and distinguished as ' Mexican ' or ' Vera Cruz ' jalap.

The jalap plant has been cultivated in India and Jamaica but these colonies do not at present compete with Mexico in the supply of the drug.

The Spaniards became acquainted with this and similar purgative Convolvulaceous plants early in the sixteenth century, and exported considerable quantities of them to Europe.

**Description.**—Jalap occurs in pieces of very varying size, most commonly about that of a hen's egg, although sometimes they attain 10 cm. or more in diameter. They vary also much in shape, being sometimes napiform, sometimes fusiform or irregularly oblong. The small are usually entire, but the larger bear gashes that have been made to facilitate the drying. Towards the lower extremity they taper off and show a fractured end where the slender part of the root has been broken off. The surface is dark brown, furrowed and wrinkled (but not conspicuously convoluted), and marked with numerous, paler, elongated transverse scars (lenticels). They are heavy and compact, and so hard as to be broken with difficulty, but softening readily in water. Internally they have a yellowish grey or dingy grey colour, and are very tough or horny, the section exhibiting irregular dark lines often concentrically arranged. These lines are due to the formation of secondary cambiums, an abnormal development that is occasionally, although not often, met with. Under the lens numerous dark resin cells are visible, especially in the cortical portion, but woody tissue is not easily discerned.

Jalap has a distinct and characteristic odour which is often ascribed to the smoke from the fire over which the roots have been dried, but which is, partly at least, inherent in the drug. The taste is at first sweetish, but afterwards disagreeably acrid.

The heat to which the drug is subjected during the drying is generally sufficiently high to gelatinise the starch, especially in the interior of the roots, where the moisture is retained longer than in the outer portions ; hence the horny and not starchy appearance of the drug. Roots obtained from cultivated plants in India and Jamaica are usually more carefully dried, and present a mealy, not horny, appearance in the interior.

**Microscopical Characters.**—Powdered jalap is characterised, when examined under the microscope, by the abundant starch which may be intact or in any stage of gelatinisation ; the grains are either simple or compound ; the former, measuring mostly $25\mu$ to $40\mu$, are rounded or ovoid and exhibit concentric striæ ; compound grains may contain 2 to 6 constituents. Numerous globules of resinous emulsion, varying in size, are evident or can be made evident by solution of iodine in potassium iodide. There are occasional sclerenchymatous cells, and fragments of vessels with areolated pits. Cluster crystals of calcium oxalate are frequent.

FIG. 183.—Jalap root. Small specimen, natural size, showing the transverse lenticels.

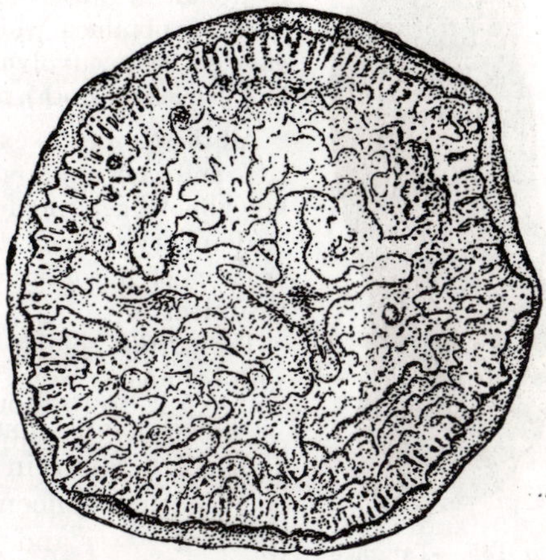

FIG. 184.—Jalap root. Transverse section. Slightly reduced. (*Pharmaceutical Journal.*)

**Constituents.**—The principal constituent of jalap is the 'resin' which can be separated by extracting the root with alcohol, concentrating the tincture, pouring it into water, washing and drying the resinous precipitate. The drug contains in addition colouring matter, sugar, starch, $\beta$-methylæsculetin, ipurganol, a phytosterin and calcium oxalate.

The resinous precipitate obtained as described is known as 'jalap resin.' The yield has varied from 2 to 22 per cent. of the drug ; 8 to 12 per cent. is often found. It is characterised by its partial solubility (about 10 per cent.) in ether.

Jalap resin is an exceedingly complex mixture, the composition of which is only imperfectly known. Treatment with various solvents in succession has given the following results:

*Petroleum Spirit*, 1·9 per cent., containing fatty acids and fats, phytosterol, cetyl alcohol.

*Ether*, 9·7 per cent., containing ipurganol; after treatment with alkali followed by acid, chiefly amorphous substances were obtained, together with a little phytosterol, cetyl alcohol, &c.

*Chloroform*, 24·1 per cent., containing β-methylæsculetin; after treatment with alkali and acid, formic, butyric and *d*-methylethylacetic acids, glucose, convolvulinolic acid ($C_{15}H_{30}O_3$), and a higher homologue were obtained.

Fig. 185.—Tampico Jalap root. Natural size.

*Ethyl Acetate*, 22 per cent., similar in composition to the chloroform extract.

*Alcohol*, 38·8 per cent., by treatment with barium hydroxide a water-soluble, non-purgative, 'hydrolysed,' resin was obtained, which, boiled with dilute sulphuric acid, gave glucose, convolvulinolic acid, ipurolic acid (also found in *I. purpurea*, Roth), formic, butyric and valeric acids.

All these extracts except the petroleum spirit are purgative, and each appears to be only partly glucosidic. Convolvulinolic acid (hydroxypentadecylic acid), and ipurolic acid are crystalline; they appear to exist in the resin in the form of glucosides which can be split up by appropriate treatment.

The portion soluble in ether is assumed to be identical with scammony resin.

Good jalap should yield not less than 10 per cent. of total resin when exhausted with alcohol, the tincture concentrated and poured into 8 volumes of water and the precipitated resin washed and dried. Much of the drug that at present reaches the market contains considerably less, and its exclusion from use in making the official preparations of jalap is ensured by the minimum requirement of 10 per cent. adopted by the British Pharmacopœia. On the other hand there seems to be little doubt that the quality of the drug has deteriorated during the last twenty-five years, from 12 to 18 per cent. being formerly the usual amount of resin contained in it. Under exceptional circumstances the quantity has risen in roots cultivated in India to upwards of 20 per cent.

**Uses.**—Jalap is a powerful stimulant of the intestinal secretion, producing in small doses a laxative effect, and in large doses active purgation. It is much used as a hydragogue cathartic.

**Varieties.**—*Tampico jalap* is afforded by *Ipomœa simulans*, Hanbury, a plant resembling *I. Purga* and growing on the eastern slopes of the Mexican Andes. The roots are exported from Tampico (a town on the Gulf of Mexico, about 200 miles north of Vera Cruz), whence it derives its name. This root, which frequently appears on the London market, is distinguished by its irregular shape and remarkable, convoluted surface, which does not exhibit the small transverse scars characteristic of true (Vera Cruz) jalap. It yields about 10 per cent. of resin, which is distinguished from the resin of true jalap by its complete solubility in ether. This resin (tampicin) is probably identical with the ether-soluble resin of jalap and scammony.

*Orizaba jalap* (light, woody, stalk, or male jalap), is produced by *Ipomœa orīzabensis*, Ledanois. (See below.)

## ORIZABA JALAP ROOT

### (Mexican Scammony, Ipomœæ Radix)

**Source, &c.**—Orizaba jalap root is obtained from *Ipomœa orizabensis*, Ledanois, a Convolvulaceous plant growing near Orizaba in the Mexican Andes (about 65 miles south-west of Vera Cruz). It produces a large, woody root which is cut into pieces of varying size, and shape and dried. The drug, which has long been known as light, woody, stalk, or male jalap has largely been imported in considerable quantities under the name of Mexican Scammony.

**Description.**—Orizaba jalap root occurs mostly in large transverse slices, 5 to 10 cm. in diameter, and 1 to 2 cm. thick, or in irregular rectangular portions of larger slices. It is greyish-black in colour and coarsely wrinkled; internally it is greyish or brownish and more or less horny. From the transverse surfaces coarse woody fibres protrude in irregular, concentric circles. The odour is slight and the taste faintly acrid. Occasionally small tapering roots 2 to 3 cm. in diameter and 8 to 10 cm. long are met with.

**Constituents.**—Like jalap and scammony root the chief constituent of Orizaba jalap root is the 'resin' it contains; other constituents of less importance are sugar, β-methylæsculetin, dihydroxycinnamic acid, fatty acids, phytosterol, starch and calcium oxalate. The amount of crude resin determined as described under 'Jalap Root' varies from 6 to 22 per cent., averaging 12 to 18 per cent.

The crude resin is a very complex mixture. It contains about 6 per cent. of fatty substances, 64·8 per cent. of resin soluble in ether, 24·8 per cent. soluble in ethyl acetate and 2·3 per cent. soluble in chloroform. The ether-soluble portion is not identical with the ether-soluble portion of jalap resin. By alkaline hydrolysis it yields ipuranol. *d*-*a*-methylbutyric acid, tiglic acid and a product which, by acid hydrolysis, yields dextrose, methylpentose, jalapinolic acid and methyl jalapinolate. The resin soluble in ethyl acetate gives similar results. The chief constituents, therefore, appear to be the glucoside and methylpentoside of jalapinolic acid and its methyl ester; these are very similar to those of scammony resin (see p. 351). The following table shows the solubility of certain Convolvulaceous resins when treated with various solvents in success.

| | Jalap | Scammony | Mexican Scamm. | Braz. Jalap | *Ipomœa purpurea* |
|---|---|---|---|---|---|
| Petroleum Spirit | 1·9 | 4·5 | 6·2 | 2·1 | 8·0 |
| Ether | 9·7 | 92·5 | 64·8 | 5·4 | 7·3 |
| Chloroform | 24·1 | 0·4 | 0·6 | 73·4 | 9·8 |
| Ethyl Acetate | 22·0 | .. | 24·8 | 14·2 | 23·8 |
| Alcohol | 38·8 | 1·8 | 2·3 | 4·7 | 49·0 |
| Acid Value | 15 | 20 | 20 | .. | .. |
| Saponification Value | 140 | 235 | 180 | .. | .. |
| Optical Rotation | −37° | −19·8° | −23° | .. | .. |

The therapeutic action of Orizaba jalap resin appears to be very similar to that of scammony resin and although the two resins are not identical the British Pharmacopœia permits both roots to be used in the preparation of scammony resin.

**Note.**—Brazilian jalap, which has lately appeared on the market, is the root of *Piptostegia Pisonis*, Martius (N.O. *Convolvulaceæ*). It occurs in the form of transverse, circular slices about 3 to 5 cm. in diameter, and 1 cm. thick, pale greyish brown, with several concentric rings and exhibiting pale resin cells; it contains 20 per cent. of resin, very little of which is soluble in ether.

The root of *Ipomœa purpurea*, Linné (N. America), contains 4·8 per cent. of a purgative resin.

## TURPETH

### (Turpethum, Radix Turpethi)

**Source, &c.**—The dried root and stem of *Ipomœa Turpethum*, Robert Brown (N.O. Convolvulaceæ), a native of India, Ceylon, the Malay Archipelago and N. Australia.

**Description.**—Turpeth occurs in pieces of very varying length and thickness; they are often 10 to 20 cm. long and 1 to 5 cm. wide, deeply wrinkled longitudinally, and of a dull grey or brown colour; sometimes the bark has been slit on one side and the central wood removed. Fracture short, the wood in old specimens in several concentric circles; vessels very large; medullary rays wide; in the bark resin cells and frequently abnormal wood bundles. Odour slight, taste nauseous but slowly developed.

**Constituents.**—Turpeth contains from 5 to 10 per cent. of resin, part of which is soluble and part insoluble in ether. The ether-soluble resin is a mixture of $\alpha$- and $\beta$-turpethein, the ether-insoluble resin is turpethín. Turpeth resin has not yet been thoroughly examined; $\alpha$-turpethein appears to be a rhamnoside; $\beta$-turpethein a glucoside. Further research will probably show a complex composition analogous to that of scammony resin, jalap resin, &c.

**Uses.**—The drug has long been used in India as a purgative.

## BELLADONNA ROOT

### (Radix Belladonnæ)

**Source, &c.**—The deadly nightshade, or belladonna, *Atropa Belladonna*, Linné (N.O. *Solanaceæ*), is a tall branching herb, attaining a height of 2 metres. It is widely distributed over central and southern Europe; in England it is confined chiefly to the southern counties, but it is cultivated in Bedfordshire, Hertfordshire, and elsewhere, and is grown for medicinal use in Germany also.

Belladonna possesses a stout, branching, tap root which for medicinal use should be collected in the autumn when about three or four years old, cut into pieces, and carefully dried. The commercial drug, much of which has been imported from Germany, though probably collected in Hungary, is often of inferior quality.

**Description.**—Good root as found in commerce, occurs in pieces varying from 15 to 20 cm. in length, and from 1 to 2 cm. in diameter. It is of a pale greyish brown colour, finely wrinkled longitudinally, usually cylindrical or gradually tapering, and often crowned with the remains of hollow aerial stems. Sometimes the roots are cut longitudinally to facilitate drying. They break with a short fracture, and should be whitish and starchy internally. The section exhibits a greyish bark separated by a dark line (cambium) from a whitish central portion (wood) in which, especially near the cambium ring, dark groups of vessels and fibres are scattered; these groups, however, seldom exhibit more than an indistinctly radiate arrangement except close to the crown of the root, where one or more rings of radiate yellowish wood may be found, the root passing imperceptibly into rhizome. The bark is not fibrous, and adheres closely to the wood. Most of the parenchymatous cells of both bark and wood contain small compound starch-grains, and some of them are filled with numerous very minute (sandy) crystals of calcium oxalate; these characters, visible only under the microscope, serve as additional means of identifying belladonna root.

Old roots become woody, and may then exhibit a prominently radiate structure, whilst roots gathered in spring contain sugar and but little starch, and hence show a shrunken outer surface and a dark, spongy interior. In both cases the drug is believed to contain less alkaloid than the young root collected in the autumn and possessing the characters given though roots that are dark internally and not starchy often contain high percentages of alkaloid.

Notwithstanding the fact that the preparations of belladonna root are standardised, the official description is framed so as to exclude such old or partially exhausted roots.

The root has an odour recalling liquorice root, and a slightly bitter taste.

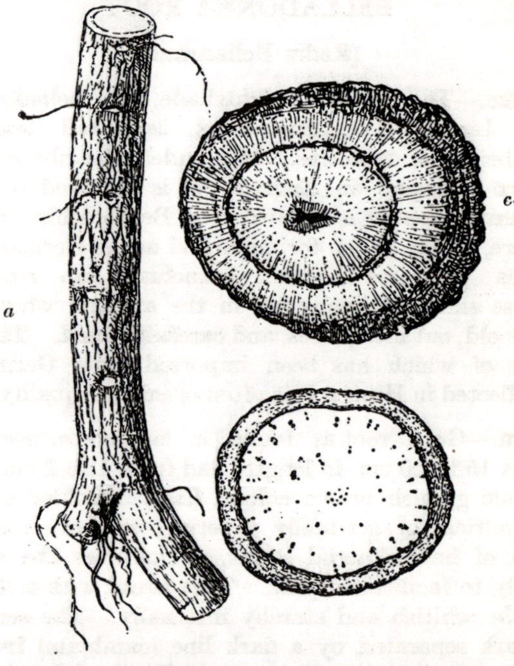


FIG. 186.—Belladonna root. *a*, portion of young root, natural size; *b*, transverse section of the same, magnified; *c*, transverse section of the upper part of the root, magnified. (Holmes).

The student should carefully observe

(*a*) The *firm*, *plump appearance* of typical belladonna root,
(*b*) The *short*, *mealy fracture*,
(*c*) The *dark cambium line*, and *distribution of the groups of vessels and fibres*;

and should compare the root with

(i) *Elecampane root*, in which oil glands are present, and the groups of vessels inconspicuous,
(ii) *Scammony root* (small pieces), which have a very characteristic structure,

(iii) *Marshmallow root*, which has a radiate wood, in which scattered groups of vessels are not visible, and a fibrous bark.

**Constituents.**—The principal constituent of belladonna root is the alkaloid hyoscyamine which appears to be accompanied by a little atropine, although it is possible that the atropine found has been produced from hyoscyamine during the process of extraction, a change which is readily effected. Traces of scopolamine appear also to be present, but belladonnine and apoatropine, which are found in the mother liquors from which the other alkaloids have been crystallised, probably do not exist as such in the root, but are formed during the various manipulations in the course of their isolation (compare p. 45).

The root also contains $\beta$-methyl-æsculetin (chrysatropic acid, scopoletin), a crystalline fluorescent principle, widely distributed throughout the natural order *Solanaceæ*, and found also in gelsemium rhizome and a number of other drugs.

The total amount of alkaloid that the root contains varies as a rule between 0·4 and 0·6 per cent., but may occasionally rise to 1·0 per cent. (Cripps). Gerrard has shown that the plant is more active when about four years old than when younger, and that the wild plant is rather richer in alkaloid than the cultivated, while Schmidt's experiments indicate that the roots of young plants contain more alkaloid than those of old plants. The Pharmacopœia therefore directs the root to be collected in the autumn, and excludes old roots by the limitation of the size and by the description of the transverse section.

**Uses.**—Belladonna acts as a local anæsthetic and anodyne, and is often applied as such externally. Internally it is given to check the sweating in phthisis, as a sedative to the respiratory nerves, to relieve spasmodic cough, and in numerous other cases.

**Adulterants.**—*Scopola Rhizome*. The rhizome of *Scopŏla carniōlica*, Jacquin (N.O. *Solanaceæ*), is not unfrequently found in the belladonna root imported from Germany. The plant grows in the Carpathian Mountains and other parts of Austro-Hungary. It produces a horizontal rhizome from which aerial stems from 30 to 60 cm. high arise bearing leaves which resemble belladonna leaves but are thinner, more lanceolate in shape, and taper more towards the base (compare p. 46). The rhizome, which is collected in large quantities and forms an important source of the alkaloids hyoscyamine, atropine, and scopolamine, may attain 10 cm. or more in length and about 2 cm. in thickness. It is nearly black in colour and tortuous, and bears on the upper surface numerous large depressed stem-scars. It contains hyoscyamine, scopolamine, and possibly also atropine, the total alkaloid (0·6 to 0·7 per cent.) somewhat exceeding in amount that present in belladonna root.

*Japanese Belladonna*, *Scopola Japonica*, Maximowicz, closely resembles scopola rhizome.

*Indian Belladonna Root*. Considerable quantities of belladonna root have recently been imported from India, the bulk having been collected in Kashmir from wild plants of *Atropa lutescens*, Jacquemont, which is closely allied to *A. Belladonna*. Indian belladonna root closely resembles the European, but may be distinguished by the section, which exhibits a radiate wood even in the small roots. The drug has been stated to contain 0·7 per cent. of alkaloid, but it is doubtful whether this is entirely mydriatic Solanaceous alkaloid.

*Phytolacca Root*. The root of *Phytolacca decandra*, Linné (N.O. *Phytolaccaceæ*), a plant indigenous to North America but naturalised in southern Europe, has been used to adulterate belladonna. It may, however, readily be detected by the transverse section which exhibits several concentric rings of wood bundles, in consequence of which large roots readily fissure longitudinally. It contains abundant acicular crystals of calcium oxalate in place of the sandy calcium oxalate of belladonna. As it attains a much larger size than belladonna root it is often cut into pieces for admixture. It contains a resinoid substance, phytolaccin, and has a purgative action.

## RHUBARB RHIZOME

(Rhizoma Rhei, Radix Rhei)

**Source, &c.**—The structure of commercial rhubarb indicates that the drug is derived from two species of *Rheum*. One of these is certainly *R. officinale*, Baillon; the other is probably *R. palmatum*, Linné, var. *tanguticum*, Maximowicz. The former occurs abundantly in central and western China at an elevation of 3,000 to 4,000 metres, and is collected chiefly in the mountains separating Tibet from the Chinese province of Szechuen and extending eastwards to Hupeh. The plant produces a rhizome which may attain a very great size, and to which large thick roots are attached. The plants are dug up towards the end of September and the roots cut off. The rhizomes are cleaned, and the crown and the bark removed, the larger being cut either longitudinally or transversely. They are dried by stringing them on cords which are then stretched from tree to tree, or on hurdles in huts in which a fire is kindled, or on heated stones. The drug is conveyed chiefly by river to Chung-King on the Yang-tse-Kiang river, and thence to Shanghai, whence it is shipped to Europe, after having undergone further trimming and sorting on its journey.

*Rheum palmatum*, Linné, var. *tanguticum*, Maximowicz, occurs in the province of Kansu. Large quantities of the drug are collected near Lake Kokonor in Tibet, and are apparently derived from this plant. It is carried over the mountains to the Han river, and thus reaches Hankow. Further quantities of rhubarb are also produced in the province of Shensi, and may possibly be obtained from a third, at present unknown, species of *Rheum*.

Practically all the Chinese rhubarb now exported passes through Hankow to Shanghai for shipment to Europe, and consists of rhizome only.

Formerly the drug was conveyed by caravan *via* Persia to the Syrian ports, whence it reached Europe and was known as Turkey rhubarb ; some was sent direct from China (China or Canton rhubarb),

FIG. 187.—Case of Chinese Rhubarb.

and some was shipped *via* India (East Indian rhubarb), the latter being the most common variety as early as 1640. Subsequently, by a commercial treaty between Russia and China, the trade was diverted to the Russo-Siberian route, and the drug was conveyed *via* Irkutsk to Moscow, still retaining its designation of Turkey rhubarb. Later, however, when the Chinese opened some of their northern ports to European trade, the drug was exported by the route which it now takes, and the trade names of Turkey and East Indian rhubarb indicate nothing more than the routes by which the drug formerly found its way to the European market.

**Description.**—The official drug occurs in pieces of very varying size and shape. Those derived from smaller rhizomes that have not been sliced occur in cylindrical, conical, or barrel-shaped pieces varying in size, but frequently from 7 to 10 cm. in length and 3 to 6 cm. in thickness, although of course they may be larger or smaller. The larger rhizomes, which are usually cut longitudinally, yield plano-convex pieces, frequently tapering more or less towards either end ; such pieces often measure 8 or 10 cm. in length or breadth and 4 cm. in thickness. These two principal forms are known as 'rounds' and 'flats' and are commonly sold in separate cases. They are

FIG. 188.—Chinese Rhubarb (Shensi flat), showing network of whitish lines.

frequently drilled, a piece of string being occasionally found in the perforation. The outer surface is dusted over with a fine bright yellow powder, and bears evidence of having been not only peeled with a knife, but carefully dressed with a file or scraped after the drug has been dried. Here and there small portions of a dark layer may still be found. These characters are common to the commercial varieties of Chinese rhubarb generally, but the following apply more particularly to the finest variety, viz. Shensi rhubarb.

If the outer surface, freed from the yellow powder that adheres to it, is examined with a lens, a number of minute, longitudinal, dark reddish brown lines and dots can be observed imbedded in a white ground mass, which latter frequently forms a delicate network, in the meshes of which the alternating white and dark red lines alluded to can be seen. The dark lines are medullary rays containing brown colouring matter; the white lines consist of bast parenchyma containing starch and calcium oxalate. Occasionally minute dark points or projections may be detected which, when carefully shaved off with a knife, exhibit radiating red and white lines; these are the remains of fibro-vascular bundles from the leaves (leaf-traces) and pass through the bark into the central portion of the rhizome.

The drug is firm, heavy, and compact, the outer surface showing little sign of shrinkage during the drying; it breaks with an uneven fracture, the fractured surface, which varies from bright pink to dull

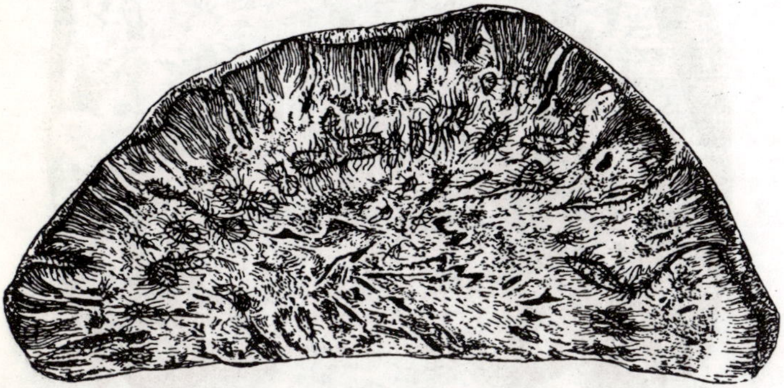

FIG. 189.—Chinese Rhubarb (Shensi flat). Transverse section, showing the remains of the bark (as a narrow exterior layer), the cambium (as a dark line), the wood (with a more or less distinctly radiate structure), and an irregular circle of starspots (abnormal fibro-vascular bundles). (Planchon and Collin.)

grey in colour, exhibiting under the lens dark reddish brown lines alternating with white ones ('nutmeg' fracture). The transverse section shows near the periphery a more or less continuous ring of large, conspicuous, star-like spots, each of which consists of dark red lines (medullary rays) radiating from a common centre through a white groundwork (parenchyma). These star-spots are fibro-vascular bundles which also traverse the inner part of the drug in varying directions. Exterior to this ring the drug exhibits a radiate appearance due to the formation of a narrow ring of secondary wood by the cambium, which is sometimes evident as a dark line about 1 mm. from the margin (compare fig. 189), but is often not seen, as not only the whole of the bark but even part of the secondary wood may have been removed by the peeling. If the peeling has been sufficiently deep to remove the bark and the secondary wood as well, then the star-spots appear on the outer surface, but until then they are indistinct. They can always be seen on the flat (inner) surface of the plano-convex pieces. Beyond the secondary wood is the cambium, usually visible as a dark line, and here and there are the narrow remains of the bark. Small cylindrical pieces are generally more favourable for examination than flat slices of a larger rhizome.

The structure of this drug is complex, but it must be remembered that it is a very fleshy rhizome, the internodes of which are so closely approximated as to be almost suppressed.

The drug possesses a characteristic odour and bitter, astringent

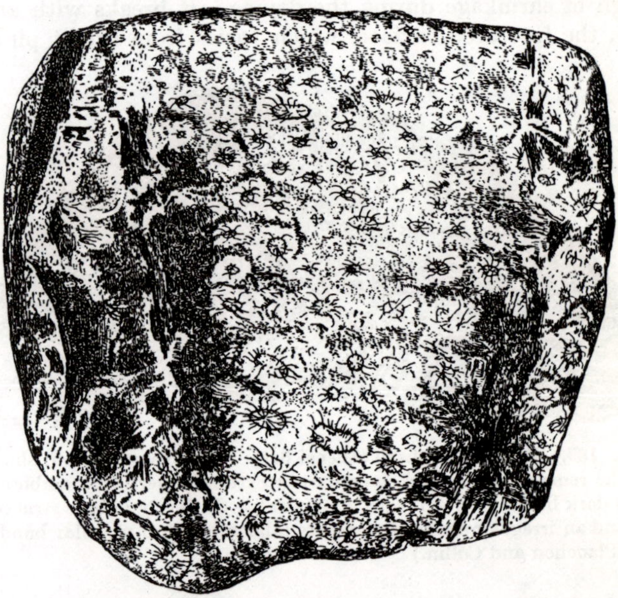

FIG. 190.—Chinese Rhubarb (Shensi flat). Flat (inner) surface of a plano-convex piece, showing irregularly distributed star-spots. (Planchon and Collin.)

taste ; when chewed it is very gritty between the teeth, a character due to the calcium oxalate, which occurs in considerable quantity in large cluster-crystals.

The student should particularly observe

(*a*) The *firm, compact, texture,*
(*b*) The *outer surface, seldom much wrinkled,*
(*c*) The *continuous ring of star-spots on the transverse section,*
(*d*) The *network of white lines on the outer surface* (not shown when the peeling has been carried too deep).

**Constituents.**—Rhubarb exerts an astringent and also a laxative action.

The chief astringent constituent is gallic acid, in addition to which a small quantity of tannin and (?) catechin is present.

The chief laxative constituent is, according to Tutin and Clewer (1911), an amorphous non-glucosidic resinous mixture. By hydrolysis with dilute sulphuric acid this yields gallic acid, cinnamic acid, rhein, emodin, aloe-emodin, emodin monomethyl ether, chrysophanic acid, and a trihydroxydihydroanthracene. The gallic and cinnamic acids are probably present as esters.

In addition to the resin the drug contains rhein, emodin, aloe-emodin, emodin monomethyl ether and chrysophanic acid in the free state and also in the form of glucosides. These substances are all derivatives of oxymethylanthraquinone and are accompanied in the drug by rheinolic acid.

Further constituents are starch, fat, dextrose, levulose, pectin, calcium oxalate, &c.

FIG. 191.—Chinese Rhubarb (Canton round), showing granular fracture. Natural size.

*Chrysophanic acid*, $CH_3.C_{14}H_5O_2(OH)_2$, is a dihydroxymethylanthraquinone, but the positions of the groups are not known with certainty.

*Emodin*, $CH_3.C_{14}H_4O_2(OH)_3$, is hydroxychrysophanic acid and is therefore a trihydroxymethylanthraquinone.

*Emodin monomethyl ether*,

$$CH_3.C_{14}H_4O_2(OH)_2O.CH_3,$$

is the corresponding methoxychrysophanic acid; it has been found in *Rumex Ecklonianus*, Meissner, and may be prepared from emodin.

*Aloe-emodin*, $C_{14}H_5O_2(OH)_2.CH_2.OH$, is the primary alcohol (hydroxymethyldihydroxyanthraquinone) corresponding to chrysophanic acid; as its name indicates, it is found in aloes.

*Rhein*, $C_{14}H_5O_2(OH)_2.CO_2H$, is the corresponding carboxylic acid, and may be obtained by the oxidation of aloe-emodin.

The total quantity of oxymethylanthraquinones, free and combined, present in rhubarb has been estimated to vary from 2 to 4 per cent. Only chrysophanic acid and aloe-emodin are laxative. The mixed glucosides of the oxymethylanthraquinones crystallise readily together; this crystalline mixture, termed by Gilson rheopurgarin, is, however, destitute of purgative action. The suggestion that has been made that the comparative medicinal value of rhubarb may be determined by estimating the total oxymethylanthraquinones present would appear, therefore, to be unacceptable.

According to Gilson, the gallic acid is present as a glucoside (glucogallin), and gallic acid, cinnamic acid, and rheosmin as another glucoside (tetrarin), but this is improbable.

The amount of calcium oxalate contained in rhubarb varies considerably, averaging about 7·3 per cent. (Fluckiger). The ash is subject to a similar variation; it has been found as low as 3·5, and as high as 43·27; usually good Chinese rhubarb yields from 7 to 13 per cent. (B.P. not more than 15). English rhapontic rhubarb yields usually less than Chinese. It occurs in very large cluster crystals, and is the cause of the gritty taste of the drug.

**Uses.**—In small doses rhubarb is a bitter stomachic and intestinal astringent; in larger doses it causes purgation, which is followed by an astringent effect due to the tannoid constituents. It is given in cases of indigestion with diarrhœa and as a mild laxative.

FIG. 192.—Chinese Rhubarb (flat High-dried). Slightly reduced.

**Varieties.**—Three varieties of Chinese rhubarb are recognised on the market, viz. Shensi, Canton, and High-dried, each of which may occur in 'flats' or 'rounds.'

*Shensi Rhubarb* is characterised by its very compact nature, bright yellow coat, distinct whitish reticulations, and 'nutmeg' fracture (fig. 189): the odour is agreeable and free from marked empyreuma, the taste slightly bitter. That which exhibits a bright pink fracture is preferred.

*Canton Rhubarb* may be distinguished from Shensi by the fracture, which is more or less uniformly granular, exhibiting no marked marbling, and by the odour and taste which are more distinctly empyreumatic and disagreeable. It is more fibrous in its nature and less compact, the coat is not so bright, and the whitish reticulations are less marked. Formerly the pieces bore a deep triangular nick, but this is now more often wanting. It is less esteemed than Shensi rhubarb.

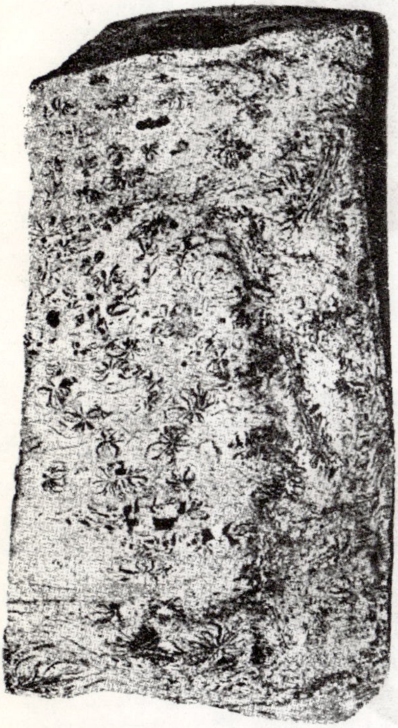

FIG. 193.—English Rhubarb (*R. officinale*), showing the star-spots on the radial section. Natural size.

FIG. 194.—English Rhubarb (*R. rhaponticum*). Rhizome. Slightly reduced.

*High-dried Rhubarb* may resemble either Shensi or Canton in the fractured surface, but the coat is much duller and rougher, and the odour and taste more empyreumatic than Canton. The rounds are often much shrunken, and frequently exhibit the remains of the bud, dark patches being often visible on the outer surface. The flats show distinct severe paring, and are often so hard as to spring when broken with the hammer; they are usually of better quality than the rounds.

*English Rhubarb.*—In England two species of *Rheum*, viz. *R. officinale* and *R. rhaponticum*, Linné, are cultivated (Oxfordshire, Bedfordshire,

&c.). The rhizomes are dried and sold separately from the roots. Those of *R. officinale* resemble the Chinese drug, but, being more spongy, shrink and wrinkle as they dry, and are softer to cut; the white reticulations are commonly absent, the white and red lines being parallel to one another; the star-spots are also fewer and more scattered. The

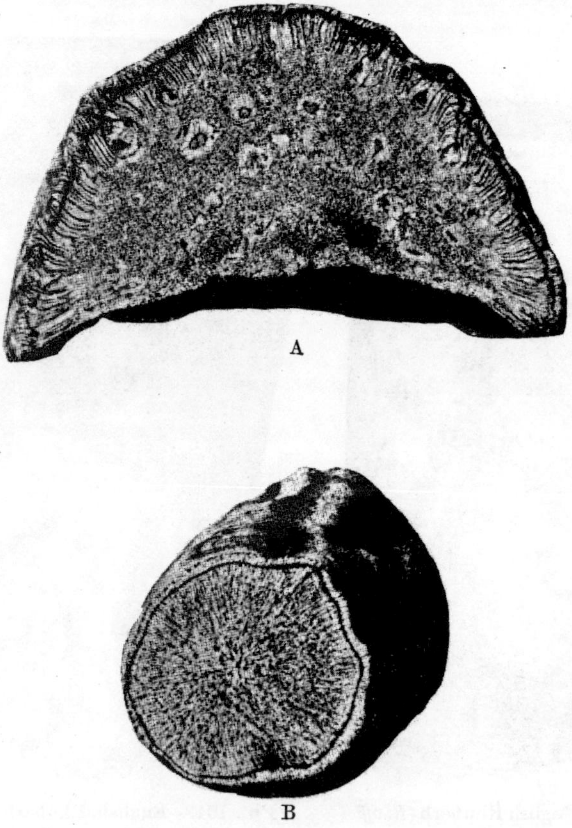

A

B

FIG. 195.—English Rhubarb (*R. rhaponticum*). A, transverse section of rhizome; B, transverse section of root. Natural size.

roots are readily distinguished by their long cylindrical shape and distinctly and entirely radiate transverse section. The constituents of this variety of English rhubarb are similar to those of the Chinese.

The rhizomes of *R. rhaponticum* are also much shrunken and usually pinkish in colour; they may be readily distinguished by the transverse section which exhibits a diffuse circle of isolated star-spots. Rhapontic rhubarb contains no emodin, aloe-emodin, or rhein; its most characteristic constituent is a crystalline glucoside, rhaponticin (rhapontin, ponticin; not an anthraquinone derivative), the presence of which can be demonstrated by the following test:

Percolate 10 gm. of the powdered drug with 60 per cent. alcohol; collect 25 c.c. of the percolate, evaporate at 80° to 7 gm., shake vigorously while still warm with 10 c.c. of ether, pour off the ethereal solution into a small flask, cork and set aside; needle-shaped crystals of rhaponticin will separate within 24 hours. This test serves to identify the drug derived from *R. rhaponticum*, and will detect it in a mixture of 1 part of rhapontic rhubarb with 3 of Chinese but in this case crystals do not separate for a few days.

Rhapontic rhubarb also contains chrysophanic acid, a crystalline substance chrysopontin, $C_{16}H_{16}O_5$, and a glucoside yielding by hydrolysis chrysorhapontin, $C_{16}H_{16}O_4$. According to Hesse the drug contains, in addition to rhaponticin and chrysophanic acid, anhydrorhapontigenin, rhabarberone, chrysarone, gluco-chrysarone, gallic acid and rhapontic acid.

*Chinese Rhapontic Rhubarb.*—This variety has latterly been imported from China; it closely resembles English rhapontic but is usually darker, often hollow in the centre, the section exhibiting alternating paler and darker concentric rings and a yellow rather than pink colour; it yields rhaponticin by the test described above.

**Allied Drugs.**—Many species of *Rumex* contain oxymethylanthraquinone derivatives and have been used as substitutes for rhubarb, e.g., *R. alpinus*, Linné, *R. obtusifolius*, Linné, &c.

*R. hymenosepalus*, Torrens, yields canaigre root, used in America for tanning (30 per cent. of tannin).

## KAVA RHIZOME

(Kava-kava: Kavæ Rhizoma)

**Source, &c.**—Kava rhizome is the rhizome of *Piper methysticum* Forster (N.O. *Piperaceæ*), a shrub indigenous to the Sandwich Islands. The plant produces a large rhizome which should be freed from roots and periderm, cut into pieces and dried.

**Description.**—Although the British Pharmacopœia requires that the periderm should be removed, much of the commercial drug is unpeeled. The larger rhizomes are cut into irregularly cubical or wedge-shaped pieces or slices about 1 to 5 cm. thick; the smaller may be entire and have numerous stout roots attached, the drug then exhibits a dark grey periderm within which is a whitish or pale brownish grey cortex and a narrow ring of whitish wood enclosing a large, dense, often discoid pith of the same colour. In the wood slender vascular bundles with distinct vessels alternate with wide medullary rays. This structure is best seen in the smaller rhizomes.

The parenchymatous medullary rays readily break down and the vascular bundles then separate into strands giving to the drug a coarsely fibrous appearance. All the parenchymatous cells are loaded with starch in simple or compound (two to three) grains. The odour is slight and agreeable; the taste at first bitterish but subsequently distinctly numbing.

**Constituents.**—Kava rhizome contains about 5·8 per cent. of resin, part of which ($\alpha$-kava resin) is soluble in petroleum spirit, the remainder ($\beta$-kava resin) being soluble in ether. In addition it contains methysticin (colourless crystals melting at about 137°; 0·3 per cent.), yangonin (colourless crystals melting at 151°; 0·18 per cent.), and pseudomethysticin (yellowish crystals melting at 113°; 0·18 per cent.); these readily crystallise from a concentrated alcoholic tincture of the drug. Further constituents are an alkaloid, kavaine, two glucosides and an abundance of starch. The medicinal activity appears to be due to the resins.

**Uses.**—Kava rhizome is employed medicinally as an antiseptic and diuretic in gonorrhœa and cystitis. The purified resin dissolved in sandalwood oil forms a proprietary preparation for gonorrhœa. It is used in the South Sea Islands for preparing an intoxicating drink, the intoxication being said to affect the power of movement while leaving the intellect clear.

## SERPENTARY RHIZOME

### (Serpentary Root, Virginian Snakeroot, Rhizoma Serpentariæ)

**Source, &c.**—Two varieties of serpentary, or, as it is sometimes termed, snakeroot, are official in the British Pharmacopœia, viz. *Virginian snakeroot* and *Texan* or *Red River snakeroot*.

*Virginian snakeroot* is obtained from *Aristolochia Serpentaria*, Linné (N.O. *Aristolochiaceæ*), a small herbaceous plant with slender perennial rhizome, growing in the United States, to the east of the Mississippi.

*Texan* or *Red River snakeroot* is the produce of *Aristolochia reticulata*, Nuttall, a rather stouter plant, growing, as its name indicates in the south-western States.

The drug is collected in the autumn and dried. It was known in England in 1632, its use having been undoubtedly learnt from the American Indians, and it found a place in the London Pharmacopœia of 1650.

**Description.**—The snakeroot at present in commerce is the variety known as Texan, and is derived from *A. reticulata*. It consists of a short, slender, rhizome about 25 mm. long and 3 mm. thick, to the flanks and lower surface of which are attached numerous long, curved, but not wiry or interlacing roots; these are comparatively stout, being frequently half the thickness of, or even as thick as, the rhizome itself, and seldom much shrivelled. The latter is usually horizontal, but not unfrequently assumes an oblique or even vertical position, and gives off from its upper surface (or from one side) numerous closely

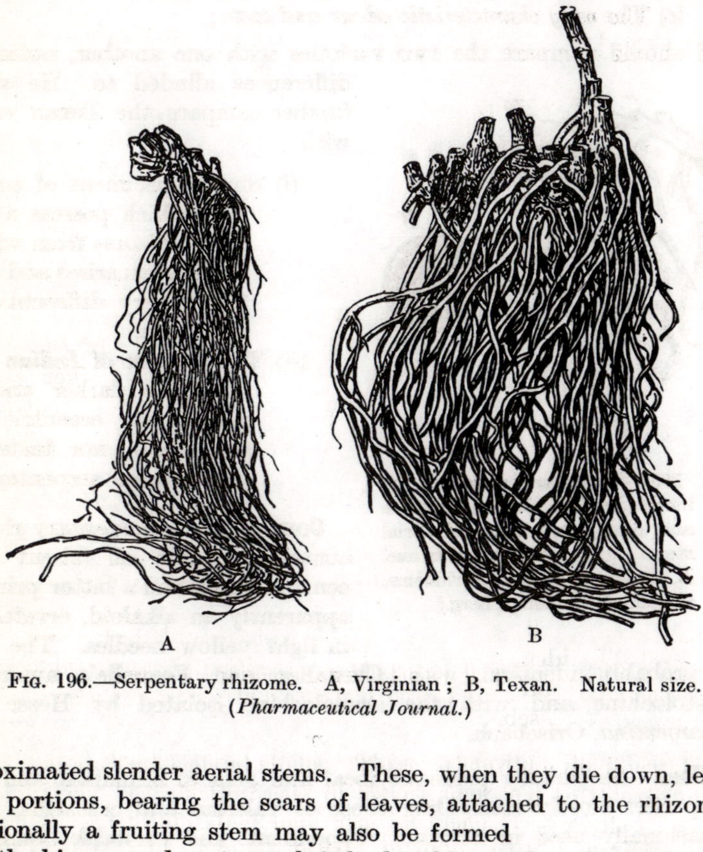

FIG. 196.—Serpentary rhizome. A, Virginian; B, Texan. Natural size. (*Pharmaceutical Journal.*)

approximated slender aerial stems. These, when they die down, leave short portions, bearing the scars of leaves, attached to the rhizome; occasionally a fruiting stem may also be formed

Both rhizome and roots are brittle, breaking with a short fracture. The former exhibits in transverse section a whitish pith that is distinctly eccentric, being much nearer to the upper than the under surface of the drug; the wood-bundles are numerous, yellow and curved, the bark yellowish brown and thin; the section of the root shows a slender yellow wood and a thick white bark.

The drug, by far the greater part of which consists of the roots, has a tolerably uniform yellowish brown colour, a characteristic camphoraceous odour, and a strong, disagreeably bitter and acrid taste.

Virginian snakeroot, which appears to be seldom imported, differs from the foregoing in having a shorter and thinner rhizome and thinner wiry, interlacing roots forming little matted masses, but resembles it in colour, odour, and taste.

The student should observe

(*a*) The *slender rhizome with portions of the stem arising from the upper surface in close succession,*

(*b*) The *eccentric position of the pith,*

(*c*) The *very characteristic odour and taste*;

and should compare the two varieties with one another, noting the differences alluded to. He should further compare the Texan variety with

(i) Small specimens of *valerian root,* which possess a short erect rhizome from which a single stem arises and which has a very different odour and taste.

(ii) The rhizome of *Indian pink,* which is darker and has neither the eccentric pith, the odour, nor taste that characterise serpentary.

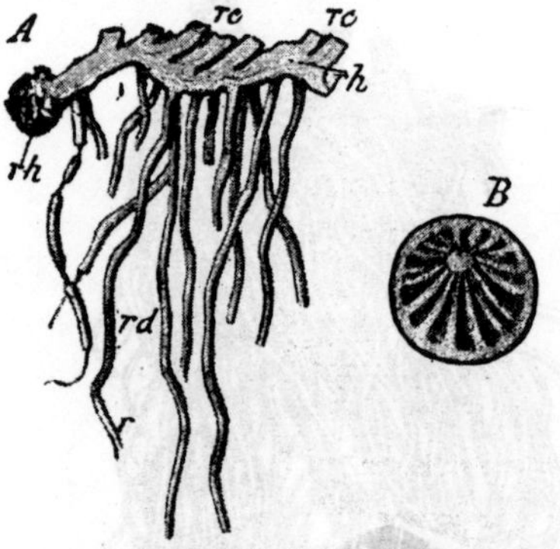


FIG. 197.—Serpentary rhizome. *A*, longitudinal section of rhizome and roots: *tc*, remains of aerial stems; *rd*, roots; *rh*, rhizome. *B*, transverse section of rhizome, magnified 3 diam. (Berg.)

**Constituents.**—Serpentary rhizome contains volatile oil (about 1 per cent.), tannin, and a bitter principle, apparently an alkaloid, crystallising in light yellow needles. The latter is probably identical with Chevalier and Fenuelle's amorphous aristolochine, and with the aristolochine isolated by Hesse from *A. argentina*, Grisebach.

**Uses.**—Serpentary possesses local and general stimulant and tonic properties closely resembling those of valerian and cascarilla. It is occasionally used in nervous, despondent, and excitable conditions, as well as in low fevers and febrile states.

## SASSAFRAS ROOT

### (Radix Sassafras)

**Source, &c.**—The sassafras, *Sassafras officinale*, T. Nees and Ebermayer (N.O. *Laurineæ*), is a tree of medium size, widely distributed over the eastern United States, extending from Canada southwards to Florida.

All parts of the plant secrete in special cells an aromatic volatile oil, but this is especially the case with the bark of the root.

**Description.**—The tree produces large, woody, branching roots covered with a dark reddish or greyish brown spongy bark, from which, however, they are often entirely freed, the bark constituting in America a separate article of trade, and realising, on account of its greater fragrance, a higher price than the wood. The latter is greyish yellow or greyish red in colour, soft, and easily cut; it exhibits in transverse section distinct annual rings traversed by thin, dark, medullary rays and containing large vessels. Both bark and wood, but especially the former, have an agreeable, fragrant odour, and an aromatic, slightly astringent taste.

**Constituents.**—The root contains about 2 per cent. of volatile oil, which is obtained from it by distillation, and employed in large quantities as a perfume and flavouring agent; its chief constituent is safrol (80 per cent.). The bark yields from 6 to 9 per cent. of oil.

**Uses.**—Sassafras is supposed to increase the action of the skin in syphilis and rheumatism, and thus to be an alterative, but its physiological action is not definitely known.

## TURMERIC

### (Rhizoma Curcumæ)

**Source, &c.**—The plant yielding turmeric, *Curcuma longa*, Linné (N.O. *Scitaminĕæ*), is apparently a native of southern Asia, though no longer known in the wild state. It is largely cultivated in India, China, Java, and other tropical countries. The rhizome has long been employed both as a spice and as a colouring agent (*Crocus indicus*). It was certainly known to Dioscorides, and described by him as a root resembling ginger, but having a yellow colour and bitter taste. During the Middle Ages it fell into disuse, and is now considered much inferior to ginger and other spices, whilst in India it is largely employed as a dye and as a condiment.

The rhizomes are dug up after the herbaceous aerial stems have died down; there is then found an upright, bulb-shaped rhizome, from which the stem has sprung and to which several cylindrical descending branches are attached. One or more of the latter, destined to produce aerial stems in the following year, will curve upwards and thicken to form bulb-shaped organs. The rhizomes so obtained are then steamed in their own juice or boiled in water for a half to one day, by which the vitality that they otherwise obstinately retain is destroyed. They are finally dried either in the sun or in an oven, and (sometimes) sorted into 'fingers' and 'bulbs,' the former being the cylindrical descending branches, the latter the erect, stem-producing ones.

**Description.**—Finger or long turmeric occurs in curved or nearly

FIG. 198.—Long Turmeric. (Pereira.)

FIG. 199.—Round Turmeric. (Pereira.)

straight cylindrical pieces bluntly tapering at each end. The outer surface is of a deep yellowish brown colour, longitudinally wrinkled and marked with transverse rings (leaf-scars). Occasionally they bear short knob-like branches, or show large circular scars where these have been broken off. They are hard and heavy, and break with a short fracture; internally they have a uniform dull brownish yellow, waxy appearance and tough horny consistence. The transverse section exhibits little of its structure beyond a paler (or sometimes darker) ring separating the stele from the cortex. This remarkable appearance of the interior of the rhizomes is due to the prolonged boiling they undergo, by which not only is the starch gelatinised and a horny consistence imparted to the drug, but the colouring matter, previously restricted to certain scattered cells, becomes uniformly diffused throughout the rhizome.

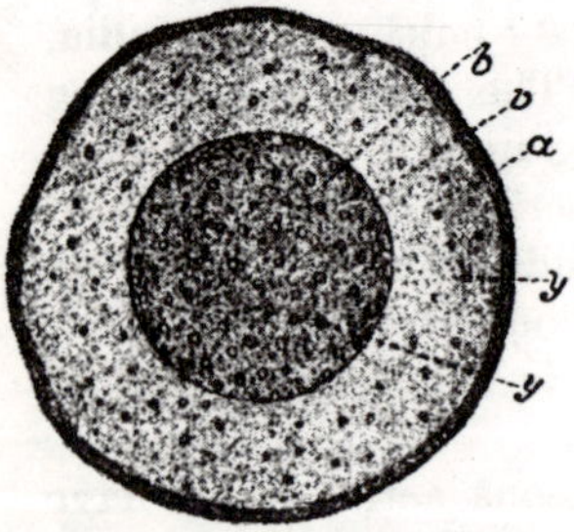


FIG. 200.—Turmeric rhizome. Transverse section. *a*, cortex; *v*, endodermis; *b*, stele; *y*, vascular bundles. Magnified 3 diam. (Berg.)

Bulb or round turmeric resembles the finger variety, but is, as its name indicates, shorter and thicker.

The drug has a characteristic aromatic odour and taste, and when chewed colours the saliva yellow.

The student should observe

(*a*) The *difference in shape* between 'bulbs' and 'fingers,'
(*b*) The *yellowish brown colour*,
(*c*) The *horny consistence and waxy appearance of the interior*.

**Constituents.**—Turmeric contains about 5 per cent. of volatile oil-resin, a crystalline yellow body, curcumin. These are confined, in the fresh rhizome, to the particular secreting cells in which they have been produced, but pass during the scalding into the surrounding tissue, the parenchymatous cells of which are filled with amorphous masses of gelatinised starch.

Curcumin, $C_{21}H_{20}O_6$, forms reddish yellow prisms melting at 183°, and readily dissolving in alcohol, forming a deep yellow solution, the colour of which is changed to reddish brown by alkalies. Evaporated after the addition of boric acid the colour is reddish brown, which alkalies then change to blue. A mixture of alcohol and sulphuric acid dissolves it with production of a brilliant crimson red colour and forms an excellent means of detecting powdered turmeric in the presence of many other substances.

**Uses.**—Turmeric is used as a condiment and colouring agent.

**Varieties, &c.**—Several commercial varieties of turmeric are known. The majority of the drug is shipped from India, that from Madras being the most esteemed. Other species of *Curcuma* (*C. angustifolia*, Roxburgh, *C. leucorhiza*, Roxburgh) have paler coloured rhizomes; these are utilised in India for the production of starch, which is known as 'East Indian Arrowroot.'

**Allied Drug.**—Zedoary (*C. Zedoaria*, Roscoe), India; circular slices of a rhizome resembling bulb turmeric; greyish or yellowish; starchy; section exhibiting numerous oleo-resin cells, hairs, and bundles without sclerenchymatous elements; taste and odour resembling ginger but less aromatic.

## GINGER

### (Zingiber, Rhizoma Zingiberis)

**Source, &c.**—Ginger is the dried rhizome of *Zingiber officinale*, Roscoe (N.O. *Scitamineæ*), a reed-like plant producing leafy stems a metre high, springing from branching rhizomes. It is a native of Asia, but is cultivated in many tropical countries, notably in the West Indies, in India, Africa, and Japan.

There can be little doubt that ginger was well known in India as a spice from the earliest times. The Greeks and Romans were well acquainted with it, and its use spread in the ninth and tenth centuries through Europe to England. It was introduced by the Spaniards into the West Indies and Jamaica.

In Jamaica the plant is largely cultivated. It is propagated by dividing the rhizome into 'fingers,' each of which contains a bud, and planting these about a foot apart. They grow rapidly and flower in the autumn. When the aerial stems wither the rhizomes are dug up, freed from the roots, and washed. They are then peeled with a narrow-bladed knife, by which the layer of cork and part of the parenchyma of the cortex are removed, after which they are again washed, and dried in the sun. The product is known as unbleached Jamaica ginger. Although the use of a revolving drum for removing much of the peel and of a drying machine for rapidly drying the rhizomes has been suggested, this machinery does not appear to have been successfully introduced.

Much ginger is found in commerce from which the cork layer has not been removed, or which has been deprived of it on the flatter sides only. Such ginger is known as coated '' or 'unscraped,' whilst that which has been completely peeled is called 'scraped.' Sometimes, too, the rhizomes are treated with sulphurous acid or chlorine, or they are dusted over with calcium sulphate or carbonate which imparts to them a whitish appearance; ginger that has been treated so is termed 'bleached' or 'limed' ginger. In addition therefore to being obtained from different countries, ginger may vary in appearance according to the way in which it has been prepared for the market. As with nutmegs, limed ginger is undoubtedly less susceptible to the attacks of insect pests. Ginger is commonly washed and limed abroad, but is also usually rewashed and heavily limed in London.

**Description.**—Unbleached Jamaica ginger, which is the official variety and the one that is most esteemed, occurs in flattened branched pieces technically termed 'races' or 'hands.' These vary in length, but average about 7 to 10 cm. From the upper surface of the main rhizome, which is usually straight, numerous branches about 3 cm. in length appear to arise and take an upward course; they are often slightly compressed laterally, enlarged near the rhizome, and, tapering abruptly, terminate in the remains of an undeveloped bud or a small depressed scar indicating the point of attachment of an aerial stem. The branches themselves also produce lateral branchlets. All these branches and branchlets arise from buds on the under surface of the rhizome in the axils of cataphyllary leaves. When the growth of the main axis is terminated by the production of an aerial shoot, one of these buds develops, curves upwards, and itself in due time produces an aerial shoot and develops a lateral branch. Each piece of the drug or 'hand' is therefore a sympodial branch system.

The colour of the drug is usually pale yellowish buff; the surface is strongly striated or even fibrous, the fibres being the leaf-traces passing through the cortex to the leaves, and laid bare by the peeling of the rhizome. It breaks with a short, mealy or sometimes resinous fracture, short scattered fibres usually protruding from the fractured surface; these are the fibro-vascular bundles in the stele and cortex.

The smoothed transverse section exhibits a large stele which is sharply delimited by a fine yellow line from a narrow cortex, both stele and cortex containing numerous yellow oil-cells. No cork layer is to be discerned. Ginger has an agreeable aromatic odour and a strong pungent taste.

The student should observe

(*a*) The *fibrous surface*,

(*b*) The *short fracture with protruding fibres* and *yellow oil-cells*.

**Microscopical Characters.**—The transverse section exhibits a large stele and a cortex free from cork. The parenchymatous cells, which are very abun-

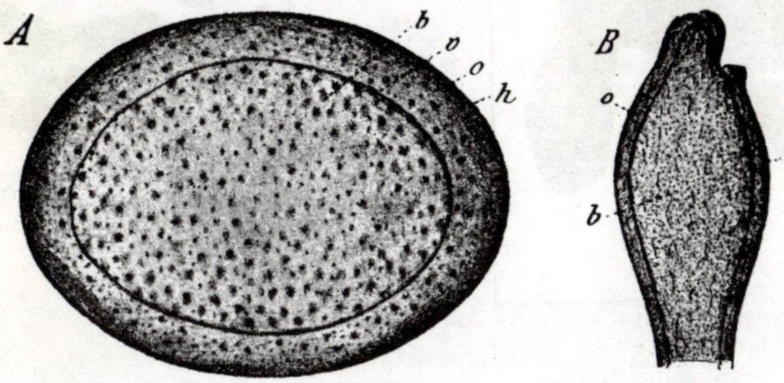


FIG. 201.—Ginger. *A*, transverse section through the rhizome, magnified 3 diam. *B*, longitudinal section through a branch, natural size. *o*, cortex; *v*, endodermis; *b*, stele. (Berg.)

dant, have very thin walls; some contain yellow oleo-resin, but most are filled with starch. The fibro-vascular bundles in the stele contain a few reticulated vessels supported by sclerenchymatous fibres, which have rather thin walls and large cavities; abutting on the vessels are small cells filled with a dark brown secretion. The starch grains are typically Scitaminaceous; they are simple, $12\mu$ to $30\mu$ long, ovoid or sack-shaped and flat, the hilum being at the extreme, pointed end.

Powdered ginger is characterised by the abundant typical starch, by the very thin walls of the parenchymatous cells, by the characteristic sclerenchymatous fibres, and by the oleo-resin cells; the latter are mostly broken, but the suberised walls are always to be found. It should contain no cork, no isodiametric sclerenchymatous cells, and no hairs. It should yield about 7 per cent. of extract to 90 per cent. alcohol, from 12 to 15 per cent. to cold water, not more than 6 per cent. of ash, and not less than 1·5 per cent. of ash soluble in hot water. The last two data are specially useful in detecting adulteration with exhausted ginger.

**Constituents.**—Ginger contains from 0·25 to 3 per cent. of a volatile oil possessing the aroma but not the pungency of the drug. The latter property is due to a yellowish oily body, gingerol, which is odourless, but has an intensely pungent taste. The drug contains in addition resin and abundance of starch. It yields from 3 to 5 per cent. of ash.

Gingerol is a mixture of homologous phenolic substances of the formulæ $C_{17}H_{26}O_4$, $C_{18}H_{28}O_5$, &c. Fixed alkalies, especially when heated, destroy its pungency but baryta water splits it up into fatty aldehydes, especially *n*-heptaldehyde and zingerone. Zingerone, $C_{11}H_{14}O_3$, is crystalline and pungent;

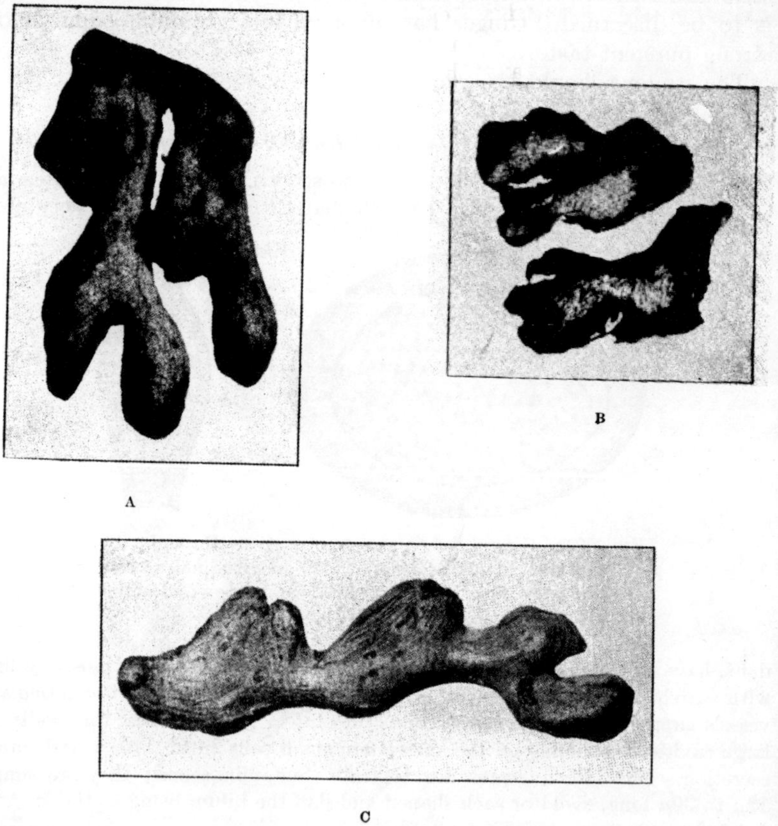


FIG. 202.—Ginger. A, Jamaica; B, African; C, Cochin. All slightly reduced.

it has a sweet odour and is allied to vanillin from which it has been synthesised; it is 4-hydroxy-3-methoxyphenylethylmethylketone. The pungency of gingerol is destroyed by boiling with 2 per cent. solution of potassium hydroxide, while that of capsaicin or paradol is scarcely affected.

**Uses.**—Ginger is largely used as a condiment, and medicinally as a carminative and aromatic stimulant.

**Varieties.**—The chief commercial varieties of ginger are Jamaica, Cochin, African, Japanese, and Bengal; an inferior grade of Jamaica ginger, obtained by allowing a part of the 'hand' to remain in the ground after the first crop has been collected and grow without further attention, is known as 'ratoon' ginger.

*Cochin ginger* occurs in both the scraped and coated varieties; the latter bears on its ventral and dorsal surfaces, but not on the lateral, portions of a reddish grey cork, coarsely wrinkled both longitudinally and transversely. The lateral surfaces, that have been freed from the cork, are striated and of a rather paler colour. The drug is usually in smaller 'hands' than the Jamaica, the branches ('fingers') are commonly shorter and thicker, and the aroma less agreeable.

*African ginger* is a coated ginger, the ventral and dorsal surfaces bearing patches of wrinkled cork of an earthy-brown colour. The cortical tissue that is exposed on the lateral surfaces is sometimes of a dingy grey colour and lighter than the cork, sometimes nearly black and then much darker. The drug in bulk is darker than Cochin ginger and appears discoloured from want of care in the preparation of it for the market. Although deficient in aroma, it is an exceedingly pungent ginger, in this respect excelling the Jamaica drug. It yields about 10 per cent. of alcoholic extract.

*Bengal ginger* is dark and partly coated; it resembles African.

*Japanese ginger* usually occurs in small flattened unscraped pieces; it is not produced by *Z. officinale*, as many of the starch grains are compound and the volatile oil differs in physical properties from that of Jamaica ginger; it has been referred to *Z. Mioga*, Roscoe.

*Ratoon ginger* is small, dull, dingy, greyish brown, and bears evidence of having been imperfectly peeled and carelessly cured. It is of inferior aroma and pungency.

Of all the commercial varieties of ginger, Jamaica is the most aromatic and African the most pungent.

## GALANGAL RHIZOME

### (Lesser Galangal, Rhizoma Galangæ)

**Source, &c.**—Galangal root is the rhizome of *Alpinia officinārum*, Hance (N.O. *Scitamineæ*), a reedlike plant, attaining about a metre in height, a native of and cultivated on the island of Hainan and the neighbouring south-east coast of China. The rhizome is dug up in the autumn, washed, trimmed, cut into pieces, and dried; during the latter process the pale colour of the fresh rhizome turns to a reddish brown. The drug is exported in bales made of split cane, plaited, and bound round with cane.

Galangal root is not now much used in England, but is still employed in some parts of Russia both as a spice and as a medicine.

**Description.**—The drug consists of a branched rhizome, about 12 mm. thick, that has evidently been cut whilst fresh into pieces about 5 or 10 cm. long. These are frequently cylindrical, but sometimes tapering or enlarged, and often branched. They are dull reddish brown, longitudinally striated or shrivelled, and bear, at intervals

FIG. 203.—Galangal root. Natural size.

of about 5mm., pale, encircling, sinuous or frilled remains of cataphyllary leaves. Here and there the broken upper end of a root remains attached to the rhizome, but these are by no means numerous. It is hard, tough, and difficult to break.

The interior of the drug has a reddish brown colour, sometimes deeper, sometimes paler than that of the exterior. The transverse section exhibits a distinct central column (stele) surrounded by a wide cortex. In both of these portions paler fibro-vascular bundles and numerous deep red resin-cells may be distinguished under the lens. The drug has an agreeable spicy odour and a strongly pungent spicy taste.

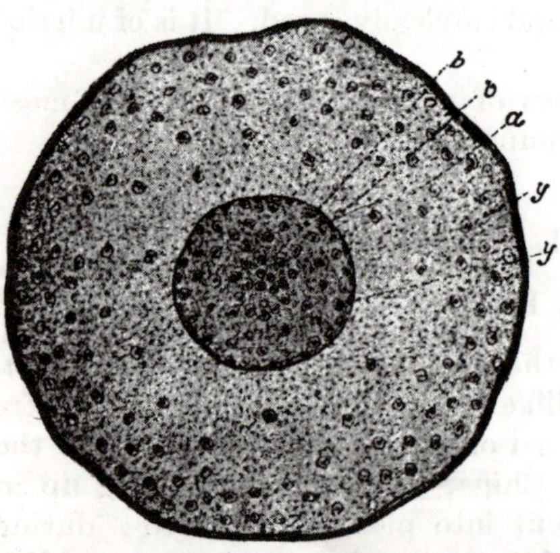

FIG. 204.—Galangal rhizome. Transverse section, magnified 3 diam. *a*, cortex; *v*, endodermis; *b*, stele; *y*, vascular bundles. (Berg.)

The student should observe

(*a*) The *reddish brown colour*,
(*b*) The *remains of the cataphyllary leaves*,
(*c*) The *pungent taste*,
(*d*) The *structure exhibited by the transverse section*.

**Constituents.**—Galangal rhizome contains a little volatile oil (cineol, methyl cinnamate), and a pungent oily body, galangol. It also contains three tasteless, yellow crystalline substances, viz. kæmpferide, galangin, and the monomethyl ether of galangin. Alpinin, which was formerly considered to be a constituent of the drug, appears to be a mixture of kæmpferide and galangin. Galangin is dioxyflavanol and has been obtained synthetically. Kæmpferide is 1.3-dioxy-4-methoxyflavonol.

**Uses.**—Galangal rhizome has stimulant and carminative properties and is used for flatulent dyspepsia.

**Varieties, &c.**—Greater galangal, the rhizome of *A. Galanga*, Willdenow, which is occasionally imported from Java, is much larger, orange-brown externally and pale buff internally; it has similar properties, but is less pungent.

## ORRIS RHIZOME

### (Orris Root, Rhizoma Iridis)

**Source, &c.**—Orris rhizome, or root as the drug is commonly termed, is derived from three species of *Iris* (N.O. *Iridĕæ*), all of which are cultivated for that purpose, viz. :—

*Iris germanica*, Linné, a handsome plant with dark blue or purplish blue flowers, distributed over central and southern Europe, extending to Africa and India, and a common garden plant in England. It is cultivated in Italy, especially in the neighbourhood of Florence and Verona, and also in Morocco.

*Iris pallida*, Lamarck, with pale bluish flowers, a native of the eastern Mediterranean countries; it is also cultivated in Italy, and yields with *I. germanica* the bulk of the drug.

*Iris florentina*, Linné, with large white flowers, also a native of the eastern Mediterranean region, and cultivated in Italy, but not so abundantly as the other two.

The rhizomes of all three species so closely resemble one another that there are no definite means of distinguishing them. They are dug up in the late summer when two or three years old, trimmed and peeled; they are dried in the sun on a kind of matting made of bamboo rods for about 5 days; they are then spread on a cool, dry, tiled floor for 8 days and finally sorted by hand. During the slow drying the rhizomes, which in the fresh state are almost inodorous and have an acrid taste, acquire an agreeable fragrance and lose their acridity.

**Description.**—Orris rhizome occurs usually in pieces from 5 to 10 cm. long and 2 to 3 cm. thick, of a dull white colour. They are often dorsi-ventrally flattened and contracted at intervals or bear one or two short lateral branches at the apex. Each of the enlargements corresponds to a year's growth of the rhizome; the branches are developed from buds after the rhizome has flowered, which may not occur for three or four years.

The under surface is marked with the small dark circular scars of roots, and on the upper surface traces of the leaves, or at least of the fibro-vascular bundles that passed into them, can be discerned, the drug showing distinct evidence of having been peeled. It is hard, heavy, and compact, and breaks with a short fracture, the interior being yellowish and horny. The transverse section exhibits a large stele containing scattered darker bundles and a comparatively narrow cortex.

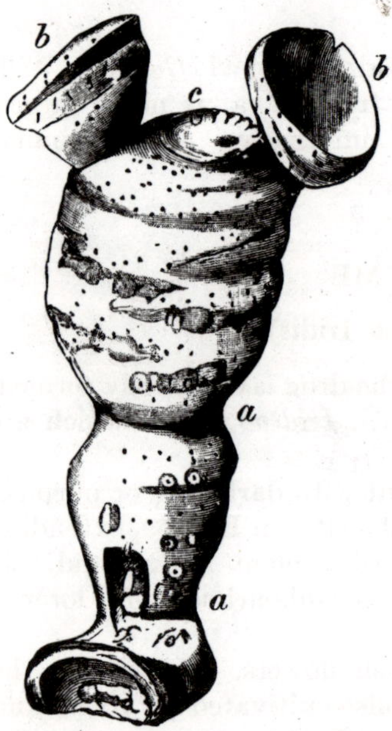


FIG. 205.—Orris rhizome. Rhizome of *Iris germanica*. *a*, constrictions indicating winter growth; *b*, *b*, young branches; *c*, scar left by previous year's flowering shoot. (Tschirch, after Hartwich.)

The drug has an agreeable aromatic odour and a slightly bitter taste.

The student should particularly observe

(*a*) The *contractions* of the rhizome,
(*b*) The *scars of leaves upon the upper* and of *roots on the under surface*,
(*c*) The characters of the *transverse section*,
(*d*) The *characteristic odour*.

**Constituents.**—By distillation with steam, orris rhizome yields about 0·1 or 0·2 per cent. of a yellowish, buttery, aromatic substance, commonly called oil or butter of orris; this consists principally (about 85 per cent.) of myristic acid together with irone, an oily liquid with a powerful odour of violets. Irone is the only aromatic constituent of the rhizome. The drug also contains a crystalline glucoside, iridin, which must be carefully distinguished from the brown, resinous, eclectic remedy of the same name; the latter is obtained from the rhizome of *Iris versicŏlor*, Linné.

**Uses.**—Orris rhizome is used as a perfume, dentifrice, &c., although formerly medicinal qualities were attributed to it.

**Varieties.**—The chief commercial varieties of orris root are the Florentine, Veronese, and Mogadore. A very inferior quality, of small size, dark colour, and little fragrance, is occasionally imported from Bombay.

*Florentine orris root* is usually nearly white in colour, carefully peeled, plump, and very fragrant.

*Veronese orris root* closely resembles the Florentine, but generally has a yellower colour, is rather less carefully peeled, often more wrinkled and more elongated; most of the pieces are pierced with a small hole at one end by which they were strung during the drying.

*Mogadore orris root* is altogether inferior to both the foregoing varieties. It is in smaller, flatter, and more shrunken pieces, which often bear at their apices the shrivelled remains of numerous concentrically arranged leaves. Patches of reddish cork are left attached to the drug, which is of darker colour and inferior fragrance.

*Iris versicŏlor*, Linné. The rhizome is narrow, dark brown, longitudinally wrinkled, and purplish internally; the taste is pungent and acrid. It is the source of the 'iridin' of commerce.

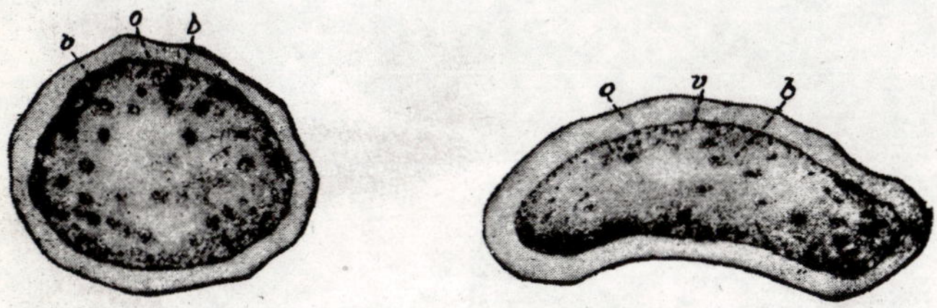


FIG. 206.—Orris rhizome. Transverse sections. *o*, cortex; *v*, endodermis; *b*, stele. Natural size. (Berg.)

## SARSAPARILLA

### (Radix Sarsæ, Radix Sarsaparillæ)

**Source, &c.**—Several varieties of sarsaparilla are imported but the one known as Jamaica sarsaparilla is the most esteemed in this country. This variety is obtained from *Smilax ornāta*, Hooker filius (N.O. *Smilacĕæ*), a climbing plant with woody stems ascending lofty trees and springing from a stout, knotty rhizome. From the rhizome slender cylindrical roots are thrown off horizontally and creep for many feet a few inches below the surface of the earth. In collecting the roots they are first laid bare and then cut off near the rhizome. After they have been dried they are made into bundles; a number of these are placed upright and bound with wire into a disc-shaped bale.

The plant is a native of Central America (Costa Rica). The root was formerly exported *via* Jamaica, hence the designation 'Jamaica' sarsaparilla, but it is now sent chiefly to New York, and thence to England.

**Description.**—Jamaica sarsaparilla occurs in bundles about half a metre long and 12 cm. in diameter, weighing about a kilogramme. Each bundle consists of numerous long, slender roots about 3 mm. in thickness, doubled up and bound loosely with one of the same roots. These usually have a dark reddish brown colour, are much shrunken and furrowed longitudinally, and bear tolerably numerous branching rootlets. They are tough and flexible, not breaking easily even when bent double. The transverse section exhibits a narrow, dark reddish brown cortex surrounding a central stele, which consists of a ring of yellowish wood with large, radially arranged vessels and a white, starchy pith.

The characters of the root are, however, somewhat variable, and it is not difficult to find roots that are paler in colour, less shrunken and more starchy than those described ; even the same root may vary

FIG. 207.—Bundles of Jamaica Sarsaparilla. Reduced.

at different points. But the dark reddish brown colour, the shrunken cortex, the presence of fibrous rootlets (technically known as ' beard ') are regarded as important characters of good Jamaica sarsaparilla. The bundles are always free from the rhizome (' chump ').

The drug has no odour, and only a slightly bitter taste.

The student should observe

(a) The *dark red shrunken cortex, which does not exhibit transverse cracks*,

(b) The *numerous wiry rootlets*,

(c) The *transverse section* ;

and should compare the drug with Indian sarsaparilla (see p. 340) which is marked with transverse cracks, is rigid and tortuous, and has a distinctive aroma.

**Constituents.**—The chief constituent in Jamaica, Honduras and probably other sarsaparillas is sarsasaponin, $C_{44}H_{76}O_{20},7H_2O$, a crystalline glucoside yielding by hydrolysis sarsasapogenin and dextrose. According to Power and Salway this is the only definite saponin-glucoside in Jamaica sarsaparilla, which also contains sarsapic acid (a crystalline dicarboxylic acid), dextrose, fatty acids, sitosterol-*d*-glucoside, resin, &c. Parillin and smilacin, formerly cited as constituents, are probably mixtures. It has been assumed that sarsaparilla is practically devoid of therapeutic value, but this has by no means been proved. Like other drugs containing saponins it possesses hæmolytic properties. Kobert has shown that the hæmolytic action of Honduras sarsaparilla is about eight times that of Jamaica. The drug contains also varying quantities of starch; it yields from 10 to 20 per cent. of aqueous extract, and about 7 per cent. of ash.

**Uses.**—Sarsaparilla has been administered as an alterative in syphilis, chronic skin diseases, and rheumatism, but great diversity of opinion exists as to its therapeutic value.

**Varieties.**—Several other varieties of sarsaparilla are imported into the English market; the following are the most important:—

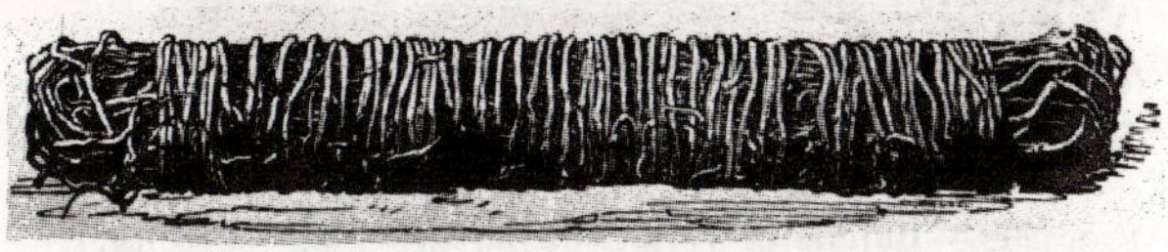

FIG. 208.—Bundle of Honduras Sarsaparilla. Reduced. (Pereira.)

1. *Honduras Sarsaparilla*, the botanical origin of which is unknown. The drug is imported from British Honduras in serons (see fig. 2, B) containing a number of bundles about 75 cm. long and 5 or 6 cm. wide, much longer and narrower therefore than the bundles of the Jamaica variety; they are sometimes closely whipped round or sometimes loosely bound with a long root. The roots are distinguished from those of the Jamaica variety by their pale yellowish or brownish colour, and by their less shrunken, more plump and starchy appearance; they have generally fewer rootlets attached and are always free from rhizome. The section exhibits a pale, starchy cortex, usually thicker than that of Jamaica sarsaparilla, but a similar stele.

This variety is largely used on the Continent, where it is generally preferred.

2. *Lima Sarsaparilla* is imported from Panama in bundles about 60 cm. long and about 7 cm. in diameter, loosely folded, bound with a root, and made into bales similar to those of Jamaica sarsaparilla, each containing about 300 bundles and weighing nearly 100 kilograms. This drug shows a close resemblance to Jamaica sarsaparilla, and indeed can only be distinguished with certainty by the different packing and by the anatomical characters of the cells of the endo-

dermis and exodermis, which in certain cases constitute a most valuable means of identifying and distinguishing these drugs.

3. *Guayaquil Sarsaparilla* is imported in rectangular pressed bales containing a number of flattish bundles about 50 cm. long and 15 cm. wide, containing the knotty rhizome and portions of the stout, round aerial stems. Sometimes the root is imported loose in bales. It has a mahogany brown colour, is usually larger than the Jamaica, not so much furrowed and with less numerous rootlets.

4. *Vera Cruz* or *Mexican Sarsaparilla* is obtained from *S. medica*, Schlechtendal et Chamisso. Both rhizome and roots are collected and dried, the drug not being made up into bales. It consists of a number of rhizomes, to each of which are attached numerous, straight dull greyish brown, shrunken roots, laid together. Latterly the roots deprived of the rhizomes have been exported. Like Lima sarsaparilla the variety is well characterised by the form of the cells of the endodermis and exodermis.

FIG. 209.—Bundle of Vera Cruz Sarsaparilla. (Pereira.)

5. *Native Jamaica Sarsaparilla* is obtained from plants presumably of *S. officinalis*, Humboldt, Bonpland et Kunth, cultivated on the island of Jamaica. This truly Jamaican sarsaparilla, commercially known as 'native' Jamaica, must be carefully distinguished from the ordinary Jamaica (or, better, Costa Rica) sarsaparilla. It arrives packed loose in bales, and is of a pale reddish or greyish brown colour. The root bears scattered, rather stout rootlets, and exhibits in transverse section a pale cortex separated by a distinct line from a rather darker stele. These characters sufficiently distinguish 'native' Jamaica sarsaparilla from the Costa Rica drug.

**Substitutes.**—A considerable variety of roots have from time to time found their way into the European markets under the name of sarsaparilla; e.g. roots of *Philodendron* sp., rhizomes of *Pteris* sp., &c.; most are readily distinguishable from the genuine drug.

## SQUILL

(Scilla, Bulbus Scillæ, Radix Scillæ)

**Source, &c.**—The squill, *Urginĕa Scilla*, Steinheil (N.O. *Liliaceæ*), is a bulbous plant indigenous to the countries bordering on the

Mediterranean, and frequently appearing in great abundance. It is one of the most ancient of medicinal plants, having been well known to the Greeks and Romans; the drug was introduced into European medicine by the Arabian physicians of the early ages.

The plant produces a large tunicated bulb, often weighing several pounds, which lies partially imbedded in the ground. Two varieties are known, the white and the red; the former is collected largely in Sicily and Malta and is preferred in England, the latter is collected in Algeria and is the variety used in France. The scales of the white

FIG. 210. Squill bulb, reduced. (Bentley and Trimen.)

squill are whitish or yellowish, whilst those of the red variety are of a dull red colour, but many intermediate forms are known.

The roots are cut off and the external thin scales stripped off until the bulb is white. Three or four incisions half an inch deep are made across the base and three or four at right angles to these. The bulb is then cut into transverse slices about 2 mm. thick which are dried in the sun.

**Description.**—The commercial drug usually consists of narrow, flattish, curved strips from 2 to 5 cm. in length and about 3 mm. thick. They frequently taper towards both ends, are of a yellowish white colour and more or less translucent. When quite dry they are brittle and can easily be powdered, but they rapidly absorb moisture from the air, becoming tough and flexible. They have only a slight odour, but a disagreeable bitter and acrid taste. Occasionally the entire bulbs are imported, but they are difficult to keep in the fresh state, as they preserve their vitality for a long time, and if allowed to remain in a warm place rapidly develop an aerial shoot.

**Constituents.**—The constituents of squill are imperfectly known. Merck (1879) separated scillitoxin, scillipicrin, and scillin, all of which exhibit glucosidal properties. Scillitoxin and scillipicrin are both amorphous and act upon the heart, the former being the more active of the two; scillin is crystalline, but is inactive. Scillain (Jarmerstedt, 1880) appears to be a purer form of scillitoxin. Waliszewski (1893) separated scillinin, scillipicrin and scillamarin. Kopaczewski (1914) isolated scillitin and scillidiuretin; scillitin (0·2 to 0·37 per cent.) is an intensely bitter, purified form of scillitoxin and probably the active constituent in the purest condition yet obtained.

The bulbs also contain mucilage, sinistrin (a carbohydrate soluble in water but insoluble in alcohol, probably identical with triticin and irisin), and calcium oxalate in bundles of long acicular crystals; the latter easily penetrate the skin when the bulbs are handled, and give rise to excessive irritation. This irritation has, however, also been referred to a volatile or unstable substance present in the drug.

**Uses.**—Squill closely resembles digitalis in increasing the vigour and diminishing the frequency of cardiac action; it is also a powerful expectorant, and is much used in chronic bronchitis and for coughs generally. In large doses it produces emesis.

## URGINEA

### (Indian Squill, Urginea)

**Source, &c.**—Indian squill is obtained from *Urginca indica*, Kunth, a plant resembling European squill but producing a smaller, tunicated bulb; it is found in sandy soil near the sea throughout India. The bulbs are collected soon after the plants have flowered, divested of their dry, outer, membranous coats, cut into slices and dried.

**Description.**—The drug occurs in curved or sickle-shaped strips, either separated or connected, several together, to a portion of the shortened axis; usually 1 to 5 cm. long and 5 to 10 mm. wide; yellowish white, fleshy, often longitudinally ribbed; tough when slightly moist but brittle and pulverisable when dry; taste bitter and acrid.

**Constituents.**—Indian squill has not yet been thoroughly examined; in all probability the chief constituents are similar to those of *Urginea Scilla*.

**Uses.**—Used in India and the Eastern Colonies in the place of European squill.

## COLCHICUM CORM

### (Colchicum Root, Radix Colchici)

**Source, &c.**—Colchicum corm is the contracted subterranean stem of the meadow saffron, *Colchicum autumnale*, Linné (N.O. *Liliaceæ*), a plant widely distributed over Europe and abundant in some parts of England in moist meadows and pastures.

The meadow saffron produces in the autumn a conspicuous reddish purple flower springing from the side of a contracted and enlarged stem (corm) situated several inches below the surface of the ground (fig. 100, p. 187). This corm has supplied the flower with the necessary materials for its growth, and has thereby been deprived of part of the reserve material which the leaves that die down in the summer have stored up in it. The proper time for collecting the corm is therefore after the early summer leaves have filled it with reserve material and died down, but before the production of the flower in the autumn has partially exhausted it.

**Description.**—The fresh corms are about 4 cm. long and 3 cm. broad, bluntly conical in shape, flattened on one side, and enveloped in an outer brown and inner reddish yellow membranous coat derived from the leaves of the previous summer. Internally the corm is firm, white, and fleshy; it has a disagreeable odour, and exudes, when cut, a bitter juice that is white and milky from the presence in it of numerous starch grains.

The corms are also cut into thin transverse slices, which are dried at a gentle heat and freed from the remains of the membranes by winnowing. They then form whitish slices about 3 cm. wide, yellowish on their outer surface and reniform in outline, the depression corresponding to the position of the flower. They break readily with a short starchy fracture. The transverse surface exhibits, when smoothed, numerous scattered darker points (fibro-vascular bundles). The drug is inodorous, but has a bitter taste.

The student should observe

(*a*) The *reniform outline of the sliced drug*,
(*b*) The *starchy fracture*,
(*c*) The *bitter taste*.

**Constituents.**—Colchicum corm contains from 0·5 to 0·6 per cent. of the poisonous alkaloid colchicine (compare p. 188). The drug also contains abundance of starch.

**Uses.**—Colchicum is chiefly used to relieve the pain and inflammation and shorten the duration of acute gout and certain gouty affections, but is liable to cause intestinal pain accompanied by vomiting and purging.

## WHITE HELLEBORE RHIZOME

### (Rhizoma Veratri Albi)

**Source, &c.**—White hellebore, *Veratrum album*, Linné (N.O. *Liliaceæ*), is an herbaceous plant with erect perennial rhizome, common in the mountains of central and southern Europe. It produces large, ovate, ribbed leaves and a flowering stem that attains a height of a metre or more. The rhizome appears to have been known and used medicinally for many years, but owing to its powerful and uncertain action it has been employed chiefly as an external application, though to a limited extent only.

The rhizome is dug up in the autumn, and the leaves, which are all radical until a flowering stem is produced, are cut off close to it. It is then usually dried entire, but is sometimes cut longitudinally into halves or quarters to facilitate drying, sometimes deprived of its roots and occasionally sliced transversely. The separation of the roots is to be deprecated, as they appear to be more active than the rhizome. When fresh the rhizome has an alliaceous odour, but this is lost by drying.

**Description.**—White hellebore rhizome (when freed from the roots) averages about 5 cm. in length and 2 cm. in thickness, and is of a dull black colour externally. The upper part is nearly cylindrical, but the lower extremity, where the rhizome gradually perishes and rots away as growth progresses, is usually bluntly conical or truncate. It is crowned with a dense leafy mass consisting of the thin, dry remains of numerous concentrically arranged leaf-bases which have been cut off level close to the rhizome ; the outer of these are coarsely fibrous, the parenchymatous tissue having perished, leaving the veins persistent. The surface of the rhizome is rough and wrinkled and shows encircling scars, in the centre of which the slender wood is distinctly visible. In the untrimmed rhizome the roots are very numerous and stout : they completely envelop the rhizome, so that the untrimmed drug is much more bulky than the rhizome alone. They are usually dull grey or yellowish in colour, and commonly show a disposition to shrivel longitudinally rather than transversely.

The rhizome frequently branches, two or even three branches springing from the same rhizome. This is caused by the production of a flowering stem ; the main axis being thus terminated, the growth of the rhizome is continued by the development of one or more of the buds that are situated in the axils of the inner radical leaves. Such a branch may grow for several years before it flowers.

The drug breaks with a short fracture, the interior being whitish, firm, compact, and starchy. The cortex is about 3 mm. thick, and separated by a wavy, brownish line (endodermis) from the stele, the latter being irregularly traversed in a characteristic manner by yellowish fibro-vascular bundles. The cortex also exhibits occasional bundles (leaf-traces), and gives off here and there a root.

The drug has a bitterish, acrid taste but little odour ; the powder is strongly sternutatory.

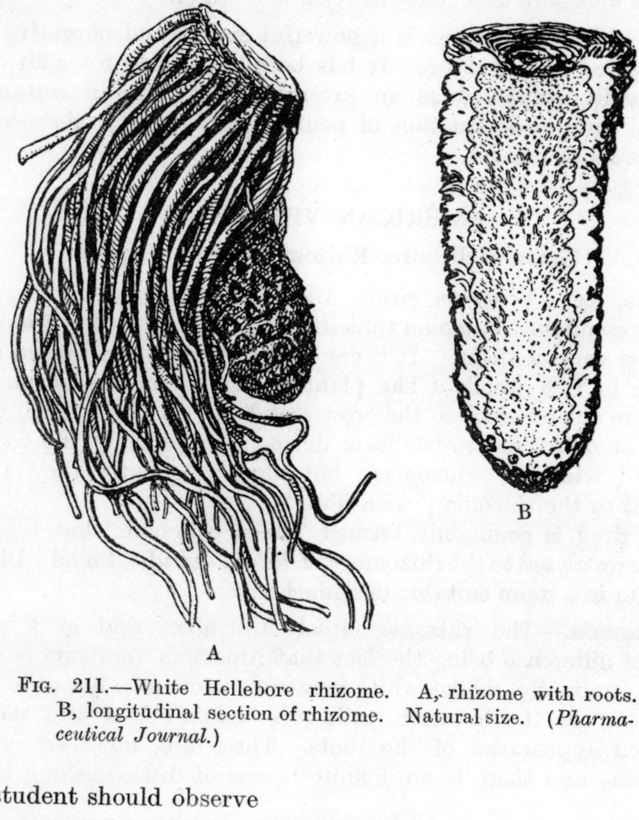

FIG. 211.—White Hellebore rhizome. A, rhizome with roots. B, longitudinal section of rhizome. Natural size. (*Pharmaceutical Journal.*)

The student should observe

(*a*) That the rhizome is *usually entire*,
(*b*) That the *roots are greyish* in colour, and *not often much shrivelled transversely*,
(*c*) The characters of the *transverse section* ;

and should compare the drug with American veratrum (see below).

**Constituents.**—White hellebore contains several alkaloids, amounting in all to 0·5 to 1·0 per cent. The most important and the most toxic of these is protoveratrine, $C_{32}H_{51}NO_{11}$ (0·03 per cent.), which is obtainable in small crystals melting at 245° to 250°. It closely resembles aconitine in its action, and is in addition a powerful sternutatory. Jervine, $C_{26}H_{37}NO_3,2H_2O$, is also crystalline and toxic, but it is less active than protoveratrine. Rubijervine and pseudojervine are also present, but are said to be inactive. Whether protoveratridine occurs preformed in the drug is doubtful. The roots appear to be somewhat richer in alkaloid than the rhizome (Bredemann, 1906), and should not therefore be discarded.

White hellebore also contains resin and starch.

**Uses.**—White hellebore is a powerful emetic and purgative when administered in full doses. It has been prescribed for gout, but is now usually employed as an external application in certain skin diseases, for the destruction of pediculi and other noxious vermin, and as a moth-powder.

## AMERICAN VERATRUM

### (Green Hellebore, Rhizoma Veratri Viridis)

**Source, &c.**—*Veratrum viride*, Aiton (N.O. *Liliaceæ*), is a plant so closely resembling *V. album* that its claim to be considered a distinct species is very doubtful. It is common in the eastern United States, growing in rich woods. The plant is dug up in the autumn, the leaves are cut off close to the crown, and the rhizome is then usually halved or quartered to facilitate drying; occasionally the roots are cut off ('trimmed' rhizomes), but more frequently they are left attached to the rhizome ('with fibre').

The drug is commonly termed 'green hellebore,' but this name is better restricted to the rhizome of *Helleborus viridis*, Linné; American veratrum is a more suitable designation.

**Description.**—The rhizome closely resembles that of *V. album*, the chief difference being the fact that American veratrum is usually cut longitudinally, whilst white veratrum is commonly entire; other characters are the brighter, yellowish brown colour and the more shrivelled appearance of the roots. These are, however, variable characters, and there is no definite means of distinguishing the two drugs.

**Constituents.**—The constituents of *Veratrum viride* are apparently identical with those of *Veratrum album*, with the (doubtful) exception that the former contains an alkaloid—cevadine—that is not found in the latter; they are present in about the same proportion.

**Uses.**—American veratrum has been recommended as a sedative, but is seldom prescribed.

## SWEET FLAG RHIZOME

### (Rhizoma Acori Calami)

**Source, &c.**—The sweet flag or sweet sedge, *Acŏrus Calămus*, Linné (N.O. *Aroidĕæ*), is a native of Eastern Europe and Central

Asia, but has become widely diffused by cultivation. It has established itself in England as a wild plant on the edges of lakes and streams. In dry summers large quantities are collected in East Prussia and Galicia.

The rhizome has long been esteemed as a most valuable medicine in India, whence probably its use spread to Europe.

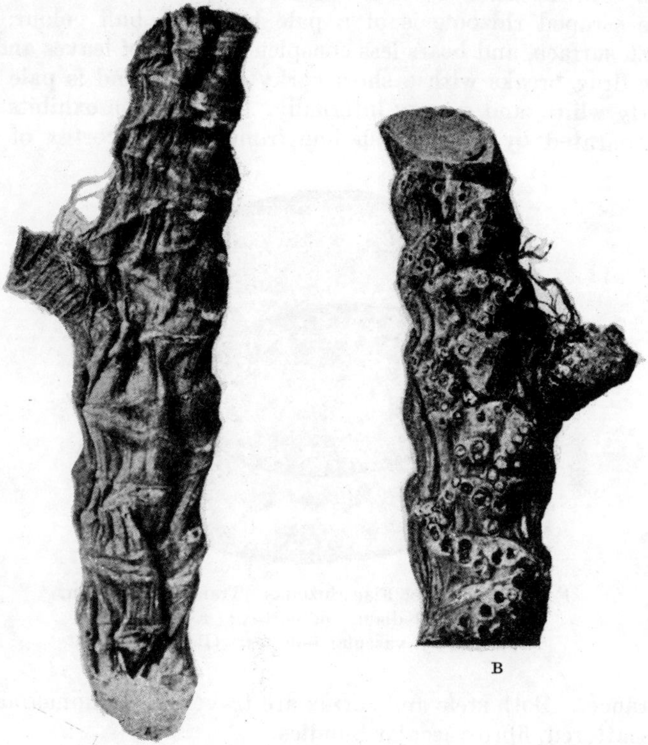

A B

FIG. 212.—Sweet Flag rhizome. A, upper surface, showing leaf-scars. B, under surface, showing root-scars. Natural size.

The long, creeping, horizontal rhizome is collected in the autumn, trimmed, cut into pieces 10 cm. or more in length, and dried. Sometimes it is partially deprived of the outer cork layer by peeling or scraping; but this is inadvisable, as the peeled rhizomes yield less volatile oil than the unpeeled.

**Description.**—The rhizome commonly occurs in pieces varying from 5 to 15 cm. or more in length and from 1 to 2 cm. in thickness. They are covered with a thin brownish cork and are much shrunken bearing deep longitudinal wrinkles. They are marked on the upper surface with large triangular leaf-scars that encircle the rhizome, springing from each side alternately. To these scars the fibrous leaf-traces are sometimes attached. The under surface bears an irregular zigzag line of small raised root-scars that are circular and exhibit a central stele surrounded by a narrow cortex. The rhizome produces an occasional slender lateral branch which is distinctly marked with the scars of leaves and roots.

The scraped rhizome is of a pale brownish buff colour, has a roughish surface, and bears less conspicuous scars of leaves and roots.

The drug breaks with a short corky fracture, and is pale brown, or nearly white, and spongy internally. The section exhibits a large stele separated by a yellowish line from a thick cortex of similar

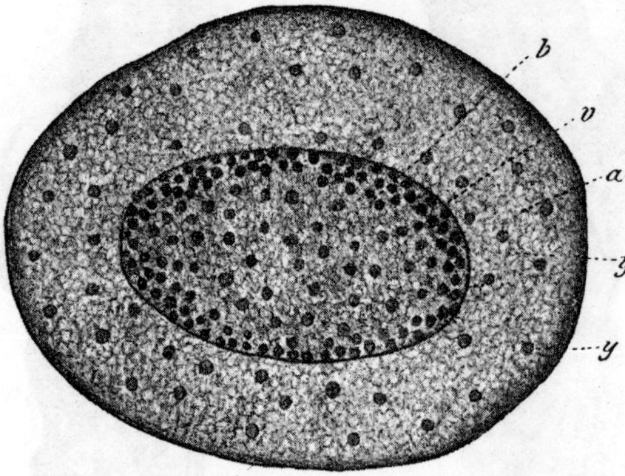


FIG. 213.—Sweet Flag rhizome. Transverse section, magnified 3 diam. *a*, cortex; *v*, endodermis *b*, stele; *y*, vascular bundles. (Berg.)

appearance. Both stele and cortex are traversed by numerous small, oval, scattered, fibro-vascular bundles.

The freshly fractured surface has an agreeable aromatic odour. The taste is disagreeably bitter and pungent.

The student should observe

(*a*) The *large triangular leaf-scars*,
(*b*) The *zigzag line of root-scars*,
(*c*) The *spongy texture*,
(*d*) The *bitter*, *pungent taste*.

**Constituents.**—Sweet flag rhizome contains from 1·5 to 3·5 per cent. of an aromatic volatile oil, the chief aromatic constituent of which is asaryl aldehyde. It contains also an amorphous bitter principle, acorin, yielding by oxidation acoretin and by treatment with acids or alkalies volatile oil and sugar. Certain alkaloidal substances (choline, trimethylamine) appear also to be present, but they require further investigation. The drug contains an abundance of starch and a little tannin.

**Uses.**—The drug has stimulant and tonic properties; it has been used for ague and for atonic dyspepsia.

## COUCH GRASS RHIZOME

### (Rhizoma Tritici, Radiz Graminis)

**Source, &c.**—The couch grass, *Agropȳron repens*, Beauvais (*Triticum repens*, Linné, N.O. *Gramineæ*), is abundant in this country, being often a troublesome weed. It produces a slender rhizome, running for a considerable distance just below the surface of the ground, giving off lateral branches, and at the nodes, which are 2 to 3 cm. apart, small fibrous roots. The rhizome, which appears to have been used by the Greeks and Romans for certain affections of the bladder, is collected, cut into pieces about 1 cm. long, and dried.

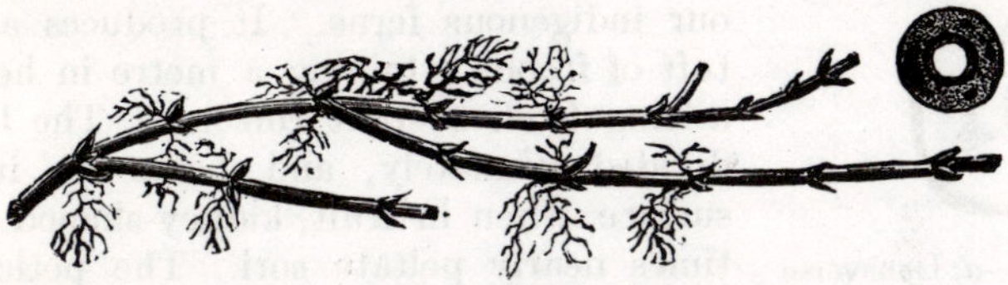

FIG. 214.—Couch Grass rhizome and transverse section; the latter magnified 3 diam. (Maisch.)

**Description.**—Couch grass is generally met with in commerce cut into short pieces. The rhizome is very slender, averaging about 2 mm. in diameter, and of a dark straw-yellow colour. The surface is quite glabrous, hard and shining and usually bears five or six rather prominent longitudinal ridges. Some of the pieces bear the nodes, at which there may be found the persistent fibrous remains of a sheathing leaf-basis, and either a few threadlike paler roots or the scars left by them. The transverse section exhibits, under a lens, a narrow pale inner ring (the stele), hollow in the centre, surrounded by a darker translucent cortex.

The drug has but little odour, and a sweetish, mucilaginous taste. The student should observe

(*a*) The *slender hollow rhizome*,

(*b*) The *translucent* (*not starchy*) *cortex*.

**Constituents.**—Couch grass contains a carbohydrate, triticin (5 per cent.), which is not very soluble in water, and yields by hydrolysis levulose; it appears to occur in the rhizomes of other Graminaceous plants, and possibly is widely diffused in the vegetable kingdom.

Mucilage, inosite, and levulose are also constituents of the rhizome, but starch is not present.

**Substitutes.**—Most of the commercial drug is imported from the Continent and frequently consists of the rhizome of *Cynodon Dactylon*, Persoon. This contains abundance of starch which affords a ready means of distinguishing it; the section is white and is blackened by solution of iodine.

**Uses.**—Couch grass has been employed as a diuretic in certain affections of the bladder.

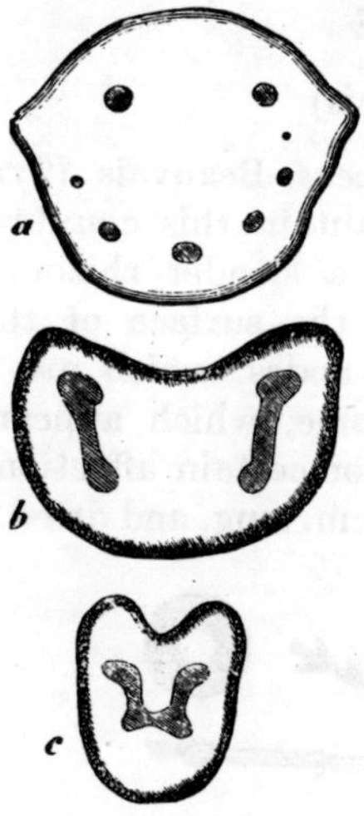


FIG. 215.—*a*, transverse section of base of petiole of *Aspidium Filix-mas*. *b*, transverse section of petiole of *A. Filix-fœmina*; *c*, the same just below the lamina. Slightly magnified. (Luerssen.)

## MALE FERN RHIZOME

### (Rhizoma Filicis Maris, Filix Mas)

**Source, &c.**—The male fern, *Dryopteris Filix-mas*, Schott (N.O. *Filicinĕæ*), is abundant in Great Britain, and one of the commonest of our indigenous ferns. It produces a circular tuft of fronds attaining a metre in height and arising from a stout rhizome. The lamina is divided pinnately, and bears on its under surface, when in fruit, kidney-shaped or sometimes nearly peltate sori. The petiole bears numerous brown, scarious scales, especially in the lower part.

The rhizome is collected in the autumn; the fronds are cut off, leaving the lower swollen portion, about 3 cm. in length, attached to the rhizome, the roots are removed, and the drug is then dried. Sometimes the rhizome is cut longitudinally to facilitate the drying. Much is imported from Germany, large quantities being collected in the Harz and Thuringian Mountains.

**Description.**—The drug occurs in pieces up to about 12 or 15 cm. in length, but usually shorter, and 4 cm. in thickness; only about one-half of this is due to the rhizome itself, the rest being formed by the bases of the petioles that have been left attached to it. These, which constitute the major part of the drug, are usually from 2 to 3 cm. in length, from 5 to 15 mm. in thickness, and enlarged near the middle. They are curved, bluntly angular or longitudinally furrowed, and of a dark brown or nearly black colour externally. They are more or less densely covered with dry, brown membranous scales. Examined under the microscope the marginal cells of these scales will be seen to be prolonged at intervals into simple hairlike processes each consisting of two parallel and contiguous cells, but the scale does not bear any glandular hairs, excepting sometimes two at its base : this is an important diagnostic character of the drug (Lauren,

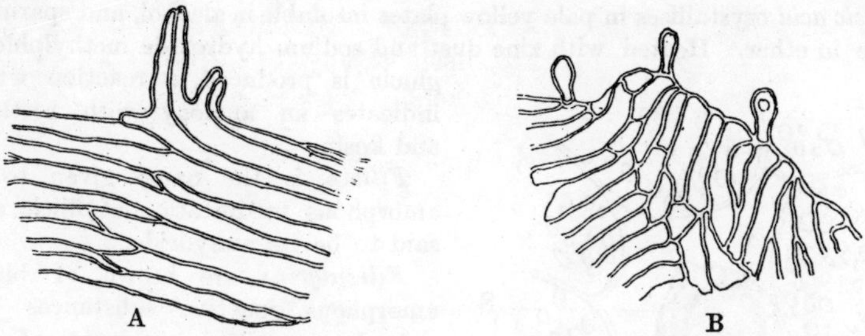


FIG. 216.—A, margin of scale from petiole of *A. Filix-mas*; B, from that of *A. spinulosum*. Magnified. (Lauren.)

1896). When broken the petioles should be (if recently dried) green internally, and the smooth section should exhibit about eight (usually seven, eight, or nine) steles arranged in a diffuse circle.

The rhizome itself should also be green internally, and exhibit in section about as many principal steles as the petiole. The drug has a disagreeable, nauseous, bitter taste.

**Constituents.**—The chief constituent of male fern rhizome is about 5 per cent. of a yellow, amorphous substance of acid nature termed filmarone, to which the vermifuge properties of the drug are apparently to be attributed. Filicic acid, aspidinol, flavaspidic acid, flavaspidinol, and albaspidin are also contained in it, as well as fixed oil (6 per cent.), filicitannic acid (7·8 per cent.), resin, and starch. Some of these constituents are formed by the remarkable secreting cells produced by the rhizome and petioles in intercellular spaces in the parenchymatous tissue. The drug yields about 4 per cent. of ash.

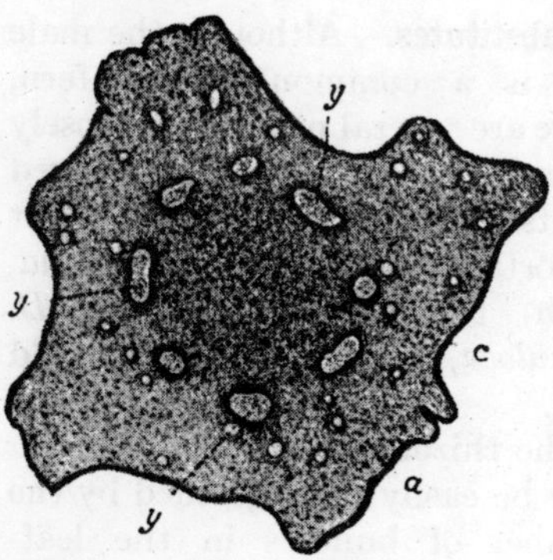


FIG. 217.—Male Fern rhizome. Transverse section, magnified 3 diam. *y*, steles. (Berg.)

*Filmarone*, which is said to be anthelmintic but not toxic, is insoluble in water, sparingly soluble in alcohol and petroleum spirit, but readily soluble in acetone, chloroform, and ether. Dissolved in acetone or alcohol it slowly decomposes into the comparatively inert filicic acid and aspidinol, both of which substances are present in the drug. This decomposition also takes place in the dried drug, as well as in the liquid extract, causing in the latter case, the formation of a crystalline deposit of filicic acid and deterioration of the preparation. Male fern is said to lose its activity when kept; a good sample should show a green and not brown colour in the transverse section.

*Filicic acid* crystallises in pale yellow plates insoluble in alcohol, and sparingly soluble in ether. Heated with zinc dust and sodium hydroxide methylphloroglucin is produced, a reaction which indicates an analogy with rottlerin and kosin.

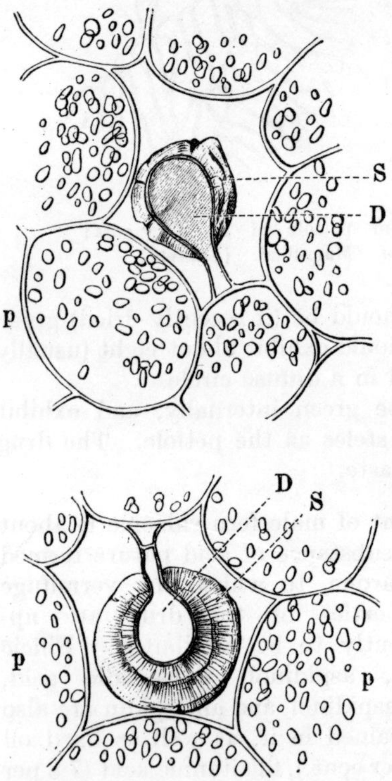


FIG. 218.—Intercellular spaces in the rhizome of Male Fern, showing the secreting glands, D, and secretion, S. p, cortical parenchyma containing starch-grains. Magnified 240 diam. (Vogl.)

*Filicin* is the name given to an amorphous modification of filicic acid said to be its anhydride.

*Filixnigrins* are brown or black, amorphous, inactive substances and are decomposition products of the other constituents of the rhizome.

Although a number of substances have been isolated from male fern rhizome our knowledge of its vermifuge constituent is still far from complete.

**Uses.**—Male fern rhizome has a powerful toxic action on tapeworms, which it kills and expels. In large doses it is an irritant poison.

**Substitutes.**—Although the male fern is a common British fern, there are several others that closely resemble it and might be gathered in its place. Of these the most important are *A. Filix-fœmina*, Roth, the lady fern, and *D. spinulosa*, O. Kuntze, the shield fern.

The rhizome of *A. Filix-fœmina* may be easily distinguished by the number of bundles in the leaf-base, for whilst the male fern has from seven to nine the lady fern has only two large ones. Moreover, the lady fern produces no secreting cells in the parenchyma of rhizome or petiole. It has decided anthelmintic properties, and contains filicic acid and possibly filmarone also. It is seldom, if ever, found in the commercial drug.

The rhizome of *D. spinulosa* is more difficult to distinguish, as it contains secreting cells similar to those of the male fern, and about the same number of bundles. According to Lauren, the character of the margin of the scales borne by the petiole is distinctive, that of *D. Filix-mas* bearing hairlike projections, whilst that of *D. spinulosa* produces glandular secreting cells. The rhizome of *D. spinulosa* is frequently found mixed with the male fern in German commerce (up to 90 per cent). It is an active vermifuge, and is largely used in Finland. It contains aspidin (polystichin), polystichinin, polystichalbin, filicic acid and probably filmarone. The extract prepared from it is said to be twice as active as that from *D. Filix-mas*.

## SUBTERRANEAN ORGANS IN LESS FREQUENT USE

*Blue Cohosh* (Caulophyllum). The rhizome and roots of *Caulophyllum thalictroides* (Linné), Michaux (N.O. *Berberideæ*), an herbaceous plant of the United States. Horizontal, about 10 cm. long and 6 to 12 m. thick, irregular, somewhat tortuous and branched, greyish brown ; on the upper surface, short knotty branches and cup-shaped depressions ; on the under surface, numerous roots about 10 cm. long and 6 to 10 mm. thick. Section whitish with numerous narrow, woody wedges surrounding a large, starchy pith. Taste sweetish, somewhat acrid. Contains an alkaloid, methylcytisine ($C_{12}H_{16}ON_2$ formerly designated caulophylline), similar in action to cytisine, but of only one-tenth the toxicity, two saponins (caulosaponin and caulophyllosaponin), resin, phytosterol, and fat (Power and Salway, 1913). Used as a diuretic, emmenagogue, and vermifuge.

*Leptandra* (*Culver's Root, Black Root*).—The rhizome and roots of *Veronica virginica*, Linné (*Leptandra virginica*, Nuttall, N.O. *Scrophularineæ*), a tall perennial herb abundant in eastern and central North America. Rhizome horizontal or oblique, dark greyish-brown, hard and woody, 10 to 15 cm. long and about 5 mm. thick ; terminated by the remains of the aerial stem and bearing such remains at distant intervals ; on the upper surface small, brown cataphyllary leaves. Roots, numerous, wiry. Transverse section of the rhizome exhibits a narrow, dark cortex, a paler ring of wood of about equal thickness and a large dark pith ; sometimes hollow in the centre ; no odour ; taste, unpleasantly bitter and slightly acrid. Contains 3.4-dimethoxycinnamic acid, mannitol, glucose, verosterol, cinnamic and *p*-methoxycinnamic acids in the form of esters. 'Leptandrin' is an indefinite mixture obtained by pouring a concentrated tincture with water, washing and drying. Used as a cholagogue. The fresh rhizome is violently cathartic and emetic, but these properties disappear on drying and keeping.

## SECTION IX

# STARCHES, GLANDS, HAIRS, &c.

## STARCH

### (Amylum)

**Source, &c.**—The official varieties of starch (Amylum, B.P.) are those obtained from wheat, *Tritĭcum satīvum*, Lamarck, from maize, *Zea Mays*, Linné, and from rice, *Orȳza satīva*, Linné (N.O. *Graminĕæ*). Wheat is largely cultivated in temperate climates, whilst maize and rice are grown chiefly in warmer or subtropical countries.

In preparing the starch from the fruits, it is essential in the first place that the cells containing it should be ruptured in order to liberate the starch grains, and in the second place that the starch grains thus liberated should be separated from other matters, both soluble and insoluble, that accompany them, especially from the nitrogenous gluten which often clings to them pertinaciously. The cells are ruptured by grinding the softened grain to a pulp, and the gluten is removed by one of the following processes, viz. :

(*a*) A mixture of crushed grain and water is allowed to undergo putrefactive decomposition ; the gluten is decomposed, lactic, acetic, and other acids being produced and removed by washing.

(*b*) The crushed grain is mixed with a dilute solution of caustic soda by which the gluten is dissolved.

(*c*) The grain is crushed and mixed with water to a dough from which the starch grains are washed by kneading it in a stream of water, leaving a mass of gluten behind.

The starch is purified by washing, straining and allowing the milky liquid containing starch grains in suspension to clear, by which a more or less complete separation into pure starch and starch mixed with varying amounts of cell-débris is effected. It is finally dried, during which the moist mass gradually splits up into angular fragments. These are then ground to form starch powder.

**Description.**—From whatever source starch is obtained it forms either a fine white powder, or angular or columnar masses easily reducible to powder. It is inodorous and quite insoluble in water, to which it should impart not more than a faintly acid (wheat and

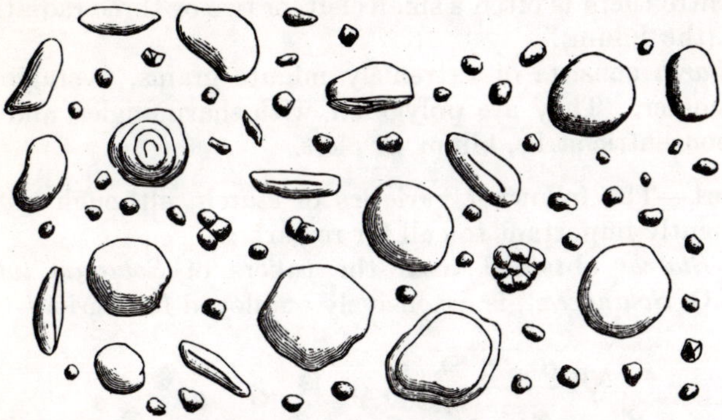

FIG. 219.—Wheat Starch. Magnified 300 diam. (Tschirch.)

potato starches) or faintly alkaline (maize and rice starches) reaction. It leaves, when incinerated, only traces of ash. Boiled with water and cooled it gives a cloudy, more or less gelatinous mixture, which is coloured deep blue by solution of iodine. Air-dry starch usually retains from 12 to 16 per cent. of moisture.

FIG. 220.—Maize Starch. Magnified 300 diam. (Tschirch.)

Although the different varieties of starch show certain differences in the temperature at which they gelatinise, the chief means of distinguishing them and ascertaining their freedom from admixture lies in the examination with the microscope.

*Wheat starch* consists of large and small grains mixed together, with few intermediate in size; the former are lenticular in shape and sometimes marked with faint concentric rings. The hilum is central but not conspicuous. The larger of the large grains average about 30 to

38 microns [1] ($\mu$) in diameter, the smaller of the large grains 15 to 25, whilst the small grains average 6 to 7.

*Maize Starch* consists of grains that are nearly uniform in size and rather smaller than the large grains of wheat starch ; (mostly $12\mu$ to $18\mu$ ; they are polygonal with blunt angles, or more or less rounded. In the centre there is often a small cleft, or two or three radiating from a centre (the hilum).

*Rice Starch* consists of extremely minute grains, averaging about $6\mu$ in diameter. They are polygonal, with sharp angles, and without evident concentric striæ, hilum, or cleft.

**Varieties.**—The following varieties of starch, although not official are sufficiently important to call for remark :—

*Potato Starch*, obtained from the tubers of *Solanum tuberosum*, Linné (N.O. *Solanaceæ*), is extensively employed for various technical

FIG. 221.—Rice Starch. Magnified 300 diam. (Tschirch.)

purposes and has been frequently used to adulterate maranta starch and certain drugs.

The grains are much larger than those of any of the official starches, varying mostly from $45\mu$ to $75\mu$ in length and from $45\mu$ to $60\mu$ in breadth; larger grains may occasionally be found and smaller ones are by no means uncommon. They have usually a flattened ovoid shape and possess a small hilum that is often situated near the *narrow* extremity of the grain, whilst the broad end frequently exhibits a sinuate outline; the concentric striæ are well marked.

*Maranta Starch*, obtained from the rhizomes of *Maranta arundinācea*, Linné (N.O. *Marantaceæ*), is known in commerce as St. Vincent, Bermuda, and Natal arrowroot, or simply as arrowroot, but as the latter term is applied to a number of other starches, it is desirable that this, the arrowroot of English commerce, should be specified as maranta starch. Thus the starch of *Curcŭma angustifolia*, Roxburgh, and *C. leucorhiza*, Roxburgh (N.O. *Scitaminĕæ*), is known

[1] A micron is the one-thousandth part of a millimetre.

as East Indian arrowroot; that of *Manĭhot utilissima*, Pohl (N.O. *Euphorbiaceæ*), and that of *Ipomœa Batātas*, Choisy (N.O. *Convolvulaceæ*), as Brazilian arrowroot; that of *Canna edŭlis*, Edwards,

FIG. 222. Potato Starch. Magnified 300 diam. (Tschirch.)

and other species of *Canna* (N.O. *Scitaminĕæ*) as Queensland arrowroot, &c.

The grains of maranta starch resemble in general appearance those of potato starch, but they are as a rule decidedly smaller,

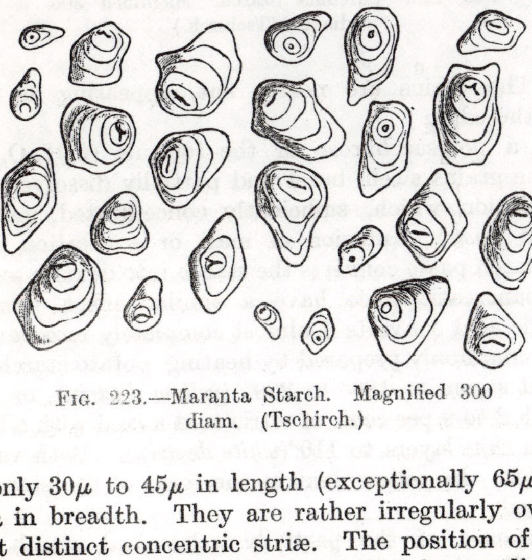

FIG. 223.—Maranta Starch. Magnified 300 diam. (Tschirch.)

averaging only 30$\mu$ to 45$\mu$ in length (exceptionally 65$\mu$ to 75$\mu$) and 24$\mu$ to 30$\mu$ in breadth. They are rather irregularly ovoid in shape and exhibit distinct concentric striæ. The position of the hilum is usually indicated by a cleft or by three or four radiating fissures, and in the majority of grains is near the broad end. The size of the grain, the character and position of the hilum sufficiently distinguish this starch from potato starch, which it otherwise resembles.

*Curcuma starch*, from the rhizomes of *C. angustifolia*, Roxburgh, and *C. leucorhiza*, Roxburgh (N.O. *Scitamineæ*), is known as East Indian arrowroot and is largely used in India, although it does not find its way, to any great extent, to this country.

The grains average about $36\mu$ to $60\mu$ in length, although both larger and smaller ones may be found. They vary considerably in shape, but typical ones are of an elongated ovate outline. The hilum is situated in a little projection from the narrow end of the grain, and is therefore remarkably eccentric ; it is often indistinct, but its position can be found by following the concentric striæ which are usually

FIG. 224.—Curcuma Starch. Magnified 300 diam. (Tschirch.)

discernible. The grains are nearly flat, appearing as rods when viewed from the edge.

Starch is a polysaccharose of the formula $(C_6H_{10}O_5)_n$. Boiled with water the grains swell, burst and partially dissolve, forming an opalescent solution which, sufficiently concentrated, gelatinises on cooling (*starch paste*). Infusion of malt or a solution of diastase acting upon starch paste converts the starch into dextrin and maltose. Saliva and pancreatic juice have a similar effect. Boiling with dilute sulphuric acid converts it almost completely into dextrose.

Dextrin is commonly prepared by heating potato starch by means of superheated steam to 180° to 200° (*yellow dextrin*), or by mixing the starch with 2 to 9 per cent. of nitric acid mixed with a little water and heating in thin layers to 110° (*white dextrin*). Both varieties are soluble in water. In yellow dextrin the form of the starch grain is but little altered.

*Tapioca* consists of the partially gelatinised starch from the rhizomes of *Manihot utilissima*, Pohl (N.O. *Euphorbiaceæ*) ; the starch is either raked on heated plates (*flake tapioca*) or pressed through a sieve and heated (*seed, pearl, bullet pearl tapioca*).

*Sago* is similarly prepared from the starch from the stem of *Metroxylon Rumphii*, Martius, or *M. Sagu*, Rottboell, (N.O. *Palmæ*). Substitutes may be prepared from other starches, e.g. potato, &c.

## LYCOPODIUM

### (Lycopodium)

**Source, &c.**—Lycopodium consists of the spores of the common clubmoss, *Lycopōdium clavātum*, Linné (N.O. *Lycopodiaceæ*), and probably other species. The clubmoss, with its long creeping stems and ascending branches covered with narrow moss-like leaves, is distributed over Great Britain and Europe generally. It produces fruiting branches covered with small scaly leaves and resembling a slender elongated cone. On the upper surface of these leaves (sporophylls) near the base are small receptacles (sporangia) filled with minute spores. These spores form the drug; they are collected by shaking the fruiting branches over a cloth in which the lycopodium collects as a fine yellow powder which is freed from extraneous matter by passing it through a fine sieve and is then packed in sacks, which are usually covered with matting. It is collected chiefly in Russia.

**Description.**—Lycopodium appears as a pale yellow, fine, very mobile powder that floats when thrown on to the surface of water. Blown into a flame it burns instantly with a brilliant flash, but when incinerated in a crucible it is slowly consumed, leaving only about 4 per cent. of ash.

Examined under a microscope lycopodium is seen to be composed entirely of minute spores, each of which is about $25\mu$ in diameter and has the shape of a triangular pyramid with a convex base. The entire surface of the spore is covered by a delicate network of projecting ridges. When crushed the spores burst and a drop of yellowish oil exudes from each. These characters are so well marked that sophistication is readily detected.

**Constituents.**—The spores contain about half their weight of fixed oil, which, however, does not make itself evident until the cellular membrane in which it is enclosed is either broken or destroyed. The oil consists principally of lycopodium-oleic acid (80 per cent.) and myristic acid (about 2 per cent.) combined with glycerin. Lycopodium-oleic acid resembles ordinary oleic acid but is not identical with it. Phytosterin, sugar, and traces of an alkaloid are also present in the drug. Pure lycopodium yields about 1 to 1·5 per cent. of ash.

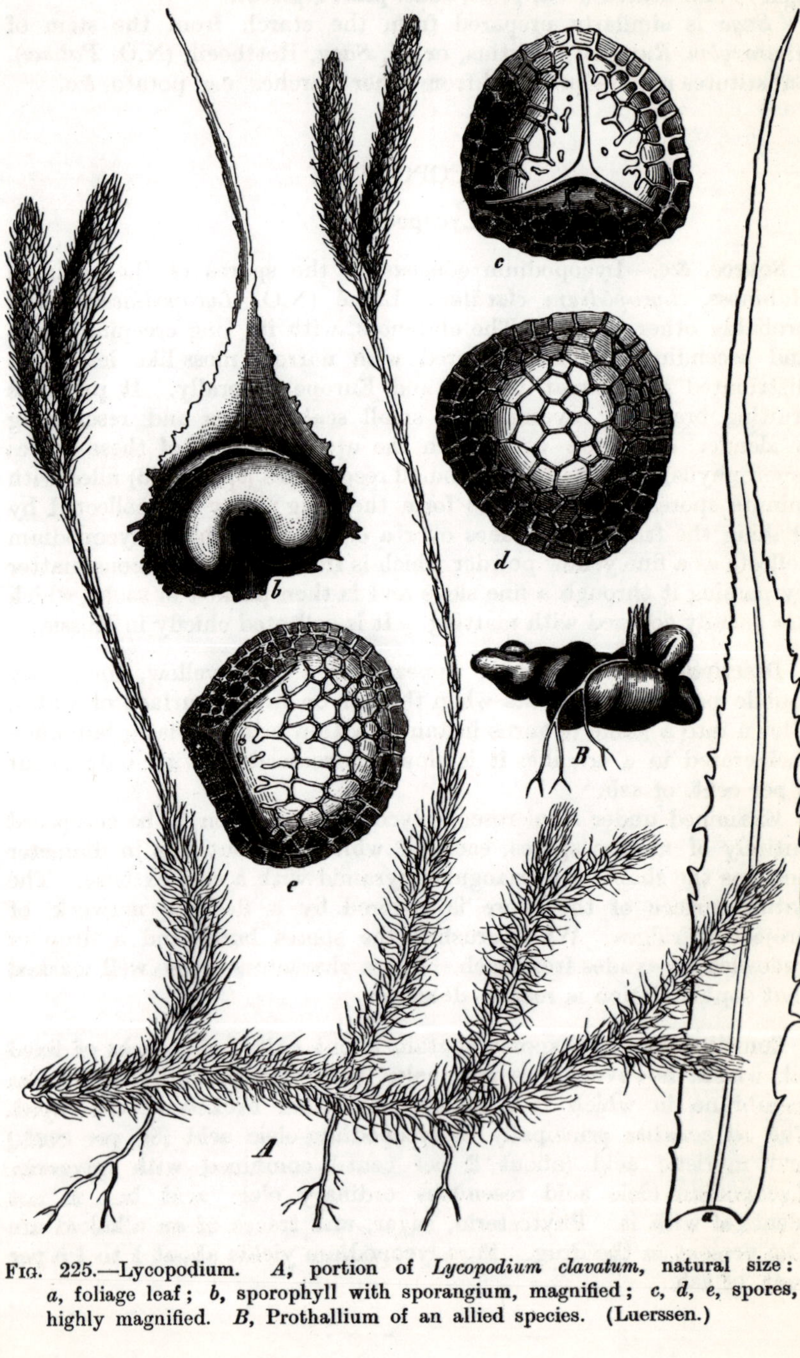


FIG. 225.—Lycopodium. *A*, portion of *Lycopodium clavatum*, natural size: *a*, foliage leaf; *b*, sporophyll with sporangium, magnified; *c*, *d*, *e*, spores, highly magnified. *B*, Prothallium of an allied species. (Luerssen.)

**Uses.**—Lycopodium is sometimes used as a dusting powder for excoriated surfaces and for preventing the mutual adhesion of pills.

**Adulteration.**—Lycopodium is frequently adulterated, the following substances, all of which are easily detected by microscopical examination, having been found:—potato starch, maize starch slightly roasted and coloured, sulphur, powdered colophony, powdered amber, dextrin, powdered boxwood, powdered talc, and pollen of various kinds, especially that of coniferous trees, which has been actually sold (in Austria) under the name of *Lycopodium hungaricum.* Pine pollen consists of ovoid grains bearing an enlargement on each side. The drug should not yield more than 4 per cent. of ash (absence of mineral matter). Sulphur may be detected by its solubility in carbon disulphide.

## LUPULIN

### (Lupulinum)

**Source, &c.**—Lupulin consists of the glands obtained from the strobiles of *Humŭlus Lupŭlus*, Linné (N.O. *Urticaceæ.*)

The cone-like, collective fruits of the hop (see p. 128) are known as strobiles, and consist of leafy stipules and bracts, the latter enfolding at their base minute fruits. Both the bases of the bracts and the fruits (to a less degree the stipules) are sprinkled over with bright shining glands which when fresh have a pale greenish yellow colour, which darkens as the hops are kept. These glands can be separated more or less completely by shaking and beating the ripe hops, and they are also detached during the manipulations to which hops are subjected in gathering and drying, and collect together with sand, debris, and other extraneous matter, on the floors of the hop-kilns. By sweeping the floors and sifting the sweepings, much, if not all, commercial lupulin is obtained.

FIG. 226.—Hop. *a*, strobile of the Hop, natural size; *b*, bract enfolding at its base a small fruit, and showing lupulin glands, natural size; *c*, fruit magnified (Tschirch.)

**Description.**—Commercial lupulin is a granular, brownish yellow powder with a strong, hop-like odour and bitter, aromatic taste. Examined under the microscope it is seen to be composed of a number of glands, each of which consists of a hemispherical layer of cells, the cuticle of the concave surface of which has been raised, dome-like, by the secretion of oil or oleo-resin between it and the cell-wall. When burst by pressure, which is very easily effected, the gland discharges a granular oily liquid.

Lupulin is generally very impure, as indeed it must be if it represents simply the sifted sweepings of the hop-floors, but the glands themselves are easily recognised under the microscope, and as easily distinguished from accidental (sand, débris of the strobiles, &c.) or intentional impurities.

**Constituents.**—Lupulin contains volatile oil, bitter principles, resin, wax, and traces of alkaloids, volatile acids, &c. (compare 'Hops,' p. 127). Pure lupulin yields to ether about 80 per cent. of its weight, and affords about 2·5 per cent. of ash. Commercial lupulin, however, often gives from 10 to 25 per cent. or even more ash, and yields from 40 to 70 per cent to ether. By keeping, lupulin gradually darkens in colour and acquires an unpleasant odour.

**Uses.**—Lupulin is occasionally employed as a stomachic tonic, and also as an hypnotic to promote sleep.

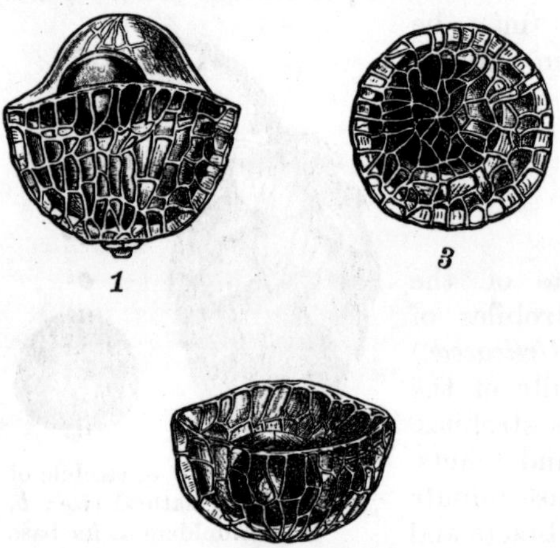

FIG. 227.—Lupulin. 1 and 2, side views; 3, seen from below. Magnified 100 diam. (Vogl.)

## KAMALA

### (Kamala)

**Source, &c.**—Kamala consists of the glands and hairs that cover the fruits of *Mallōtus philippinensis*, Müller Argoviensis (N.O. *Euphorbiaceæ*), a small tree widely distributed throughout India, Ceylon, the Malay Archipelago, Australia, &c.

The drug, which has probably been used in India for many centuries as a dye-stuff, was known to the Arabian physicians of the tenth century, and at the present time still retains in the Indian bazaars its Arabic name, *wars*. It was introducd into European medicine as a vermifuge about 1858. It is collected chiefly in Orissa (south-west of Calcutta), Bengal, and Bombay.

The tree produces three-celled capsular fruits about the size of a large pea, and more or less completely covered with a red powder. These fruits are gathered, dried, and thrown into a basket, where they are shaken and rubbed with the hands; the red powder covering them is detached, and, falling through the basket, is caught on a cloth placed beneath it. This powder, which consists of the stalked glands and stellate hairs from the surface of the fruits, constitutes the drug.

**Description.**—Kamala is a fine, granular, mobile powder of a dull red or madder colour, without odour and almost tasteless, floating when thrown on to the surface of water. Alcohol, ether, chloroform, and caustic alkalies are coloured deep red by it, but water has little action on it. That it is not a homogeneous powder can easily be seen by gently shaking it, when a greyish portion (hairs) will aggregate on the surface.

FIG. 228.—Fruits of *Mallotus philippinensis.* Natural size. (Vogl.)

Examined with a microscope, after the removal of most of the colouring matter by solution of potash, kamala will be seen to consist of glands and hairs. The former, which are much smaller than lupulin glands, are of a depressed globular shape; they are filled with a deep red resin, and contain a number of club-shaped secreting cells radiating from a common centre. The hairs are thick-walled, curved, and usually arranged in small groups.

**Constituents.**—The most important constituent of kamala is rottlerin (from *Rottlera*, a former generic name of the tree), $C_{33}H_{30}O_9$, which crystallises in thin salmon-coloured plates. By treatment with hot caustic alkalies rottlerin yields methylphloroglucin; by reduction with caustic soda

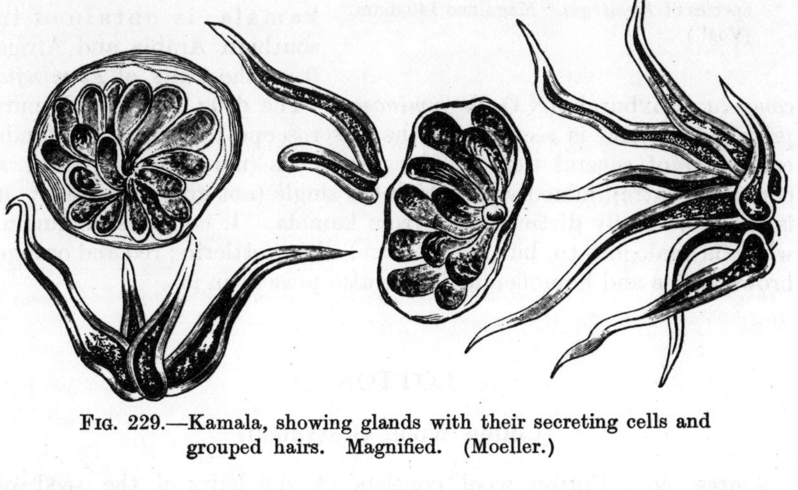

FIG. 229.—Kamala, showing glands with their secreting cells and grouped hairs. Magnified. (Moeller.)

and zinc dust dimethylphloroglucin is produced. The same substances may be obtained from kosotoxin and also from filmarone by similar means, thus showing a remarkable analogy between these three vermifuge substances, all of them being derivatives of phloroglucin. The drug also contains a yellow crystalline substance, a red and a yellow resin, and wax. Isorottlerin (Perkin) appears to be impure rottlerin.

If quite pure, kamala yields about 1·5 per cent. of ash, but this amount is usually exceeded by the commercial drug, even when of good quality, from which from 3 to 5 or even 10 per cent. may be obtained.

**Adulteration, &c.**—Kamala is often grossly adulterated with ferric oxide or with a ferruginous sand, or with brick dust, inferior qualities of the drug yielding 50 per cent. or even more ash. Its quality may be roughly judged by throwing a little on to the surface of water; kamala will float, but most adulterants will sink. Substitutes for kamala consisting of ground safflower (florets of *Carthămus tinctōrius*, Linné), dyed starch, &c., have been observed, but are easily detected by the microscope.

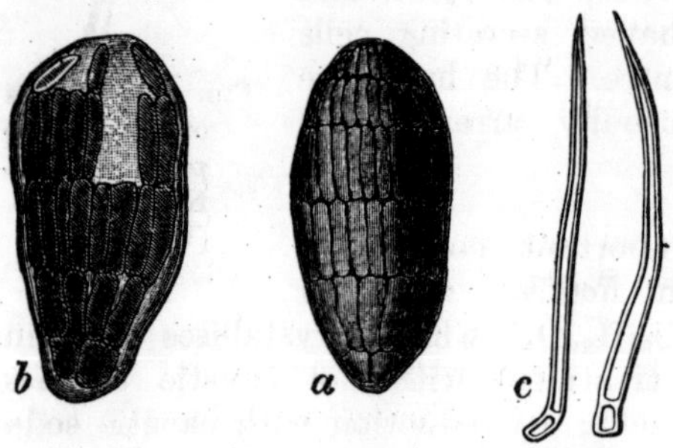


FIG. 230.—Glands (*a* and *b*) and hairs (*c*) of a species of *Flemingia*. Magnified 140 diam. (Vogl.)

**Uses.**—Kamala is an efficacious remedy for tapeworm, expelling the worm and producing free purgation.

**Substitute.**—*True wars* or *wurus*, a drug analogous to kamala, is obtained in southern Arabia and Africa from the fruits of *Flemingĭa congesta*, Roxburgh (N.O. *Leguminosæ*). The drug has a dull purplish colour, and is seen under the miscroscope to consist of glands composed of several tiers of elongated cells (not radiating from a common centre), mixed with which are single (not grouped) hairs; it is therefore easily distinguished from kamala. It contains flemingin, which is analogous to, but not identical with, rottlerin; red and orange brown resins and homoflemingin are also present in it.

## COTTON

### (Cotton Wool, Gossypium)

**Source, &c.**—Cotton wool consists of the hairs of the seed of *Gossypĭum barbădense*, Linné (N.O. *Malvaceæ*), and other species of *Gossypium*.

The cotton plants, of which many varieties exist, are either herbs, shrubs, or small trees. They produce three to five-celled capsular fruits containing numerous seeds covered with a woolly mass of long white or yellowish hairs. These are separated from the seeds and freed from impurities by machinery specially designed for that purpose.

Each of the hairs thus separated is covered with a thin cuticle by which certain waxy and fatty substances are secreted; the presence of these is objectionable, as they repel moisture, and consequently the wool will only slowly absorb watery fluids. To remove them, the cotton wool is usually boiled under pressure with a dilute caustic alkali, after which it is washed, bleached by chlorinated lime and hydrochloric acid, again washed and then dried; finally the fibres are loosened by machinery, and separated by a current of air, from which they are collected as a fleecy wool. The wool thus freed from the wax and fat naturally present in it possesses much better absorbent properties and is in most cases more suitable as a surgical dressing (absorbent wool).

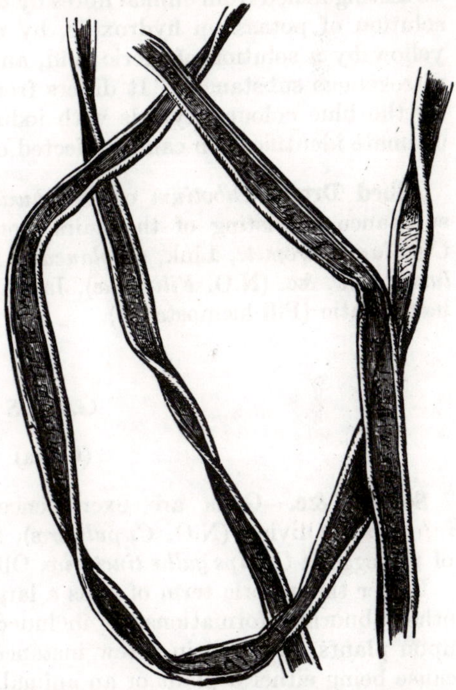

FIG. 231.—Cotton. Magnified. (Tschirch.)

**Description.**—Each of the soft white filaments of which cotton wool consists is a single hair from the surface of the seed. They attain as much as 5 cm. in length, and appear, when examined under the microscope, as flattened, twisted bands with slightly thickened rounded edges. Cotton wool should readily sink in water, showing that the waxy coating with which it is naturally provided has been removed, as directed by the Pharmacopœia; it should not, however, communicate to water either an acid or an alkaline reaction, as might be the case if it had not been completely freed from the acids and alkalies commonly used to remove the wax. Cotton wool is inodorous, tasteless and insoluble in water but almost completely soluble in an ammoniacal solution of cupric oxide, the cuticle remaining undissolved. Solutions of iodine colour cotton wool yellow, which is changed to deep blue by sulphuric acid. It burns easily, leaving less than 1 per cent. of ash.

**Constituents.**—Cotton wool consists principally of cellulose $(C_6H_{10}O_5)_n$, associated with traces of inorganic matter, albuminoids, &c. The fatty matter is partly a wax soluble in alcohol and ether, and melting at 86°, partly a mixture (apparently) of stearic and palmitic acids.

Cotton wool is best identified by microscopic examination. It may be distinguished from animal fibres by being insoluble in hot 5 per cent. solution of potassium hydroxide, by not being stained permanently yellow by a solution of picric acid, and by containing only traces of nitrogenous substances. It differs from many other vegetable fibres by the blue colour it yields with iodine and sulphuric acid, but its ultimate identification can be effected only by the microscope.

**Allied Drug.**—*Cibotium* or *Penghwar Djambi* is a brownish woolly substance consisting of the hairs from the bases of the fronds of *Cibotium Barometz*, Link, *C. glaucum*, Hooker and Arnold, *Alsophila lurida*, Bl., &c. (N.O. *Filicineæ*), Java, Sumatra, &c.; it is used as a hæmostatic (Pili hæmostatici).

## GALLS

### (Galla)

**Source, &c.**—Galls are excrescences on the twigs of *Quercus infectoria*, Olivier (N.O. *Cupuliferæ*), resulting from the deposition of the eggs of *Cynips gallæ tinctoriæ*, Olivier.

Under the generic term of galls a large variety of excrescences and other abnormal formations are included, that are produced not only upon plants, but also in a few instances upon animals, the exciting cause being either a plant or an animal. The plants that induce the formation of galls are exclusively fungi, but a variety of animals are capable of producing them, the principal being species of *Cynips* and *Aphis*. Thus the true oak-apple, the bedeguar of the rose, the oak gall, oak spangles, &c., are all varieties of galls produced by various insects.

The official drug is a particular variety of gall produced by a particular insect upon a particular plant. This variety of gall is known commercially as 'Aleppo' galls or 'Turkey' galls; they are collected in Asiatic Turkey, especially in the province of Aleppo.

The wasp that produces these galls is *Cynips gallæ tinctoriæ*, Olivier. Of this insect the female only is known, reproduction taking place parthenogenetically. By means of her ovipositor the wasp deposits an egg between the rudimentary leaves near the growing-point of young shoots of the oak. The larva emerges from the egg and wounds the delicate tissue with its mandibles, simultaneously introducing into the tissue a secretion that stimulates a rapid development of tissue. The new tissue thus formed assumes an concentric arrangement and within this the larva lives, feeding upon starch produced by the cells. Arrived at maturity the larva passes into the pupa stage, from which the wasp emerges and, boring through the gall with its mandibles, escapes. The gall is therefore probably to be regarded as a metamorphosed shoot. Very remarkable is the fact that no development of tissue takes place until a glandular secretion passes from the mandibles of the larva into the surrounding tissue, which is always merismatic in nature; this formation of tissue continues as long as the exciting substance is supplied, but should the larva perish it at once ceases and the growth of the gall is arrested.

Galls should be collected preferably before the insect escapes; after that has happened they become lighter in weight and are less esteemed.

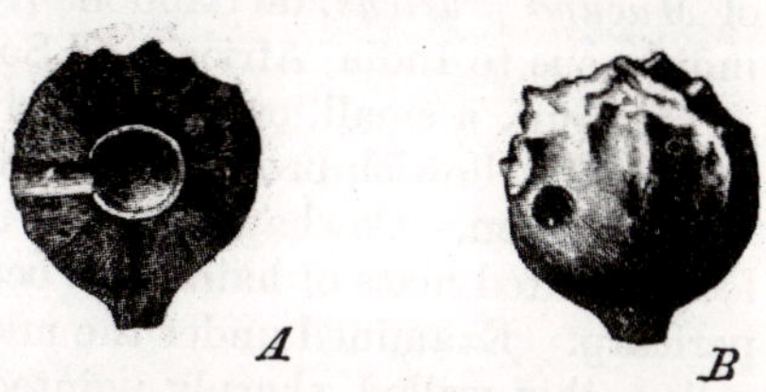

FIG. 232.—Galls. *A*, showing interior; *B*, exterior, both with hole bored by the insect. (Vogl.)

**Description.**—Aleppo galls are nearly spherical in shape and vary from 12 to 20 mm. in diameter. They are hard and heavy, and bear, especially in the upper portion, short, bluntly pointed projections. They are of a bluish green or olive green colour externally, yellowish or brownish white within. There is usually a small cavity in the centre, in which the remains of the larva or of the gall-wasp may be found. They have no odour, but an intensely astringent taste followed by a slight sweetness.

**Constituents.**—Galls contain as principal constituent from 50 to 70 per cent. of tannic acid, which, to distinguish it from other varieties of tannic acid, is termed gallotannic acid. They contain also a little gallic acid (2 to 4 per cent.), ellagic acid, cyclogallipharic acid, sugar and starch.

Gallotannic acid, the 'tannic acid' of the British Pharmacopœia, is a pale yellow amorphous substance yielding bluish black precipitates with solutions of ferric salts. Its aqueous solution darkens when exposed to the air with simultaneous formation of gallic acid, $C_7H_6O_5,H_2O$, and sometimes also of ellagic acid, $C_{14}H_6O_8,2H_2O$.

**Uses.**—Galls are used medicinally as a local astringent chiefly in the form of a suppository or ointment. They find an extensive application technically in tanning and dyeing, in the manufacture of ink, &c.

**Varieties.**—*White galls* are the galls collected after the escape of the gall-wasp; they are rather larger than the 'blue' galls, rather lighter in weight, and yellowish in colour. They are less esteemed, and are considered to contain less gallotannic acid, which, however, does not from analyses appear always to be the case.

*Chinese* and *Japanese galls* are produced by *Aphis chinensis*, Bell., on *Rhus semiālata*, Murray (N.O. *Anacardiācěæ*); they are of a very irregularly lobed shape, reddish brown in colour, hollow and covered with a thick, grey, velvety down. They are largely used in the manufacture of gallotannic acid, of which they contain about 70 per cent.

## COWHAGE

### (Mucuna)

**Source, &c.**—Cowhage consists of the hairs attached to the fruit of *Mucuna pruriens*, de Candolle (N.O. *Leguminosæ*), a climbing plant indigenous to India, Africa, and South America.

The fruit, a small, curved, nearly black legume, is densely covered with stiff yellowish brown hairs which form the commercial drug.

**Description.**—Cowhage appears in commerce as a yellowish brown, loosely felted mass of hairs with occasional small black portions of the pericarp. Examined under the microscope it is seen to consist of one-celled, thin-walled, sharply pointed hairs which are 2 or 3 mm. long, and bear numerous short, often recurved prominences.

**Constituents and Uses.**—The hairs contain a little tannin, but their action as a vermifuge is purely mechanical.

# SECTION X

## DRIED LATEX

Latex is the name given to the milky liquid contained in laticiferous tissue. This tissue may take the form of elongated, branching cells (*Apocynaceæ*, *Asclepideæ*, *Urticaceæ*, most *Euphorbiaceæ*) or of superposed cells with intact transverse walls (many *Convolvulaceæ*) or of superposed cells the transverse walls of which have been perforated, thus forming laticiferous 'vessels' (*Papaveraceæ*, many *Compositæ*, *Campanulaceæ*, *Sapotaceæ*, some *Convolvulaceæ* and *Euphorbiaceæ*, &c.). In addition to water, salts, proteids, pectin, gum, fat, wax, enzymes, &c., the latex frequently contains substances of medicinal or technical value, as, for instance, alkaloids (*Papaver*), caoutchouc (*Hevea*, *Ficus*, &c.), guttapercha (*Palaquium*, *Mimusops*), acrid principles (*Euphorbia*), &c. It is readily exuded by the plant on incision and may thus be collected and dried (opium, lactucarium, euphorbium) or further treated for the isolation of one or more of its constituents (caoutchouc, guttapercha).

## OPIUM

**Source, &c.**—Opium consists of the dried latex of the opium poppy, *Papaver somniferum*, Linné (N.O. *Papaveraceæ*). The drug was apparently known in very remote times, as both the Greeks and Romans were well acquainted with it, and with the manner in which it was collected then, as it is now, from the unripe capsule of the opium poppy. The physicians of the Arabian school probably introduced the drug into India as well as into Europe. It was originally used as a medicine, the practice of opium-eating having originated probably in Persia.

Opium is collected principally in Greece, Serbia, Asia Minor, Persia, India, China, and in small quantities in Australia. The cultivation of the opium poppy has been experimentally carried on in France, Germany, and other countries, where, however, the expense of the necessary labour and land has been so great as to render it unprofitable, although opium of very high quality has been obtained. Indeed, the highest percentage of morphine ever observed in the drug (22·88) was found in a sample collected near Amiens, in France.

All parts of the plant yield on incision a white latex, but the unripe capsule is especially rich, and from it alone opium is obtained. The wall of the capsule contains an elaborate branching and anastomosing system of laticiferous vessels that accompany generally the fibrovascular bundles; these vessels are filled with a white, milky secretion, and, as they are in open communication with one another, a considerable area of the system is drained of its latex when the vessels are incised at any particular point.

Although the methods adopted in the different countries for collecting

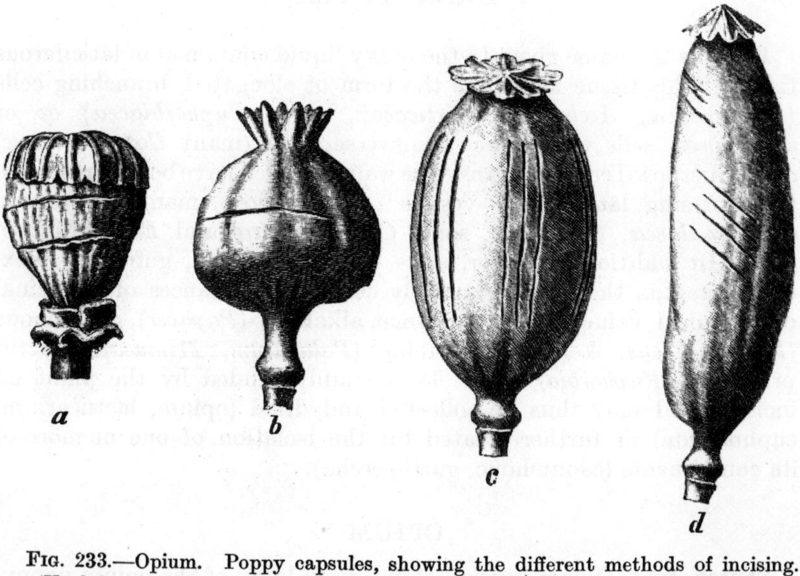

FIG. 233.—Opium. Poppy capsules, showing the different methods of incising. (Vogl, from specimens in the Museum of the Pharmacological Institute, Vienna.)

the opium vary in their minor details, the principal features remain the same; in India alone a particular method of treating the opium is adopted. Soon after the petals have fallen, and whilst the capsule is still unripe, being about 4 cm. in diameter, incisions are made in the wall, great care being taken that they do not penetrate to the interior of the fruit, for then both opium would be lost and the seeds prevented from ripening; the latter have a commercial value, for they contain a fixed oil which they yield when submitted to pressure.

The incisions are sometimes limited to a single transverse one extending round the capsule (Asia Minor); sometimes they are oblique or vertical and two or three together (India); or the method adopted varies in different districts of the same country. Sometimes the capsules are cut once only, sometimes they are subjected after a short lapse of time to a second and third cutting. This is usually done in the afternoon, and on the following morning the latex that has exuded and partially dried is scraped off with a knife. The scrapings are united, and as soon as sufficient has been collected it is formed into a cake which is further dried by exposure to the sun.

These cakes are wrapped in poppy leaves (Turkey; Bulgaria) or paper (Persia), and are then ready for exportation.

**Varieties of Opium.**—1. *Turkey.*—This variety of opium is obtained in both south-eastern Europe and Asia Minor. In the former, Macedonia produces increasing quantities of opium which are exported

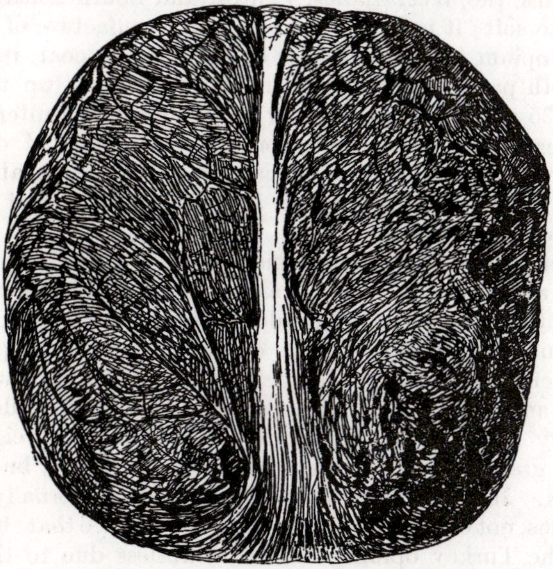

FIG. 234.—Turkey Opium. Slightly reduced. (*Pharmaceutical Journal.*)

from Salonica, and the cultivation is being pushed forward in Bulgaria; in Asia Minor, the central and north-western districts yield the bulk of the drug, which is sent chiefly from Constantinople, Smyrna and Salonica.

Turkey opium occurs in flattened cakes varying commonly from $\frac{1}{4}$ to 1 kilogram in weight, occasionally much larger (5 kilograms). They are covered with a greyish green leaf of the poppy plant, the broad midrib of which is usually conspicuous, and are sprinkled over with the brown, triangular, winged fruits of a species of *Rumex*. These are thrown amongst the cakes by the opium merchants to prevent them from sticking together when packed in cases ('chests') for shipment.

When fresh the drug is plastic; internally, it has a rich reddish brown colour of varying depth, and is coarsely granular or more or less smooth according to the variety (see below). It has a strong characteristic and not unpleasant odour and a bitter taste.

Several subvarieties of Turkey opium are recognised on the market. The most important of these are the Karahissar, Ghiveh, and Boghaditz opiums, which are largely used by pharmacists. Salonica, Tokat, and Malatia opiums are known as 'soft shipping' opiums; as the name indicates, the paste is usually, though not necessarily, soft; they are used by the morphine makers and are also exported to the United States, the West Indies, Central and South America. Yerli opium is also soft; it is largely used for the manufacture of morphine.

Salonica opium is characterised by its rougher coat, its (usually) dark, smooth paste, its very high moisture content (up to, or even exceeding, 35 per cent.), and its very high morphine content (usually 18 to 22 per cent. of the dried opium). 'Druggist's' opium has a smoother coat than the Salonica variety, contains about 22 to 26 per cent. of moisture and about 13 to 16·5 per cent. of morphine, calculated on the dried opium.

2. *Persian Opium* is produced most largely in the neighbourhood of Ispahan and Shiraz in the west and near Meshed in the north-east. After collection the opium is kneaded, often with added gum, into a homogeneous mass, which is then divided into brickshaped masses about 15 cm. long; after further drying these are wrapped in red paper and tied round with red or sometimes yellow string. Occasionally it is formed into bluntly conical masses weighing from 200 to 400 grammes each or short sticks or flat cakes, but these are seldom seen. It is usually of a dark brown colour internally and quite homogeneous, not exhibiting the granular appearance that characterises much of the Turkey opium; this is doubtless due to the method of preparation.

The bulk of the opium normally imported into England consists of the Turkey and Persian drugs; the latter is largely exported to South America and elsewhere for smoking, but when sufficiently cheap is used for the manufacture of morphine.

3. *Indian Opium.*—The chief opium-producing districts in India are the central districts of the Ganges, including Behar, Benares, and Patna, and the tablelands of Malwa which lie to the north-east of Bombay, most of the drug being produced by licensed cultivators and purchased by the Government. That which is destined for use in India is dried in the sun until the moisture in it is reduced to 10 per cent., and then made into flat cakes (Bengal abkari opium) or round balls (Malwa opium) and wrapped in oiled paper. For the Chinese market it is formed into round balls which are enveloped in cases made of dried poppy petals, opium, and water, each cake with its shell resembling a Dutch cheese in size and shape.[1]

[1] Details of the process will be found in the *Pharmacographia Indica.*

During the war large quantities of Indian opium were imported into this country in the form of a uniform, black, paste, in flat cakes of 2 lb. wrapped in waxed paper.

**Constituents.**—Amongst the numerous constituents of opium the most important are the alkaloids it contains. Of these no fewer than twenty-one have been reported, but the existence of some of these in the drug is open to considerable doubt. Chief of them, both in its medicinal importance and in the quantity in which it exists, is morphine; narcotine and codeine are of secondary importance, and next to these come thebaine, narceine, and papaverine.[1]

Morphine, $C_{17}H_{19}NO_3,H_2O$ (or $C_{17}H_{17}NO<^{OH}_{OH}+H_2O$), forms colourless crystals, very slightly soluble in cold water but readily in solutions of caustic alkalies or alkaline earths; it is almost insoluble in cold ether, chloroform, or benzene. Heated with hydrochloric acid in a sealed tube to 140°–150°, it is partly converted into apomorphine ($C_{17}H_{17}NO_2$). Morphine is a powerful hypnotic, but apomorphine, in addition to a hypnotic action, produces a powerful emetic effect.

Codeine, $C_{18}H_{21}NO_3$, or methyl morphine ($C_{17}H_{17}NO_2<^{OH}_{OCH_3}$) is usually obtained from the mother liquors from which morphine has been crystallised, but can also be prepared by methylating morphine. It forms large rhombic crystals soluble in 80 parts of water and readily soluble in chloroform. It is a strong base, and is liberated from its salts by fixed alkalies. but not by ammonia. It has only a mild hypnotic action.

Narcotine, $C_{22}H_{23}NO_7$ or $C_{19}H_{14}(OCH_3)_3NO_4$, crystallises readily in rhombic prisms or needles, and may be extracted from the residue left after treating opium with water; it is a weak base and has little or no narcotic action, hence it has also been called 'anarcotine.'

All these are well-defined, crystalline alkaloids, and together constitute in good dry opium about one-fifth of the weight of the drug.

Amongst the other constituents attention must be directed to meconic acid, $HC_7H_3O_7,3H_2O$, a crystalline organic acid that exists to the extent of about 5 per cent. combined with morphine; this well-characterised and easily identified acid is important in toxicological investigations as corroborative of the presence of opium.

Meconin and meconoiosin are two indifferent substances and exist in small quantity only.

Mucilage, sugar, wax, caoutchouc, and salts of calcium and magnesium are also contained in opium, but starch, tannin, oxalic acid, and fat, all of which are common constituents of plants, are not found in it, and their presence therefore indicates adulteration of the drug.

The proportion of water contained in freshly imported Turkey

[1] For a complete list and a table of the reactions by which they may be distinguished, Henry's *Plant Alkaloids*, p. 199, may be consulted.

opium varies from 7 to 35 per cent. Good Turkey opium yields from 4 to 8 per cent. of ash and from 55 to 65 (usually over 60) per cent. of dry aqueous extract (calculated in each case upon the dry opium).

The quality of opium is, however, determined almost exclusively by the proportion of morphine it contains. Good Turkey opium yields from about 12 to 22 per cent. of anhydrous morphine (calculated upon the dry drug), but is subject, of course, to considerable variation. Persian opium is less rich in morphine, and contains, when of good quality, from 8 to 12 per cent; in exceptional cases 16 and even more per cent. has been recorded. Indian opium varies from 3 to 15 per cent., the bulk ranging from 7·5 to 10 per cent. Efforts have been and are being made to raise the morphine content of this variety. Chinese opium has been shown to contain from 4·3 to 11·2 per cent. The alkaloid certainly exists in combination with acids (probably partly with meconic, partly with sulphuric acid), and is easily and completely extracted by water.

Narcotine varies usually from about 2 to 8 per cent., although the latter figure is occasionally exceeded. It appears to occur in larger proportion in Indian and Persian opium than it does in Turkey, and to exist chiefly in the free state, it being a weak base. Codeine exists in proportions ranging from 0·3 to 4·0 per cent. in combination with acids. Thebaine varies from 0·2 to 0·5 per cent.; narceine from 0·1 to 0·4 per cent. The remaining alkaloids constitute together about 1 per cent. of the drug.

For the preparation of tincture and extract of opium which are standardised to contain 1 per cent. and 20 per cent. of morphine respectively, any suitable variety of opium may be used provided it contains, when dry, not less than 7·5 per cent. of anhydrous morphine, but for other official uses opium must contain between 9·5 and 10·5 per cent. of morphine, and opium containing more than 10 per cent. may be diluted with opium containing between 7·5 and 10 per cent. or with milk sugar. Hence any opium containing over 7·5 per cent. of morphine may be officially used, the strength being adjusted by suitable blending. This permission was apparently given with the view of favouring the importation of the best varieties of Indian opium. Turkey opium, on the other hand, is occasionally diluted to contain about 10 per cent. of morphine before exportation.

**Adulteration.**—The necessarily high price of the drug and its nature invite adulteration. Stones, small shot, pieces of lead, and such substances have been found in opium. Gum, grape must, sugary fruits, &c., are said to be added to it. Persian opium frequently contains added gum and sometimes also mineral matter (clay). The percentage of morphine, the ash, moisture, and residue insoluble in water all give valuable indications, the last-named excluding the sugary pulp of apricots and other fruits. The insoluble residue will contain portions of the outer epidermis of the poppy capsule, but not of the inner (absence of powdered capsules); such fragments are more frequent in Turkey than in Indian opium, the former, obtained usually from a horizontal incision, requiring more scraping than the latter, in which the opium collects at the bottom of a vertical incision. Starch should not be present in appreciable quantity; it is, however, regularly found in Persian opium (Mjoen, 1895), and traces are said to be frequently present in Turkey opium (Cæsar and Loretz, 1898) but not in quantity sufficient to constitute an adulteration.

**Assay.**—Opium may be assayed by the official process.

**Uses.**—Opium is one of the most valuable of drugs. It is unexcelled as an hypnotic and sedative, and is frequently administered to relieve pain and calm excitement. It is also used as an astringent in diarrhœa and dysentery, and as a sedative in certain forms of cough, dyspnœa, &c.

## LACTUCARIUM

**Source, &c.**—Lactucarium consists of the dried latex of *Lactuca virosa*, Linné (N.O. *Compositæ*), and other species of *Lactuca*. *Lactuca virosa* is a native of central and southern Europe, but is cultivated in England.

All the plants belonging to the sub-order *Liguliforæ*, in which the genus *Lactuca* is included, contain a system of laticiferous vessels forming an anastomosing network penetrating to all parts of the plant (compare fig. 177). They are especially numerous in the bast of the stem, but are present also in the pith. Hence when the stem of *L. virosa* is wounded a free exudation of latex takes place in the form of white milky fluid of intensely bitter taste. This latex when dried constitutes the drug. It is collected in Germany near Zell on the Mosel, where most of the commercial drug is produced, by cutting the stem off about a foot below the summit; the latex which exudes is taken off by the finger and transferred to a china cup. It soon coagulates and is then removed from the cup by warming and gently tapping, when it falls out in the shape of a blunt cone which may be dried entire or cut into four pieces and dried. Thin slices are daily taken off the stem and the collection of lactucarium thus proceeds during the summer. The annual production is only about 150 kilograms.

**Description.**—German lactucarium occurs in hard, opaque, irregular, angular pieces, sometimes flat or curved on one or two of their sides, of a dingy brown or reddish brown colour. In the interior they are paler; if quite fresh creamy white and of the consistence of wax;

but the colour changes by exposure to a dull brown, the drug becoming at the same time hard. They possess a strong characteristic odour, different from but recalling that of opium, and a bitter taste.

Lactucarium is tough and difficult to powder. Boiled in water it softens to a plastic mass, but only very little of it is soluble. The cooled and filtered liquid should not be coloured blue by iodine, indicating the absence of starch, which has been found in factitious lactucarium, but, as Composite plants contain no starch, this should not be present in the genuine drug. Lactucarium is only partially soluble in alcohol and ether.

**Constituents.**—It appears doubtful whether lactucarium possesses any particular therapeutic action, although it has been used as a sedative. Lactucin, lactucic acid (crystalline), and lactucopicrin (amorphous) are three bitter principles, the re-examination of which is to be desired. Lactucerin (lactucone), constituting about one-half of the drug, and extracted from it by boiling alcohol, is a crystalline, tasteless, inert, waxy substance, yielding by saponification acetic acid and lactucol (or, according to Hesse, $\alpha$- and $\beta$-lactucerol).

In addition to these constituents, that are apparently peculiar to lactucarium, various other substances commonly found in the latex of plants have been detected, such as caoutchouc, albumen, mannite, certain inorganic substances, &c.

The alkaloid hyoscyamine, which Dymond detected in both wild and cultivated lettuce, especially when the flowering stage is reached, and to which possibly any sedative action of the fresh plant might be due, could not be found in lactucarium.

**Uses.**—Lactucarium has been used as a sedative, but is now not much employed.

## GUTTAPERCHA

**Source, &c.**—Guttapercha is obtained from the latex of several species of *Palaquium* and *Payena* (N.O. *Sapotaceæ*), the principal being *P. oblongifolium*, Burck, *P. borneense*, Burck, *P. Leerii*, Burck, and *P. Treubii*, Burck, all of which are stately trees indigenous to the Malay Archipelago. They contain in the bast, as well as in the cortical parenchyma of the stem, extending into the mesophyll of the leaves, numerous superposed, elongated laticiferous cells which are filled with a granular latex. It is this latex that forms, when properly prepared, guttapercha. To that end the trees with a trunk about 30 cm. in diameter are felled, the branches stripped off, and transverse or oblique channels cut in the bark. Into these the latex is discharged, and in them it coagulates. This coagulation is apparently due to the coagulation of an albumin in the latex by which the separate particles of guttapercha are entangled and retained. The coagulated latex is then scraped from the incision, kneaded under hot water to free it from accidental impurities, beaten with mallets, and finally made into cakes which are bought up by Chinese merchants.

At present Sumatra yields the bulk of commercial guttapercha, but the destructive and wasteful method of collection has led to numerous endeavours to cultivate the trees and obtain from them the guttapercha by a more rational method. These endeavours have met with some success. Burck has shown that by making V-shaped incisions in the living tree large quantities of guttapercha can be obtained, and that the tapping can be continued for three or four years without apparent injury to the tree. It has also been found that the leaves contain more guttapercha than the stem-bark, and that it can be extracted by boiling the dried and crushed leaves with a suitable solvent, such as toluene or petroleum spirit; from the latter solution the pure guttapercha is deposited on cooling (Obach's process). Not only could the trees be preserved by this means, but the stools of trees that have already been felled send up shoots, the leaves of which could be utilised, although the stems are too young to fell. The guttapercha obtained by this means is, however, not equal to that produced by incision, and the process has not yet established itself.

The yield by felling has been variously computed at about 300 to 600 gm. from each tree, whereas by tapping 1,400 gm. has been obtained annually; the leaves are said to yield 10 per cent. of their weight.

**Description.**—Crude guttapercha occurs in commerce in lumps or blocks of very variable, often large size, of a brown or a greyish brown colour externally, reddish yellow or reddish grey internally, and of laminated or fibrous nature. It has (especially when rubbed), an odour that is not disagreeable and it is flexible though scarcely elastic. It often contains mechanical impurities, from which it is freed by slicing and washing it and pressing it whilst plastic through wire gauze, or by softening in hot water and rolling into strips; these are then torn by machinery into shreds and the shreds kneaded into dark-brown lumps (purified guttapercha).

White guttapercha is prepared by dissolving purified guttapercha in chloroform, decolorising the solution with charcoal, filtering, and precipitating with alcohol.

Purified guttapercha is firm, tough, and flexible, but scarcely elastic; it can be cut easily with a knife, and at a temperature of 45°–60° it softens and can be rolled or drawn. It is soluble in chloroform and carbon disulphide in the cold; turpentine and benzene dissolve it when warmed, alcohol and ether only partially, whilst caustic alkalies and dilute acids have no action upon it. It is a very bad conductor of heat and electricity, and upon the latter fact depends its extensive use as an insulator.

**Constituents.**—The principal constituent of guttapercha is a hydrocarbon, gutta, $C_{30}H_{48}$; with this are associated several crystalline substances which have been grouped into three series, viz. albans, fluavils, and albanans. Guttapercha also contains tannin, salts, substances resembling sugar, and an unstable substance, guttan.

Gutta is white when pure, and has been obtained in minute crystals; on exposure to the air it gradually assumes a reddish colour.

The albans are soluble in boiling alcohol, but not in cold; they are crystalline and are distinguished by the prefixes $\alpha$, $\beta$, $\gamma$, &c.; they yield by hydrolysis with alcoholic potash the corresponding alcohol, albaresinol, together with cinnamic acid.

The fluavils are soluble in cold alcohol; they are also crystalline, are similarly distinguished, and yield an alcohol and cinnamic acid when saponified.

The albanans are insoluble both in cold and hot alcohol.

As the albans, &c., from different varieties of guttapercha are not identical, these substances are further distinguished by a prefix indicating their source; thus those obtained from commercial (Sumatra) guttapercha are termed 'sumalbans,' 'sumafluavils,' &c.

The quality of commercial guttapercha is judged by exhausting with ether or petroleum spirit and determining the resins (albans, fluavils, &c.) which dissolve, then removing the gutta with chloroform, and weighing the insoluble residue.

**Uses.**—Apart from the varied technical uses of guttapercha a solution of that substance in chloroform is used to form a protective covering on the skin or as a means of applying various remedies.

**Allied Drug.**—*Balata* is a similar substance obtained from *Mimusops globosa*, Gaertner, by tapping and coagulating the latex either by standing (sheet balata) or by boiling (block balata); it closely resembles guttapercha both in physical characters and chemical composition.

## CAOUTCHOUC

**Source, &c.**—Caoutchouc is a substance contained in the latex of a large number of plants, especially of those belonging to the natural orders *Euphorbiaceæ*, *Urticaceæ*, and *Apocynaceæ*, but it is by no means confined to these. Indeed, it is so widely distributed that in all cases the latex of plants appears to contain caoutchouc or some similar substance as a constant constituent. Many secrete it in such quantity as to form suitable sources for its commercial extraction. Commercial caoutchouc, however, always consists of caoutchouc associated with fat, resin, &c., and is best termed 'rubber.'

Pará rubber, which is regarded as the best variety, is obtained from *Hevēa brasiliensis*, Müller Argoviensis, and probably other species (N.O. *Euphorbiaceæ*), a large tree widely distributed in the basin of the Amazon and the low ground flanking its tributaries, the Madeira,

Purus, Jurna, &c. The tree is also largely cultivated in Ceylon, the Straits Settlements, &c.

In the collection of the latex, spiral, herring-bone or isolated incisions are made in the bark of the trunk and the latex collected. This latex consists of globules of caoutchouc with small quantities of other bodies held in suspension by proteid substances ; it is treated in various ways for the production of rubber.

In Brazil the small tin cups in which the latex has been collected are emptied into calabashes which the collector carries to his hut. Here he has a pole, about the length of a broomstick and thickened

FIG. 235.—Sampling Pará rubber ; in the centre balls of Pará rubber cut transversely, showing the central hole.

for a few decimetres at a little distance from one end, supported horizontally on a cross-bar. Under the thickened portion a clay dome, or sheet iron cone, is placed in which a fire is kindled on to which palm nuts are thrown. The latex is emptied into a large basin and a little is poured by means of a dipper on to the thickened part of the pole. This is now rotated in the dense smoke of the burning nuts. The latex is rapidly coagulated by the acetic acid in the smoke to a layer about 1 mm. thick, the creosote and other tarry substances from the smoke acting as preservatives. The coating and coagulation is repeated again and again until a short, thick, cylindrical mass weighing usually from 3 to 30 kilograms, but sometimes much more, has been obtained. The pole is then removed and the ball hung up to dry. In this way 'ball' Pará rubber is obtained. Sometimes a paddle which may be dipped into the latex, is used instead of a pole and the rubber removed by splitting it along one side; this gives 'spade,' 'paddle,' or 'biscuit' Pará rubber. Pará rubber, therefore, consists of the latex dried after coagulation, and contains in addition to caoutchouc all the other constituents of the latex (except water). The coagulation is probably due to precipitation of part of the albumin by the acetic acid of the smoke. The latex contains about 40 per cent. of caoutchouc, 2·8 per cent. of proteids, and 55 per cent. of water.

**Description.**—Pará rubber occurs in commerce in elastic, rounded, cylindrical, or flattened masses of varying size and shape, and of characteristic odour. Externally they are of a uniform brownish black colour, but internally they exhibit under a lens a large number of very thin yellowish layers separated by dark lines, each layer representing a coat of coagulated latex. The drug is insoluble in water, alkalies, or dilute acid, but chloroform, oil of turpentine, carbon bisulphide, and benzene make it swell and become soft and gelatinous, the caoutchouc dissolving and leaving a gelatinous substance undissolved. Digestion with alcohol removes from Pará rubber about 1·5 per cent. of resin, but inferior rubbers may yield much more.

**Constituents.**—Pará rubber contains from 40 to 60 per cent. of the hydrocarbon caoutchouc (caoutchouc-gutta) which when quite pure is white and corresponds in composition to the empirical formula $(C_{10}H_{16})_n$; it is a very unstable body and rapidly oxidises. In Pará rubber it is associated with 30 to 50 per cent. of the gelatinous substance which probably protects the caoutchouc from change; small quantities of resin, fat, colouring and mineral matter are also present. Good rubber always contains a considerable amount of the gelatinous substance and only small quantities of resin, &c.; in poor rubbers the reverse is the case.

**Varieties.**—*Ceylon* or *Plantation Rubber.*—This now forms a very important commercial article, the production of which is rapidly increasing. The latex is coagulated by allowing it to stand for a few hours, with or without the addition of a little acid, and the coagulum is put into a washing-machine consisting of a pair of very heavy corrugated cylinders over which water is poured. The rubber globules are united by this treatment and come from the machine in the form of thin, crinkled sheets (crêpe rubber) which are hung up to dry. Sometimes the coagulum is passed through a machine which presses the rubber into small vermiform pieces; these are afterwards submitted to heavy pressure which converts them into 'block' rubber. Such rubber is of course free from the proteids and other substances present in the Pará rubber. It usually occurs in large thin sheets of a pale yellow or brownish black colour or in translucent dark brown or golden brown blocks about 4 cm. thick.

The keeping properties of rubber are much increased by combining it with sulphur (vulcanisation). This may be effected by heating the rubber with sulphur, or by means of a solution of sulphur chloride in carbon disulphide. Ordinary soft rubber usually contains from 2 to 4 per cent. of sulphur, hard rubber (ebonite) from 20 to 30 per cent.

The description of the numerous other commercial varieties of rubber would be far beyond the scope of this work, but the following may be briefly noticed as illustrating the various methods of preparation and the varying appearance of this important and interesting substance :—

*Assam Rubber* from *Ficus elastica*, Roxburgh (N.O. *Urticaceæ*) ; the latex is coagulated by boiling and formed into dark red lumps which are compacted into larger masses.

*Mangabeira Rubber* from *Hancornia speciosa*, Müller Argoviensis (N.O. *Apocynaceæ*) ; collected in Eastern Brazil, the latex being coagulated with alum or salt water. It occurs in large, flat, pinkish white pieces often very wet from presence of salt water.

*Central American Rubber* from *Castilloa elastica*, Cervantes (N.O. *Urticaceæ*) ; the latex is coagulated with salt or sodium bicarbonate or with an infusion of various plants ; the masses of coagulum are pressed together into slabs composed of black, wavy sheets.

*West African Rubber*.—The latex of *Funtumia* (*Kickxia*) *elastica*, Stapf (N.O. *Apocynaceæ*), is coagulated by boiling or by pouring it into cavities in the trunk of the tree. That of *Landolphia florida*, Bentham (N.O. *Apocynaceæ*), is smeared by the natives over their bodies, and after drying rolled into balls.

*Mozambique Rubber* from *Landolphia Kirkii*, Dyer (N.O. *Apocynaceæ*), occurs in balls or in sausage-shaped masses consisting of strings of rubber wound round a little wooden spindle.

*Ceara Rubber* from *Manihot Glaziovii*, Müller Argoviensis (N.O. *Euphorbiaceæ*). The bark is shaved off until the latex trickles down the trunk ; it is allowed to dry, peeled off in strings and rolled into balls.

## EUPHORBIUM

### (Euphorbium)

**Source, &c.**—Euphorbium is the dried latex of *Euphorbia resinifera*, Berg (N.O. *Euphorbiaceæ*), a plant about a metre in height, resembling in appearance and habit a cactus, and common in the mountainous districts of Morocco, especially on the spurs of the mountains in the neighbourhood of the town of Morocco, where the drug is principally collected.

The plant produces green, fleshy, quadrangular stems and branches which bear small, scaly leaves supported by two persistent, thorny stipules ; the flowers are small and borne on short peduncles ; the fruit consists of three carpels, and resembles typical Euphorbiaceous fruits.

Both the cortex and the pith of the plant contain long, branching, laticiferous cells ; these when wounded discharge their latex in the form of milky drops, the exudation in rainy seasons being very copious. This dries to a resinous mass which is collected in the autumn by

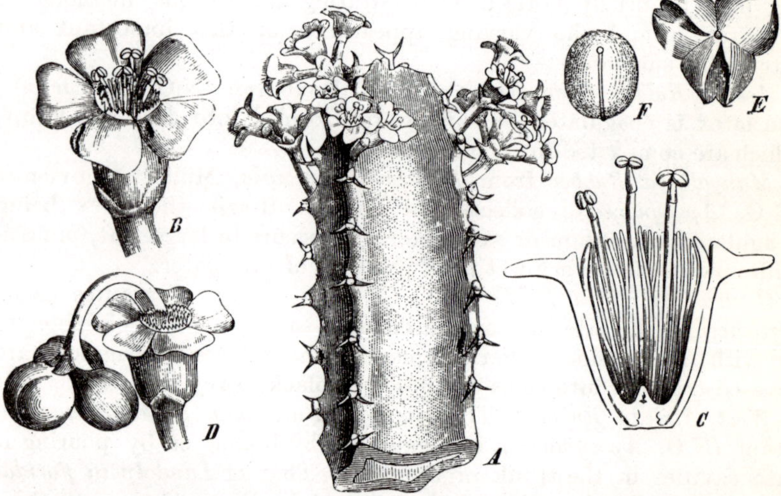


FIG. 236.—*Euphorbia resinfera.* *A*, flowering branch, natural size. *B* and *C*, staminate, *D*, pistillate flower, magnified. *E*, fruit, and *F*, seed, magnified. (Luerssen, after Berg and Schmidt.)

the poorer class of Arabs and brought to Morocco for sale. It is exported chiefly from Mogadore. So acrid is the drug that the faces of persons handling it have to be protected by cloths.

**Description.**—The drug occurs in dull yellow or brown pieces seldom exceeding 1·5 cm. in width, often mixed with fragments of the quadrangular, thorny stem and other débris. Many of the pieces have evidently solidified round a pair of stipules and are pierced by holes corresponding to them, or sometimes even include the stipules themselves ; some are pierced with single holes and retain the fruits or flowers or portions of them. The fruits are characterised by their shape, each consisting of three nearly separate, one-celled, compressed, keeled carpels attached at the apex and base to a central axis.

The drug itself is very brittle, breaking readily between the fingers; internally it is of a dull yellowish brown colour. It has no particular odour, but an extremely acrid taste and the powder, if inhaled, produces violent sneezing. It is partially soluble in 90 per cent. alcohol (62 per cent.), in ether (56 per cent.), and in water (32 per cent.), but almost completely in glacial acetic acid. Petroleum spirit dissolves about 36 per cent., and the solution, carefully poured over sulphuric acid containing one drop of nitric acid in 20 c.c., develops a blood-red zone; this reaction is characteristic of euphorbium.

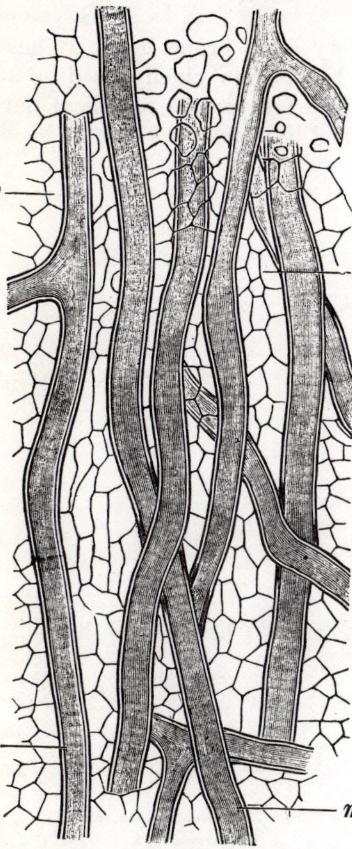


FIG. 237.—Laticiferous cells in the stem of *Euphorbia resinifera*. *p*, parenchyma; *m*, laticiferous cells. Magnified. (Tschirch.)

**Constituents.**—Euphorbium consists chiefly of euphorbone (40 per cent.), euphorbo-resene (20 per cent.), euphorbic acid (0·7 per cent.), calcium malate (25 per cent.) together with an intensely acrid substance which has not yet been isolated, vegetable débris, &c.; it yields about 5 per cent. of ash.

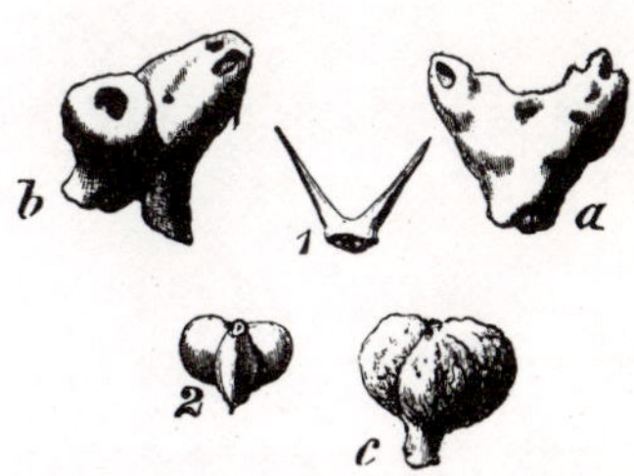


FIG. 238.—Euphorbium. 1, pair of stipules; *a*, and *b*, stipules covered with gum-resin; 2, fruit; *c*, fruit covered with gum-resin. Natural size. (Vogl.)

Euphorbone crystallises from acetone in colourless, odourless, and tasteless acicular crystals melting at 115°–116°.

Euphorbo-resene is yellowish brown, amorphous, and tasteless.

The acrid substance to which the physiological action of the resin is due is soluble in water, alcohol, and ether, but so far all attempts to purify it or isolate it in crystalline form have failed; it has been obtained as an amorphous mass of intensely and persistently acrid taste.

Many species of *Euphorbia* yields an acrid, milky juice analogous in composition to the above (Hencke, 1886).

**Uses.**—Euphorbium possesses very acrid properties. It has been employed as a drastic purgative, but is now chiefly used as a vesicant in veterinary practice, and also for various technical purposes such as an antifoul for coating ships' bottoms.

## SECTION XI

## DRIED JUICES

A number of plants exude on incision liquids that have been contained in certain specialised cells. These liquids differ in their composition from the latex contained in laticiferous tissue and also from the oil, oleo-resin, or oleo-resinous emulsions secreted in cells, glands, or ducts. Some of them contain substances of medicinal value, as, for instance, various species of *Aloe*, *Eucalyptus*, &c.; these yield when dried drugs that, for want of a better term, have been designated 'dried juices.' In many cases tannin forms the chief constituent as it does in the various kinos; in others, particular substances not produced in other natural orders or even genera are found, such as aloin.

### ALOES

(Aloe)

**Source, &c.**—Aloes is officially described as the juice that flows from the transversely cut leaves of certain species of *Aloe* (N.O. *Liliaceæ*) evaporated to dryness.

The aloe plants are indigenous to eastern and southern Africa, but have been introduced into other tropical countries, as for instance the West Indies, and will flourish even in southern Europe. They produce spikes of yellow or red flowers and large fleshy leaves resembling those of the American agave, an ornamental plant commonly grown in this country and often erroneously called an aloe.

The leaf which is fleshy and mucilaginous in its interior, contains near the epidermis a row of isolated fibrovascular bundles each of which is surrounded by an endodermis. The cells of the pericycle and sometimes of the neighbouring parenchyma are unusually large, and filled with a viscous, yellow liquid, the aloetic juice. When the leaves are cut from the plant this juice slowly drains from the cells containing it. Probably the transverse walls of the elongated and axially arranged cells give way under the pressure exerted upon them by the surrounding tissue, and thus the aloetic juice is drained from cells situated at a considerable distance from the incision.

Little or nothing flows from the mucilaginous tissue in the centre of the leaf, and as it contains no aloetic juice no effort is or should be made to obtain it; in fact, its presence would lower the quality of the aloes.

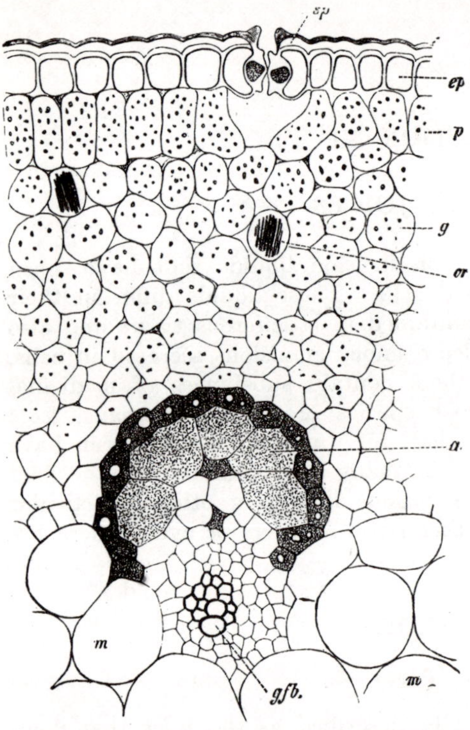


FIG. 239.—Transverse section near the margin of an Aloe leaf. *ep*, epidermis; *gfb*, vascular bundle, the pericyclic cells, *a*, of which are much enlarged and contain a yellow secretion (aloes); *p*, palisade; *g*, parenchyma; *sp*, stoma; *cr*, calcium oxalate crystals; *m*, mucilaginous parenchyma. (Magnified.) (Tschirch.)

The drug is obtained chiefly from Eastern and Southern Africa and from the West Indies. The means by which it is prepared are probably practically identical in all three countries. The leaves are cut from the plant and placed so that the juice that drains from the aloetic cells may be received in any convenient vessel. This juice is then concentrated, either (probably) by allowing it to evaporate spontaneously (Socotrine aloes) or by boiling it until it has reached the desired consistence, when it is poured into boxes or gourds (West Indian aloes) or transferred it to kegs (Socotrine aloes) or skins (Zanzibar aloes), or cases or barrels (Cape aloes) and allowed to solidify.

The manner in which this operation is performed has a remarkable influence on the physical characters of the aloes. If the juice is rapidly concentrated, the concentration carried as far as practicable, and the resulting aloes quickly cooled, a drug is obtained that breaks with a vitreous or glassy fracture, and in small splinters is quite homogeneous and transparent even when examined under the microscope. Such an aloes is termed a 'vitreous,' 'lucid,' or 'glassy' aloes. But if the evaporation is carried on slowly and not quite so far as in the preceding case, a drug is obtained that is opaque and exhibits when examined under the microscope minute prismatic crystals embedded in a transparent resinous mass. Such an aloes is termed 'hepatic' or 'livery.' The crystals that it contains are aloin, and the reason why in this case the aloin crystallises, whilst in the vitreous aloes it does not, is probably to be found partly in the fact that the conditions under which evaporation takes place are more favourable to crystallisation, and partly in the conversion of the crystalline aloinin to an amorphous modification (see below). Nevertheless every variety of aloes not only may, but actually does, occur in both the vitreous and the opaque modification.

**Varieties of Aloes.**—The chief commercial varieties of aloes may be conveniently considered under three heads, viz.: East African, South African, and West Indian.

## East African Aloes

1. *Socotrine Aloes.*—This variety is stated in the British Pharmacopœia to be derived from *Aloe Perryi*, Baker, and probably other species. It is brought by Arab traders from the east coast of Africa, the island of Socotra (600 miles west of Aden), and possibly also from the coast of Arabia, to Bombay, whence it is exported to Europe. It arrives usually in kegs or tins, or occasionally in large barrels, and commonly has a pasty, semi-liquid, or even treacly consistence. It is then of a brownish yellow colour and quite opaque, but if not too viscid it separates on standing into a clear, dark brown, supernatant liquid and a dark yellow sediment which, under the microscope, is seen to consist of minute, prismatic crystals (aloin). The odour of the fresh aloes is usually remarkably unpleasant, but this may change when the aloes is kept under certain (not at present precisely known) conditions to a rather agreeable fragrance that has been compared to myrrh and saffron.

As this variety of aloes contains a varying amount of water, and is not, as imported, in a suitable condition for use in pharmacy, it must be dried at a gentle heat, best by exposing it on wooden trays in a warm room. It then forms hard, dark brown or sometimes, if dried at too high a temperature, nearly black masses, breaking with a dull, waxy, uneven, often porous fracture, and possessing a strong, characteristic unpleasant odour and an extremely bitter, nauseous taste. It should be almost entirely soluble in alcohol, and yield about half its weight to cold water. Small splinters are opaque, but mounted in a drop of almond oil and examined under the microscope they exhibit numerous minute prismatic crystals (of aloin) embedded in a transparent, dark yellowish brown mass. This aloes therefore belongs to the class of aloes known as livery or hepatic aloes. From other aloes it may be distinguished by moistening a little of the

28

powder with nitric acid, when a reddish or yellowish brown colour is produced.

2. *Zanzibar Aloes*, which is sometimes regarded as a variety of Socotrine, is commonly poured into skins which are then packed in cases ('monkey-skin' aloes). It is usually hard, but is sometimes soft, and then the contents of the case become compacted into a solid mass. It has a liver-brown colour, and a dull waxy, but nearly smooth and even fracture. In this respect it differs from the foregoing, which breaks with an uneven fracture, as it does also in its odour, which is characteristic and strong but not disagreeable.

Both the foregoing varieties of aloes are usually hepatic, a fact that points to a slow concentration of the aloe juice, possibly by spontaneous evaporation. Both occur also in the vitreous variety, but East African aloes is seldom imported in this condition.

### South African Aloes

3. *Cape Aloes*.—Large quantities of aloes of excellent quality are exported from Mossel Bay to Cape Town in Cape Colony. It is obtained from *Aloe ferox*, Miller, a species widely distributed in South Africa and often occurring in considerable quantity. The aloe juice is collected in goat-skins covering a hollow in the ground around which the cut leaves are piled; from these it is transferred to tins, and finally boiled down over a naked fire in iron pans. This variety of aloes, which is preferred in Austria and Germany, occurs in masses of a dark reddish brown or nearly black colour, often with a greenish tinge, and breaking with a clean, glassy fracture. Thin splinters are perfectly transparent, of a reddish brown or amber-yellow colour, and exhibit no trace of crystals when examined under the microscope. It belongs therefore to the class of glassy, lucid, or vitreous aloes, and may be easily distinguished from all other glassy aloes by its very distinctive sour odour and by the pale yellow colour of its powder.

Recently the aloe juice has been allowed to undergo slight fermentation and then dried in shallow troughs exposed to the sun's heat. This aloes is known as 'Crown' and also as 'Uganda' aloes, the latter being a very misleading fancy name; it occurs in bricks or fragments of a yellowish brown colour and bronze-gold fracture; it has the odour of Cape aloes, but is distinctly an hepatic aloes. It is now practically obsolete.

4. *Natal Aloes*.—The botanical source of this variety of aloes, which is seldom imported, has not yet been definitely ascertained (possibly *A. succotrina*, Garsault).

Natal aloes is almost always opaque, and has a characteristic dull greenish black or dull brown colour. In odour it closely resembles Cape aloes, and it may therefore be easily distinguished from all other opaque aloes by this character alone. It yields when scraped a pale greyish green or sometimes pale yellowish powder, and when this is mixed with a little sulphuric acid and the vapour of nitric acid blown over it, a deep blue coloration is produced.

These last three characters are so distinctive that Natal aloes is one of the easiest to identify.

## West Indian Aloes

5. *Barbados Aloes.*—But very little aloes is now produced in the island of Barbados, and that which bears this name is in reality prepared in the Dutch islands of Curaçao, Aruba, and Bonaire; it is therefore, often and more appropriately, termed Curacao aloes. The plant which yields it appears to be *Aloe vera*, Linné, or possibly *A. chinensis*, Baker.

Curaçao or Barbados aloes arrives usually in boxes, sometimes in gourds. It occurs in the vitreous as well as in the livery variety; the former, commercially known as 'Capey' Barbados, may become opaque on keeping, a change to be ascribed to the slow crystallisation of the aloin. It is also by no means uncommon to find packages that are partly filled with glassy, partly with opaque, aloes; such differences in the appearance are probably due to slight differences in evaporating the juice.

Livery Curaçao aloes of good quality varies in colour from yellowish or reddish brown to chocolate brown, lower grades being black and occasionally burnt. It breaks with a dull, waxy, even fracture, small splinters exhibiting under the microscope numerous minute crystals of aloin. It closely resembles Zanzibar aloes, and, in fact, can be distinguished (apart from chemical tests) only by its distinctive odour.

Vitreous Curaçao aloes, which is not a highly esteemed variety, is distinguished from the livery by its transparency. It has usually, in small fragments, a garnet-red colour; in other respects it resembles the foregoing.

The student should have little difficulty in identifying the foregoing varieties of aloes. Two only (Cape and Curaçao) are commonly met with in the vitreous form, and they are easily distinguished both by their colour and by their very different odour. Five (Socotrine, Zanzibar, Natal, Curaçao, and Uganda) are hepatic. Of these the odour of the Natal distinctly recalls that of Cape, whilst the greyish green or pale yellowish brown powder is characteristic. Uganda aloes is readily identified by its bronze-gold colour. Both Natal and Uganda aloes are practically obsolete. Socotrine aloes is remarkable for its unpleasant odour and uneven, porous, fracture. Zanzibar and hepatic Curaçao are very similar in appearance, but differ in odour; these are the only two likely to be confused, and if necessary a chemical test must be applied.

**Chemical Reactions.**—(*a*) General Reactions. The following are the most important general reactions for aloes :—

*Schönteten's Reaction.*—Dissolve 0·1 gramme of aloes in 10 c.c. of boiling water and add 0·5 gramme of powdered borax ; a green fluorescence is produced. This reaction is due to barbaloin ; all aloes except Natal (see below) respond to it.

*Bornträger's Reaction.*—Dissolve 0·1 gramme of aloes in boiling water, cool, add 10 c.c. of benzene, shake vigorously, separate the benzene solution and shake it with solution of ammonia ; the alkaline solution is coloured red. This reaction is due to aloe-emodin ; all aloes except Natal, respond to it.

*Chrysammic Acid Reaction.*—Heat 1·0 gramme of aloes with 20 c.c. of nitric acid in a dish on a waterbath for 2 hours replacing the liquid lost by evaporation ; then dry on the waterbath ; the resulting brown powder dissolves in ammoniacal water with violet coloration.

(*b*) Special Reactions :—

*Natal Aloes*, which is quite distinct from all other aloes, is readily identified by Histed's reaction :—Dissolve 0·1 gramme of aloes in a few drops of sulphuric acid on a porcelain slab and gently blow the vapour of fuming nitric acid over the surface ; a green, then red, and finally deep blue colour is developed.

*Curaçao Aloes.*—To 10 c.c.of a *freshly prepared*, 0·1 per cent. aqueous solution of the aloes add 1 drop of a 5 per cent. solution of copper sulphate, a trace of sodium chloride and a few drops of alcohol ; a fine wine-red colour permanent for 24 hours is produced. Of the other varieties of aloes, Cape aloes alone may give a pale colour which, however, rapidly fades. The reaction is due to isobarbaloin (see below).

*Cape Aloes.*—Drop a few drops of nitric acid on a little of the crushed aloes ; a permanent green colour is gradually produced.

*Socotrine and Zanzibar Aloes.*—With the nitric acid test, a reddish brown colour is produced. Under the same conditions Curaçao aloes gives a crimson colour.

**Constituents.**—The principal constituent of all the foregoing varieties of aloes (with the exception of Natal aloes) is the pale yellow, crystalline, glucoside, barbaloin (formerly called socaloin, zanaloin, capaloin, &c., according to its origin). In Curaçao and in true Barbados aloes the barbaloin is accompanied by isobarbaloin which is crystalline and isomeric with barbaloin but easily distinguished from barbaloin, as it yields the cupraloin test (see above). Socotrine and Zanzibar aloes contain no isobarbaloin, and Cape aloes traces only.

The crystalline aloin is accompanied by an amorphous aloin, $\beta$-barbaloin, which may be produced by heating barbaloin for about three hours to 160°–165° ; it is isomeric with barbaloin, and constitutes part at least of the water-soluble substances other than barbaloin and isobarbaloin. β-barbaloin is particularly abundant in Cape and Uganda aloes.

Other constituents of aloes are resin and aloe-emodin, in addition to water-soluble substances (other than the aloins) of which nothing definite is known. The resin of Curaçao aloes consists of barbaloresinotannol combined with cinnamic acid; that of Cape, Uganda, Natal, and probably Zanzibar and Socotrine aloes consists of capaloresinotannol combined with paracumaric acid.

Aloe-emodin is a decomposition product of barbaloin and occurs in small proportion only.

Various formulæ have been proposed for barbaloin of which that of Léger, $C_{21}H_{20}O_9$, is probably the most correct. It is sparingly soluble in water, more readily in alcohol and acetone. Its aqueous solution slowly acquires a green fluorescence when saturated with borax. It is a glucosidal methylanthraquinone derivative, yielding by prolonged standing of an alcoholic solution, or by treatment with sodium peroxide (but not with mineral acids), aloe-emodin and *d*-arabinose. It is isomeric with frangulin (compare p. 245), but the latter is hydrolysed by mineral acids yielding rhamnose.

The proportion in which the aloins are present in the respective aloes is not accurately known. Tilden (1872) and Treumann (1880) obtained from 20 to 25 per cent. of aloin from Curaçao aloes. Carr and Reynolds (1907) found in a number of specimens from 12·6 to 27·9 per cent. As no exact assay process is known, and as there appears to be considerable loss in its extraction, it may be assumed that good Curaçao aloes may contain upwards of 30 per cent. of crystallisable aloins. Socotrine and Zanzibar aloes appear to contain less. Cape aloes has yielded over 9 per cent.

Natal aloes differs materially from other aloes in its composition. It contains nataloin, homonataloin, and a resin consisting of nataloresinotannol (probably identical with capaloresinotannol) combined with paracumaric acid. Nataloin does not yield a fluorescent solution when dissolved in a solution of borax.

**Uses.**—All the varieties of aloes have more or less powerful purgative action, Cape aloes being the strongest and Natal aloes the weakest, all of them acting with remarkable slowness. Aloes is one of the most valuable purgatives in certain forms of constipation, as it improves the digestion and does not lose in activity by repetition.

**Varieties.**—In addition to the above-described varieties *Mocha aloes* is occasionally imported from Bombay in tin-lined cases; it is a black, brittle, glassy aloes of very strong odour and inferior quality, *Jafferabad aloes,* which is sent from that port to Bombay, is also nearly black; it does not enter into English commerce.

## KINO

### (East Indian, Malabar, Madras, or Cochin Kino)

**Source, &c.**—The term 'kino' or 'kano' was first applied to a red astringent substance imported from West Africa about the middle of the eighteenth century. This was obtained from *Pterocarpus erinaceus*, Poiret. It was, however, soon replaced by other similar astringent drugs, and the term 'kino' became a generic one. The official drug, although often spoken of as 'kino,' is better termed Malabar or Cochin kino. Malabar kino is the juice obtained from incisions in the trunk of *Pterocarpus Marsupium*, Roxburgh (N.O. *Leguminosæ*), evaporated to dryness. The bast of the tree, which grows in southern India and Ceylon, contains, according to v. Höhnel, numerous comparatively wide and short tubular cells arranged in axial rows; these cells are filled with a red astringent liquid, which flows from them when they are wounded. Vertical incisions, with oblique lateral ones running into them, are accordingly made in the bark; the juice that flows is collected in small cups made of leaves, or in other convenient receptacles, and soon dries in the sun to a dark mass that readily breaks up into small angular grains. It is sometimes boiled before it is evaporated, an operation that modifies the subsequent behaviour of the drug. Kino is occasionally imported as a treacly liquid which can easily be dried.

**Description.**—Kino occurs in small, glistening, angular grains that appear quite black and are remarkably free from dust. When thin laminæ or the edges of the grains are examined they are seen to be transparent and of a dark ruby-red colour. They are hard and brittle, breaking with a vitreous fracture and yielding a brownish red powder. The drug is odourless, but has, when chewed, an astringent taste, and adheres to the teeth, colouring the saliva red.

In cold water kino is only partially (from 80 to 90 per cent.) soluble; it dissolves to a greater extent in hot water, and is almost entirely soluble in alcohol. The aqueous solution turns green on the addition of a ferrous salt, violet with an alkali, and throws down a precipitate (kinotannic acid) when acidified with a mineral acid.

The student should particularly observe

(*a*) The *brilliant, apparently black, colour*,

(*b*) The *absence of any dust*;

and should compare the drug with red gum.

**Constituents.**—The principal constituent of kino is kinotannic acid, of which it contains from 70 to 82 per cent. The published assays vary from 50 to 96 per cent., but this variation is probably due partly to variation in the drug, partly to variation in the process adopted for the assay, and partly to the gradual production of oxidation products intermediate between the soluble kinotannic acid and the insoluble kino-red. This change from kinotannic acid to kino-red commences immediately the juice is exposed to the air, as evinced by a darkening in colour; it is continued slowly in the dry drug which becomes duller and less soluble, and proceeds rapidly in solutions of the drug, which may gelatinise owing to the formation of insoluble kino-red. It is caused by the presence of an oxydase enzyme, and may be prevented by destroying the activity of the enzyme by boiling the juice or the solution of the drug. Hence the boiling of the juice before evaporation, as required by the British Pharmacopœia, is a rational procedure.

In addition to kinotannic acid and kino-red the drug contains about 10 to 15 per cent. of moisture, and small quantities of pyrocatechin (catechol), gallic acid, and mineral constituents (ash 1·5 per cent.).

**Uses.**—Kino is a powerful astringent; it is given internally for diarrhœa and dysentery and is also used externally.

**Varieties.**—Kinos have been obtained from numerous plants belonging to various natural orders, including *Leguminosæ*, *Myrtaceæ*, *Polygonaceæ*, *Myristicaceæ*, and *Saxifragaceæ*. Of these kinos the following may be briefly mentioned :—

1. *Red gum* (see below).
2. *Butea gum* (see below).
3. *Eucalyptus* or *Botany Bay kino* from various species of *Eucalyptus* (Australia), the most suitable being *E. calophylla*, R. Brown, the tannin of which does not gelatinise. The drug occurs in irregular dark red pieces.
4. *African kino* from *Pterocarpus erinaceus*, Poiret, in West Africa. It closely resembles the official drug.
5. *Jamaica Kino* is an extract obtained by evaporating a decoction of the leaves, wood and bark of *Coccoloba uvifera*, Linné (N.O. *Polygonaceæ*).

## RED GUM

### (Eucalyptus Kino, Gummi Eucalypti)

**Source, &c.**—Eucalyptus Kino, or as it is commonly termed 'red gum,' is a variety of Australian kino obtained from *Eucalyptus rostrata*, Schlechtendal (N.O. *Myrtaceæ*), and other species (*E. marginata*, Smith, *E. amygdalina*, Labillardier, &c.). They are all Australian trees, *E. rostrata* forming large forests on the banks of the Murray River in New South Wales and yielding a valuable timber. This species is usually preferred as the source of red gum for medicinal use, because the tree is gregarious, cannot easily be mistaken for others, and yields freely a drug of good quality. The gum, which is secreted in cavities in the wood, or sometimes between the bark and the trunk of the tree, forming carbuncles, is obtained by making an incision and inserting a trough-shaped piece of tin by which the treacly liquid as it drains from the cut is carried into buckets or tins. In a few days it dries into a solid mass which soon becomes friable, breaking up into very dark fragments; or it may be evaporated by boiling, and the official drug is probably prepared by this method. The yield of each tree is very variable, the average being about a litre, some yielding none, others as much as 18 litres (Maiden, 1897).

**Description.**—Red gum is seen in commerce in small pieces, about the size of a pea or less. They are of a dark reddish brown colour, opaque, and more or less dusty, but thin laminæ are transparent and ruby-red, the powder being pale reddish in colour. It is somewhat tough, and has when chewed an astringent taste, colouring the saliva red and adhering to the teeth. Cold water should dissolve from 80 to 90 per cent. According to Brownscombe (1899) good qualities should yield not less than the latter percentage.

The student should observe

(*a*) The *reddish colour*,

(*b*) The *dusty appearance*

of the drug.

**Constituents.**—Red gum contains about 47 per cent. of kinotannic acid (Maiden, 1897), which is undoubtedly its principal constituent. There is also present kino-red, a gelatinisable tannin glucoside, catechin, pyrocatechin, and about 15 per cent. of moisture, the remainder consisting of substances not at present exactly known. According to Smith (1904), eucalyptus kinos contain two tannins giving with ferric chloride a violet and a green reaction respectively; the former gelatinises readily but the latter does not.

**Uses.**—Red gum is not so powerful an astringent as kino, but its action is said to be slower and more prolonged.

## BUTEA GUM

### (Bengal Kino, Buteæ Gummi)

**Source, &c.**—Butea gum is the juice obtained by incising the stem of *Butea frondosa*, Roxburgh (N.O. *Leguminosæ*), dried.

**Description.**—The drug usually occurs in small, irregular, angular fragments to one side of which dull, buff coloured portions of the cortex and cortex of the stem sometimes adhere. When fresh it is ruby-red, transparent in small fragments, and brittle; but on keeping it becomes dull, nearly black, opaque and tough. It is readily reduced to a reddish powder and has an astringent taste. It is partially soluble in water and in alcohol.

**Constituents.**—The chief constituent is kinotannic acid (15 to 62 per cent.); the insoluble matter may vary from 10 to 46 per cent.

**Uses.**—Similar to those of kino.

## ELATERIUM

**Source, &c.**—Elaterium is the feculence that is deposited, on standing, by the juice of the nearly ripe fruit of the Squirting Cucumber, collected and dried. This plant, *Ecballium Elaterium*, Richard (N.O. *Cucurbitaceæ*), is a rough, prostrate, trailing plant, common in southern Europe, particularly in the countries bordering on the Mediterranean. It is cultivated to a limited extent in England, the bulk of the commercial drug being imported from Malta. The plant and the drug under notice have long been known, the process described by Dioscorides for making elaterium being almost identical with the method now adopted.

If the fruits, which resemble small hairy gherkins about 3 to 5 cm. long, are allowed to ripen, they are forcibly separated from the peduncles, the seeds and juice being simultaneously ejected. They are therefore collected before they are quite ripe, sliced, and pressed; the slightly turbid juice is allowed to stand, during which it becomes more turbid, and throws down a deposit which is collected, drained, and dried. This forms the elaterium of commerce.

**Description.**—Elaterium occurs in thin, opaque, curved pieces about 2 mm. thick, pale green in colour if fresh (presumably from the presence of a little chlorophyll), but becoming greyish green and finally yellowish grey by keeping. It is light and friable, breaking readily with a short, very finely granular fracture exhibiting minute crystals when examined with a lens. It has a slight odour, and bitter, acrid taste.

**Constituents.**—The active constituent of elaterium, $\beta$-elaterin, has not yet been obtained quite pure. It is an extremely powerful purgative and is accompanied in elaterin by an inactive crystalline substance, $\alpha$-elaterin. Commercial elaterin is a mixture of inactive $\alpha$-elaterin (60 to 80 per cent.) and active $\beta$-elaterin. English elaterium yields approximately 20 to 27 per cent of commercial elaterin, Maltese from 14 to 17 per cent. The drug contains in addition fat, inorganic matter, &c. Both $\alpha$-elaterin and $\beta$-elaterin are contained preformed in the juice of the fruit and are not, as has been believed, liberated by the action of an enzyme.

**Adulterations.**—Elaterium is liable to admixture with chalk and starch. It should yield not more than 8 per cent. of ash.

**Uses.**—Elaterium is a powerful hydragogue cathartic, producing numerous very watery motions; it is almost entirely used as a hydragogue purgative in dropsy and uræmia.

## SECTION XII

## EXTRACTS

Under this heading are grouped together drugs prepared by evaporating aqueous decoctions of various plants.

### GAMBIER

### (Pale Catechu, Terra Japonica, Catechu)

**Source, &c.**—Gambier is an extract prepared from the leaves and young shoots of *Uncaria Gambier*, Roxburgh (N.O. *Rubiaceæ*).

This plant is a climbing shrub indigenous to the Malay Archipelago and largely cultivated on the small islands between Singapore and Sumatra, as well as in British North Borneo, and on other islands of the Archipelago. The drug was introduced into Europe towards the end of the eighteenth century, but was probably used in India at much earlier times for chewing with betel leaf (the leaf of *Piper Betle*, Linné).

The leaves and young shoots of the shrub are collected and boiled with water; the decoction is evaporated to a syrup in copper pans (iron pans would discolour it) and cooled; during the cooling the workman works a stick up and down in an oblique direction in the liquor, by which crystallisation is induced and a mass of the consistence of soft clay is obtained. This is usually conveyed in a moist state, often dripping with the mother liquor that drains from the crystalline mass, to Singapore, where it is cut into cubes from 2 to 3 cm. in diameter and dried. Occasionally it is formed into strips, plates, or small round discs, or sometimes it is imported in large blocks, but cube gambier is the form in which it is usually employed in pharmacy.

**Description.**—Gambier, as observed, is usually seen in the form of tolerably regular cubes, measuring from 2 to 3 cm. each way; it is light in weight, and of a dull, dark reddish brown colour externally, which, however, varies slightly, even on different sides of the same piece. The cubes break easily, and internally are of a pale cinnamon-brown colour, porous and friable. The drug has no odour; the taste is at first bitter and astringent but afterwards sweetish.

Gambier of good quality is almost entirely soluble in boiling water, and yields not less than 80 per cent. to alcohol.

The filtrate from the alcoholic solution, made strongly alkaline with solution of potash and shaken with petroleum spirit, imparts to the latter a brilliant green fluorescence, a reaction which is characteristic of gambier.

A little of the powdered drug mounted in water and examined under the microscope exhibits numerous minute acicular crystals (of catechin), but should be free from starch. The residue left after extraction with alcohol may also be tested for starch. The ash should not exceed 5 per cent.

**Constituents.**—Gambier consists principally of catechin and catechutannic acid, these two substances in varying proportions constituting together in good specimens over 60 per cent. of the drug; the percentage of catechin varies from 7 to 33 per cent., that of catechutannic acid from 22 to 50. These figures, however, vary with the care with which the drug has been prepared. Brown substances, rubinic and japonic acids, of unknown chemical nature are also present.

Catechin, $C_{15}H_{14}O_6,4H_2O$, forms white, silky, acicular crystals with an astringent taste; it is sparingly soluble in cold but freely in boiling water, the solution giving an intense green colour with ferric salts. In the presence of caustic alkalies and water, catechin readily absorbs oxygen and yields a black dye; with carbonated alkalies the colour is red. Cotton may be dyed brown by steeping in a hot solution of gambier containing a little copper sulphate and then immersing it in a hot solution of potassium dichromate; the catechin is probably first converted into catechu tannic acid and then oxidised to japonic acid. Gambier also contains small quantities of a second catechin, $C_{15}H_{14}O_6$, differing from the foregoing in its melting point and crystalline form.

Catechutannic acid has been obtained as a reddish, amorphous substance easily soluble in cold water and in alcohol, the former being used to extract it from the drug. It is apparently produced from catechin by loss of a molecule of water, and itself yields an insoluble red substance, catechu-red, when boiled with water or with dilute mineral acids. These three substances, catechin, catechutannic acid, and catechu-red, appear therefore to stand in close relation to one another, and the relative proportion in which they occur in the drug depends largely upon the care with which it is manufactured, well-prepared gambier containing most catechin. Hence commercial gambier may contain as much as 50 per cent. of catechutannic acid, and as little as 7 per cent. of catechin.

Trimble (1888) found in three samples of gambier the following composition:—

| | |
|---|---|
| Catechin . . . . . . | 7 to 19 per cent. |
| Catechutannic acid . . . . | 33 to 47 „ „ |
| Gum . . . . . . . | 10 to 16 „ „ |
| Ash . . . . . . . | 3 to 4 „ „ |
| Moisture . . . . . . | 9 to 11 „ „ |
| Colouring matter, &c. . . . . | 4 to 28 „ „ |

Other investigators have obtained larger proportions both of catechin (up to 33 per cent.) and catechutannic acid (up to 50 per cent.).

Other constituents of the drug are catechu-red, quercetin, and gambier-fluorescin, a fluorescent substance which can be removed from an alkaline solution of gambier by shaking with petroleum spirit. This fluorescent substance is absent from black catechu and many similar extracts, and forms therefore a valuable means of identifying gambier (Dieterich, 1897).

**Uses.**—Gambier is employed medicinally as a local astringent in the form of a lozenge or as a general astringent in diarrhœa ; its use for these purposes is, however, insignificant compared with the quantities consumed in the dyeing and tanning industries ; for the former, a mixture of catechin and catechutannic acid is said to give the best results.

## CUTCH

### (Catechu, Black Catechu, Catechu nigrum)

**Source, &c.**—Cutch is an extract prepared from the heartwood of *Acacia Catechu*, Willdenow (N.O. *Leguminosæ*), a tree of medium size common in India and Burma. It yields a valued timber, and also an astringent bark that is used for tanning.

This drug must be carefully distinguished from the foregoing (gambier), which is official under the name of catechu. The two substances are quite distinct, and that obtained from *Acacia Catechu* is to be regarded as true catechu. To avoid confusion it would be well to adhere to the terms 'gambier' and 'cutch' for the two drugs, thus avoiding the use of the term 'catechu,' which has been applied to both.

Cutch has long been used in India as a masticatory, but it was not introduced into Europe till the latter half of the seventeenth century.

To obtain the drug the tree is felled, the bark and sapwood stripped from the trunk, the dark red heartwood cut into chips and boiled in water in earthen pots. The decoction is then strained and boiled down in iron pots with continual stirring until it attains the consistency of syrup. When sufficiently cool to handle, the extract is spread upon leaves arranged within a wooden frame or mould and left for the night. In the morning the cutch is dry, and forms brick-like masses weighing about 20 kilograms, which are broken up for the market.

**Description.**—Cutch occurs in nearly black masses, the outer portions of which are hard and brittle, but the interior often still soft. It breaks easily, the fractured surface having a dull gloss, and containing a number of small cavities. It yields a dull brown powder and has no odour, but an astringent and subsequently sweetish taste. When macerated with cold water it forms a brown magma, which exhibits under the microscope numerous minute crystals similar to those found in gambier. Boiling water dissolves it almost entirely but on cooling deposits a crystalline sediment.

**Constituents.**—Cutch closely resembles gambier in chemical composition. It contains as principal constituents catechutannic acid and acacatechin, but the former is usually present in much larger quantity than the latter. Acacatechin differs from the catechin of gambier in its formula, $C_{15}H_{14}O_6,3H_2O$, and melting point. The drug contains in addition catechu-red and small quantities of quercetin, but is free from the fluorescent substance that is present in gambier. Although the two drugs, gambier and cutch, as usually found on the market have little resemblance to one another, this is due solely to the manner of preparation, the syrupy liquor being in the case of gambier allowed to crystallise. This method is sometimes pursued in India with cutch, and the resulting drug, 'katha,' then closely resembles gambier, but can always be distinguished from it by the absence of the fluorescent substance.

Trimble (1888) found three samples of cutch to have the following composition :—

| | |
|---|---|
| Catechin . . . . . . | 2 to 10 per cent. |
| Catechutannic acid . . . | 25 to 33 ,, ,, |
| Gum . . . . . . . | 20 to 29 ,, ,, |
| Ash . . . . . . . | 2 to 3 ,, ,, |
| Moisture . . . . . | 12 to 25 ,, ,, |
| Colouring matter, &c. . . . . | 21 to 25 ,, ,, |

Good cutch contains from 10 to 12 per cent. of catechin, whereas katha contains about 50 per cent.

**Uses.**—Cutch is employed chiefly in the dyeing and tanning industries, especially in the former, the amount of colouring matter it contains often rendering it more suitable for dyeing than gambier.

**Varieties.**—Similar extracts are also prepared from other substances and called 'cutch'; thus mangrove cutch is obtained from the bark of *Ceriops candolleana*, Arnold (N.O. *Rhizophoreæ*), which contains 42 per cent. of tannin.

## CURARE

**Source, &c.**—Under the name of curare several (at least three) varieties of a dark, extract-like mass appear in commerce. They are all arrow poisons prepared by tribes of Indians in the valleys of the Amazon and Orinoco and their tributaries. The manner in which these extracts are prepared, and the ingredients, vegetable or animal, that enter into them are only imperfectly known. The bark of various species of *Strychnos* (*S. toxifera*, Bentham; *S. Castelnæi*, Weddel; *S. Gubleri*, G. Planchon, *S. Crevauxii*, G. Planchon, &c.) appear to be essential constituents. In the bark of these plants considerable quantities of poisonous alkaloids are present.

**Description.**—Curare has been imported in gourds, in small earthen pots, and in bamboo tubes, but gourd curare is now no longer a commercial article. It has the appearance of a very dark brown or nearly black extract resembling black catechu, often containing small cavities. That imported in bamboo is dark brown and granular, the broken fragments frequently exhibiting crystals sufficiently large to be visible to the naked eye. It has little or no odour, but a very bitter taste.

All these varieties of curare are poisonous when injected subcutaneously, but when administered by the mouth they are harmless, producing, it is said, the effect of a stomachic tonic. The degree of toxicity varies not only in the different varieties but in different specimens of the same variety, and the strength, therefore, of each parcel must be determined before it can be used medicinally. Bamboo curare yields to water about 84 to 88 per cent., gourd curare 34 to 75 per cent., pot curare 50 to 87 per cent. (Böhm, 1898). These figures suffice to show the extreme variability of the drug.

**Constituents.**—Gourd curare contains the alkaloids curarine and curine; curarine is extremely toxic, but curine is less so. Bamboo curare contains tubocurarine and curine. Pot curare contains protocurarine, protocurine, and protocuridine. Urarine has also been reported; it is said to possess the typical curare action.

**Uses.**—Curare has been employed as a remedy for hydrophobia and chorea; it has also been found useful for tetanus, but it would appear desirable to abandon the use of the crude drug in favour of that of its active alkaloids.

# SECTION XIII

## GUMS

Under this name a number of substances resembling in their properties acacia gum, tragacanth, or cherry gum have been grouped together. They are all insoluble in alcohol, ether, &c., but dissolve or at least swell in water, yielding either viscous adhesive solutions or gelatinous mixtures.

When submitted to carefully regulated hydrolysis with dilute mineral acids various sugars are produced, but the gum is not entirely converted into sugars, about 20 per cent. resisting such treatment. This residue has proved in each gum investigated to be an organic acid with which the various sugars separated during the hydrolysis have been combined. Hence the gums must be regarded as consisting of glucosidal acids of high molecular weight. In some gums these acids exist almost wholly in the free state, but in the majority of gums they are partly combined with potassium, magnesium, or calcium in the form of salts.

Amongst the sugars that have been obtained from gums are the pentoses, arabinose, xylose, and tragacanthose, and the hexose, galactose. In addition to the glucosidal organic acid gums contain mineral matter (up to 5 per cent.) together with small quantities of sugars and of nitrogenous substances.

Gums are yielded by trees and shrubs belonging to a number of natural orders, but especially *Leguminosæ*, *Rosaceæ*, *Rutaceæ*, *Anacardiaceæ*, *Combretaceæ*, and *Sterculiaceæ*. They are produced by the conversion of the cell-walls of the tissues into gum (gummosis), doubtless by means of a diastasic enzyme of the origin of which nothing definite is known. Typical gums (acacia gum, cherry gum) are formed as a protective coating after the infliction of injury on the tree, and are to be regarded as pathological products. Mucilages, on the other hand, are normal products of the plant and are secreted in certain cells. Tragacanth occupies an intermediate position as it is not a pathological product.

The so-called artificial gum (dextrin) produced from starch differs essentially from the gums in being entirely converted into dextrose by dilute mineral acids; it is strongly dextrorotatory, most gums being slightly lævorotatory.

## ACACIA GUM

### (Gum Arabic, Acaciæ Gummi)

**Source, &c.**—Acacia gum is a dried exudation from the stem and branches of various species of *Acacia* (N.O. *Leguminosæ*), especially of *Acacia Senegal*, Willdenow, a small tree attaining a height of 5 or 6 metres, and growing freely both in Western Africa (Senegambia) and in Eastern Africa (the upper Nile districts), possibly also in Central Africa, forming forests of considerable extent. The first formation of gum appears to take place in the cambium or the young bast tissue produced from the cambium and to be induced by a wound inflicted on the stem. Such wounds may be produced by animals (ants, beetles, &c.), or by the cracking of the bark during the process of growth, or intentionally with the object of producing gum. Bacteria, moulds and other organisms may thus gain admittance to the gum-forming tissues, but whether these take an initial part in the formation of the gum is more than doubtful.

The best gum is produced near Kordofan from trees specially cultivated and worked for gum. Strips of bark 0·5 to 1 metre long and from to 2 to 8 cm. wide are removed from the trunks and large branches, care being taken not to injure the cambium. About two or three weeks afterwards the gum is collected, and the orchard is then picked over about every four days until the rain sets in, when the exudation of gum ceases. Gum is also spontaneously exuded from wild trees, but this is usually rather darker in colour and not so valuable. The trees begin to yield gum when they are about three years old and continue to yield for about fifteen years.

The gum as collected is in translucent tears which, 'ripened' by exposure to the sun, develop cracks and become friable. Native girls clean it by picking out pieces of bark and sifting out the sand. It is then packed into sacks and sent from the nearest rail-head (El Obeid) to Port Sudan for shipment. Large quantities are sorted and packed in Omdurman.

In some districts another acacia yielding smaller quantities of gum (*A. Seyal*, Delile) occurs. The gum, which is called talk or talka gum, is collected by the natives with that of *A. Senegal* (hachab gum). As talka gum is of inferior quality it has to be separated by picking.

**Description.**—Kordofan (hachab) gum, which is the best variety, occurs in rounded or ovoid tears, varying in size from a pea to a hazel nut or even larger. They are often quite white, but sometimes show a yellowish tinge, and are opaque from the presence in the outer part of the tears of small fissures. In consequence of these they easily break up into a number of small, transparent, angular fragments with glistening, vitreous surfaces. The drug is practically inodorous, and has a bland, mucilaginous taste. Whilst the finest qualities are white, or have at most only a yellowish tinge, inferior grades have a decided yellow or reddish or brownish red colour and then contain traces of tannin.

Acacia gum is insoluble in alcohol, but dissolves freely in water, forming a translucent, viscid, but not glairy or ropy liquid, that feebly reddens litmus paper. A 10 per cent. aqueous solution of good qualities is slightly lævorotatory, and when boiled with an equal volume of Fehling's solution throws down a slight but distinct deposit of cuprous oxide. Solution of lead acetate produces no precipitate, but subacetate produces a copious white one; whilst a saturated solution of borax forms with a strong solution of gum a clear, translucent jelly. Inferior (brown) gum usually contains tannin which may be detected by solution of ferric chloride.

**Constituents.**—Acacia gum consists almost entirely of a glucosidal acid of high molecular weight, which has been termed arabic acid, combined with potassium, magnesium, and calcium; by hydrolysis each molecule yields two molecules of the sugar arabinose and four of galactose together with an organic acid to which the name of arabic acid has also been given, but which is better termed isogeddic acid, as it is isomeric with the corresponding acid, geddic acid, obtained from geddah gum. The glucosidal acid of acacia, is therefore, a diarabinan-tetragalactan-isogeddic acid, the termination 'an' indicating the anhydride of the corresponding sugar. This acid can be obtained from the gum by acidifying an aqueous solution with a mineral acid, dialysing it until the mineral constituents are removed, and fractionally precipitating with alcohol. Whilst moist it dissolves in water, but the dried acid only swells in water, dissolving on the addition of an alkali.

Gum acacia also contains an oxydase enzyme, and hence readily turns powdered guaiacum resin, or the tincture diluted with water, blue. It loses about 14 per cent. of moisture when dried at 100° and yields from 2·7 to 4·0 per cent. of ash. It contains further a small percentage of nitrogen, but this does not enter into the composition of the gum itself (distinction from gelatin, &c.); it is probably due to the enzyme, from which the gum cannot be entirely freed.

**Uses.**—Acacia gum is used medicinally as a demulcent and as a means of suspending oils, resin, &c., in aqueous fluids.

**Varieties, Impurities, &c.**—Gum-yielding acacias are widely distributed over tropical and subtropical countries, and furnish large quantities of gums which, though in most cases unsuitable for medicinal use, are extensively employed for certain technical purposes.

**Varieties.**—*Sennaar gum*, *Gedaref gum*, *Ghezireh gum*, *Talka gum*, *Somali gum*, *Aden gum*, &c., are varieties of East African gums and are considered inferior to Kordofan gum; the latter gives a lævorotatory solution, whereas many of the other East African gums give a dextrorotatory solution.

*Senegal gum* is exported from the Senegal river to Bordeaux. It is derived from *A. Senegal*, but may be distinguished from Kordofan gum by being (usually) slightly more coloured, less fissured and by containing vermiform tears.

*Mogadore gum* (*A. gummifera*, Willdenow); is mostly dark in colour and but little fissured; occasional white fissured tears are probably Sudan gums.

*Indian* (*Acacia*) *gums* include Amrad gum (*A. arabica*), Amritsar gum (*A. modesta*), &c. The flora of the deserts of Sind resembles that of the African. The gums are often in large tears, varying in colour from yellow to dark brown, and are used for calico printing, &c. (For ghatti gum see below.)

Cape gum (*A. horrida*, Willdenow) and Australian gum (*A. dealbata*, Link; *A. pycnantha*, Bentham) find application in various industries.

Many of these gums form glairy, ropy solutions with water, and when diluted throw down gelatinous deposits of gum that has swelled but not dissolved. An acacia gum suitable for pharmaceutical use should be free from both these characters, and should further give no reaction for starch (which might be present as an adulterant of powdered gum) or for tannin, which is present in certain inferior varieties of gum (such as Australian); a 10 per cent. aqueous solution should be slightly lævorotatory (absence of dextrin, from which an artificial gum has been prepared, certain sugars, &c.).

## GHATTI GUM

### (Gummi Indicum)

**Source, &c.**—Ghatti gum is obtained from *Anogeissus latifolia*, Wallich (N.O. *Combretaceæ*), a large tree indigenous to India and Ceylon.

**Description.**—The gum occurs in vermiform or rounded tears, the best qualities being almost colourless, the inferior yellow to dark brown. The surface is dull and the fracture uniform and glassy, not exhibiting cracks. Its aqueous solution gives only a slight precipitate with solution of lead subacetate (that of acacia gum gives a copious one). With water it forms a nearly colourless mucilage about twice as viscous as that made with the same proportion of acacia.

**Constituents.**—The constituents of ghatti gum are, as far as is known, similar to those of acacia.

**Uses.**—Ghatti gum is admirably adapted for pharmaceutical use; it has excellent emulsifying properties.

## TRAGACANTH GUM

### (Tragacantha)

**Source, &c.**—Tragacanth gum is a dried, gummy exudation from the stem of *Astragalus gummifer*, Labillardière (N.O. *Leguminosæ*), and other species of *Astragalus*.

These plants are small, branching, thorny shrubs, about a metre in height, and are natives of southern and eastern Europe and especially of Asiatic Turkey and Persia, where they form one of the most characteristic forms of vegetation. When the stem is incised a gum exudes and dries, the form that it assumes being dependent on the form of the incision, vertical slits yielding flat, ribbon-shaped pieces and punctures vermiform tears. It is produced by the transformation of the cell-walls of the pith and medullary rays into gum, which easily absorbs water, and, swelling, exerts considerable pressure on the surrounding tissue. When, therefore, the stem is wounded the gum is forcibly pressed out, a piece about 2 cm. long being exuded in half an hour; it carries with it the starch grains present in the cells in a more or less unaltered condition, and these are therefore a natural constituent of the drug.

Of the influences that induce gummosis of the cell-wall, nothing definite is known.

In some districts accidental wounds by grazing cattle appear to suffice for the production of tragacanth. In Asiatic Turkey it is collected chiefly in Anatolia and shipped from Smyrna; in Persia the Baktiari mountains, south of Ispahan, and the neighbouring districts yield much tragacanth, which is conveyed from the Persian Gulf ports to Bombay, and thence to Europe. The former variety is known as Smyrna, the latter, which alone is official, as Syrian or Persian.

**Description.**—*Syrian* or *Persian tragacanth* occurs in thin, flattened curved, ribbon-shaped flakes of a translucent, horny appearance and nearly colourless or faintly yellowish. The flakes are often 3 cm. long and 1 cm. wide, and are marked with numerous concentric ridges, conveying the impression that the gum has been exuded in successive portions. The flakes break with a short fracture, are odourless and almost tasteless. Soaked in cold water they swell considerably, forming a gelatinous mass, but only about 8 or 10 per cent. dissolves.

*Smyrna tragacanth* occurs in similar pieces in which, however, the ribbon-like character is less pronounced, and which are more opaque than the Persian, so that the two, when placed side by side, can easily be distinguished. Moreover, Smyrna tragacanth contains sufficient starch to give a decided reaction with the iodine test, whereas Persian responds only very faintly.

**Constituents.**—That portion of tragacanth that is soluble in water consists chiefly of a polyarabinan-trigalactan-geddic acid and yields by hydrolysis arabinose, galactose, and geddic acid; the insoluble part of the gum is termed bassorin; it is converted by baryta water into isomeric $\alpha$- and $\beta$-tragacanthan-xylan-bassoric acids which yield by hydrolysis tragacanthose, xylose, and bassoric acid. Tragacanth contains also water, traces of starch, cellulose, and nitrogenous substances, and yields about 3 per cent. of ash.

**Uses.**—Tragacanth is chiefly employed medicinally as a means of temporarily suspending insoluble powders in mixtures, and to give the requisite firmness to pill-masses. Very large quantities are used as a thickening agent in calico printing.

**Varieties, &c.**—In addition to the tragacanth described (flake tragacanth), much is imported of a very inferior quality ('hog' gum, Caramania gum), the botanical origin of which is doubtful; such gum is usually in tears or irregular vermiform pieces, and darker in colour. Sometimes it is whitened with lead carbonate before being used to adulterate the finer qualities.

Vermicelli tragacanth is in thin vermiform pieces often of good colour; tragacanth of this nature is collected in northern Morea from *A. cylleneus*, Boissier et Heldreich.

*Indian Tragacanth* (Sterculia gum, Karaya gum) is obtained from *Sterculia urens*, Roxburgh (N.O. *Sterculiaceæ*), and possibly other species. It occurs in irregular, striated, often vermiform, whitish, or pale brownish or pinkish brown pieces, here and there with fragments of bark attached. It has a distinctly acid odour, forms a transparent, colourless jelly with water and assumes at most a slight brownish colour when boiled with 5 per cent. solution of potash (tragacanth turns canary-yellow). The powder has been used to adulterate powdered tragacanth. Scoville gives the following test:—Shake 2 grammes, moistened with alcohol, with 50 c.c. of water till homogeneous; add 2 grammes of borax dissolved in 50 c.c. of water, shake vigorously and stand over night. If Indian tragacanth is present the mucilage will be stringy. The presence of sclerenchymatous cells (from adhering bark) in the powder points to the presence of Indian tragacanth. It is largely used in India as a substitute for tragacanth.

Powdered acacia may be detected in powdered tragacanth by making a mucilage 1·5 in 50, adding guaiacol 0·5 in 50 of water and 1 drop of hydrogen peroxide. Pure tragacanth remains colourless; if acacia is present a brownish colour is developed (Cæsar and Loretz).

# SECTION XIV

# RESINS

Under the term resin a number of substances possessing certain properties in common are grouped together, but it is impossible to define the group as sharply as those of the fats, waxes, &c. The most important of these properties is insolubility in water, but ready solubility in alcohol, the alcoholic solution leaving on evaporation a film of transparent varnish. Resins are hard, amorphous, brittle solids (occasionally semi-solids), which soften when heated, and finally melt to a clear, adhesive fluid; they contain a large percentage of carbon, but only little oxygen, and burn with a smoky flame; they are free from nitrogen. Many are soluble in ether or in chloroform, but only a few dissolve in petroleum spirit.

Although most resins, and indeed all resins of medicinal or technical importance, are natural products, it is possible to produce resins artificially. Various volatile oils, especially those containing terpenes, yield resins by oxidation, and resinous bodies are also formed by the action of potassium hydroxide on aldehydes, &c.

Natural resins are commonly produced by special cells and discharged, together with other products, into a cavity formed by the separation of the cells from one another; in this way the oleo-resins of the *Coniferæ*, the gum-resins of the *Umbelliferæ*, &c., are produced. The tissue intervening between oleo-resin ducts frequently breaks down, so that branching cavities are formed; in this case the cell-walls appear to be converted into resin or oleo-resin, but the quantity thus produced is seldom large.

Guaiacum resin, on the other hand, is secreted in the vessels and cells of the heartwood, completely filling them and thus obstructing the passage of water. The resin of Indian hemp is secreted by external, stalked glands, whilst shellac is probably produced by the lac insect.

In addition to the secretion ducts normally present in the plant, others may be formed as the result of injury, and this may also take place even if the plant produces normally no such ducts. The number of the ducts thus formed may be very large and produce large quantities of oleo-resin which, discharged over the wound, forms a temporary protection for it. This flow of oleo-resin, which is termed 'secondary flow' to distinguish from the 'primary flow' from ducts normally present, is the source of most of the oleo-resins of technical importance.

None of the resins are simple bodies; all are mixtures, and most of them are mixtures of complex nature.

The following are the chief classes of substances that have been isolated from the resins:

1. *Resinotannols*: aromatic, amorphous, brownish resin-alcohols allied to the tannins; they occur partly free, but more generally combined with aromatic acids or with umbelliferone in the form of esters (tannol-resins).

2. *Resinols*: crystalline, colourless resin-alcohols, also occurring partly free, partly in the form of esters (resinol-resins).

3. *Resin-acids*: these are partly crystalline, and mostly occur free; they combine with alkalies to form soaps, and with metals to form esters that often crystallise readily.

4. *Resenes*: these are indifferent bodies, and neither esters nor acids.

5. *Glucoresins*: these yield sugar by hydrolysis.

6. *Various acids*, more particularly belonging to the aromatic series, such as benzoic, cinnamic, paracumaric, salicylic, &c., which occur partly free and partly combined with the resin-alcohols.

The resins dealt with in this volume may be conveniently grouped according to their constituents, thus colophony, Burgundy pitch, sandarac, and amber are all of Coniferous origin, and present many points of analogy. Guaiacum contains chiefly resin-acids, and in mastich the principal constituent is a resene. Benzoin and dragon's blood contain resin-alcohols (resinols) together with aromatic acids. In shellac fatty resins are present while araroba, consisting almost entirely of crystalline substances, is not, strictly speaking, a resin, although it may conveniently be appended.

## COLOPHONY

### (Amber Resin, Resin, Resina)

**Source, &c.**—Colophony is the residue left after the distillation of the oil of turpentine from the crude oleo-resin of various species of *Pinus* (N.O. *Coniferæ*).

Most Coniferous trees contain in the wood a branching system of schizogenous secretion ducts filled with a viscid oleo-resin, which is exuded when the ducts are punctured. The amount thus yielded is, however, small. Very much larger quantities, practically the entire yield, are obtained by hacking the bark of the tree. In the new wood produced by the cambium immediately after such injury large numbers of oleo-resin ducts are formed, from which abundant oleo resin is poured out over the wound; after a time this formation of ducts and discharge of oleo-resin diminishes, and eventually ceases, but may be continued by repeating the injury which caused their production. The flow may therefore be continued for some considerable time. Hence the oleo-resin is not a normal ('physiological') but an abnormal ('pathological') product.

The term 'common' turpentine is practically restricted to the oleo-resin obtained in America, as the English market is almost exclusively supplied from that source. The bulk is obtained from *Pinus palustris*, Miller, the long-leaf pine, but *P. Tæda*, Linné, the loblolly pine, *P. echinata*, Miller, the short-leaf pine, and *P. cubensis*, Grisebach, the Cuban pine, all yield a considerable quantity. These trees, especially the long-leaf pine, form extensive forests in the southern and south-eastern United States, extending from Texas to North Carolina.

The oleo-resin is collected in the following manner:

In the winter, when no oleo-resin flows, cavities are cut in the trunk of the tree near the base; they slope inwards and downwards, and are destined to receive the turpentine. In the spring triangular incisions are made above the cavity or 'box,' the bark and part of the young wood being removed. The turpentine soon rapidly exudes and collects in the box, from which it is removed by a dipper. After eight or ten days the flow diminishes, but may be increased by cutting a strip of bark above the triangular incision; this process of hacking is repeated until the autumn, when the flow of turpentine gradually ceases. The last portions that are slowly exuded partially dry before they reach the box, and form a white incrustation on the hacked surface. This incrustation is removed and forms the drug known in America as 'scrape,' and in England as 'gum thus' or 'American' or 'common' frankincense.

The crude turpentine is removed by a dipper from the boxes to barrels for transportation to the stills. These are of copper, and set in brick furnaces. Water is added and the whole warmed, any chips of wood, &c., that float to the top being skimmed off. The head is then luted on, and the heat increased. At first, water and oil of turpentine distil over, subsequently oil of turpentine alone. Water is occasionally added to prevent the resin from charring. After the distillation has been stopped the melted resin is run through wire strainers into barrels.

The finest resin is that obtained from the tree in the first year, when the crude turpentine yields about 80 per cent. of it. After that the proportion of oil of turpentine in the oleo-resin gradually diminishes, whilst that of the resin increases, but the latter become darker and darker in colour. Tschirch has shown that the yield of turpentine is greatly increased if the trees are wounded before the spring.

So enormous is this industry in the United States that it is computed that at least 800,000 acres of virgin forest are newly invaded annually to supply the turpentine stills in operation.

**Description.**—The resin thus obtained should form pale amber-coloured, transparent, glassy masses, very brittle and easily powdered. It is rather heavier than water, the specific gravity varying from 1·070 to 1·085 (Dieterich). It has a faint terebinthinate odour and taste. At about 80° it softens, but it does not completely melt until the temperature exceeds 100°. It is soluble in alcohol, ether, and chloroform, and leaves when incinerated but little ash. Freshly

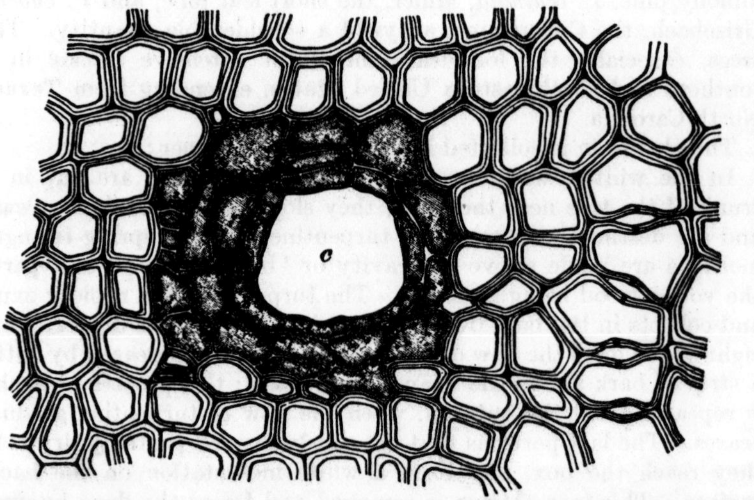

FIG. 240.—Transverse section through the wood of *Pinus maritima*, showing an oleo-resin duct, *c*. Magnified. (Tschirch.)

powdered colophony is almost entirely soluble in petroleum spirit, but becomes much less soluble by long keeping. A thin film of the resin increases in weight by exposure to the air, a change probably due to oxidation.

**Constituents.**—According to Tschirch and Studer (1903) American colophony has the following approximate composition:

| | |
|---|---|
| α-Abietic acid . . . . . . | 30 per cent. |
| β-Abietic acid . . . . . . | 22 ,, |
| γ-Abietic acid . . . . . . | 32 ,, |
| Resene . . . . . . . | 5 ,, |
| Volatile oil . . . . . . | 0·5 ,, |
| Bitter principle . . . . . | trace |

The abietic acids are unstable crystalline isomeric acids to which the formula $C_{20}H_{28}O_2$ is assigned.

The α- and β-acids are removed from an ethereal solution of the resin by shaking with solution of ammonium carbonate; the -acid by subsequent shaking with sodium carbonate solution, the resene remaining in the ethereal liquid. The acid value[1] of colophony varies from about 136 to 180 and the saponification value from 157 to 200. Although the saponification value is uniformly higher than the acid value, the resin is nevertheless free from esters.

Colophony may be identified by its high acid value and by the following reactions: 0·1 gramme dissolved in 10 c.c. of acetic anhydride and cooled assumes a bright wine-red colour on the cautious addition of sulphuric acid; 0·1 gramme warmed with 2 c.c. of methyl sulphate gives a rose or violet coloration; a solution of 0·1 gramme in petroleum spirit shaken with a 0·1 per cent. aqueous solution of cupric acetate is coloured bright emerald green.

**Uses.**—Colophony has stimulant and diuretic properties. It is used chiefly as an ingredient of ointments and plasters.

**Varieties, &c.**—Resin is obtainable in commerce in various grades ranging from 'water-white' to nearly black. Opaque resin is made by melting common resin with water.

*Bordeaux turpentine* is obtained chiefly from *P. maritĭma*, Poiret, in the south-western departments of Landes and Gironde. A vertical incision is made through the bark and about 1 cm. deep into the wood and an earthen pot is fixed at the bottom, in which the turpentine is collected. The cut is gradually lengthened until it is about 3 metres long, then the opposite side of the tree, and finally the edges of the cicatrix of the first cut are incised, the tree remaining productive for many years. A product similar to 'scrape' is also obtained; it is termed 'galipot.'

The resin consists of pimarinic, pimaric, and α- and β-pimarolic acids. The volatile oil is distinguished from American oil of turpentine by being strongly lævorotatory.

*Venice turpentine* is obtained from the larch, *Larix europœa*, de Candolle, in France and Southern Tyrol, by boring into the stem in the spring, and collecting the oleo-resin that exudes. It is a yellowish, slightly turbid, viscid liquid, with bitter aromatic taste. The resinous

[1] The acid value is the number of milligrammes of potassium hydroxide necessary to neutralise the acids present in one gramme of the resin. It is determined by dissolving a weighed quantity of the resin in alcohol and titrating with alcoholic solution of potassium hydroxide, using phenolphthalein as indicator.

The saponification value is the number of milligrammes of potassium hydroxide necessary to neutralise the acids and also to saponify the esters present. It is determined by boiling a weighed quantity of the resin with an excess of alcoholic volumetric solution of potassium hydroxide and titrating back with sulphuric acid.

The ester value is the number of milligrammes of potassium hydroxide necessary to saponify the esters present. It is obtained by deducting the acid value from the saponification value.

portion consists chiefly of α- and β-larinolic acids. A factitious mixture of colophony and turpentine is commonly substituted for it.

*Oil of turpentine* is chiefly produced by the foregoing method, Lower grades are obtained by the steam-distillation of the wood, roots, stumps, sawdust, &c. (wood turpentine), and also as a by-product in the manufacture of sulphite cellulose. For pharmaceutical use it is freed from resinous and other impurities by redistillation. A comparatively small amount of the commercial oil is imported from France. American oil is usually dextrorotatory, though it may be slightly lævorotatory. French is strongly lævorotatory. It consists chiefly of *d*- and *l*-pinene. Specific gravity, 0·860 to 0·870; refractive index at 25°, 1·465 to 1·480. It should distil almost entirely between 156° and 180° leaving no appreciable residue. It may be adulterated with illuminating petroleum or with rosin oil.

Colophony distilled at 80° to 250° yields 'rosin spirit' which is colourless, insoluble in water and in alcohol; about one-half consists of hydrocarbons boiling below 120°. At about 300° the resin yields 'rosin oil' a viscous, colourless to dark brown oil consisting of a mixture of hydrocarbons of high boiling point together with abietic acid.

## BURGUNDY PITCH

### (Pix Burgundica)

**Source, &c.**—Burgundy pitch is a resinous exudation obtained from the stem of *Picĕa excelsa*, Link (N.O. *Coniferæ*), melted and strained.

Burgundy pitch is collected chiefly in Finland, smaller quantities being obtained in the Black Forest and in the Jura mountains. Incisions are made through the bark into the outer layers of wood, and the oleo-resin that slowly collects between the bark and the wood is, after the lapse of some months, scraped out of the incisions, melted under water, and strained.

**Description.**—True Burgundy pitch is an opaque, yellowish brown or dull reddish brown substance, hard and brittle, but gradually taking the form of the vessel in which it is kept. It is strongly adhesive, breaks with a conchoidal fracture, and has a very agreeable aromatic odour, especially when heated. The taste is sweet and aromatic without bitterness.

It is soluble in twice its weight of glacial acetic acid, and readily soluble in alcohol.

**Constituents.**—Burgundy pitch contains resin and a little volatile oil. The latter appears to consist chiefly of α- and β-picea-pimarolic acids, small quantities of picea-pimarinic and picea-pimaric acids, and resene (juro-resene).

**Uses.**—Burgundy pitch is used in the preparation of a few plasters.

**Varieties, &c.**—Much of the Burgundy pitch of commerce is a factitious mixture of resin, turpentine, and palm oil; it may be distinguished from the genuine by its incomplete solubility in twice its weight of glacial acetic acid ('Pharmacographia').

## AMBER

### (Succinite, Succinum)

**Source, &c.**—Amber is the fossil resin of certain extinct Coniferous trees, chiefly of *Pinus succinifĕra*, Conwentz. The term is a generic one, several varieties of amber having been distinguished, but that known as Baltic amber, or, better, succinite, is the only one of commercial importance.

The amber-yielding trees probably formed, in the early part of the period geologically known as the oligocene and belonging to the tertiary formation, extensive forests over the northern parts of Scandinavia. They contained oleo-resin ducts similar to those of Coniferous trees of the present day, and in addition frequently produced abnormal quantities of oleo-resin as the result of injuries, to which they appear to have been especially liable. The oleo-resin, hardened by exposure and liberated by the gradual destruction of the trees, was carried, towards the latter end of the oligocene period, from the position occupied by the forest and deposited in a bed of blue earth of considerable extent near the eastern shores of the Baltic. From this bed of earth which lies below the sea-level the amber is now recovered chiefly by mining, but some little is washed by the sea from the exposed surface of the stratum and thrown up by the waves on the shore.

Succinite is also occasionally found on the east coast of England, whither it has been brought by the sea.

**Description.**—Succinite occurs in pieces varying much in size and shape, usually with rounded edges and covered with a dark crust. Internally it is transparent, translucent, or quite opaque (cloudy amber), and of varying shades of yellow or brown. It exhales when warmed a slight but characteristic odour, and is almost tasteless. It is hard, breaking with a bright, conchoidal fracture that is sometimes glassy, sometimes opaque. It is partially soluble (about 30 per cent.) in alcohol, ether (about 20 per cent.), and chloroform (about 20 per cent.); it melts at 280° to 290°, at the same time decomposing and yielding water, succinic acid, and various tarry products.

**Constituents.**—That portion of succinite which is soluble in alcohol consists of free succino-abietic acid (itself a complex mixture the chief constituent being succino-abietinolic acid together with succoxyabietic acid, succino-silvic acid and succino-abietol) associated with a little bornyl succino-abietate. The part insoluble in alcohol, succinin, is a compound of succinic acid, with a resin-alcohol, succino-resinol. Succinite contains in addition traces of sulphur and of inorganic substances. By destructive distillation it yields a dark tarry oil accompanied by water; the former, separated from the watery fluid and redistilled, forms the genuine yellow oil of amber of commerce, which may be distinguished from a common substitute made from colophony by its higher specific gravity (0·926 to 0·930) and optical rotation (+ 23° to + 26°).

## GUAIACUM RESIN

### (Resina Guaiaci)

**Source, &c.**—Guaiacum resin is the resin obtained from the stem of *Guaiacum officinale*, Linné, or *Guaiacum sanctum*, Linné (N.O. *Zygophyllĕæ*). The bulk of the resin of commerce is produced in the following rather crude way from the trunk of the tree, the heartwood of which, as already stated (see 'Guaiacum Wood'), contains from 20 to 25 per cent. of resin: A log of the wood is supported in a horizontal position above the ground by two upright bars. Each end of the log is then set on fire, and, a large incision having been previously made in the middle, the melted resin runs out therefrom in considerable abundance ('Pharmacographia'); or one end of a log of wood is raised, and fire applied to it, when the melted resin will run out of a groove cut in the other end, and may be received in potsherds (block resin).

It is said that the resin may also be obtained in the form of tears by incisions made into the trunk, but it is more than doubtful whether the tear resin of commerce is so produced.

**Description.**—Guaiacum resin is usually seen in large masses of dark colour, often more or less covered with a greenish powder. The resin breaks easily with a clean, glassy fracture, thin splinters viewed by transmitted light being transparent, and varying in colour from yellowish green to reddish brown. The powder is greyish, but becomes green by exposure to light and air. It has a slightly acrid taste, and, especially when warmed, a somewhat balsamic odour. It is freely soluble in alcohol, chloroform, and solution of caustic potash, incompletely in ether, but only slightly soluble in petroleum spirit, carbon disulphide, or benzene.

The resin in tears forms rounded masses, attaining 2 to 3 cm. in diameter, usually covered with a greenish powder, and exhibiting the characters already detailed.

The commercial drug is never completely soluble in alcohol. The residue, which in the case of tear resin is about 1·5 per cent., and in good samples of the lump averages about 7·5 per cent. (Evans, 1898; not over 10 per cent., B.P., 1914), may in exceptional cases amount to as much as 25 per cent. It consists chiefly of fragments of vegetable débris, gummy matter, &c.

Guaiacum resin is easily identified by its remarkable reaction with oxidising agents. This is best seen by dissolving a little of the resin in alcohol and adding a drop of dilute solution of ferric chloride; the liquid instantly assumes a deep blue colour which is destroyed by reducing agents, but restored by oxidising agents.

The student should observe

(*a*) The *varying colour* that the resin exhibits when viewed by transmitted light,

(*b*) The *greenish powder* with which it is often more or less covered,

(*c*) Its *characteristic*, though not powerful, *odour and taste*.

**Constituents.**—Guaiacum resin consists chiefly of resin-acids and has approximately the following composition :—

| Constituent | |
|---|---|
| α-Guaiaconic acid, β-Guaiaconic acid | 70·00 per cent. |
| Guaiaretic acid | 11·25 ,, |
| Guaiac-β-resin | 15·00 ,, |
| Guaiacic acid, guaiac-yellow, vanillin, saponin, &c. | 2·25 ,, |

*α*-Guaiaconic acid is a colourless amorphous substance, probably a mixture, one constituent of which is changed by oxidising agents to deep blue guaiac-blue; sulphurous acid reduces this to a colourless substance, which may again be oxidised.

*β*-Guaiaconic acid is colourless and crystalline. Guaiaretic acid is light brown, amorphous, and insoluble in ether. Guaiac-β-resin is brown and amorphous, and appears to be chiefly a decomposition product of the guaiaconic acids; it contains the substance that yields guaiac-blue by oxidation.

**Uses.**—The action of guaiacum is that of a local stimulant or, in large doses, irritant. It has been employed locally in the form of the lozenge, and has also been given in chronic gout and rheumatism.

**Adulterants, &c.**—Guaiacum resin is normally so cheap that it offers little inducement for adulteration. Colophony which has been mixed with it may readily be detected by shaking the freshly powdered resin with petroleum spirit and shaking the filtrate, which should be colourless, with 0·1 per cent. solution of cupric acetate; if colophony is present the petroleum spirit will assume a vivid green colour. The acid value is about 90, but the determination is unsatisfactory, as the end-reaction is indefinite. Guaiacum resin in other resins may generally be detected by the ferric chloride reaction.

*Peruvian guaiacum* is brownish and aromatic, and is used in perfumery; its botanical origin is unknown.

## ARAROBA

### (Goa Powder)

**Source, &c.**—Araroba, or, as it is often termed, Goa powder, is a substance found in cavities in the trunk of *Andīra Ararōba*, Aguiar (N.O. *Leguminosæ*), a large tree common in the damp forests of Bahia (Brazil).

This remarkable substance is found filling longitudinal fissures in the trunk of the tree. From careful microscopical examinations that have been made of the fragments of wood picked from the crude drug, it would appear that the walls of both parenchymatous and prosenchymatous cells, as well as those of the vessels, undergo a complete change, by which they are converted into a yellowish, powdery, substance, araroba. This change is observable first in the secondary thickenings of the cell-wall, but afterwards the cell-walls themselves break-down, thus forming cavities of considerable size which are filled with the araroba. The exact nature of the change of the influences that induce it, and the manner in which it is effected, is unknown; it appears to be a pathological not a normal physiological process.

Araroba is collected by felling the tree, sawing the trunk into lengths, and splitting these longitudinally. The yellowish powder is then scraped out with the axe, by which means numerous splinters of wood and other débris are simultaneously removed. It is exported in that crude condition, and may be purified by sifting it as free as possible from fragments of wood, drying, and powdering it. The drug appears to have been long known to the natives of Brazil as a cure for certain skin diseases. In 1864 Kemp drew attention to the Goa powder that was used in India for similar purposes, and this was proved in 1875 to be identical with the araroba of the Brazilians.

**Description.**—The crude drug, as imported, consists of a brownish yellow or umber-brown powder mixed with numerous small and large fragments of wood. Microscopical examination of the powder shows the presence of numerous minute prismatic crystals, and granular, amorphous matter accompanied by vegetable débris. The smoothed transverse surfaces of the larger fragments of wood show thin, yellow, medullary rays, vessels, and here and there yellow masses (of araroba). As already observed, it is sometimes purified by simply freeing it from the débris present, drying and powdering; but the purification is more generally effected by grinding the crude drug, drying it, boiling it with benzene, and filtering; the hot benzene solution deposits on cooling a yellow, crystalline powder which forms the chrysarobin of commerce. This is a yellow, crystalline, tasteless, and inodorous powder, soluble in hot chloroform and benzene, almost entirely soluble in hot alcohol, very sparingly and incompletely soluble

FIG. 241.—Araroba. Portion of a transverse section of the wood of *Andira Araroba*, near to a cavity filled with araroba. *S*, vessels; *l*, wood fibres; *p*, wood parenchyma; *m*, medullary rays. Nearly all the elements contain a dark substance, which in the upper part forms a dense mass, *x* (omitted from the illustration). Magnified. (Vogl.)

in petroleum spirit, and practically insoluble in water. Hot solution of potash dissolves it almost entirely. To solution of ammonia it imparts at first a slight pink tinge; this colour, however, quickly deepens owing to oxidation of the chrysarobin (which is insoluble in ammonia) to chrysophanic acid (which is soluble).

Purified araroba melts when heated, gives off yellow fumes, and finally burns, leaving not more than 1 per cent. of ash.

Crude araroba, as imported, often contains from 15 to 30 per cent. of water, which appears to be added to prevent the irritating dust from rising; it may yield from 50 to 75 per cent. of purified araroba.

**Constituents.**—Purified araroba varies considerably both in the constituents present and the proportion in which they occur. The following table may suffice to indicate its approximate composition (Tutin and Clewer, 1913) :—

| | |
|---|---|
| Chrysophanol (chrysophanic acid) . . | 4·7 to 6·7 per cent. |
| Emodin monomethyl ether . . . | about 1·6 to 2·2 „ |
| Chrysophanolanthranol ('chrysarobin') | about 26·0 to 62·0 „ |
| Anthranol of emodin monomethyl ether . | Small amount |
| Monomethyl ether of dehydroemodin-anthranol . . . . . . . | 13·4 to 41·1 per cent. |
| Ararobinol . . . . . . . | about 4 „ |
| Emodin . . . . . . . . | trace |
| Inseparable mixture of substances and amorphous matter . . . . . | about 12·0 to 30·0 per cent. |

*Chrysophanolanthranol*, often called chrysarobin, $C_{15}H_{12}O_3$, crystallises in yellow leaflets melting at 199°. It is insoluble in aqueous ammonia, but is converted by oxidation into chrysophanol which is soluble.

*Chrysophanol* (chrysophanic acid), $C_{15}H_{10}O_4$, crystallises in deep orange-coloured leaflets melting at 191°.

*Monomethyl ether of dehydroemodinanthranol*, $C_{16}H_{12}O_4$, crystallises in pale yellow needles melting at 265°.

*Ararobinol*, $C_{23}H_{16}O_5$, crystallises in yellow crystals without definite melting point.

There is no definite knowledge as to which constituent or constituents its medicinal value is to be attributed.

**Uses.**—Purified araroba has been used successfully in ringworm, psoriasis, and other skin diseases; it acts apparently by destroying low vegetable organisms.

## BENZOIN

### (Benzoinum)

Benzoin occurs in several well-defined commercial varieties, one only of which, viz., Sumatra benzoin, is official.

The trees from which benzoin is obtained do not contain any special secreting cells or ducts, nor is normally any benzoin produced; the drug is a purely pathological product, the formation of which may be induced by injury to the tree.

#### Sumatra Benzoin

**Description.**—This, the official variety, is obtained from *Styrax Benzoin*, Dryander (N.O. *Styraceæ*), a tree indigenous to Sumatra and Java. The drug is obtained by hacking the bark of the tree with an axe. After the infliction of such injury the cambium rapidly produces new wood in which a ring (or sometimes two rings) of oleo-resin ducts are formed. By the breaking down of the tissue intervening between the ducts large schizolysigenous ducts are produced in which considerable quantities of oleo-resin are secreted. These ducts open on to the wounded surface over which their secretion is discharged. Similar ducts are also formed in the bark. The benzoin thus formed partly accumulates between the bark of the tree and the trunk and partly exudes from the incisions ; it is allowed to get firm and is then collected, bought by Chinese traders, softened by heat or hot water, and packed in oblong boxes, being often, it is said, mixed with vegetable débris in the course of repacking.

Sumatra benzoin occurs in masses consisting of opaque, white tears embedded in a dull, greyish-brown, or sometimes reddish-brown matrix. It possesses an agreeable, though not very powerful odour, recalling storax, and a slightly acrid taste. When cautiously heated it melts and evolves whitish irritating fumes of benzoic and cinnamic acids. When a little of the crushed resin is warmed with alkaline solution of potassium permanganate benzaldehyde is evolved, indicating the presence of more than traces of cinnamic acid in the drug.

Sumatra benzoin is frequently contaminated with pieces of bark and other débris, especially in the angles and round the sides of the box in which it is packed.

Good Sumatra benzoin should not contain more than 10 per cent. of substances insoluble in alcohol, whilst inferior qualities often yield up to 30 per cent. ; it should also not afford more than 5 per cent. of ash.

**Constituents.**—Sumatra benzoin consists principally of two alcohols combined with cinnamic acid, and associated with free benzoic and free cinnamic acid. Of the two alcohols, òne, benzoresinol, is probably identical with the siaresinol of Siam benzoin ; the nature of the other is not definitely known ; it has been stated to be sumaresinotannol, but this alcohol may have been produced from the alcohol originally present (? lubanol) by oxidation during the extraction (compare Siam benzoin). The drug contains, in addition, traces of benzaldehyde, vanillin (1 per cent.), phenylpropyl cinnamate, styrol, and styracin, all of which combine to produce its particular fragrance.

The total amount of aromatic acids present in the commercial drug, freed from vegetable débris and moisture, averages 26 per cent., free cinnamic acid 10·3, combined cinnamic acid 7·3, free benzoic acid 6·5 and combined benzoic acid 2·5.

## SIAM BENZOIN

**Description.**—This variety of benzoin is collected in the Siamese province of Luang Prabang, in the neighbourhood of the Mekong

River. Part is exported to London *via* Singapore, but much reaches the market by way of France. It appears in commerce either as separate tears or in the form of masses composed of tears more or less firmly cemented together by a dark reddish brown, transparent resin.

The tears vary considerably in size, but seldom exceed 5 cm. in length, by 1 cm. in thickness; usually they are much smaller. They are flattened, or sometimes, if large, concavo-convex, this shape being evidently caused by the resin accumulating between the trunk and bark of the tree. They are opaque, brittle, and milky-white internally, but are usually covered with a thin coating of brownish resin, which increases as the tears are kept, being produced from the opaque tear by oxidation.

Block Siam benzoin consists of small tears cemented together by a glassy, reddish-brown, transparent or translucent resin, which gives them a peculiar varnishy appearance. In this case also the proportion of the red, transparent resin increases as the drug is kept until it becomes its most prominent feature.

Both varieties of Siam benzoin are characterised by their agreeable fragrant odour, recalling vanilla; they are almost entirely soluble in alcohol and in ether, yielding only about 2 per cent. of insoluble residue. When quite pure Siam benzoin affords only traces of ash.

Good commercial samples of Siam benzoin should not yield more than 3 per cent. of substances insoluble in alcohol or more than 1 per cent. of ash.

**Constituents.**—The white tears of Siam benzoin are crystalline and consist chiefly of the benzoate of a resin alcohol, lubanol. When warmed to about 50° the white tears become yellow, red and brown and lose their crystalline character, a change due to oxidation. In addition the drug contains siaresinol benzoate which also is crystalline and an amorphous benzoate which in alkaline solution, readily oxidises.

Further constituents are vanillin and an oily aromatic liquid, probably an ester of benzoic acid. Siam benzoin contains about 39 per cent. of total aromatic acids (about 23 per cent. free and 16 per cent. combined), 36 per cent. being benzoic acid and 3 per cent. cinnamic acid (Cocking and Kettle, 1914). This proportion of cinnamic acid is so small that it cannot be detected by the official test (heating with solution of potassium permanganate).

### Penang Benzoin

Two distinct varieties of Sumatra benzoin have been known under this name. The one, now generally termed storax-benzoin, has a very agreeable odour resembling storax; the other, known as 'glassy Penang' or simply 'Penang' benzoin, is distinguished by its glistening, glassy fracture and slight odour. Both varieties are packed like Sumatra benzoin.

### PALEMBANG BENZOIN.

This variety, which is not official, is produced in Sumatra, it is said from *S. Benzoin.* It is commonly imported in tins, four of which are packed in a wooden case. It differs markedly in appearance from both Siam and Sumatra benzoins, consisting principally of a translucent, greyish-brown or reddish resinous mass in which a few scattered, opaque, white tears are embedded. It breaks with an irregular fracture, the fractured surface being uneven and often exhibiting small cavities. The odour, which is not strong, recalls that of Sumatra benzoin. Its constituents have not yet been accurately investigated; it appears to contain benzoic but no cinnamic acid. It is not official, and is used only for the preparation of benzoic acid.

**Uses.**—Benzoin taken internally acts as a carminative expectorant and diuretic ; externally it is stimulant and antiseptic.

## DRAGON'S BLOOD

### (Sanguis Draconis)

**Source, &c.**—Dragon's blood is a resinous secretion produced on the fruits of *Dæmonōrops propinquus*, Beccari, *D. ruber*, Martius, and probably other species. The two species named were formerly included in *Calamus Draco*, Willdenow (N.O. *Palmæ*) ; they are climbing palms with long, flexible stems, and are indigenous to Sumatra and Borneo.

The plant produces numerous small fruits about the size of a cherry, covered with hard, yellowish, imbricated scales, which overlap one another from apex to base. From between these scales a red resin, probably produced in the pulp of the fruit, exudes and more or less completely covers the fruit. The fruits are shaken together in a basket, and the separated resin mixed with water, pressed into moulds, and then melted; or it is made into a cake which is wrapped in a cloth, steeped in hot water and pressed to form a solid block. It is said to be nearly always mixed with the milky juice of *Garcinia parviflora*, Miquel (Treub, 1891).

**Description.**—Dragon's blood occurs in lumps of very varying size and shape. They are often large rounded masses, sometimes weighing several kilograms, bearing the impress of sacking or reed-matting, or they may be rounded, flattened cakes 10 cm. or more in diameter and about 5 cm. in thickness ; sometimes the cakes are smaller and thinner. Occasionally it is imported in sticks about 20 to 25 cm. long and 2 to 3 cm. thick, or 30 cm. long and

1·5 cm. thick, each carefully wrapped in a palm leaf. These varieties are known as 'lump,' 'saucer,' 'reed,' &c., dragon's blood.

Good samples of the drug usually have a dull, dark red colour, and are more or less covered, where the pieces have rubbed against one another, with a crimson powder. They are brittle and friable, breaking with a glossy but irregular, uneven fracture, minute fragments being translucent and of a deep garnet-red colour.

The drug yields when crushed a bright crimson powder, has no odour, and is practically tasteless, breaking up when chewed into a fine gritty powder. Inferior qualities are duller in colour and tougher. They yield a duller crimson or even brick-red powder, and exhibit less powder on the surface of the lumps. Such specimens frequently contain numerous fragments of the fruit scales, which are easily seen when the drug is broken, or are left when it is exhausted with alcohol.

Tears, in which form the drug is now seldom seen, give a glassy, conchoidal fracture, thin flakes being of a clear garnet-red colour.

**Constituents.**—Dragon's blood consists principally of a red resin (56·8 per cent.), a compound of dracoresinotannol with benzoic and benzoylacetic acids. Other constituents are a white, amorphous dracoalban (2·5 per cent.), a yellow resinous dracoresene (13·58 per cent.), vegetable débris (18·4 per cent.), and ash (8·3 per cent.).

It is frequently considerably adulterated both with earthy matter and with fragments of the scales of the fruits, the amount of residue insoluble in alcohol amounting sometimes to as much as 40 per cent. of the drug.

**Uses.**—Dragon's blood is chiefly used for colouring varnishes, &c.

**Varieties.**—The term 'dragon's blood' has also been applied to several other resins resembling Sumatra dragon's blood in appearance. They may be distinguished by their insolubility in benzene and carbon disulphide. The only one of these that appears in commerce is Socotrine dragon's blood which is occasionally imported from Bombay and Zanzibar and is technically termed 'Zanzibar drop' dragon's blood. It is obtained from *Dracæna Cinnabari*, Balfour. It occurs in small tears or fragments seldom exceeding 2 cm. in length with a vitreous fracture, thin splinters being of a ruby red colour. It does not when heated evolve an odour of benzoic acid, and contains no scales similar to those found in Sumatra dragon's blood.

## MASTICH

### (Mastiche)

**Source, &c.**—Mastich is a resin obtained from a broad-leaved variety of *Pistacia Lentiscus*, Linné (N.O. *Anacardiaceæ*), a shrub or small tree indigenous to the countries bordering on the Mediterranean.

The resin, which has been known from the earliest times. and was formerly much more highly prized than it is now, is collected on the island of Scio in the Grecian Archipelago, and also in Cyprus, and possibly on other islands, but is exported only from Scio.

The bark of the tree, which contains a circle of oleo-resin ducts in the bast, is punctured with a small instrument resembling a chisel ; the oleo-resin exudes in the form of small tears which in a few days become dry and hard. It is then collected, that taken from the tree itself forming the best qualities, whilst that which has dropped upon the ground is inferior.

**Description.**—Mastich occurs in small hard tears about the size of peppercorns. The majority are pear-shaped, ovoid, or nearly globular ; sometimes, but not often, they are elongated and resemble small stalactites. When fresh they are nearly colourless and quite clear and bright, but by keeping and handling they become pale yellow in colour and acquire a dull, dusty surface. They are brittle, breaking with a clear, glassy, conchoidal fracture, the interior of the tears being quite transparent. When chewed the tears break up at first into a sandy powder, which subsequently agglomerates into a plastic mass. The drug has an agreeable, rather aromatic odour, and a slight agreeable taste, both of which though, not pronounced, are characteristic.

The student should observe

(*a*) The preponderance of *rounded or pear-shaped tears*,
(*b*) The *characteristic odour*,
(*c*) The formation of a *plastic mass when the resin is chewed* ;

and should compare the drug with

*Sandarac resin* (see below).

**Constituents.**—Mastich has approximately the following composition:

| | | |
|---|---|---|
| α- and β-Masticonic acid, amorphous, soluble in alcohol | 38·0 | per cent. |
| α-Masticoresene, soluble in alcohol | 30·0 | ,, |
| β-Masticoresene, insoluble in alcohol | 20·0 | ,, |
| α- and β-Masticinic acids | 4.0 | ,, |
| Volatile oil | 2·0 | ,, |
| Masticolic acid, crystalline | 0·5 | ,, |

It consists, therefore, chiefly of resin acids and resenes associated with about 2 per cent. of volatile oil (chiefly *d*-pinene).

**Uses.**—Mastich was formerly employed as a stimulant, and was also used in the manufacture of varnishes. For the latter purpose it has been superseded by other cheaper resins, whilst as a medicine it is obsolete.

**Varieties.**—Bombay or East Indian mastich is obtained from *P. Khinjuk*, Stokes (and probably other species); it is darker than genuine mastich, less vitreous, more soluble in alcohol and less soluble in oil of turpentine. It may also be distinguished by the acid value (103 to 109) which is much higher than that of genuine mastich (45 to 67).

## SANDARAC

### (Gum Juniper, Sandaraca)

**Source, &c.**—Sandarac is a resin obtained from *Callĭtris quadrĭvalvis*, (N.O. *Coniferæ*), a small tree about 7 metres high, growing on the mountains in the north-west of Africa. It is usually obtained by incision, the tears when sufficiently hard being collected and exported, chiefly from Mogadore.

**Description.**—Sandarac occurs in small tears about the same size as mastich, but usually of an elongated, more or less cylindrical or stalactitic form, several of which are sometimes united into a small, flattened mass. Globular or pear-shaped tears are comparatively rare in it, and by this means it can be distinguished at sight from mastich.

The tears, as usually seen in commerce, have a dull dusty surface and a pale yellowish colour; they are brittle, breaking with a glassy conchoidal fracture, and displaying a clear, transparent interior, in which, as in amber, small insects are occasionally embedded. The resin has a slight terebinthinate odour and a terebinthinate, slightly bitter taste; when chewed it breaks up between the teeth into a sandy powder which, unlike mastich, shows no disposition to agglomerate into a plastic mass.

It is completely soluble in alcohol and ether, partially only in chloroform, carbon disulphide, and oil of turpentine.

The student should observe

(*a*) The *preponderating stalactitic form*,
(*b*) The *terebinthinate odour*,
(*c*) The *indisposition to form a plastic mass between the teeth*;

and should compare this drug with

(i) *Mastich* (see p. 468),
(ii) *Olibanum* (see p. 481).

**Constituents.**—Sandarac consists of resin associated with traces of volatile oil, bitter principle, &c. The chief constituent of the resin is (optically) inactive pimaric acid (sandaraco-pimaric acid, 85 per cent.), obtainable in acicular crystals melting at 170°; other constituents are sandaracinic acid (2·3 per cent.), amorphous callitrolic acid (10 per cent.), and sandaracoresene. Callitrolic acid is easily converted into the lactone which is insoluble in alcohol.

**Uses.**—Sandarac is chiefly used in the manufacture of varnishes; it is paler in colour than shellac, and is therefore more suitable for light woods.

**Varieties.**—*Australian sandarac*, from *C. verrucōsa*, Robert Brown, is occasionally imported. The tears are softer, larger, and more aromatic than those of African sandarac, which it otherwise resembles. Its composition is similar, but it contains more volatile oil and more inactive pimaric acid.

## SHELLAC

### (Lac, Lacca)

**Source, &c.**—Shellac is a resinous exudation that encrusts the bodies of *Coccus* (*Tachardia*) *Lacca*, Kerr (Order *Hemiptera*, Family *Coccidæ*).

These minute insects live upon plant juices sucked up by a proboscis that penetrates the succulent tissues of the host. Twice or sometimes three times in the year the larvæ emerge from the dead bodies of the females, crawl away and establish themselves in new situations. They are minute creatures, about 0·5 mm. long, of an orange-red colour with fully formed feelers and powerful legs, but no visible separation of the body into head, thorax, or abdomen. They puncture the young twigs with their probosces and suck up the juices; they become fixed, and their legs, being no longer required, drop off; a resinous secretion forms around their bodies and gradually more or less encrusts the twigs. At the period of maturity the males, which undergo a complete metamorphosis, escape from their pupæ, impregnate the females and shortly afterwards die. The females rapidly increase in size, assume a bright red colour, develop viviparous larvæ and die. New colonies are established by cutting off twigs with the gravid females and tying them on to various trees.

The principal trees visited by the lac insects are *Aleurītes laccifera*, Willdenow (N.O. *Euphorbiaceæ*), *Ficus religiosa*, Linné (N.O. *Urticaceæ*), *Schleichĕra trijuga*, Willdenow (N.O. *Sapindaceæ*), *Butēa frondosa*, Roxburgh (N.O. *Leguminosæ*). *Acacia arabica*, Willdenow (N.O. *Leguminosæ*), and *Cajānus indica*, Sprengel (N.O. *Leguminosæ*), are cultivated for that purpose.

Whether the resin is secreted by the insects or produced by the plant as a result of the irritation caused by the insects is at present not definitely known. The resin shows, in its chemical composition, a variation from vegetable resins, and this seems to indicate that it owes its origin, in part at least, to a change effected by the insects in the constituents of the plant.

After the secretion of the resin the twigs are broken off, and form without further preparation the stick lac of commerce. Stick lac therefore consists of the twigs of the trees coated with a granular brownish or reddish resin which is frequently 7 mm. thick. Enclosed in this resin are the bodies of the insects, which contain a valuable colouring matter. The latter is extracted by breaking the lac from the twigs, crushing it, and exhausting it with water and sometimes subsequently with dilute soda solution. The aqueous solution contains a red colouring matter, laccaic acid, allied to, but not identical with, carminic acid; it is evaporated to dryness, and the residue, pressed into cakes, constitutes the lac dye of commerce. After the resin has been crushed and freed from colouring matter, it is spread out on floors to dry and bleach; it then forms brownish grains which are known as seed lac. This still contains various débris of the insects, &c.

It is melted with a little orpiment and resin (2 to 5 per cent.) and pressed whilst hot through a cloth. The strained lac thus obtained is spread into sheets about 3 mm. thick which are softened by heat and stretched to thin sheets. These sheets, broken up and sorted, constitute the shellac of commerce. Garnet lac is a dark coloured variety, and button lac is in the form of thin discs.

Lac is prepared in various parts of India, particularly in Bengal and Assam, and is exported chiefly from Calcutta.

**Description.**—Shellac usually occurs in thin, brittle, translucent, leafy flakes devoid of odour and taste. The colour varies from brownish yellow (orange shellac) to deep reddish brown (garnet shellac), the palest being considered the best. It melts when heated, evolving a characteristic odour.

**Constituents.**—A recent investigation of stick lac (Farner, 1899) resulted in the separation of the following constituents :—

| | | |
|---|---|---|
| Wax | 6·0 | per cent. |
| Colouring matter (laccaic acid) | 6·5 | „ |
| Resin | 74·5 | „ |
| Residue (sand, vegetable and animal débris) | 9·5 | „ |
| Moisture and loss | 3·5 | „ |

Of the resin 35 per cent., composed principally of fatty acids, was soluble in ether containing alcohol. The part insoluble in ether (65 per cent.) consisted of a resinotannol combined with aleuritic acid, a crystalline acid belonging to the fat acid series. A colouring matter erythrolaccin, crystallising in golden yellow crystals, is also present.

Shellac consists principally of the resin (90 per cent.) and wax and other substances (10 per cent.) Orpiment, although used in its preparation, seldom occurs in shellac. Adulteration with colophony may be detected by dissolving in alcohol, precipitating the resin with water, filtering, drying, and digesting with petroleum spirit; if colophony is present the petroleum spirit solution will assume an emerald-green colour when shaken with a 0·1 per cent. aqueous solution of cupric acetate. The presence of colophony is also indicated by the iodine value which should not exceed 12 in pure shellac (colophony 136 to 180).

Bleached shellac is made by dissolving shellac in hot solution of sodium carbonate and adding chlorinated lime, which forms chlorinated soda and deposits calcium carbonate. The resin is then precipitated, collected, 'pulled' under water, twisted into sticks, and kept under water.

**Uses.**—Shellac is largely used in varnishes, but is not employed medicinally.

The following resins are largely used for various technical purposes :

*Zanzibar copal* (gum animi) ; from *Trachylobium Hornemannianum*, Hayne, (N.O. *Leguminosæ*) ; a fossil resin dug up on the East Coast of Africa ; varies much in size ; pale yellow to deep reddish-brown or greenish-red ; transparent or semi-transparent ; surface warty or longitudinally striated ; consists chiefly of trachylolic acid (80 per cent.) isotrachylolic acid and resene.

*American copal*, from *Hymenœa Courbaril* (Brazil), Linné ; pale brown transparent and brittle, of agreeable odour.

*Australian copal* (gum Kauri) ; from *Agathis australis*, Stender ; mostly fossil, pale yellow or greenish yellow, conchoidal vitreous fracture, balsamic odour.

*West African copal* ; from *Copaifera Guibourtiana*, Benth.

*Manilla copal* (=East Indian Dammar) ; from *Dammora orientalis*, Lambert (N.O. *Coniferæ*) ; fragments of large masses ; yellow, slight but distinctly aromatic odour.

The copals are entirely soluble in alcohol, but only partially in benzene, chloroform, &c.

*Dammar*, probably from a species of *Shorea* (N.O. *Dipterocarpeæ*) ; imported from Singapore ; pale yellow, transparent or translucent nodules or stalactitic masses coated with white powder ; fracture vitreous, often exhibiting air bubbles.

*Gum accroides* (grass-tree gum ; black-boy gum), from *Xanthorrhœa hastilis*, Robert Brown, and *X. australis*, Robert Brown (N.O. *Liliaceæ*, Australia). The resin of the former is yellow and contains xanthoresinotannol combined with paracumaric acid, and also free paracumaric and cinnamic acids ; that of the latter is red and contains erythroresinotannol combined with paracumaric acid, free paracumaric acid but no cinnamic acid. The resin is a natural exudation covering the stem and leaf bases of the plant.

## SECTION XV

## GUM-RESINS

The gum-resins consist, as their name indicates, chiefly at least of resin and gum. With these constituents, however, there are always associated small quantities of other substances such as volatile oil, bitter principle, enzyme, &c. They are secreted in schizogenous or schizolysigenous ducts, being formed in the resinogenous layer of the walls of the tapetal cells, and discharged into the ducts in the form of milky liquids which are exuded when the ducts are punctured.

The resins contained in the gum-resins exhibit in general the characters detailed on pp. 453–4.

The gum of most of the gum-resins resembles, but is not identical with acacia gum, very possibly it consists of two of more glucosidal acids in varying proportions. It is always accompanied by an enzyme from which it has never yet been freed ; it therefore always contains nitrogen.

### GAMBOGE

### (Cambogia)

**Source, &c.**—Gamboge is a gum-resin obtained from *Garcinia Hanburii*, Hooker filius (N.O. *Guttiferæ*), a tree of moderate size found in Cambodia, Siam, and the southern parts of Cochin China.

The bark of the tree contains in the cortex, as well as in the bast, secretory ducts filled with a yellow, resinous emulsion, the two systems of ducts being connected by transverse canals at the nodes.

The gamboge is obtained by making, in the rainy season, a spiral cut in the bark from the height of about 3 metres down to the ground. The emulsion wells out and trickles down the incision into a hollow bamboo placed to receive it. From this it is transferred to smaller bamboos ; these are set aside until, in about a month, the gamboge has solidified. It is removed from the bamboo by drying over a fire until the bamboo cracks and can be stripped off. The drug is sold to local collectors, who convey it to Bangkok or Saigon, whence it is exported to Europe, usually *via* Singapore.

It is occasionally formed whilst soft into cakes of various shapes or into thick sausage-like masses, which are wrapped in leaves, the impression of which they bear on their surface (Saigon gamboge).

**Description.**—The finest qualities of gamboge occur in rolls, 3 to 5 cm. in thickness, and from 10 to 20 cm. in length, nearly cylindrical, solid or hollow in the centre, and marked externally with longitudinal striations derived from the inner surface of the bamboos in which they have been dried. The drug breaks easily, with a smooth, uniform, conchoidal fracture, the freshly fractured surface having a dull gloss and being of a rich reddish yellow or brownish orange colour. It is easily reduced to a bright yellow powder, with little odour, but with an acrid taste.

Thin splinters mounted in oil and examined under the microscope exhibit a ground-mass of gum in which numerous minute granules of resin are scattered accompanied by occasional crystals of calcium oxalate and starch grains derived from the incised tissues.

Rubbed with the wet finger gamboge instantly forms a yellow emulsion. It is almost completely dissolved by the successive action of alcohol and water. The yellow emulsion yielded with water becomes nearly clear and deep orange red on the addition of ammonia.

**Constituents.**—Gamboge consists essentially of a mixture of 70 to 80 per cent. of resin, with 15 to 25 per cent. of gum.

The resin, formerly known as cambogic acid, is soluble in alcohol, ether, chloroform, benzene, petroleum spirit, &c., as well as in solutions of alkaline hydroxides and carbonates; from its alkaline solutions it is precipitated by acids. From it three organic acids have been separated, viz. $\alpha$-, $\beta$- and $\gamma$- garcinolic acids, the last named being characterised by the red colour of even a very dilute alkaline solution.

The gum is analogous to acacia gum; it is lævorotatory and contains an oxydase enzyme.

**Uses.**—Gamboge produces purging and in large doses vomiting. It has been employed as a hydragogue cathartic, but is now seldom used as a medicine.

**Adulterants.**—The chief adulterants are starch, inorganic matter (such as sand, &c.), and vegetable débris. These are all easily detected by their insolubility in alcohol and water used successively, or in dilute ammonia.

**Varieties.**—*Pipe gamboge*, as above described, is the best variety.

*Inferior gamboge* breaks with a dull, rough, granular fracture, and the fractured surface, which often exhibits small cavities, is of a dark brownish colour.

*Lump* or *cake gamboge* consists of pipe gamboge bent and pressed whilst soft so as to form a cake ; or it may occur in irregular lumps which are frequently soft in the interior and often contain abundant visible impurity in the shape of sand, small stones, &c.

*Saigon gamboge* is occasionally exported from Saigon in short, thick, cylindrical cakes wrapped in palm leaves.

*Indian gamboge* is obtained in India from *G. Morella*, Desrousseaux, and resembles Siam gamboge in its essential qualities ; it is used as an equivalent of gamboge in India and the Eastern Colonies.

## SCAMMONY

### (Scammonium)

**Source, &c.**—Scammony is a gum-resin obtained by incision from the living root of *Convolvulus Scammonia*, Linné (N.O. *Convolvulaceæ*).

The scammony plant is a twining plant indigenous to the eastern Mediterranean and resembles the common bindweed, but is much larger. It produces a root attaining as much as a metre in length and 10 cm. in thickness, from which a number of slender aerial stems spring. The gum-resin is collected in Asiatic Turkey, chiefly near Smyrna and Aleppo.

The abnormal structure of the root has been already alluded to (see Scammony Root). In the cortex as well as in the parenchyma surrounding each of the columns of wood are cells containing a resinous emulsion. These cells are arranged in vertical rows, and when an incision is made the contents drain from a series of cells by the breaking of the transverse walls. In collecting the drug the workman clears away the earth surrounding the upper part of the root so as to leave 10 or 15 cm. of the root exposed. This is then cut off in a slanting direction 5 to 10 cm. below the crown, and a mussel shell is stuck into it just beneath its lowest edge so as to receive the milky sap which instantly flows out. The shells are usually left till the evening, when they are collected and the cut parts of the roots scraped with a knife so as to remove any partially dried drops of juice ('Pharmacographia'). If the gum-resin were then dried at once a drug of fine golden brown colour and translucent appearance would be produced ; but as a rule the contents of the shells are collected, and daily added to, until a sufficient quantity has accumulated ; this is then softened, mixed into a homogeneous mass, and allowed to dry. During the long standing in a more or less moist condition the scammony undergoes a fermentative change, and the drug when dry has a dark colour and porous appearance when broken. It is also frequently adulterated whilst in the soft state.

**Description.**—Scammony is usually imported in flattened cakes about 1·5 cm. thick and 10 or 15 cm. in diameter. It varies in colour from dark grey to dark brown or nearly black, the surface being often covered with a grey powder. It breaks very readily, and the freshly exposed surface is glossy, resinous, often more or less porous, and of a dark brown or nearly black colour. Thin fragments, however, when viewed by transmitted light, are seen to be brown in colour and translucent. It is very easily reduced to an ash-grey powder, and readily yields a milky emulsion when rubbed with water. Scammony has a characteristic cheesy odour and an acrid taste.

Very fine qualities of the drug which, however, are scarce, differ in being of a golden brown colour, almost transparent in thin fragments, and less porous.

The student should observe its *brittle nature*, *translucent brown colour* in thin fragments, and *characteristic odour* ; it yields a milky emulsion when rubbed with the wet finger, a character that distinguishes it from scammony resin, which does not emulsify when so treated.

**Constituents.**—Pure scammony contains about 88 or 90 per cent of resin identical with that obtained from the root (see p. 351) and entirely soluble in ether ; the residue consists principally of gum. But scammony of such purity as this is seldom met with in commerce. Good qualities yield from 70 to 80 per cent. of resin to ether, and are known as ' virgin ' scammony.

**Uses.**—Scammony is used as a brisk cathartic and vermifuge, especially for children.

**Adulteration.**—The high price of scammony and its plastic nature before it is dried invite and facilitate adulteration. Starch, chalk, and earthy matter are frequently used to increase its weight, and occasionally other resins are fraudulently mixed with it.

Adulteration with inorganic matter is readily detected by incineration. Good scammony should not yield more than 3 per cent. of ash. Starch is easily recognised by microscopical examination or by the iodine test, but traces of the starch of scammony root are often present in genuine scammony, and caution must be used in concluding that the drug has been adulterated. Most foreign resins may be detected by dissolving the drug in hot solution of potassium hydroxide ; on the addition of an acid, scammony resin is not precipitated whereas most foreign resins are.

Inferior scammony is usually tough and has a dull fracture, small splinters being opaque instead of translucent. Factitious scammony that was practically free from resin soluble in ether has even been offered for sale as genuine scammony.

## MYRRH

### (Myrrha)

**Source, &c.**—Myrrh is a gum-resin obtained from the stem of *Commiphŏra Myrrha*, Holmes, and probably other species (N.O. *Burseraceæ*). It is collected chiefly in Somaliland (in the north-east of Africa), brought down to the coast and sent to Aden, whence it is shipped to Europe, either direct or *via* Bombay. Some myrrh is said also to be collected in the south cf Arabia. Several other species of *Commiphora* are found in Arabia and north-eastern Africa yielding gum-resins that more or less resemble the official myrrh in appearance and odour.

These plants are shrubs or small trees, and, like most other plants belonging to the same natural order, they contain numerous schizogenous ducts in the bark, in which an oleo-gum-resin is secreted. In the case of the species from which the official myrrh is derived the tissue intervening between the ducts frequently breaks down and thus lysigenous cavities of considerable extent are produced which, together with the ducts that remain intact, are filled with a granular secretion. When the bark is wounded, therefore, the secretion is discharged in considerable quantity. It is first yellowish white and fluid, but soon hardens to a firm, reddish brown mass which constitutes the commercial drug. The secretion also exudes from fissures that are formed in the bark by natural causes ; indeed the bulk of the drug is said to be naturally exuded.

It not unfrequently contains other gums and gum-resins from which it is freed by hand picking.

**Description.**—Myrrh occurs in irregular rounded tears, or lumps composed of agglutinated tears, varying in size from small grains up to masses nearly as large as the fist, pieces about the size of a walnut being of common occurrence. They have a reddish yellow or reddish brown colour and a rather rough, dull, dusty surface. They break fairly easily, the fractured surface having a rich brown or reddish brown colour and translucent, unctuous, granular appearance, often exhibiting whitish spots or veins ; thin splinters are translucent or almost transparent. The drug has an agreeable aromatic odour and an aromatic, bitter and acrid, but not unpleasant taste. Triturated with water it yields a yellowish emulsion. If one gramme of the coarsely powdered drug is shaken for a few minutes with 10 c.c. of ether and 2 c.c. of the ethereal solution allowed to evaporate to a thin film in a small porcelain dish, the residue is coloured instantly deep violet-black by the vapour of bromine, or violet by nitric acid diluted with an equal volume of water. The resinous film obtained by evaporating tincture of myrrh will also yield similar reactions.

The student should observe

(*a*) The *unctuous, granular* (not uniform or vitreous) *fracture*,

(*b*) The *translucent* (not opaque) *appearance of thin fragments*,

(*c*) The *aromatic, bitter taste.*

The taste and the violet colour-reaction are, perhaps, the most characteristic features of myrrh.

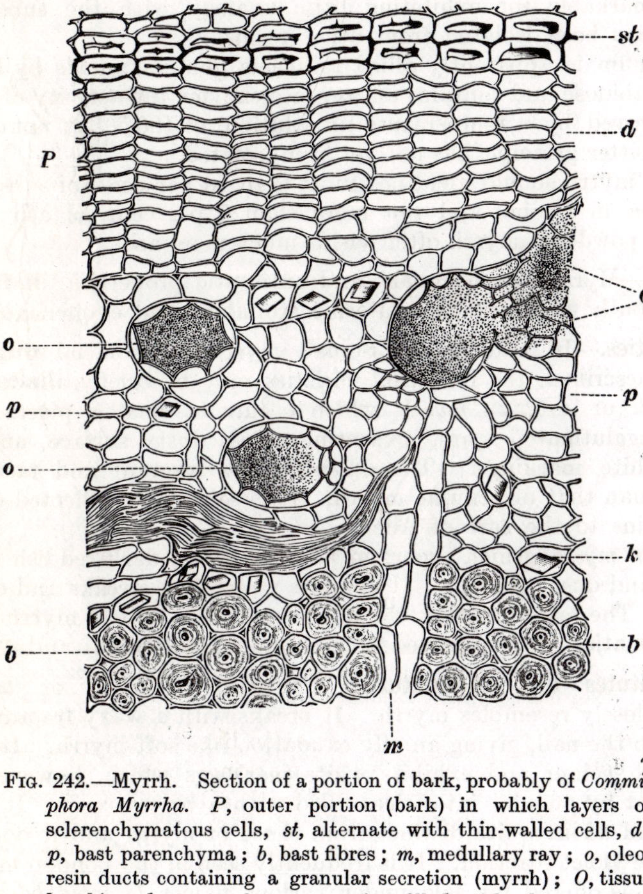


Fig. 242.—Myrrh. Section of a portion of bark, probably of *Commiphora Myrrha*. *P*, outer portion (bark) in which layers of sclerenchymatous cells, *st*, alternate with thin-walled cells, *d*; *p*, bast parenchyma; *b*, bast fibres; *m*, medullary ray; *o*, oleo-resin ducts containing a granular secretion (myrrh); *O*, tissue breaking down to form a cavity filled with the secretion. Magnified 280 diam. (Vogl.)

**Constituents.**—Myrrh consists of a mixture of resin, gum, and volatile oil. The latter can be obtained by distillation with water to the extent of 2·5 to 6·5 per cent. (Schimmel). The resin occurs in amounts varying from 25 to 35 per cent., the remainder of the drug consisting of gum, moisture, and various impurities.

Alcohol dissolves the volatile oil and resin. The volatile oil is yellowish and viscous and resinifies with great rapidity. Both the volatile oil and the resin give the characteristic violet reaction.

The resin is not entirely soluble in ether. The insoluble (smaller) portion is separable into three free resin-acids ($\alpha$-, $\beta$- and $\gamma$-commiphoric acids), a combined resin-acid (commiphorinic acid), and two phenolic resins ($\alpha$- and $\beta$-heerabo-myrrholol). The portion soluble in ether contains $\alpha$- and $\beta$-myrrhololic acids. These constituents are remarkable for exhibiting little analogy with the substances that have been isolated from other resins.

The gum is apparently allied to acacia gum ; it yields by hydrolysis arabinose and contains an oxydase enzyme the activity of which is destroyed by a temperature (in solution) of 100,° but not of 90°.

The bitter principle has not yet been isolated.

Good myrrh should yield not more than 70 per cent. of substances insoluble in alcohol and not more than 5 per cent. of ash ; commercial powdered myrrh often yields much more ash.

**Uses.**—Myrrh has stimulant and antiseptic properties ; it is used as a mouth wash and as a uterine stimulant and emmenagogue.

**Varieties.**—In addition to Somali myrrh (the official drug), as above described, the following varieties may be briefly alluded to :

*Fadhli* or *Arabian myrrh*, which occurs in smaller pieces made up of agglutinated tears, presenting a less dusty surface, and free from white markings. The odour is less fragrant and taste less bitter than that of genuine myrrh. It is said to be collected on the mountains to the east of Aden.

*Yemen myrrh*, which occurs in large pieces of dark reddish brown colour and dusty surface. It exhibits no whitish streaks and exudes no oil. The taste is bitter, the odour resembles that of myrrh but is less aromatic. It is exported from Makullah to Bombay and Aden.

**Substitutes and Admixtures.**—*Perfumed bdellium* or *bissabol* which closely resembles myrrh. It breaks with a waxy fracture and yields to the nail, giving an oily exudation like soft myrrh. It has a yellowish colour and exhibits white markings which, however, are traversed by angular interstices filled with a brown resin. It has a taste and odour quite distinct from those of myrrh and it does not yield the violet reaction. It is frequently seen in the London market, where it is offered for sale under various names (*scented bdellium*). It is probably derived from *C. erythrœum*, var. *glabrescens*, Engler.

*Opaque bdellium*, a very hard, yellow ochre-coloured, opaque gum-resin with but a slight odour and a bitter taste. Portions of a papery bark are frequently found associated with it. The tincture (1 in 6) assumes an intense greenish black colour with solution of ferric chloride.

*African bdellium*, in hard pieces, translucent in thin layers, and red when viewed by transmitted light. The fracture is dull and slaty, the margins possessing a powdery appearance; it has a bitter taste and an odour recalling pepper. The tincture gives no precipitate with ferric chloride.

*Indian bdellium*, which occurs in large irregular masses of a dark reddish brown colour. The fractured surface resists the nail, and is covered with characteristic, minute, shiny points of resin which also appear on the outer surface. The odour is feeble and cedar-like; it appears to be developed only on keeping. The taste is slightly acrid and devoid of bitterness.

*Gum hotai*, liver-coloured, opaque masses, is sent in large quantities to Bombay; used for washing the hair. It contains an acid resin and a saponin.

## OLIBANUM

### (Frankincense, Olibanum)

**Source, &c.**—Olibanum or, as it is sometimes termed, frankincense (to be carefully distinguished from American frankincense) is a gum-resin obtained from *Boswellia Carterii*, Birdwood, and other species of *Boswellia* (N.O. *Burseraceæ*).

These plants are small trees that grow in southern Arabia and in Somaliland. Like the trees that yield myrrh they contain schizogenous ducts in the bark, in which an oleo-resin is secreted. The Somalis incise the bark and collect the gum-resin as soon as it has sufficiently dried. The drug is conveyed to Aden and thence to Bombay, whence it is exported to Europe.

**Description.**—Olibanum occurs in small tears varying from 0·5 to 3 cm. in length and usually ovoid, pear-shaped, or club-shaped, but sometimes stalactitic in form, occasionally agglutinated into small masses. They are usually of a pale yellowish colour, frequently with a greenish, bluish, or reddish tinge, semi-translucent and covered with a dull white dust, the surface of the tear being dull even after the dust has been removed. They are brittle, breaking easily between the fingers; internally they are opalescent and translucent, the fractured surface being dull and waxy.

The drug has a fragrant, balsamic odour and an aromatic, slightly bitter taste, and soften to a plastic mass when chewed. Triturated with water it yields a whitish emulsion.

The student should observe

(*a*) The *fragrant odour*,

(*b*) The *opalescent*, *waxy interior of the tears*.

**Constituents.**—Olibanum consists principally of resin (60 to 70 per cent.), gum (27 to 35 per cent.), and volatile oil (5 to 7 per cent.).

These constituents have been further investigated, with the following results (Halbey, 1898):

| | | | |
|---|---|---|---|
| Soluble in alcohol, 72 per cent. | Boswellic acid, free | 33·0 | per cent. |
| | ,, ,, combined | 1·5 | ,, |
| | Olibanoresene | 33·0 | ,, |
| | Volatile oil | 7·0 | ,, |
| | Bitter principle | 0·5 | ,, |
| Insoluble in alcohol, 28 per cent. | Gum (arabic acid with Ca and Mg) | 20·0 | ,, |
| | Bassorin | 6·0 | ,, |
| | Vegetable Débris | 2·0 | ,, |

According to this analysis the resin consists principally of a resin acid (boswellic acid) and an indifferent resin (olibano-resene) in about equal proportions. The gum consists mainly of arabin, with which is associated a little bassorin. The volatile oil is yellowish and fragrant; it contains pinene, dipentene, and phellandrene, but the aromatic constituent is not yet known.

**Uses.**—Olibanum is used chiefly in the manufacture of incense and as an ingredient in plasters and fumigating pastilles.

## AMMONIACUM

(Ammoniacum)

**Source, &c.**—Ammoniacum is a gum-resin exuded from the flowering and fruiting stem of *Dorēma Ammoniacum*, D. Don (N.O. *Umbelliferæ*), and probably other species, distributed throughout Persia and extending into southern Siberia.

The drug is collected chiefly in central Persia.

The stems of the ammoniacum plants contain, especially in the cortex, numerous, large, schizogenous ducts full of a milky secretion. In the summer, when the plant is fruiting, it is visited by numbers of beetles, which puncture the stem and cause an abundant exudation of the secretion in the form of milky drops, some of which harden on the stem, whilst others drop on to the ground. It is collected, sorted, and exported from the Persian Gulf ports.

**Description.**—Ammoniacum occurs in commerce in two forms—viz. *tear ammoniacum* and *lump ammoniacum*, the former being alone official.

The tears are small, rounded or nodular masses varying usually from 0·5 to 3 cm. in diameter. When fresh they are of a pale dull yellow colour, which, however, darkens by keeping. They are hard and brittle when cold, but soften when warmed. Internally the tears are opaque, and vary in colour from milky white to pale brownish yellow, the freshly fractured surface having a waxy lustre. The drug has a characteristic but not alliaceous odour, and a bitter, acrid taste. Triturated with water it forms a white emulsion, which is coloured deep orange red by a solution of chlorinated soda, yellow by solution of potash, and, transiently, faintly violet by ferric chloride. The latter reaction, although not strong, should be noted, as it depends upon the presence of a constituent (salicylic acid) that is absent from other similar gum-resins, and is characteristic therefore of ammoniacum ; a more distinct coloration is obtained by shaking the emulsion with asbestos, filtering clear and adding to the filtrate a drop of solution of ferric chloride.

Lump ammoniacum consists of agglutinated, whitish, yellowish grey or bluish grey tears, mixed with varying quantities of extraneous substances, such as stones, dirt, stems and other débris of the plant and occasionally the broad, flat mericarps of the fruit, the presence of which indicates the time at which the drug has been collected. The substance of the tears agrees with the description of the tears already given.

Good qualities of lump ammoniacum consist of tears varying in size from a pea to a hazel-nut or even larger, with a little intervening dark-coloured ground substance, and but few pieces of stem, fruits, &c. Intermediate forms composed of more or less agglutinated tears also occur.

If about 1 gramme of the drug is boiled for a few minutes with 20 c.c. of equal volumes of hydrochloric acid and water, and the solution filtered into water made strongly alkaline with ammonia, no blue fluorescence should be developed. This reaction indicates the absence of umbelliferone (which, in alkaline solution, exhibits a strong blue fluorescence) and distinguishes Persian ammoniacum from galbanum, asafetida, and African ammoniacum, all of which yield umbelliferone. Free umbelliferone may also be tested for by adding an alcoholic tincture of ammoniacum to an alcoholic solution of ammonia.

The student should observe

(*a*) The *clean, hard tears*, which do not crumble between the fingers,
(*b*) The *characteristic odour*,
(*c*) The *positive result of the test for salicylic acid*,
(*d*) The *negative result of the test for umbelliferone* ;

and should compare the drug with selected tears of galbanum, which are much softer, have a distinctive odour, and yield a positive result with test for umbelliferone.

**Constituents**.—Ammoniacum consists of volatile oil 0·1 to 1·0 per cent., resin (about 65 to 70 per cent.), gum (about 20 per cent.), moisture (2 to 12 per cent.), ash (1 per cent.), and insoluble residue (3·5 per cent.).

Luz (1893) found that the resin could be separated into two portions, one consisting of a resene not attacked by caustic alkalies, and the other and larger portion decomposed by long-continued boiling with solution of caustic potash into salicylic acid and a resin-alcohol, ammoresinotannol, which was obtained as a brownish powder. The resin chiefly consists therefore of ammoresinotannol combined with salicylic acid.

The gum is allied to acacia gum.

Luz was unable to detect sulphur in either the volatile oil or the resin, although that element had previously been reported present in ammoniacum.

The drug contains also traces of free salicylic acid, which is the source of the violet colour produced when ferric chloride is added to an aqueous emulsion.

Good qualities of the drug yield about 3 per cent. of ash and 65 per cent. of resin.

**Uses.**—Ammoniacum is a stimulant, and, being excreted by the bronchial mucous surfaces, stimulates and disinfects the secretion. It is used as a disinfectant expectorant in chronic bronchitis with profuse discharge, and in plasters as a stimulant to the skin.

**Varieties, &c.**—The official or Persian ammoniacum is distinguished from African ammoniacum, said to be obtained in Africa from *Ferŭla communis*, Linné, var. *brevifolia*, by the orange red colour it yields with solution of chlorinated soda, and also by yielding a negative result with the tests for umbelliferone.

## GALBANUM

### (Galbanum)

**Source, &c.**—Galbanum is a gum-resin obtained from *Ferŭla galbaniflŭa*, Boissier and Buhse (N.O. *Umbelliferæ*), and probably from other species.

These plants are, like those yielding ammoniacum, large Umbelliferous plants indigenous to and widely distributed over Persia. Two varieties at least of the drug are well recognised—viz. Persian, which is soft and contains fruit and stalks, and Levant, which is dried and contains slices of the root, seldom fruits or stalks. The latter is the variety at present usually met with.

Like the ammoniacum plants the galbanum plants contain, especially in the cortical portion of the stem and root, numerous schizogenous ducts that secrete a milky gum-resinous fluid.

Part of the drug is apparently obtained by natural exudation from the stem, but part is certainly produced by laying bare the root, cutting the stem off near the crown, and collecting the juice that exudes and hardens, successive slices of root being removed at intervals of several days. The former procedure would probably yield the tears that are found in commercial galbanum, whilst the slices of root found in the drug indicate the latter method as the one by which most of the drug is obtained. It is exported chiefly from the Persian Gulf ports.

**Description.**—Galbanum occurs in distinct tears, in small agglutinated masses, and in lump form. The tears are rounded or irregular in form, and, though they vary in size, are usually about as large as, or rather larger than, a pea. Externally they are yellowish brown or orange brown in colour, and often rough and dirty. They are not, like the tears of ammoniacum, hard, but are so soft that they can usually be squeezed flat between the finger and thumb, becoming ductile and sticky. They break easily with a granular, irregular fracture, and are opaque, yellowish, and soft internally. Sometimes the tears are more or less translucent and of a bluish green colour.

Thin transverse slices of the root are commonly found mixed with commercial galbanum ; they are usually about 2 or 3 cm. in diameter and frequently bear on one side the dried secretion derived from the freshly cut surface.

The drug has a characteristic, not exactly unpleasant, aromatic odour, and a rather disagreeable, aromatic, and bitter taste. An alcoholic tincture poured into alcoholic solution of ammonia yields a brilliant blue fluorescence, indicating the presence of free umbelliferone.

Galbanum occurs also in lumps, corresponding to lump ammoniacum ; these consist of yellowish or bluish green or brownish tears embedded in a brownish mass and mixed with slices of root and various foreign substances.

The student should observe

(*a*) The *soft yellowish brown tears*, occasionally *bluish green*,
(*b*) The *characteristic odour*,
(*c*) The *positive result of the test for umbelliferone* ;

and should compare the drug with ammoniacum.

**Constituents.**—Galbanum consists, apart from extraneous substances, of volatile oil (about 5 to 10 per cent.), resin (about 60 per cent.) and gum (about 20 per cent.), the residue being made up of inorganic matter (about 2 per cent., sometimes much more) and moisture (from 1 to 10 per cent.)

Conrady found (1894) in a good specimen of commercial galbanum :

| | |
|---|---|
| Volatile oil . . . . . . . . | 9·5 per cent. |
| Resin . . . . . . . . . | 63·5 ,, |
| Gum and impurities . . . . . . | 27·0 ,, |
| Ash . . . . . . . . . | 8·25 ,, |

The resin boiled with solution of potassium hydroxide, yielded galbaresinotannol and umbellic (dioxycinnamic) acid; the latter, however, is not contained in the drug itself, but is formed from umbelliferone, the anhydride of umbellic acid which is first split off from the

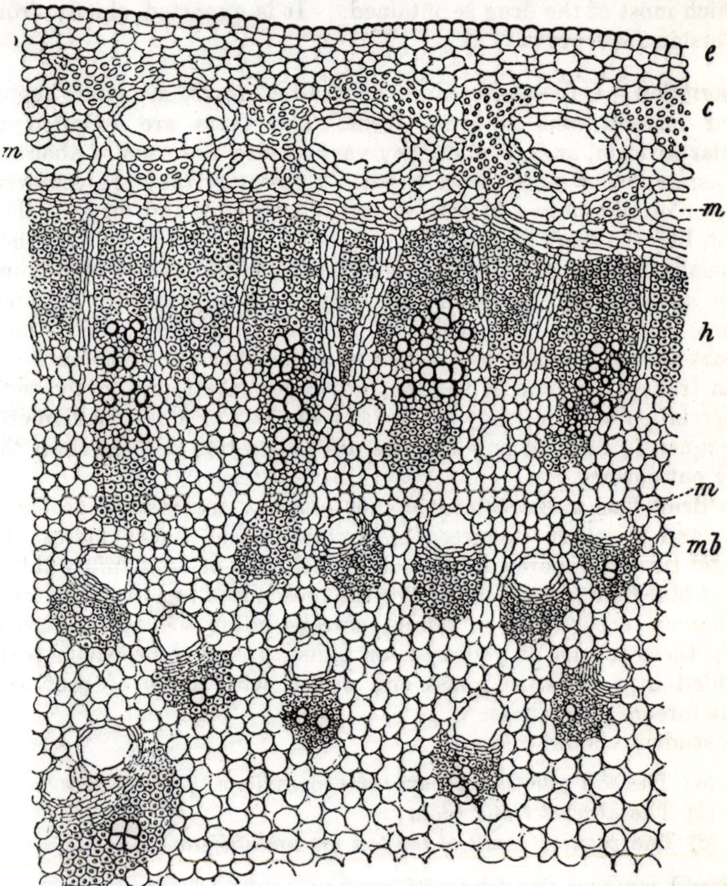


FIG. 243.—Transverse section through a portion of the stem of *Ferula galbaniflua*, showing the distribution and structure of the gum-resin ducts. *e*, epidermis; *c*, collenchyma; *m*, gum-resin ducts; *h*, wood; *mb*, bundles in pith; magnified. (Tschirch.)

resin and then converted into umbellic acid. The galbaresinotannol was obtained as a brown powder.

Umbellic acid yields in the cold with solution of potassium hydroxide and chloroform an intense green colour. Umbelliferone when boiled with solution of potassium hydroxide and chloroform yields a similar coloration, as does also galbanum itself. Galbanum also assumes a violet red colour when warmed with hydrochloric acid; the constituent to which this characteristic reaction is due is not known.

Good qualities of the drug should yield about 40 per cent. of substances insoluble in alcohol, about 10 per cent. of moisture, and give on incineration not more than about 7 per cent. of ash.

**Uses.**—Galbanum is used chiefly as a stimulant in plasters.

## ASAFETIDA

### (Asafetida)

**Source, &c.**—Asafetida is a gum resin obtained from the root of *Ferŭla fœtida*, Regel, *F. rubricaulis*, Boissier (N.O. *Umbelliferæ*), and probably other species. These plants, like those yielding ammoniacum and galbanum, are large Umbelliferous plants growing in eastern Persia and western Afghanistan. In the cortex of the stem, and especially in that of the root, there are numerous large, schizogenous ducts filled with a milky, gum-resinous emulsion; these when wounded discharge their contents, which then gradually acquire by evaporation a firmer consistence.

Part of the drug is certainly collected in a manner similar to that in which part at least of commercial galbanum is obtained—viz. by laying bare the root of the plant and cutting off the stem close to the crown; the emulsion that flows from the cut surface is allowed to harden, for which purpose the root is protected by a dome-like covering of sticks and leaves; the hardened gum-resin is then scraped off, a slice of the root cut off, and the juice again allowed to exude, and so the process is repeated. Some is probably obtained by incising the stem.

Herat and Kandahar are the centres of the asafetida trade. The drug is exported from Bunder Abbas and other ports on the Persian Gulf, partly also from Bombay, mostly in large tin-lined cases, but a small quantity arrives as a pasty mass in tins or hides.

**Description.**—Asafetida occurs in three forms, viz.: paste, tear and mass (block or lump). Paste and tear are the purer forms, but the bulk of the drug is mass.

The tears, some of which are separate, some more or less agglutinated together, are rounded or flattened, and vary from 1 to 3 cm. in diameter. They are of a dull yellow or sometimes dirty grey colour; some darken on keeping, finally becoming reddish brown, but others retain their original colour for years, thus indicating some difference in the drug. Probably the red variety is derived from *F. fœtida*, the white from *F. rubricaulis*. When fresh they are usually tough at ordinary temperatures, becoming harder when cooled and softer when warmed. Internally they may be yellowish or milky white, translucent or opaque ; the freshly exposed surface, may gradually, pass through a very characteristic change of colour, becoming first pink, then red, and finally reddish brown (*F. fœtida*), or may remain nearly white (*F. rubricaulis*). The drug has an intense, penetrating, persistent, alliaceous odour, and a bitter, acrid, alliaceous taste.

Mass asafetida consists of the tears agglutinated into a more or less uniform mass and mixed with varying quantities of extraneous substances such as stones, slices of the root, earthy matter, calcium carbonate, calcium sulphate, &c. ; it is generally much inferior to the tears.

Asafetida contains no free umbelliferone (compare 'Galbanum'), but yields it when boiled with hydrochloric acid ; a reaction which serves to distinguish it from a variety of the drug met with in Bombay yielding no umbelliferone. From galbanum the tears of asafetida may also be distinguished by the green colour the freshly fractured surface assumes when it is touched with nitric acid diluted with an equal volume of water, or by the bright red or brownish red colour with sulphuric acid, changing to violet when the acid is washed off with water.

**Constituents.**—Asafetida consists principally of volatile oil, resin, and gum.

Good samples yield from 10 to 17 per cent. of volatile oil, from 40 to 64 per cent. of resin, about 25 per cent. of gum, and 1·5 to 10 per cent of ash. The amount of mineral matter in mass asafetida may rise to 60 per cent, or exceptionally even more ; fine tears may contain as little as 1·5 per cent.

The resin consists of ferulic acid combined with asaresinotannol; although the drug contains no free umbelliferone, it easily yields that substance by the action of sulphuric or hydrochloric acid on the ferulic acid and on the resorcin produced simultaneously from the resin.

The volatile oil contains pinene together with various disulphides, $C_7H_{14}S_2$, $C_{11}H_{20}S_2$, $C_{10}H_{18}S_2$, &c., the percentage of sulphur varying from 17 to 37.

**Uses.**—Asafetida is a powerful nervine stimulant, and is used in the nervous disorders of hysteria. It has also a well-marked stimulant action on the bowel, and is employed to expel flatulence and relieve constipation. Much of it is exported to the continent and also to the United States, where is is used on the cattle ranches.

## SECTION XVI

## OLEO-RESINS

The drugs that are grouped together under the above heading are mixtures of resins with volatile oils or oily liquids. They are secreted in schizogenous or schizolysigenous ducts which may be of either normal (Canada turpentine, copaiba) or pathological (balsam of Peru, storax) origin.

All of the constituents may vary considerably in their composition. Those oleo-resins which contain benzoic or cinnamic acid are frequently termed ' balsams.'

### CANADA TURPENTINE

### (Canada Balsam, Terebinthina Canadensis)

**Source, &c.**—Canada turpentine is an oleo-resin obtained by incision from the balsam fir, *Abies balsāmea*, Miller, and the hemlock spruce, *A. canadensis* (Linné), Miller (N.O. *Coniferæ*), trees widely distributed over the northern United States and Canada, extending to Hudson's Bay. The drug, which has long been known, is collected in Lower Canada, especially in the province of Quebec.

The tree contains schizogenous oleo-resin ducts, as most Coniferous trees do, but they are restricted to the bark, none occurring normally in the wood. In addition, however, to these secretion ducts, cavities are formed which fill with oleo-resin and produce blisters on the smooth trunk of the tree. From these blisters the oleo-resin is obtained by puncturing them with the pointed spout of a can which serves to receive the turpentine.

**Description.**—Canada turpentine is a clear, transparent liquid, about as viscid as honey, and of a pale yellow or greenish yellow colour, often exhibiting a slight greenish fluorescence. By keeping it becomes more viscid, and finally it gradually dries to a hard resin which remains transparent and shows little disposition to crystallise, a quality that renders it particularly valuable as a medium in which to preserve microscopical preparations. It has an agreeable balsamic odour and a rather bitter and acrid taste. It is completely soluble in chloroform, benzene, and ether, but only partially in alcohol.

**Constituents.**—Canada turpentine consists approximately of 16 to 24 per cent. of volatile oil mixed with from 70 to 80 per cent. of resin.

The volatile oil consists chiefly of *l*-pinene.

Of the resin about 20 per cent. is composed of an indifferent resene, canado-resene, which is remarkable for its insolubility in alcohol. A further 20 per cent. of the resin is amorphous canadinic acid. The remaining 60 per cent. consists of two amorphous resin acids, *α*- and *β*-canadinolic acids, associated with 0·5 per cent. of crystalline canadolic acid.

The bitter principle, which is soluble in water, has not yet been isolated.

**Uses.**—Canada turpentine is extensively used as a microscopic mountant. For this purpose the resin, obtained by heating the turpentine until the volatile oil is driven off, is dissolved in xylol, or some other suitable solvent. It is also used for cementing lenses.

**Note.**—*Oregon balsam* is a similar oleo-resin obtained from *Abies Menziesii*, Lindley, on the Pacific slope; it is thinner than Canada balsam and yields a sticky film on evaporation.

## COPAIBA

### (Copaiva, Balsam of Copaiba)

**Source, &c.**—Copaiba is an oleo-resin obtained from the trunk of *Copaifera Lansdorfii*, Desfontaines (N.O. *Leguminosæ*), and other species of *Copaifera*.

The trees from which the official oleo-resin is obtained are large trees attaining over 100 metres in height, indigenous to Brazil and the north of South America. The drug, which was highly esteemed by the natives of Brazil and had probably long been used by them as a medicine, was introduced into Europe about the beginning of the seventeenth century.

The oleo-resin is contained in anastomosing, schizogenous secretion ducts that form an extensive network in each zone of the secondary wood of both stem and root, extending throughout the entire length of the zone, resembling therefore in arrangement the laticiferous tissue of the dandelion. These ducts are formed in the young wood and rapidly attain their normal diameter, which is often very considerable; at the level of the insertion of the branches a number of lateral ducts connect zone with zone. In addition to these schizogenous ducts lysigenous cavities also appear to be formed by the breaking down of the cell walls and their probable transformation into resinous or oleo-resinous substances.

The oleo-resin is collected by cutting in the trunk of the tree near the base a cavity sloping inwards and downwards, and penetrating to the centre of the trunk, resembling the ' box ' made in the trunk of the turpentine trees. Into this cavity the oleo-resin is discharged; it is transferred to barrels and other vessels for exportation.

The large size of the secretion ducts, and their extensive distribution in each zone of wood throughout the entire length of the tree, render the amount of oleo-resin that may be secreted by each tree very considerable. Even as much as 48 litres is said to have been obtained from a single tree, others again yielding but little.

The drug is exported from the seaports on the northern coast of South America—viz. Para, Maranham, Maracaibo, Bahia, Cartagena, &c., these towns giving their names to the commercial varieties of the drug. As these differ in the percentage of volatile oil and of resin and in the composition of the latter the following typical commercial varieties may be described at length.

## Maracaibo Copaiba

**Description.**—Maracaibo copaiba is a clear, viscous, brownish-yellow liquid with a slight but distinct green fluorescence. It possesses a characteristic aromatic odour and an unpleasant, acrid and rather bitter taste. It is miscible in all proportions with chloroform, carbon disulphide, and benzene, and also with an equal volume of petroleum spirit, but with larger proportions of the latter a slight precipitation takes place; with absolute alcohol it behaves similarly. The specific gravity varies from 0·980 to 0·999, or even slightly higher. The proportion of volatile oil varies from 42 to 53 per cent.

**Constituents.**—Maracaibo copaiba consists of a mixture of resin and volatile oil with which traces of a bitter principle and fluorescent substance are associated.

The volatile oil is invariably lævorotatory, the rotation in 100 mm. tube varying from — 5° to — 35°, usually, however, it does not exceed — 25°; its specific gravity varies from 0·896 to 0·910 and its boiling-point from 250° to 275°; it does not respond to the test for gurjan balsam (see below). These details are important indications of the purity of the oleo-resin.

The resin contains a small proportion of copaivic acid (removed by ammonium carbonate from ethereal solution), but consists chiefly of $\beta$-metacopaivic acid (extracted by sodium carbonate after removal of the copaivic acid); with these there is associated a small quantity of two indifferent copaibo-resenes and of crystalline illurinic acid (compare African copaiba).

The bitter principle and fluorescent substance have not yet been obtained in a pure state.

### Pará Copaiba

**Description.**—Pará copaiba is a thin, clear, bright yellow liquid quite free from any fluorescence. The specific gravity varies from 0·917 to 0·980, but is usually low. In accordance with the greater fluidity of this variety the percentage of volatile oil is high, viz. from about 55 to 90 per cent. In other characters the drug resembles the Maracaibo variety.

**Constituents.**—Pará copaiba also consists essentially of volatile oil and resin.

The resin consists of amorphous resin-acids associated with crystalline parácopaivic acid (2 per cent.), homoparácopaivic acid, and two resenes.

The volatile oil appears to be identical with that of Maracaibo copaiba, but the resins are certainly not.

### African Copaiba

African copaiba, the botanical source of which is not known, is imported from the Niger basin in West Africa. It is a rather dark yellow, slightly fluorescent oleo-resin possessing an aromatic, piperaceous odour and frequently depositing crystals on standing. The specific gravity varies from 0·985 to 1·000. It contains about 40 per cent. of volatile oil and 60 per cent. of resin (including the crystalline substance).

**Constituents.**—The oil boils at 260° to 275° and differs essentially from the oil of South American copaiba in being dextrorotatory, the rotation in 100 mm. tube being about 10° 21′.

The crystalline deposit consists of illurinic acid identical with that obtained from Maracaibo copaiba. The remainder of the resin consists of amorphous resin-acids, fluorescent substance, &c.

### Adulterants of Copaiba

The British Pharmacopœia (1914) recognises any variety of South American copaiba containing about 45 per cent. of volatile oil, and therefore excludes thin Pará varieties containing much over that proportion but admits varieties of the Maracaibo type. The drug is, however, not infrequently sophisticated to bring it within the official limits.

The following are the chief adulterants and the means of detecting them.

*Fixed vegetable oils*, such as castor oil, render the resin left after evaporation of the volatile oil (best at a temperature of about 120° to 130°) tough or pasty, whereas it is usually (in Maracaibo copaiba) hard and brittle. The presence of fixed oil can also be detected by the high ester value; copaiba resin consists almost entirely of resin-acids and has a very low ester value (seldom over 15; fixed oils about 190).

*Volatile oils*, such as turpentine, may often be detected by distilling off the volatile oil in a current of steam and determining its specific gravity (0·896 to 0·910), boiling-point (250° to 270°), and optical rotation (− 7° to − 35°) (sp. gr. of turpentine 0·850 to 0·880; b.pt. 155° to 165°). The optical rotation is important; a dextrorotatory oil would indicate the presence of African copaibas (compare also the official tests for Oleum Copaibæ).

*Colophony.*—Colophony may be added to thin copaibas without making them suspiciously viscous. Copaiba should form a transparent solution with one-third of its volume of solution of ammonia.

*Paraffin Oil.*—Five gm. of the copaiba boiled with 15 c.c. of 95 per cent. alcohol for one minute and cooled should not separate oily drops (of paraffin oil).

*Gurjun balsam*, an oleo-resin obtained by incision from the trunk of *Dipterocarpus turbinatus* (N.O. *Dipterocarpeæ*) and other species, large trees indigenous to eastern India and Burma, is used both as a medicine and for various technical purposes. It somewhat resembles copaiba in odour and taste, but is usually darker in colour and fluorescent. It contains from 40 to 80 per cent. of volatile oil together with neutral and acid resins and has therefore a composition analogous to that of copaiba. Its presence in copaiba may be recognised by adding 4 drops to a mixture of 5 c.c. of glacial acetic acid and 4 drops of nitric acid; a purple or reddish coloration indicates gurjun balsam. It may also be detected by dissolving 2 drops of the balsam in 20 drops of carbon disulphide and adding a drop of a freshly prepared and cooled mixture of nitric and sulphuric acids; if gurjun balsam is present a violet coloration will be produced, but it has been shown that a similar colour has been occasionally yielded by genuine balsams. A similar test may be applied to the volatile oil separated from the copaiba by steam distillation; not more than a faint violet colour should be produced.

Sophistication by mixing a cheap thin copaiba with a more valuable thick copaiba is difficult to detect.

**Uses.**—The active principles of copaiba are absorbed into the blood, the volatile oil, at least, being excreted by the kidneys, bronchi, and skin; hence copaiba produces along the whole genito-urinary tract, as well as in the bronchi, a stimulant and disinfectant action, increasing the mucous secretion and exciting expectoration. It is now chiefly employed in inflammatory affections of the bladder and urethra, and occasionally in chronic bronchitis. The resin is inert or nearly so.

## ELEMI

### (Manila Elemi, Resina Elemi)

**Source, &c.**—The term 'elemi' is a generic one and is applied to a number of oleo-resins having certain physical properties in common. Tschirch enumerates no fewer than 46 varieties, all of which are derived from plants belonging to the natural orders *Burseraceæ* or *Rutaceæ*. The most important and formerly official variety is exported from Manila and distinguished as Manila elemi.

Manila elemi is an oleo-resin obtained from a species of *Canarium*, probably *C. luzonicum*, Gray (N.O. *Burseraceæ*). It is obtained by making incisions in the tree and promoting the flow of oleo-resin by the application of heat. When first obtained it is probably a clear, viscid, honey-like liquid which, however, rapidly becomes crystalline. It is exported chiefly from Manila, and arrives in this country in tins or lead-lined cases, in a more or less solid condition.

**Description.**—Elemi when fresh and of good quality is pale yellow in colour, soft and granular, resembling a crystalline honey, but on keeping it gradually becomes darker in colour, firmer and finally hard. It is soluble in alcohol and ether, partially only in petroleum spirit. It has a fragrant, balsamic odour recalling fennel and mace, and a spicy, rather bitter taste. Under the microscope it is seen to contain an abundance of acicular crystals.

**Constituents.**—Manila elemi, in the soft condition in which it is usually imported, contains about 20 to 30 per cent. of volatile oil associated with amorphous and crystalline resin acids, bitter principle, &c.

The volatile oil (sp. gr. 0·87 to 0·91) consists chiefly of terpenes; the aromatic constituents have not been examined.

The resin contains *α*- and *β*-manelemic acids (soluble in solution of ammonium carbonate), *α*- and *β*-manamyrin (insoluble in ammonium carbonate and in alcohol), maneleresene (insoluble in ammonium carbonate, soluble in alcohol), bryoidin, and bitter principle. The composition of the drug may be seen from the following table:

| | |
|---|---|
| *α*-Manelemic acid | 5 to 6 per cent. |
| *β*-Manelemic acid | 8 to 10 ,, |
| *α*- and *β*-Manamyrin | 20 to 25 ,, |
| Maneleresene | 30 to 35 ,, |
| Volatile oil | 20 to 25 ,, |
| Bryoidin | about 1 ,, |
| Bitter principle and impurities | 6 to 8 ,, |

*α*-Manelemic acid is crystalline and forms crystalline salts; *α*- and *β*-manamyrin are both crystalline, as is also bryoidin.

**Uses.**—Elemi has been employed in the form of ointment, as a stimulant and antiseptic application. It is now seldom prescribed.

**Allied Drugs.**—*East African Elemi*, from *Boswellia Frereana*, Birdwood (Somaliland); in stalactitic masses; fragments pale amber yellow.

*Brazilian Elemi*, from *Protium heptaphyllum*, March (N.O. *Burseraceæ*); small brown nodules, almost free from odour, mixed with fragments of bark.

*Yucatan Elemi*, from *Amyris Plumerii*, de Candolle (N.O. *Burseraceæ*); yellow translucent pieces curved on one side, very aromatic.

All these varieties of elemi resemble Manila elemi in composition.

## BALSAM OF TOLU

### (Balsamum Tolutanum)

**Source, &c.**—Balsam of Tolu is a balsam obtained by making incisions in the trunk of *Myroxylon Toluifera*, Humboldt, Bonpland, and Kunth (N.O. *Leguminosæ*).

The tree is a native of Colombia, and occurs plentifully in the forests near the river Magdalena and its tributary the Cauca. The balsam, which receives its name of Tolu from a small town near Cartagena, on the northern coast of Colombia, is collected by cutting a V-shaped notch in the bark, and fixing below it a gourd into which the balsam flows. Many such incisions at varying heights may be made on the same tree, which, however, is much exhausted by the tapping. The contents of these gourds are emptied into skin bags and conveyed to the coast, where the balsam is transferred to tins for exportation. It is shipped chiefly from Savanilla and Cartagena.

Although the twigs of the tree contain schizogenous secretion ducts, these are soon thrown off and no new ones are formed. The bark of the trunk, from which the balsam is obtained, contains no secretory tissue, and the balsam is probably secreted in ducts formed in the new wood, a change induced by the incisions made in the tree (compare the production of colophony).

**Description.**—Balsam of Tolu when freshly imported is a soft, tenacious, yellowish brown, resinous mass, not soft enough to flow, but taking the form of the vessel in which it is kept. By keeping, it gradually hardens to a brownish, and, especially in cold weather, brittle and easily powdered mass, which, however, readily softens when warmed. It has an agreeable, fragrant, though not powerful odour, an acidulous balsamic taste, and adheres to the teeth when chewed. A small piece warmed and pressed into a thin film between two glass slides exhibits, when examined by the microscope, colourless crystals embedded in a transparent mass and accompanied by a little vegetable débris.

It is easily soluble in alcohol, acetone, and chloroform, but only partially soluble in carbon disulphide, yielding to the latter principally cinnamic acid. The solution obtained by gently warming the balsam with carbon disulphide leaves when evaporated about 25 per cent. of crystalline residue consisting chiefly of cinnamic and benzoic acids. The British Pharmacopœia requires that the balsam should contain at least 25 per cent. of free aromatic acids, but it has been shown that 20 per cent would be a more reasonable requirement. Cocking and Kettle (1918) found an average of about 36 per cent. of total aromatic acids, free and combined, of which about 8 per cent. was free benzoic and 12·8 free cinnamic acid, 7 per cent. combined benzoic, and 8 per cent. combined cinnamic acid.

**Constituents.**—Tolu balsam was examined by Oberländer (1894), who found it to contain the following constituents: about 7·5 per cent. of an oily liquid (consisting of benzyl benzoate with a little benzyl cinnamate), traces of vanillin, free aromatic acids, principally cinnamic, and resin. The resin, amounting to about 80 per cent. of the drug, yielded by saponification an alcohol (toluresinotannol), and cinnamic acid, with which was associated a little benzoic acid.

Distilled with water good fresh balsam of Tolu yields from 1·5 to 3·0 per cent. of a very aromatic volatile oil containing tolene, styrol, and free benzoic and cinnamic acids.

**Uses.**—Tolu balsam is used chiefly as a pleasant ingredient in cough mixtures. It possesses antiseptic properties due to the cinnamic and benzoic acids contained in it.

**Adulterants.**—The chief adulterants of balsam of Tolu are colophony and balsam that has previously been used in making syrup of Tolu and hence deprived of most of its free cinnamic acid and aromatic constituents.

The acid value of genuine balsam ranges from 107·4 to 147·2, and the saponification value from 170 to 202.

Colophony may be detected by exhausting the balsam with carbon disulphide and evaporating the filtered solution. Pure balsam gives about 25 per cent. of crystalline residue; if colophony is present the residue is resinous and gives with concentrated sulphuric acid a green colour; a petroleum spirit extract of the residue shaken with an equal volume of a 0·1 per cent. solution of cupric acetate assumes a bright green colour if colophony is present.

Exhausted balsam may be detected by the deficient amount of substances soluble in carbon disulphide and of aromatic acids.

## BALSAM OF PERU

### (Balsamum Peruvianum)

**Source, &c.**—Balsam of Peru is a balsam exuded from the trunk of *Myroxylon Pereiræ*, Klotsch (N.O. *Leguminosæ*), after the bark has been beaten and scorched.

The tree grows in the forests of that part of San Salvador (Central America), near the Pacific Ocean, and known as the Balsam Coast, and probably also in other parts of Central America, as, for instance, in Honduras (Dieterich), the drug having received the name 'Peruvian' from the fact that it was originally sent from San Salvador to Callao, the port of Lima, and thence to Spain.

As in the case of *Myroxylon Toluifera*, the bark of young twigs contains secretion-ducts which, however, are soon thrown off, after which no fresh ones are formed. The formation of the balsam is induced by certain treatment to which the bark is subjected, and it is, therefore, not a normal secretion of the tree, but a pathological product. Of the exact nature of the changes induced we have no knowledge, but probably, as has been proved for colophony and for storax, schizogenous secretion-ducts are formed in the new wood after the infliction of injury.

The following is a brief account of the production of balsam of Peru. The bark of the tree is first beaten with the handle of an axe or with a stone over a space about 25 × 15 cm. and the cork removed. After five days a little balsam exudes, which is soaked off by rags. Now the wounded place is heated by torches and after about a week a more abundant flow takes place which is again absorbed by rags. The place is subsequently again cut and scorched and finally the scorched bark is itself detached. The wood thus laid bare is subjected to similar treatment lasting in all about six weeks. When the secretion of balsam ceases the bark immediately above the exhausted spot is treated.

The rags are collected and when sufficient have accumulated the balsam is pressed out and boiled with water which is poured off carrying impurities with it. The bark removed from the trees is also boiled with water and the balsam thus extracted mixed with the rag-balsam. The balsam is then filled into tin canisters with screw openings holding about 27 kilos each; these are conveyed on mules to Acajutla and Belize, whence the balsam is shipped to Hamburg, &c.

**Description.**—Balsam of Peru, as seen in commerce, is a rather viscid, oily liquid, resembling common black treacle, but rather thinner, and not of an adhesive or glutinous nature; it appears black in bulk, but in thin layers it is dark reddish brown and transparent. It has a fragrant, balsamic odour, and although it has no marked taste it produces, when swallowed, a burning sensation in the throat.

Balsam of Peru is heavier than water, its specific gravity varying within narrow limits—viz. from 1·137 to 1·158, being usually between 1·140 and 1·158, and this forms a valuable means by which adulteration can be detected, for many liquids that might be used for that purpose are lighter than water, and would appreciably depress the gravity. It is soluble in chloroform, and also in an equal volume of 90 per cent. alcohol, but with a larger proportion of the latter the mixture becomes turbid. It is practically insoluble in water, that liquid removing from it only a little cinnamic acid.

Its physical characters are so well marked, especially the odour and taste, that the drug is easy to recognise, but the detection of adulteration, especially with inferior qualities of the drug, which appears to be more or less regularly practised, is a more difficult problem.

**Constituents.**—Balsam of Peru consists essentially of an oily fluid portion mixed with a dark resin. The fluid portion (cinnamein) constitutes from 56 to 66 per cent. of the commercial drug, and consists of benzyl benzoate and benzyl cinnamate in the proportion of about three of the former to two of the latter, although this varies. Both esters are colourless, crystalline aromatic bodies which readily liquefy on heating.

The resinous portion, amounting to about 28 per cent. of the commercial drug, is composed of an alcohol, peruresinotannol, combined with cinnamic and a little benzoic acid.

The drug also contains an alcohol, peruviol (= nerolidol) which possesses a sweet odour and taste, traces of vanillin, and free cinnamic acid.

**Uses.**—Balsam of Peru is antiseptic and disinfectant. It is chiefly used as an external application in certain skin diseases and for bed sores. It has also been employed as a stimulant and disinfectant expectorant in bronchitis.

**Adulterants.**—Balsam of Peru is, from its nature and high price, liable to adulteration, chiefly with such liquids as alcohol, fixed oils, turpentine, copaiba, gurjun balsam, and the like. Any such admixture lowers the specific gravity, and can generally be detected by this means. Alcohol can be removed from the balsam by shaking it with water, which with the genuine drug should cause no appreciable diminution in volume. The presence of copaiba and of gurjun balsam can be detected by taking advantage of the insolubility of the resin of balsam of Peru in carbon disulphide. One part of carbon disulphide makes a clear mixture with three parts of balsam of Peru, but on the further addition of nine parts of carbon disulphide the resin should separate leaving a clear yellow non-fluorescent liquid which on evaporating should leave a yellowish brown oil, and this warmed to 150° should not evolve an odour foreign to balsam of Peru.

The assay of the balsam may be effected by dissolving 1 gm. of the balsam in ether, removing the resin by shaking this solution with N/2 solution of sodium hydroxide, evaporating the ethereal solution, and drying and weighing the residue ; it should amount to not less than 57 per cent. Each gramme of this residue (cinnamein) should require not less than 42 c.c. N/10 alcoholic solution of potassium hydroxide for saponification, corresponding to a saponification value of at least 235 (the saponification value of benzyl benzoate being 264·3, and of benzyl cinnamate 234·0). Full details of the test will be found in the British Pharmacopœia (1914), p. 61.

Recently a factitious substitute, 'perugen,' has been made by mixing synthetically prepared benzyl benzoate ('peruscabin') with storax, benzoin, and balsam of Tolu. It has an odour quite distinct from that of genuine balsam.

## STORAX

### (Styrax)

**Source, &c.**—Storax is a balsam obtained from the trunk of *Liquĭdambar orientălis*, Miller (N.O. *Hamamelidĕæ*), a tree of medium size forming forests in the south-west of Asiatic Turkey.

Neither the bark nor the wood of the tree possesses the agreeable odour of storax, and under normal conditions this substance is not produced in any part of the plant. In the early summer, incisions are made or the bark is beaten but not so vigorously as to kill it ; a formation of storax takes place, and the balsam soaks into the wounded bark, which is stripped off in the autumn. From the bark thus saturated the balsam is obtained by pressing it, the residue being subsequently mixed with boiling water (or boiled with water) and again pressed. The liquid balsam thus obtained forms the storax of commerce, whilst the pressed bark was formerly an article of commerce, under the name of Cortex Thymiamatis. The latter, coarsely ground and mixed with storax, formed 'Styrax calamitus,' under which name at the present time a factitious mixture is generally sold.

Although the bark of the tree contains secretion ducts these do not take part in the production of storax, which is secreted in schizogenous ducts in the young wood ; these, by the breaking down of intervening tissue, form schizolysigenous cavities from which the balsam exudes into the wounded bark. The secretion is therefore purely pathological, and it is produced in the young wood, subsequently finding its way thence into the bark, with which it is removed when the latter is stripped off.

**Description.**—Crude storax is a greyish, semi-fluid, viscid substance with an agreeable, aromatic, balsamic odour and a sharp, pungent taste. It is rather heavier than water, and contains usually vegetable débris, amongst which numerous bast fibres may be found. By drying it loses from 17·4 to 25·8 per cent. of water.

The crude drug is purified by dissolving it in three or four times its volume of hot alcohol, filtering, and removing the alcohol by gentle evaporation, care being taken to lose as little of the volatile constituents as possible. From 13 to 18 per cent. of the crude storax is insoluble in alcohol, the residue consisting principally of vegetable débris associated with inorganic matter ('Pharmacographia'). According to Evers (1896) storax contains from 6 to 9 per cent. of a greyish substance insoluble in alcohol, probably a resin ester of cinnamic acid.

Thus purified, storax is a semi-transparent, yellowish brown, semi-liquid balsam, entirely soluble in alcohol, ether, chloroform, and carbon disulphide. It should have a specific gravity of 1·109 to 1·114 at 100°.

**Constituents.**—Purified storax consists of a resin mixed with an oily liquid.

The resin is composed of storesinol partly free, partly combined with cinnamic acid. Storesinol is a white odourless amorphous substance which, however, forms a crystalline potassium compound.

The oily liquid contains styrol (phenylethylene, a colourless aromatic liquid), ethyl cinnamate, phenylpropyl cinnamate (odourless liquid) cinnamyl cinnamate (= styracin, odourless crystals), vanillin, and free cinnamic acid.

The approximate composition of storax, which, however, varies considerably, may be seen from the following table :

| | |
|---|---|
| Free cinnamic acid | 17 to 23 per cent. |
| Esters of cinnamic acid | 24 to 25 ,, |
| Water | 14 to 21 ,, |
| Resin | 36 ,, |

Purified storax may contain as much as 47 per cent. of cinnamic acid (free and combined), and is, in this respect, one of the richest drugs known.

**Uses.**—Storax is a local and remote stimulant and antiseptic, resembling in these respects balsam of Peru, benzoin, &c. It is now seldom used.

**Adulteration.**—The quantitative composition of storax is so variable that its purity can with difficulty be ascertained.

The acid value should vary between 60 and 90 and the ester value between 100 and 146. It should contain not less than 20 per cent. of total cinnamic acid when tested by the official process. Storax of good quality yields often from 25 to 30 per cent. of acid.

Much of the storax imported, especially from Trieste and Marseilles, since 1907 was very deficient in cinnamic acid, the percentage dropping to 2 or 3. Such balsam had been apparently deprived of some of its most valuable constituents (particularly cinnamyl alcohol and cinnamic acid) which were in demand for the perfumery trade.

**Allied Drugs.**—American storax or sweet gum is a transparent, yellowish, viscous liquid obtained from *L. Styraciflua*, Linné. It has been found to contain cinnamein 22·86 per cent., resin esters 34·76 per cent., resin acids 2·11 per cent., free cinnamic acid 12·63 per cent., total cinnamic acid 28·02 per cent. It has been recommended as a substitute for the Turkish.

## SECTION XVII

## TARS AND EMPYREUMATIC OILS

These are viscous or oily substances produced by the destructive distillation of organic substances such as wood, coal, &c.

### TAR

### (Wood Tar, Stockholm Tar, Pix Liquida)

**Source, &c.**—Tar or, as it is better termed, wood-tar (*Pix liquida*), to distinguish it from coal-tar (*Pix carbonis*), is a bituminous liquid obtained from the wood of *Pinus sylvestris*, Linné (N.O. *Coniferæ*), and other species of *Pinus* by destructive distillation.

Various methods have been adopted for the destructive distillation of wood. The simplest consists in constructing a mound of the material and covering it with earth, leaving spaces by which air can be admitted and through which the tar produced can drain off. The wood is fired, and, the heat being carefully controlled, charcoal is left as a residue in the mound, whilst a tarry and an aqueous liquid are collected.

More modern apparatus consists of either upright or horizontal iron retorts arranged singly or in a battery of about twenty. These are heated from without, and the products of distillation may be ranged under three heads :

(i) Gaseous products analogous to coal-gas ; these are conducted back to the hearth and burnt, serving to economise fuel, or they are stored in a gasometer until required for some such purpose.

(ii) Watery products ; water containing acetic acid, methyl alcohol, acetone, &c. ; these pass over and are condensed. They serve as a source of acetic acid, methyl alcohol, &c.

(iii) Tarry products ; these separate from the watery liquid on standing.

Coniferous wood is chiefly employed, as it yields more tar than the wood of foliage trees (beech, birch, &c.). From the former 15 to 20 per cent. of tar may be obtained, from the latter 6 to 8 per cent. For further details see Thorpe, 'Dictionary of Applied Chemistry': Wood, Destructive Distillation of.

**Description.**—Tar is a dark-brown or blackish, semi-liquid substance, with a peculiar aromatic odour and a bitter, pungent taste; it is heavier than water, its specific gravity varying from 1·02 to 1·15. By keeping, it becomes thicker, and acquires a granular appearance, due to the formation of minute crystals (probably pyrocatechin, resin acids, &c.). Water shaken with it acquires a yellow colour and acid reaction (distinction from coal-tar, which imparts to water an alkaline reaction). The filtered aqueous liquid is coloured red by very dilute (0·1 per cent.) solution of ferric chloride, (distinction from juniper-tar oil). Tar is completely soluble in ten volumes of alcohol.

**Constituents.**—The following are the chief constituents of tar: benzene, toluene, xylene, and styrolene (styrol); phenol, cresol, guaiacol and its homologues; pyrocatechin and paraffin. The most important of these, as far as the medicinal activity is concerned, are probably pyrocatechin, phenol, cresol, guaiacol, and their homologues.

Tar from Coniferous woods, which alone is official, is especially rich in guaiacol and its homologues, beech-tar in guaiacol and pyrogallol derivatives, birch-tar in guaiacol and benzophenol derivatives.

**Uses.**—Tar is used as an external stimulant and antiseptic in certain skin diseases; given internally in the form of pills or syrup, it acts as a disinfectant and deodorant of offensive discharges from the bronchial tubes.

## COAL-TAR

### (Pix Carbonis)

**Source, &c.**—Coal-tar is obtained by the destructive distillation of coal in iron or clay retorts. The gaseous products that are formed during the process constitute ordinary illuminating gas; the liquid products are a watery solution of ammonium carbonate, sulphide, &c., and a heavy, dark, viscid, tarry liquid, coal-tar. The retort contains a porous, carbonaceous residue, coke.

**Description.**—Coal-tar is a nearly black, viscid liquid, with a strong, penetrating, disagreeable odour; it is heavier than water, its specific gravity varying from 1·1 to 1·2. It is only slightly soluble in water, to which it imparts an alkaline reaction (distinction from wood-tar). It consists of a large number of substances, amongst which benzene, toluene, xylene, phenol, cresol, naphthalene, anthracene, and pitch may be mentioned. For further details see Martin, 'Industrial Chemistry, Organic,' p. 415.

**Uses.**—The chief medicinal use of coal-tar is as an antiseptic application in certain skin diseases.

## JUNIPER-TAR OIL

### (Huile de Cade, Oil of Cade, Oleum Cadinum)

**Source, &c.**—Juniper-tar oil is obtained by the destructive distillation of the wood of *Juniperus Oxycedrus*, Linné (N.O. *Coniferæ*); it is prepared chiefly in the mountains in the south-east of France, in the departments of Alpes Maritimes, Var, and Gard.

Selected portions of the trunk of the tree are cut into small pieces, and placed in a crude distilling apparatus consisting of an iron pot inverted over a concave stone slab, from the centre of which a pipe leads downwards. The smaller pieces of wood are used to heat the iron pot, and the distillate which passes down the delivery pipe is collected and allowed to stand for several weeks. At the end of this time it will have separated into three distinct layers, viz. a heavy, tarry mass, an aqueous liquid, and a lighter dark oily layer. The latter, which is true huile de cade, is separated and conveyed to neighbouring towns (Nismes, Avignon, &c.) for sale.

More recently the distillation has been conducted in a brick kiln about 20 feet long and 6 feet high, the bottom of which slopes downwards and has a gutter to carry off the tar. The kiln is filled with the wood which is then fired and the openings closed. The distillation lasts several days.

**Description.**—Juniper-tar oil is a dark reddish brown or nearly black liquid, with a tarry, but by no means disagreeable, odour, and a bitter, acrid taste. It is less viscid than wood-tar, having an oily consistence. The specific gravity is usually rather less than that of water (about 0·990), but in some (old) specimens it is rather higher. It is completely soluble in ether and chloroform, partially in cold but almost completely in boiling alcohol. Water shaken with it dissolves but little, acquiring a yellowish colour and an acid reaction (distinction from coal-tar).

**Constituents.**—The composition of juniper-tar oil is but very imperfectly known. The chief constituent appears to be a sesquiterpene, $C_{15}H_{24}$, boiling at 250° to 260°. Cadinene, $C_{15}H_{24}$, boiling-point 272° to 275°, is also present, but in small quantity. The tar also contains guaiacol, ethylguaiacol, propylguaiacol, creosol, &c., together with acetic acid and its homologues.

**Uses.**—Juniper-tar oil is used as an application in various skin diseases.

**Adulteration.**—Pine-tar is best detected by the following test :—Shake 1 c.c. of the oil vigorously with 15 c.c. of petroleum spirit and filter ; to 10 c.c. of the filtrate add 10 c.c. of a 5 per cent. solution of copper acetate ; allow separation to take place and dilute 5 c.c. of the upper layer with 10 c.c. of ether. Pure oil of cade gives a pale, brownish yellow colour, pine-tar an intense green.

When fractionally distilled, from 68 to 80 per cent. should distil between 150° and 300° (pine-tar about 15 per cent.).

**Allied Drug.**—Birch tar (Oleum Rusci, Oleum Betulae empyreumaticum) is a nearly black tarry liquid obtained in Russia by destructive distillation of birch wood (*Betula alba*, Linné). It possesses the characteristic odour of Russian leather and has been used in skin diseases.

## SECTION XVIII

## FIXED OILS, FATS AND WAXES

Fixed oils and fats are substances of either vegetable or animal origin, and consist of various organic acids of high molecular weight belonging chiefly to the stearic, oleic, linolic or ricinoleic acid series combined with glyceryl; they yield by saponification with caustic alkalies the alcohol glycerol (glycerin), the alkali combining with the organic acids to form soaps.

Waxes are solid or, occasionally, liquid substances resembling the fats and oils but differing in yielding monatomic alcohols instead of glycerol by saponification, which is readily effected by boiling alcoholic solutions of caustic alkalies, but not by aqueous.

### POPPY OIL

(Oleum Papaveris)

**Source, &c.**—The oil expressed from the seeds of *Papaver somniferum*, Linné. (Compare 'Poppy Capsules.')

**Description, &c.**—Pale to golden-yellow; specific gravity 0·924 to 0·927; iodine value about 133 to 143 (or 157 for oil extracted by petroleum spirit); dries to a hard varnish. Consists of the glycerides of linolenic, linolic, oleic and palmitic acids. Used by artists.

### EXPRESSED OIL OF MUSTARD

(Oleum Sinapis Expressum)

**Source, &c.**—The oil expressed from the seeds of *Brassica sinapioides*, Roth, and of *Brassica alba*, Boissier. (Compare 'Black Mustard Seeds.')

**Description, &c.**—Brownish-yellow; specific gravity 0·914 to 0·920; congealing point $-17°$; iodine value 96 to 110 (black mustard) or 92 to 97 (white mustard). Constituents similiar to those of rape oil; used as a mildly stimulating embrocation; also for lubricating, burning and soap-making.

## RAPE OIL

### (Oleum Rapae, Colza Oil)

**Source, &c.**—The oil expressed from the seeds of *Brassica campestris*, Linné, *B. Napus*, Linné, and other species of *Brassica* (N.O. *Cruciferæ*).

**Description, &c.**—Bright yellow, rather viscous; specific gravity 0·913 to 0·917; iodine value 98 to 103; congeals at − 4° to − 6° Contains the glycerides of arachic, erucic, linolic and linolenic acids. Largely used as a lubricant and illuminant.

## CHAULMOOGRA OIL

### (Gynocardia Oil, Oleum Chaulmoogræ)

**Source, &c.**—The soft fat expressed from the seeds of *Taraktogenos Kurzii*, King (N.O. *Bixineæ*), Burma. It was formerly supposed to be derived from the seeds of *Gynocardia odorata*, Robert Brown, which closely resemble those of *T. Kurzii*.

**Description.**—The seeds are brownish-yellow. The oil is a soft fat at ordinary temperatures, melting to a brownish oil in warm weather, with a characteristic odour and acrid taste. Melting-point 22° to 30°. Specific gravity at 45° about 0·940; saponification value 198 to 213; iodine value 96 to 104; acid value 21 to 27; partially soluble in cold alcohol, almost entirely in hot. Contains palmitic acid, chaulmoogric acid (crystalline, $C_{18}H_{32}O_2$, melting-point 68°), and also higher homologues of chaulmoogric acid, but the composition of the fat has not yet been satisfactorily elucidated.

**Uses.**—Used for leprosy, psoriasis and rheumatism, both externally and internally.

## COTTON-SEED OIL

### (Oleum Gossypii)

**Source, &c.**—The oil expressed from the seeds of various species of *Gossypium* (N.O. *Malvaceæ*).

The seeds are black, ovoid to rounded, 7 to 9 mm. long, 3 to 5 mm. wide; the section shows folded cotyledons with scattered, dark secretion cells.

They contain about 20 to 25 per cent. of fixed oil and about 20 per cent. of proteids. Crushed and pressed they yield about 13 per cent. of oil, the residual cake constituting a valuable cattle food. Cooled and pressed the oil yields a quantity of solid fat (chiefly palmitin), which is used in the manufacture of margarine, and a clear yellowish oil. They also contain a crystalline yellow substance of phenolic nature (gossypol) which passes into the (cold pressed) oil, but is extracted from it by alkalies which are always used in refining the oil.

**Description, &c.**—The crude oil expressed from the seeds is dark in colour, but after refining by a partial saponification it is pale yellow, odourless and of agreeable nutty taste. Specific gravity 0·918 to 0·925; begins to congeal at about 12°; iodine value 102 to 108; consists chiefly of the glycerides of oleic, palmitic, linolic and stearic acids. It may be identified by heating in a water-bath a mixture of 2 c.c. of the oil with 1 c.c. of amyl alcohol and 1 c.c. of a 1 per cent. solution of sulphur in carbon disulphide when a deep red colour is developed except in the case of oil which has been further purified by heating to 200°. Cotton-seed oil is largely used for making margarine and the lower qualities for soap.

## OIL OF THEOBROMA

### (Cocoa Butter, Oleum Theobromatis, Butyrum Cacao)

**Source, &c.**—The solid fat expressed from the seeds of *Theobroma Cacao*, Linné (N.O. *Sterculiaceæ*). See p. 153.

**Description, &c.**—Hard, brittle, pale yellow; softens at about 25°; melting-point 30° to 33°; specific gravity 0·990 to 0·998; iodine value 35·5 to 37·5. Consists of stearic, palmitic, arachic, oleic and small quantities of other acids combined with glycerol, partly as triglycerides, partly as mixed glycerides in which two fatty acids are combined with one glyceryl group. Chiefly used in pharmacy as a basis for suppositories, pessaries, &c.

**Note.**—*Shea butter* is a greyish fat of characteristic aromatic odour obtained from the seeds of *Butyrospermum Parkii*, Klotschy (N.O. *Sapotaceæ*), North Africa. *Kokum butter* from the seeds of *Garcinia indica*, Choisy (N.O. *Guttiferæ*), is largely used in India. *Illipe* or *mahua butter* is obtained by expression from the seeds of *Bassia latifolia*, Roxburgh, and *Bassia tallow* from the seeds of *B. longifolia*, Willdenow (N.O. *Sapotaceæ*); both are used for making candles and soap. *Chinese vegetable tallow* is the fat on the outside of the seed of *Stillingia sebifera*, Mich. (N.O. *Euphorbiaceæ*); it is largely used in China for making candles.

## LINSEED OIL

### (Oleum Lini)

**Source, &c.**—The oil expressed from the seeds of *Linum usitatissimum*, Linné (N.O. *Lineæ*). (Compare 'Linseed,' p. 149.)

**Description.**—Yellowish-brown, with characteristic odour; specific gravity 0·930 to 0·940; iodine value not less than 170; unsaponifiable matter not more than 1 per cent.; does not congeal at temperatures higher than $-20°$; exposed to the air in thin layers gradually hardens to a transparent varnish; consists of the glycerides of linolic, linolenic, isolinolenic, oleic, stearic, palmitic and myristic acids; may be identified by its odour, high iodine value, and by its hardening to a varnish; may be adulterated with mineral oil, rosin oil and fish oils; mineral oil may be detected by the rise in unsaponifiable matter; rosin oil by shaking with alcohol, evaporating the alcoholic solution, dissolving the residue in acetic anhydride and adding sulphuric acid, a red colour indicating rosin oil; fish oils are difficult to detect (compare Allen, 'Commercial Organic Analysis').

Linseed oil is used in pharmacy in liniments and as a laxative.

**Note.**—Boiled oil is linseed oil which has been heated with litharge or other suitable 'driers' to about 150°; metallic salts of the fatty acids are formed and cause the oil to dry more rapidly.

## ARACHIS OIL

### (Nut Oil, Ground Nut Oil, Pea Nut Oil, Oleum Arachidis)

**Source, &c.**—The oil expressed from the seeds of *Arachis hypogæa* Linné (N.O. *Leguminosæ*). The plant is an annual, attaining 1 m. in height; the lower flowers develop fruits which bury themselves in the earth and there ripen; a single plant may produce as many as 100 fruits. These contain usually from one to three reddish brown seeds the chief constituents of which are fixed oil (about 45 per cent.), starch and proteids (about 27 per cent.). The seeds are imported chiefly from West and East Africa and from India. The press-cake forms a valuable cattle food.

**Description, &c.**—The fixed oil is pale yellow or greenish yellow with a nutty odour and taste. Specific gravity 0·916 to 0·921; iodine value 83 to 110; consists chiefly of the glycerides of arachic, stearic, lignoceric, oleic, hypogæic and linolic acids. It may be identified by saponifying 1 c.c. with 15 c.c. of N/1 alcoholic solution of potash, allowing the mixture to stand for 24 hours at 15·5°, heating on a water-bath for 3 minutes and again setting aside, when crystals of (impure) potassium arachate separate.

## ALMOND OIL

### (Oleum Amygdalæ)

**Source, &c.**—The oil expressed from the Bitter or Sweet Almond. See p. 161.

**Description, &c.**—Pale yellow, nearly inodorous. Specific gravity 0·915 to 0·920; iodine value 93 to 100; does not congeal above $-18°$.

Consists of the glyceride of oleic acid and a little linolic acid; it contains no stearic acid. 'Kernel' oil, which may be used to adulterate almond oil, is obtained from apricot kernels (or possibly peach kernels); it may be detected by shaking 1 c.c. with 5 c.c. of a cooled mixture of equal parts by weight of sulphuric acid, fuming nitric acid and water (Bieber's test).

## OLIVE OIL

**Source, &c.**—Olive oil is the oil expressed from the pericarp of the ripe fruit of the olive tree, *Olea europœa*, Linné (N.O. *Oleaceœ*).

The olive is a small tree widely distributed by cultivation, especially in the countries bordering on the Mediterranean, and has been introduced into America (California), where it promises so well and is increasing so rapidly that that country will probably soon be independent of Europe for her supply of olive oil.

The tree produces a small, ovoid, drupaceous fruit about 2 or 3 cm. in length, which, whilst unripe, is green in colour and in this state is pickled and used as a table relish. As the fruits ripen the cells of the mesocarp become filled with a fixed oil, the colour at the same time changing from green to purple. They are collected in the winter and spring (December to April), the ripe sorted out, crushed, and subjected to a moderate pressure. The crude oil that flows from the press is run into tubs and mixed with water; the latter removes colouring matter and other impurities from the oil, and the oil, after it has completely separated from the water by standing, is skimmed off and filtered; in this way the finest quality of olive oil (virgin oil) is obtained. The marc is again ground, mixed with hot water, and again pressed, this time more strongly, and thus a second quality of oil is obtained. The residue still contains a little oil which may be obtained by a third pressing or by extraction with solvents; such oil is usually used for soap-making, &c. Sometimes all the fruits are thrown into heaps and allowed to ferment; on pressing, the whole of the oil is obtained, but it is of inferior quality, and is used principally for technical purposes.

**Description.**—For medicinal use only the first pressings, obtained without heat, should be employed. Such oil has a pale yellow or greenish yellow colour, a slight characteristic odour, and a bland taste without rancidity. Its specific gravity varies from 0·915 to 0·918. It is liquid at ordinary temperatures, but when cooled to 10° it often assumes a pasty consistence, from deposition of solid fats, and at 0° it becomes a nearly solid, granular mass.

**Constituents.**—Olive oil consists chiefly of olein and a little palmitin, together with linolein and traces of arachin, the palmitin and arachin separating out in the solid form when the oil is cooled.

By saponification these compounds yield respectively oleic, palmitic, linolic, and arachic acids together with glycerin.

**Uses.**—Olive oil has nutritive and laxative properties. Applied externally it is emollient and soothing.

**Adulterants.**—Valuable indications of purity are to be found in the specific gravity (0·915 to 0·918) and in the iodine value (79 to 87). The free acid present should not exceed 1 per cent., calculated as oleic acid, lower qualities may contain up to 5 per cent. of acid, and oils for technical purposes as much as 30 per cent. Cotton-seed oil, a frequent adulterant, is best detected by Halphen's (Bevan's) test which consists in warming 2 c.c. of the oil mixed with 1 c.c. of amylic alcohol and 1 c.c. of a 1 per cent. solution of sulphur in carbon disulphide for ten minutes in a water-bath when no red colour should be developed.

Sesamé oil [from the seeds of *Sesămum indicum*, Linné (N.O. *Pedalineæ*), India] is tested for by mixing 2 c.c. with 1 c.c. of hydrochloric acid containing 1 per cent. of sucrose, shaking for half an hour, adding 3 c.c. of water, again shaking and setting aside, when the aqueous liquid should not become pink.

The oil expressed from the seeds of the tea plant (tea seed oil) closely resembles olive oil and is said to be used for adulterating it.

For details of other tests for ascertaining the purity of olive oil, reference should be made to works on analytical chemistry, such as Allen's 'Commercial Organic Analysis,' or Lewkowitsch, 'Chemical Technology and Analysis of Oil, Fats and Waxes.'

## SESAMÉ OIL

### (Gingelly Oil, Teel Oil, Benne Oil, Oleum Sesami)

**Source, &c.**—Sesamé oil is the oil expressed from the seeds of *Sesămum indicum*, Linné (N.O. *Pedalineæ*), an annual plant about 1 m. high, extensively cultivated in India, China, Japan and most tropical countries. The seeds are small, 1·5 to 4 mm. long, 1 to 2 mm. broad and 1 mm. thick, yellowish white, reddish, brown or black according to the variety, obovate, slightly pointed, with four narrow longitudinal ribs. They contain about 50 per cent. of fixed oil and about 15 per cent. of proteids but no starch.

**Description, &c.**—Pale to golden yellow. Specific gravity 0·921 to 0·924 ; iodine value 103 to 114 ; congeals at about − 5° ; consists of the glycerides of oleic, linolic, palmitic, stearic and myristic acids ; contains about 0·2 to 0·5 per cent. of sesamin (acicular crystals, melting at 118°), also sesamol (phenolic) and a substance producing the characteristic colour reaction ; may be identified by shaking

2 c.c. of the oil with 1 c.c. of a (freshly prepared) 1 per cent. solution of sucrose (or furfurol) in hydrochloric acid when a pink colour results.

**Uses.**—As a substitute for olive oil; for making margarine and also in the preparation of additive compounds with iodine and bromine.

## CROTON OIL

### (Oleum Crotonis)

**Source, &c.**—The oil expressed from the seeds of *Croton Tiglium*, Linné (N.O. *Euphorbiaceæ*). See p. 184.

**Description, &c.**—A brownish yellow to dark reddish brown, viscous, slightly fluorescent oil with a disagreeable odour and intensely acrid taste, blistering the skin and mucous membrane. Specific gravity 0·940 to 0·960; iodine value 102 to 106; miscible with half its volume of absolute alcohol; thickens slightly but does not solidify, either partially or completely when shaken with half its volume of fuming nitric acid and the same proportion of water. It consists of the glycerides of stearic, palmistic, myristic, lauric, valerianic, tiglic and other acids and contains the acrid principle croton-resin.

**Uses.**—An extremely powerful cathartic, counter-irritant and vesicant.

## CASTOR OIL

### (Oleum Ricini)

**Source, &c.**—The oil expressed from the seeds of *Ricinus communis*, Linné (N.O. *Euphorbiaceæ*). See p. 182.

**Description, &c.**—Nearly colourless, viscous, slight odour, acrid unpleasant taste. Specific gravity 0·958 to 0·970; iodine value 83 to 90; acetyl value about 150; soluble in all proportions in absolute alcohol and in 3·5 parts of alcohol (90 per cent.); consists of the glycerides of ricinoleic, isoricinoleic, stearic and dihydroxystearic acids. Its freedom from admixture with other fixed oils is shown by the petroleum spirit test: 10 c.c. with 7 c.c. of petroleum spirit form a clear mixture at 15·5°; on adding a further 3 c.c. of petroleum spirit the mixture becomes turbid, on warming to 21° clear, and on cooling to 18° again cloudy.

**Uses.**—A mild purgative; on account of its viscosity largely used, especially in warm climates, as a lubricant.

**Note.**—Turkey red oil is obtained by allowing sulphuric acid to run slowly into castor oil, kept cool, washing with water and solution of sodium sulphate and adding ammonia until a clear water-soluble liquid is formed; the ricinoleic acid is converted into ricinoleo-sulphuric acid. It is largely used in dyeing cotton, imparting to the fabric a better lustre.

## COCO NUT OIL

### (Coprah Oil, Oleum Cocois, Oleum Cocois Nuciferæ)

**Source, &c.**—The fat obtained by expression from the kernels of the coco nut, *Cocos nucifera*, Linné, and *C. butyracea* (N.O. *Palmæ*), cultivated in southern India, Ceylon, South America and other tropical regions. The coco nuts are collected, broken open, the kernels removed and dried (= coprah) ; they are then ground and the fat pressed out of the warmed powder by hydraulic presses ; on cooling it sets to a white solid. The press-cake is used as cattle food or manure. The fibrous outer part of the nut yields coir fibre which is extensively used for mats, &c.

**Description, &c.**—A soft white fat of characteristic odour and bland taste but readily becoming rancid. Specific gravity about 0·903 at 100° ; melting-point 21° to 25° ; iodine value 8 to 9·5 ; saponification value about 258. Consists of the glycerides of lauric, myristic, palmitic, stearic, oleic, caproic, caprylic and capric acids.

**Uses.**—As a lubricant and for soap-making. Cooled and pressed it yields a liquid oil and solid stearin (coco nut stearin, melting-point about 29°) which has been used as a suppository base.

## COD-LIVER OIL

### (Oleum Morrhuæ)

**Source, &c.**—Cod-liver oil is the oil extracted from the fresh liver of the cod, *Gadus morrhŭa*, Linné (Phylum *Chordata*, Sub-phylum *Craniata*, Class *Pisces*, Order *Teleostĕi*).

The cod inhabits the North Atlantic Ocean in great numbers, leaving the deeper seas and approaching the coasts, chiefly of Norway and Newfoundland, towards spawning time—that is, from January to April. During that time immense quantities of the fish are taken both by nets and lines. The livers are cut out whilst the fish is quite fresh, the healthy ones selected, and, after the removal of the gall bladders, subjected to a gentle heat, usually about 70° (according to the British Pharmacopœia not exceeding 85°). The oil that separates is drawn off and exposed to a low temperature (about − 5°), at which a considerable quantity of solid fat separates. This is removed by filtration and pressure, and the oil thus purified forms the finest medicinal oil.

Inferior qualities of the oil are obtained from the residual livers (to which are added the unhealthy and injured livers previously rejected), by subjecting them to a higher temperature. Much oil, also of inferior quality but suitable for many technical purposes, is obtained by keeping the livers until they are partially decomposed, skimming off the oil that has separated and heating and pressing the residue ; such oil is usually of a brownish colour.

After the livers have been removed and the offal separated the fish are dried, and form an important article of commerce. The chief seats of the cod fishery are the shallow banks off Newfoundland, and the Loffoden Islands near the north-west coast of Norway.

**Description.**—Cod-liver oil should be of a pale yellow colour, and have a slight, fishy, but not rancid odour. Its specific gravity should vary from 0·920 to 0·930 ; it is readily soluble in ether and chloroform, but sparingly in alcohol.

**Constituents.**—Cod-liver oil consists chiefly of jecolein and therapin associated with palmitin and probably other fat-acids combined with glyceryl. Traces of biliary acids, alkaloids (morrhuine, aselline), cholesterol, iodine, &c., are also present.

Jecolein and therapin consist of jecoleic and therapic acids combined with glyceryl. Jecoleic acid is a very unstable acid belonging to the oleic acid series.

**Uses.**—Cod-liver oil is employed as a nutritive and is a food rather than a drug.

**Adulterants.**—Cod-liver oil is liable to adulteration with other fish-liver oils and with seal oil, the detection of which is exceedingly difficult. The following characters have been suggested for a pure cod-liver oil for medicinal use :—

| | |
|---|---|
| Colour . . . . . | Pale yellow |
| Odour . . . . . | Slightly fishy, not rancid |
| Taste . . . . . | Bland, not rancid |
| Specific gravity . . . | 0·925 to 0·931 |
| Saponification value . . | 179 to 198 |
| Free fat-acid calculated as oleic | Not over 1·5 per cent. |
| Melting-point of fat-acids . . | 23° to 26° |
| Iodine value . . . . | 154° to 170° |
| Refractive index (20°) . . | Not under 1·4790 |

If three drops of nitric acid (sp. gr. 1·4) are added to 15 drops of cod-liver oil and the mixture vigorously stirred, it should develop a bright, rose-red colour.

**Note.**—Whale oil (*Balæna* sp.), seal oil (*Phoca* sp.), dolphin oil (*Delphinus* sp.), and shark oil (*Carcharias* sp.) are used as illuminants, lubricants, for leather dressing and soap making. Large quantities of whale oil are converted by hydrogenation into bland edible fats of varying degrees of hardness. (Compare Allen's 'Commercial Organic Analysis,' Vol. II.)

## SUET

### (Sevum)

**Source, &c.**—The official suet is mutton suet, obtained from the abdomen of the sheep, *Ovis aries*, Linné (Phylum *Chordata*, Sub-phylum *Craniata*, Class *Mammalia*, Order *Ungulata*). It is purified by thoroughly crushing it so as to break the membranous vesicles in which the fat is contained, melting, and straining. During the cooling it should be stirred, so as to prevent the constituents of higher melting-point separating in a more or less granular form.

**Description.**—Suet should be white, smooth, and uniform in appearance, and possess a slight characteristic odour, but be free from rancidity. It melts at about 45° to 49°, and has a specific gravity of 0·948 to 0·953 (at 15°).

**Constituents.**—It consists principally of stearin and palmitin (about 80 per cent.), associated with olein (about 20 per cent.). The acid value should not exceed 2; saponification value 192 to 195; iodine value 33 to 46; refractive index at 60° 1·4490 to 1·4510; melting-point 45° to 50°.

## LARD

### (Adeps)

**Source, &c.**—Lard is the purified fat from the abdomen of the hog, *Sus scrofa*, Linné (Phylum *Chordata*, Sub-phylum *Craniata*, Class *Mammalia*, Order *Ungulata*).

The abdominal fat of the hog is obtained in the form of flat, leafy masses known as 'flare.' These should be first washed to free them from any salt that may have been used to preserve them, then stripped as far as possible of external membrane, and hung in a current of air for a few hours to dry. They must then be crushed or comminuted in any suitable manner, such as by beating in a stone mortar or passing through a mincing machine, in order to break the membranous vesicles and liberate the fat contained in them. If this were not done, either the fat would be retained in the vesicles or so high a temperature would have to be applied that it would acquire an unpleasant taste and odour. The crushed fat is exposed to a temperature which should not exceed 57° (in order to avoid the injurious effect of too great a heat), and when completely melted strained through fine muslin and gently stirred till cool, avoiding any form of beating which would introduce air into the melted fat and favour the development of rancidity. If not stirred the lard is liable to assume a granular condition, from the crystallisation of the constituents of higher melting-point (stearin and palmitin).

**Description.**—Lard is a uniform, soft, white, homogeneous, fatty substance melting at about 38° and having at 15° a specific gravity of about 0·934 to 0·938. Odour slight, fatty, but not rancid or otherwise disagreeable; entirely soluble in ether. Acid value not over 1·2; saponification value 192 to 198; iodine value 62 to 63; unsaponifiable matter not over 0·5 per cent.; refractive index at 60° 1·4530 to 1·4550.

**Constituents.**—Lard consists of about 40 per cent. of stearin and palmitin mixed with about 60 per cent. of olein, but these proportions are subject to a little variation, and with them both melting-point and specific gravity.

**Adulterants.**—Lard is liable to contain common salt, which is often added to preserve it for domestic use; it may be tested for chlorides by boiling with water, cooling, filtering the aqueous liquid, and adding silver nitrate and nitric acid. Starch, which might be added to give it a whiter appearance, could also be detected in the filtrate by solution of iodine. Sesame oil may be detected by the test detailed under 'Olive Oil' (p. 510).

But the most frequent adulterant of lard is cotton-seed oil, which has been found in American lard, large quantities of which are imported. It may be detected by the tests described under 'Olive Oil' and by a rise in the iodine number which should not exceed 60. It should be noted that lard obtained from hogs fed upon cotton cake may give a positive result with Halphen's test.

## BEESWAX

### (Cera flava; Cera alba)

**Source, &c.**—Beeswax is the wax separated from the honeycomb of the hive bee, *Apis mellifica*, Linné, and possibly other species of *Apis* (Phylum *Arthropoda*, Class *Insecta*, Order *Hymenoptera*).

Wax is a material secreted by the bee on the under surface of its body, and employed to form the walls of the cells of the honeycomb. After the separation of the honey the residual wax is purified by melting with water, separating, and straining; it then forms the yellow wax of commerce. White wax is obtained by exposing thin bands of yellow wax for several weeks to the action of air and sunlight, occasionally watering it, and if necessary remelting, to promote the bleaching by exposing fresh surfaces to these influences. Yellow wax is also largely bleached by chemical means, such as the action of chromic acid.

Wax is imported from Jamaica, California, Chili, Egypt, Syria, Madagascar, Morocco, &c.

**Description.**—Yellow wax is a yellowish or brownish yellow solid, with an agreeable, honey-like odour, breaking with a granular fracture, and not unctuous to the touch. It is readily and entirely soluble in hot oil of turpentine, partially soluble in alcohol. It is practically insoluble in water and in boiling aqueous solution of sodium hydroxide, the two latter liquids after filtration being neither turbid, nor becoming turbid on the addition of hydrochloric acid.

The specific gravity of wax varies from 0·958 to 0·970; melting-point from 61° to 64°; refractive index at 80° 1·4380 to 1·4420. These limits are narrow, and the specific gravity and melting-point often afford very valuable information as to the purity of the sample under examination.

**Constituents.**—Beeswax consists principally of melissyl palmitate (myricin), with which is associated free cerotic acid, an aromatic body cerolein, and probably melissyl stearate.

Cerotic acid, $C_{25}H_{51}COOH$(?), is an acid belonging to the formic acid series the formula of which is not definitely established. It crystallises from alcohol in granular colourless crystals melting at 78°–79°. Ceryl cerotate, $C_{25}H_{5}COOC_{26}H_{53}$, occurs in Chinese wax, the produce of *Coccus cerifera*, Fabricius (*C. pela*, Westwood), and in wool fat.

Melissyl palmitate, $C_{15}H_{31}COOC_{30}H_{61}$, is the palmitic ester of melissyl alcohol; the latter occurs in colourless crystalline needles melting at 85°.

**Adulterants.**—Beeswax is liable to adulteration with solid paraffin, with various fats and waxes of vegetable or animal origin, with resin, stearic acid, &c.

Paraffin and bodies belonging to this class are not attacked by hot, concentrated sulphuric acid, whereas beeswax is entirely destroyed; hence if 5 grammes of beeswax are heated for fifteen minutes with 25 grammes of concentrated sulphuric acid to 160°, and the mixture afterwards cooled, rinsed with alcohol and then extracted with ether, the latter solution should leave no appreciable solid residue on evaporation. A better test for paraffin (and other foreign waxes) is Weinwurm's which is performed as follows:

Saponify 5 gm. of the wax by boiling it with 15 c.c. of N/1 alcoholic solution of potassium hydroxide and 15 c.c. of absolute alcohol, evaporate to dryness, dissolve the residue in 20 c.c. of glycerin on a water-bath, and add 80 c.c. of boiling distilled water; a clear, or at least translucent, solution should be obtained.

Resin (colophony) would be readily dissolved by cold alcohol, in which genuine beeswax is sparingly and only partially soluble.

Soap would be removed by hot water, and the filtrate would become cloudy when acidified with hydrochloric acid from separation of the fat acid.

Stearic acid, colophony, Japan wax (obtained from the fruits of various species of *Rhus* (N.O. *Anacardiaceæ*), tallow, and all fats easily saponified, would be converted into soap by the action of boiling aqueous solution of sodium hydroxide which does not appreciably attack wax ; the solution after separation and filtration would then be rendered turbid by hydrochloric acid from separation of the free fat acids.

Starch (which might be added to white wax) would be detected by boiling with water and applying the iodine test. Inorganic substances, such as kaolin, would be detected in the ash, and by their insolubility in oil of turpentine.

Foreign colouring matters may be detected by boiling the wax with alcohol for five minutes, cooling thoroughly for several hours and filtering ; the filtrate should be almost colourless.

Further information respecting the purity of wax may be obtained by determining the acid value, which should not be less than 17·9 and the ester value (69·6 to 76·1) as directed in the Pharmacopœia ; the ratio of the acid value to the ester value should be about 1 to 4.

**Note.**—*Japan Wax* is the fat secreted in the mesocarp and cotyledons of the fruit of *Rhus succedanea*, Linné, and other species of *Rhus* (N.O. *Anacardiaceæ*), Japan. The fruits are husked between millstones, crushed and boiled with water ; the fat is skimmed off, purified by melting and straining, and then poured into moulds. It is a firm, white solid, consisting chiefly of palmitic acid and its glyceride. Melting-point about 50° to 56°. It is not a true wax, but a fat.

*Carnauba Wax* is the wax secreted on both surfaces of the leaves of *Copernicia cerifera*, Martius (N.O. *Palmæ*), South America. The leaves are dried and spread on cloths ; the wax is separated by brushing and beating, and then melted and poured into moulds. Consists chiefly of melissyl (myricyl) cerotate, melissyl alcohol, carnaubic acid. Melting-point 83° to 86°. Used in candle making, boot polishes, &c.

*Chinese Insect Wax, Pela,* is produced by *Coccus cerifera*, Fabricius (N.O. *Hemiptera*) on the twigs of *Ligustrum lucidum* (N.O. *Oleaceæ*) or *Fraxinus chinensis*, Roxburgh (N.O. *Oleaceæ*). The bark with its coating of wax is stripped from the tree, the wax melted in hot water, skimmed off, purified by re-melting and poured into moulds ; colourless or pale yellowish, crystalline, almost odourless and tasteless, melts at about 81° to 83° ; consists almost entirely of ceryl cerotate.

## SPERMACETI

### (Cetaceum)

**Source, &c.**—Spermaceti is a solid wax obtained from the head of the sperm whale, *Physēter macrocephălus*, Linné (Phylum *Chordata*, Sub-phylum *Craniata*, Class *Mammalia*, Order *Cetacea*), and probably other species. Part of the spermaceti of commerce is obtained from the bottle-nosed whale, *Hyperoodon rostratus*.

The sperm whale inhabits the Pacific, Atlantic and Indian Oceans. Its head is of enormous size, occupying about one-third of the animal, which varies from 50 to 70 metres in length. In a special, large, cylindrical organ in the upper region of the huge jaw and above the right nostril an oily liquid, crude sperm oil, is secreted. After the whale has been captured this cavity is emptied of its oil, which, on cooling, deposits a quantity of crystalline matter. This, the spermaceti, is separated by pressure and purified by re-melting, and washing with dilute solution of sodium hydroxide to free it from the last traces of oil ; the spermaceti separated from the soap thus produced, and from excess of free alkali, forms as it cools crystalline masses.

**Description.**—Spermaceti occurs in translucent, crystalline masses, pearly white in colour and unctuous to the touch ; it has but little odour or taste. Specific gravity, 0·95 to 0·96 ; melting point 46° to 50° ; acid value not more than 1 ; saponification value 125 to 136 ; iodine value 3 to 44 ; refractive index at 80° about 1·4330. It is insoluble in water and cold alcohol, but soluble in ether, chloroform, and boiling alcohol, crystallising from the latter solvent on cooling.

**Constituents.**—Spermaceti consists principally of cetyl palmitate, $C_{15}H_{31}.COOC_{16}H_{33}$, together with a small proportion of esters of other fat acids.

Cetyl alcohol, $C_{15}H_{33}OH$, which can be obtained from spermaceti by saponifying with alcoholic solution of potassium hydroxide, diluting with hot water, filtering from the soluble potassium palmitate and crystallising from hot alcohol, forms brilliant, colourless crystals melting at 49·5°.

**Adulterants.**—Spermaceti may be adulterated with stearic acid, stearin, tallow, and paraffin wax. Stearic and other fat acids raise the acid value ; stearin, tallow, and paraffin wax are insoluble in boiling alcohol ; stearin and tallow raise the saponification value but paraffin wax lowers it.

**Note.**—*Ambergris* is a substance formed in the intestine of the whale, and found there or floating in the sea. It occurs in masses of varying size (1 to 20 or more pounds), greyish in colour, veined, brittle, with an agreeable, persistent odour recalling musk.

## WOOL FAT

### (Lanolin, Adeps Lanæ)

**Source, &c.**—Wool fat is a fatty or, more correctly, waxy substance secreted by the hairs constituting the fleece of the sheep *Ovis aries*, Linné (Phylum *Chordata*, Sub-phylum *Craniata*, Class *Mammalia*, Order *Ungulata*).

If a few threads of raw sheep's wool are examined under the microscope little masses of a fatty substance may be seen adhering to them. This is the crude, natural, wool fat ; part of it is soluble in water and is removed during the first cleansing process which consists in steeping the fleeces in water; part is insoluble in water and can subsequently be removed by benzene. acetone, or other suitable solvent, forming, after evaporation, a brownish grease. It is also removed when the fleeces are scoured with soap and water, the second cleansing process. When the emulsion thus produced is acidified the wool fat is separated together with the fat-acids produced by the decomposition of the soap. These fat-acids can be converted into the corresponding calcium salts and the wool fat separated by treating the product with acetone; the acetone solution, evaporated to dryness, yields crude wool fat which has to be purified by suitable means.

Wool fat may also be extracted by scouring the fleeces with hot water, and allowing the emulsion thus produced to stand, when impure wool fat rises as a cream. This can be cleansed by repeatedly mixing with water and separating by centrifugation, the resulting wool fat being subjected to a final process of purification.

**Description.**—Purified wool fat is a yellowish, tenacious, unctuous solid with a characteristic odour. It melts at about 42° and is soluble in acetone, benzene, and the usual fat solvents. It may be distinguished from other fats by its solubility in boiling alcohol and also by the following test (for cholesteryl alcohol): Dissolve 0·1 gramme in a mixture of 5 c.c. of chloroform and 0·5 c.c. of acetic anhydride; pour gently upon 5 c.c. of sulphuric acid; a purplish brown ring, passing into green is developed at the surface of contact.

**Constituents.**—Wool fat consists chiefly of cholesteryl and isocholesteryl alcohols combined with lanoceric, lanopalmitic, carnaubic, myristic, a little oleic, and possibly also palmitic and cerotic acids.

**Uses.**—Wool fat is largely used as an emollient and for promoting the absorption of drugs by the skin.

**Adulterants.**—The most probable adulterants of wool fat are mineral fats (soft paraffin) or animal and vegetable fats and oils. Wool fat, like most waxes, is not readily attacked by boiling, aqueous solution of potassium hydroxide, but may be saponified by boiling, or heating under pressure, with an alcoholic solution of the same, the saponification value varying from 90 to 102. Mineral fats are not attacked by either aqueous or alcoholic solution of potassium hydroxide, and their presence would lower the saponification value. Animal and vegetable fats and oils would by the same treatment be saponified and raise the saponification value; they would also be saponified by aqueous alkalies which would not attack wool fat or mineral fats. Glycerin can be detected by shaking the wool fat with hot water and evaporating the aqueous solution.

## SECTION XIX

## SACCHARINE SUBSTANCES

### MANNA

### (Manna)

**Source, &c.**—Manna is strictly a generic term applied to the saccharine exudations from a number of different plants belonging to various natural orders, but when not otherwise specified is understood to mean the saccharine exudation from the stem of the manna ash, *Fraxinus Ornus*, Linné (N.O. *Oleaceæ*), a small tree widely distributed over southern Europe and cultivated especially in Sicily for the production of manna.

The trees are considered fit for yielding manna when they are about ten years old. Every day a transverse or oblique incision is made through the bark on one side of the stem; the saccharine liquid that exudes flows down the stem in favourable seasons and dries, but in rainy weather it drops from the trunk and is caught upon cactus leaves (more strictly branches), placed beneath it, yielding an inferior quality. In the following year the tree is cut upon the opposite side, and in the succeeding year again on the first side. The stem is then exhausted, the tree is cut down, and from the stool two or more shoots are allowed to grow, which in ten years are again ready for tapping. In good seasons about 500 grammes of manna are obtained from each stem.

**Description.**—The finest qualities of manna, known in commerce as 'flake manna,' are in pieces about 10 or 15 cm. long and 2 or 3 cm. wide, which are more or less conspicuously three-sided, one of the sides (that which has been next the stem) being concave and smooth. It is yellowish white in colour and very brittle, even friable, exhibiting when broken an indistinctly crystalline structure. It has a slight agreeable odour and a sweet taste.

Inferior qualities of manna are of a darker, brownish yellow colour, and composed of broken flakes agglutinated into a more or less sticky, gummy mass.

**Constituents.**—Manna consists principally of mannite (mannitol), $C_6H_8(OH)_6$, of which it may contain from 40 to 60 per cent. Mannite is a well-defined, crystalline substance belonging to the class of alcohols (a hexahydric alcohol), and is widely diffused throughout the vegetable kingdom. It is associated in manna with two sugars, viz. manninotriose (6 to 16 per cent.) and manneotetrose (12 to 16 per cent.). Each molecule of manninotriose yields by hydrolysis two molecules of galactose and one of dextrose; each molecule of manneotetrose yields two of galactose and one each of dextrose and levulose. Dextrose, mucilage, inorganic substances, a minute quantity of a fluorescent substance, fraxin, and about 10 per cent. of moisture also occur in manna.

FIG. 244.—Portion of the stem of a Manna tree, showing the incisions and adhering manna. Reduced. (Vogl.)

**Uses.**—Manna is used medicinally as a gentle laxative.

**Allied Drugs.**—*Australian manna* obtained from *Myoporum platycarpum*, Robert Brown (N.O. *Polygaleæ*), in brownish masses containing about 89 per cent. of mannite.

*Alhagi manna* from *Alhagi Maurorum*, Tournefort (N.O. *Leguminosæ*), Persia, in brownish, brittle, crystalline or amorphous masses; consists chiefly of sucrose; contains no mannite.

*Tabashir* or *Bamboo manna* secreted in the surface of the stem of *Bambusa stricta*, Roxburgh (N.O. *Gramineæ*), almost entirely sucrose (not to be confounded with the tabashir which occurs in the interior of the stem and is mainly silica).

*Oak manna*, secreted in the leaves and acorn cups of *Quercus Vallonea*, Klotschy (N.O. *Cupuliferæ*), consists of sucrose and dextrose; contains no mannite.

*Tamarisk manna*, from *Tamarix gallica*, var. *mannifera*, Ehrenberg (N.O. *Tamariscineæ*), contains sucrose, 55 per cent.; levulose, 25 per cent.; no mannite.

*Briançon manna* secreted by the leaves of *Larix europœa*, de Candolle (N.O. *Coniferæ*) contains the sugar melezitose.

## HONEY

### (Mel)

**Source, &c.**—Honey is a saccharine substance deposited by the hive bee, *Apis mellifica*, Linné (Phylum *Arthropoda*, Class *Insecta*, Order *Hymenoptera*), and other species of *Apis* in the cells of the honeycomb.

The nectar secreted by the nectaries of numerous flowers contains as its principal constituent cane-sugar (sucrose). This nectar is sucked up by the bee through an air-tight tube formed from its ligula and labial palps. It is carried through the œsophagus into the crop or honey-sac, becoming mixed on its way with the salivary secretion from special glands. During its stay in the honey-sac, the sucrose of the nectar is converted into invert-sugar (dextrose and levulose) by the enzyme, invertase, which is contained in the salivary secretion. Arrived in the hive, the bee empties the honey-sac by regurgitation into a cell of the honeycomb. From the cells of the honeycomb the honey is separated by cutting and draining, or by centrifugation, or by pressure with or without heat.

Some quantity of honey is produced in England, but the chief sources of supply are California, Chili, and Jamaica.

**Description.**—Honey, when fresh, is a viscous, transparent liquid, becoming semi-solid on standing from crystallisation of the dextrose contained in it. It varies in colour from nearly white to reddish brown. It has an agreeable odour and a sweet, slightly acrid taste, both odour and flavour being to a great extent dependent upon the nature of the flowers from which the nectar was collected.

**Constituents.**—Pure honey consists chiefly of dextrose and levulose together with water in which these are at first dissolved. It also contains small quantities of volatile oil, formic acid, sucrose, dextrin, proteids, wax, pollen grains, and often fragments of dead insects, &c. It yields from 0·3 to 0·8 per cent. of ash containing traces only of sulphate and chlorides, and usually exhibits slight dextro- or lævorotation ($+ 3°$ to $- 3°$).

**Uses.**—Honey is largely used as a demulcent and sweetening agent as well as for its nutritive properties.

**Varieties.**—Honey obtained from heather and clover is considered to have the finest flavour, while that from Coniferous plants and species of *Eucalyptus* is the least agreeable. Some indication of the source of the honey may be obtained from the identity of the pollen grains contained in it. Jamaica honey is often dark in colour, while Australian honey usually has an unpleasant eucalyptus flavour.

**Adulterants.**—The most common adulterants are sucrose and commercial glucose both of which produce dextrorotation in the honey. The presence of calcium sulphate, which may be tested for in the usual way either in the honey or in the ash, indicates commercial glucose. Pure honey should show at most a slight turbidity when mixed with three or four volumes of alcohol (absence of dextrin, a frequent constituent of commercial glucose).

## SECTION XX

# CAMPHOR

**Source, &c.**—Camphor is a crystalline substance contained in the volatile oil of the camphor tree, *Cinnamomum Camphóra*, Linné (N.O. *Laurineæ*), a large tree indigenous to the island of Formosa, and to Japan and China. It is cultivated also in Ceylon, the Federated Malay States, India, Burma, Java, East and West Africa, California, &c. The leaf, petiole, bark, wood, and pith of the tree contain numerous oil cells. In these a volatile oil is secreted which, by oxidation, yields camphor. It is also found in the solid state in cavities in the trunk, probably as the result of slow volatilisation and oxidation of the oil. The yield of oil varies from 2 to 6 per cent. and that of the camphor from 0·2 to 3 per cent., old trunks containing the most.

**Production.**—Camphor is usually obtained by passing a current of steam through the chipped wood of the trunk and root, a primitive form of still being used for the purpose. From the pasty mixture of solid camphor and liquid volatile oil obtained the camphor is separated by pressing. The oil is exported under the name of camphor oil; by fractional distillation is yields light camphor oil (boiling-point, 175° to 200°) consisting of pinene, phellandrene, cineol and dipentene, and heavy camphor oil (boiling-point 200° to 275°) containing safrol and eugenol, and utilised for the production of the former on the large scale. The crude camphor is purified by mixing it with lime, charcoal, &c., and subliming it in large glass flasks (bomboloes); the camphor sublimes in the upper part of the flask and is obtained in the form of a thick ring by cracking the flask. It may also be sublimed into flat cakes or, by passing the vapour into large cooled chambers, into small crystals (flowers of camphor) which may be pressed into rectangular cakes by hydraulic pressure.

**Description.**—Camphor occurs in small, colourless crystals or in translucent, fibrous or crystalline masses with a characteristic penetrating and pungent, bitterish taste, followed by a sensation of cold. It burns readily with a bright, smoky flame, and is volatile at ordinary temperatures. It is a ketone, yielding isoborneol by reduction and camphoric acid by oxidation. With vanillin-hydrochloric acid it gives a reddish blue or bluish green colour which appears to be due to impurities derived from the plant during the distillation.

**Uses**.—Camphor is used externally as a mild rubefacient, internally as a carminative and antiseptic.

**Varieties**.—*Synthetic camphor* is obtained from pinene by converting it into pinene hydrochloride from which isobornyl acetate and isoborneol are successively prepared; isoborneol is then converted by reduction into camphor. Synthetic camphor may be distinguished from natural camphor by its optical inactivity (natural camphor is dextrorotatory) and by its not yielding the colour reaction with vanillin-hydrochloric acid. It can be resolved into dextro- and lævo-camphor.

*Borneo camphor* is obtained from *Dryobalanops aromatica*, Gaertner (N.O. *Dipterocarpeæ*), Borneo. The camphor is secreted normally in ducts in the pith, but similar ducts are probably formed pathologically in the young wood. Cavities filled with the camphor occur in the pith and from these the camphor is scraped out. It consists almost entirely of dextro-borneol. It is employed in China and Japan for embalming and also for various religious ceremonies.

*Ngai camphor* is obtained from *Blumea balsamifera*, de Candolle (N.O. *Compositæ*), Burma, by distillation of the leaves. It consists of lævo-borneol and is used in China as a medicine and for religious ceremonies.

# SECTION XXI

## ANIMALS AND ANIMAL GLANDS AND SECRETIONS

Organotherapy, or the treatment of disease by means of animal secreting organs or of preparations derived from them, has been continually in use for some 5,000 years, but it has recently been scientifically studied and considerably extended. Before proceeding to this section of the work it will be desirable to take a very brief survey of the animal kingdom, more especially as several animal substances have already been described.

Various systems of zoological classification have been from time to time proposed, but the following one is readily intelligible and well adapted for the object in view. By it the animal kingdom is divided successively into Phyla, Classes, Orders, Families, Genera, and Species. The following outline embraces the chief Phyla, Classes, and Orders, and indicates their most important characters.

*Phylum I.—Protozoa.* This phylum includes the simplest and most primitive animals, each consisting of a minute, single cell. Loose colonies are sometimes formed by division or budding, but differentiated tissues are not developed. Most of them live in water, and some form calcareous or siliceous skeletons. They are grouped into three classes, viz. the *Rhizopoda*, which are predominantly amœboid, the *Sporozoa*, which are usually encysted, and the *Infusoria*, which are actively motile by means of cilia or flagella. The Rhizopoda include such animals as the Foraminifera, the accumulated calcareous skeletons of which form most chalks and limestones. The Sporozoa are mostly parasitic, e.g. the malarial parasite. The Infusoria move actively and are found in great numbers in stagnant water, vegetable infusions, &c.

*Phylum II.—Porifera.* These are mostly passive, vegetative, marine animals which form colonies and exhibit slight division of labour, sexuality, &c. Familiar examples are the ordinary sponges in which very minute pores on the surface (inhalant canals) lead into chambers lined with flagellated cells. The movement of the flagella draws a current of water with food particles through the sponge, the waste products being discharged through larger exhalant openings. The Porifera have no body cavity, but often secrete a horny (bath sponge), calcareous, or siliceous skeleton.

*Phylum III.—Cœlentera.* Multicellular animals which, like the Porifera, have no body cavity; they are usually radially symmetrical and develop stinging cells. The most familiar examples are the sea-anemones, jelly-fishes, and corals, the last-named being colonies of individuals which secrete a calcareous skeleton.

*Phylum IV.—Platyhelminthes* or *Flat-worms.* In this group the primitive radial symmetry is lost; the body is flattened but no true body cavity is formed. Here the first indications of a head and brain are to be discerned. To this phylum belong the Trematodes or Flukes (liver fluke of the sheep) and the Cestodes or Tapeworms, all of which are parasitic.

*Phylum V.—Nemerteæ* or *Ribbon Worms.* These are the first animals possessing an open gut and closed blood-system; they are mostly marine and, for the purpose in hand, unimportant.

*Phylum VI.—Nematohelminthes.* Thread-like worms, mostly parasitic and provided with a distinct body cavity, a well-developed alimentary canal, a mouth, and an anus. To this phylum belong such well-known parasites as the round-worm (*Ascaris*), the thread-worm (*Oxyuris*), the Guinea worm (*Dracunculus*), *Trichina*, &c.

*Phylum VII.—Annelida.* These exhibit a well-developed cœlom and a distinct segmentation of body which is visible externally. The Annelida include the *Chaetopoda* or worms with bristles (earth-worm, lob-worm) and the *Hirudinea* or worms devoid of bristles (leech).

*Phylum VIII.—Echinoderma.* The echinoderms are radially symmetrical animals which exhibit a tendency to form calcareous skeletons. Familiar examples may be found in the sea-cucumbers, sea-urchins, star-fishes, &c.

*Phylum IX.—Arthropoda.* This important phylum consists of bilaterally symmetrical animals exhibiting numerous segments and bearing paired, jointed appendages; the sexes are always separate. The chief classes are *Crustacea*, *Myriopoda*, *Insecta*, and *Arachnoidea*.

The *Crustaceans* include the crabs, lobsters, cray-fish, shrimps, &c.; they mostly live in water and breathe by gills.

The *Myriopods* embrace the centipedes, millipedes, &c.

The *Insects* are usually winged, breathe by a system of air-tubes and undergo a metamorphosis. They are subdivided into eighteen orders of which the following are the most important for the purposes of the present work.

(a) *Hymenoptera.*—Ants, bees, wasps, &c.; they have four transparent wings, are usually provided with mandibles, and are furnished with a sting or with an ovipositor.

(*b*) *Lepidoptera.*—Butterflies and moths; they have four scaly wings, a suctorial mouth, and no ovipositor.

(*c*) *Diptera.*—House-flies, gnats, midges, &c.; they have one pair of transparent wings and a suctorial mouth; the larva is a maggot.

(*d*) *Coleoptera.*—Beetles; these have one pair of wings and one pair of hardened wing-cases; they are provided with biting mandibles.

All the members of the foregoing orders undergo a complete metamorphosis from larva to perfect insect.

(*e*) *Hemiptera.*—Aphides, Coccus insects, bugs, &c.; these have piercing, suctorial mouths and either two pairs of wings or none (the male Coccus insect alone undergoes complete metamorphosis).

The *Arachnoidea* include the scorpions, spiders, mites, ticks, &c.

*Phylum X.—Mollusca.* The Molluscs (snails, mussels, oysters, cuttlefishes, &c.) are unsegmented, have no appendages, and have a soft body secreting an outer shell (in the cuttlefish an inner, calcareous shell).

*Phylum XI.—Chordata* (*Vertebrata*). These have a dorsal supporting axis and a dorsal, tubular nervous system. The most important classes are the *Pisces*, the *Amphibia*, the *Reptilia*, the *Aves*, and the *Mammalia*.

The chief sub-classes of the *Pisces* are (i) *Elasmobranchii* or Cartilaginous Fishes such as the skate, shark, dog-fish, &c., and (ii) *Teleostomi*, which have a more or less bony skeleton. The most important order of Teleostomi is *Teleostei*, or true Bony Fishes, which includes the cod, herring, salmon, and most of the modern fishes.

The class *Amphibia* are those vertebrates which exhibit a transition from aquatic to terrestrial life. The frog, toad, and newt are types.

The class *Reptilia* includes the tortoises, turtles, lizards, snakes, alligators, &c.

The class *Aves* includes the divisions *Ratitæ* or Running Birds (ostrich) and the *Carinatæ* or Flying Birds.

Of the class *Mammalia* the following Orders belonging to Sub-class *Eutheria* or placental animals may be mentioned: (i) *Xenarthra* (sloths, anteaters); (ii) *Nomarthra* (pangolins); (iii) *Sirenia* (dugong, sea-cow); (iv) *Ungulata* (hoofed animals such as the ox, hog, sheep, musk-deer, &c.); (v) *Cetacea* (whales, dolphins, &c.); (vi) *Rodentia* (rats, beavers, &c.); (vii), *Carnivora* (lions, tigers, bears, &c.); (viii) *Pinnipedia* (seals, walruses); (ix) *Insectivora* (hedgehogs, shrews, &c.); (x) *Chiroptera* (bats); (xi) *Prosimiæ* (lemurs); (xii) *Anthropoidea* (monkeys, baboons, apes, man).

# ANIMALS

## LEECH

### (Hirudo)

**Source, &c.**—Leeches are aquatic worms, and are bred in ponds for medicinal use, chiefly in Germany (near Hanover) and in the south of France (near Marseilles). Two varieties are official—viz. the speckled leech, *Hirudo medicinalis*, Linné, and the five striped or Australian leech, *Hirudo quinquestriata*, Schmarda (Phylum *Annelida*, Class *Hirudinea*, Order *Gnathobdellida*); they may be distinguished by the ventral surface, which in the former is of a greenish yellow colour, spotted with black, whilst in the latter it is not spotted.

Although the dorsal surface of the leech is marked with numerous (about 100) annulations, the body is not divided into distinct segments. It tapers towards each extremity, and is provided at each with a sucking-disc by which it can attach itself to any object. The anterior disc, which is smaller than the posterior, contains three jaws radiating from a common centre; each jaw is furnished with a number of minute teeth, and resembles a portion of a circular saw. The animal attaches itself by means of its anterior sucker to the skin, which is thereby slightly raised; the three jaws, by a saw-like movement, produce three slits which unite to form the characteristic triradiate cut, and the leech gorges itself with blood; it then relinquishes its hold, and drops from the skin. The blood which it has drawn is so slowly digested that a single meal will last for several months.

Although the leech is hermaphrodite—that is, both sexes are united in the same individual—it is incapable of self-fertilisation. It reproduces itself by means of eggs, and the young require about five years to arrive at maturity.

The quantity of blood that a leech will draw is not large (from 4 to 8 c.c.), but the flow of blood from the cut is often continued for some time; this appears to be due partly at least to a substance, hirudin, which is secreted by the salivary glands of the leech and injected into the cut. The flow of blood can also be maintained by fomentation.

**Uses.**—Leeches are used to reduce inflammation by withdrawing blood. Hirudin, which retards the coagulation of blood, has been suggested for use in thrombosis and other conditions in which the blood shows a disposition to clot too readily.

## CANTHARIDES

### (Spanish Flies, Cantharis)

**Source, &c.**—Cantharides are the dried beetles, *Cantharis vesicatoria*, Latreille (*Lytta vesicatoria*, Linné, Phylum *Arthropoda*, Class *Insecta*, Order *Coleoptera*).

Cantharides, or, as they are frequently termed, Spanish flies, are widely distributed over southern Europe; they are gregarious and inhabit chiefly ash trees, privets, elders, &c. They are collected in the very early morning before sunrise (whilst they are unable to use their wings) by shaking them from the trees on to cloths placed beneath; they are killed by exposing them to the fumes of ammonia, acetic acid, or burning sulphur, or by stove heat; they are then dried, preferably by stove heat. When fresh they possess a powerful, disagreeable odour, which diminishes by keeping.

FIG. 245.—Spanish blistering beetle, *Cantharis vesicatoria*. Slightly enlarged. (Maisch.)

They are collected in southern Russia, Galicia, Roumania, and also to a much smaller extent in Italy and Spain.

**Description.**—Cantharides are about 20 to 25 mm. long, about 7 mm. broad, smooth, and of a shining green or coppery green colour. The wing-cases are long and narrow, and conceal two transparent, brown, membranous wings. Each insect possesses three pairs of legs and one pair of antennæ.

**Constituents.**—The principal constituent of cantharides is a definite crystalline body, cantharidin, which, although almost insoluble in cold water and only sparingly soluble in alcohol, dissolves readily in acetic ether, chloroform, and benzene. With caustic alkalies (potassium or sodium hydroxide) it unites to form soluble salts, and there is evidence to show that in the beetles it exists partly as free cantharidin, partly in the form of salts soluble in water. The soft parts of the insect are the chief seat of cantharidin. Good cantharides contain from 0·5 to 0·8 per cent. or occasionally as much as 1 per cent.

In addition to this substance the beetles contain about 12 per cent. of fixed oil.

**Assay.**—Moisten 20 gm. of finely powdered cantharides with 3 c.c. of hydrochloric acid, transfer to a Soxhlet apparatus and exhaust with benzene. Distil off the benzene and boil the residual fatty matter for 10 minutes with 100 c.c. of water acidified with hydrochloric acid, using a reflux condenser. Transfer the hot aqueous solution to a capacious separator, and repeat the boiling with four successive portions of 50 c.c. of water, unite the aqueous solutions, and shake out with successive portions of 30, 30, 20, and 20 c.c. of chloroform. Recover the chloroform by distillation from a tared flask, dry the residue at 60°, and wash first with 10 c.c. of a mixture of equal volumes of absolute alcohol and petroleum spirit, and finally with petroleum spirit until the latter leaves no appreciable residue on evaporation. Dry at 60° and weigh.

**Uses.**—Cantharides possess rubefacient and vesicant properties; given internally, the drug acts as an irritant poison.

**Varieties.**—The number of species of Coleoptera known to possess vesicating properties is very large, but only a few are employed for that purpose. The most important of these are Chinese blistering beetles (Chinese cantharides), large quantities of which are regularly imported. They are derived from two species, viz.:

1. *Mylabris sidæ*, Fabricius (*M. phalerata*, Pallas). This varies from 12 to 30 mm. in length, and from 5 to 10 mm. in breadth. It is black in colour, but the wing-cases are traversed by three broad, brownish yellow bands. These bands bear black, bristly hairs which, however, are not readily seen in the commercial drug, as most of them have been broken off. The beetles inhabit China, Bombay, Assam, &c.

FIG. 246.—Chinese blistering beetle, *Mylabris sidæ* (= *M. phalerata*, Pall.). (Maisch.)[1]

2. *Mylabris cichorii*, Fabricius, on the average smaller than the foregoing; it varies from 10 to 15 mm. in length, but is marked with similar yellow bands, which, however, are usually brighter in colour. The chief distinction of this species lies in the yellow downy pubescence with which the yellow bands are covered, the hairs on the black bands being black. It inhabits China and eastern India.

Chinese blistering beetles contain from 1 to 1·2 per cent. of cantharidin and form a useful source of this substance.

In addition to these, several other blistering beetles find their way occasionally to London, e.g. *M. lunata*, Pallas, and *M. bifasciata*, Oliver, from South Africa, *Epicauta gorhămi*, Mars, from Japan, &c.

*C. quadrimaculatus* (Mexican cantharides) have been imported into Hamburg; they resemble *M. cichorii* but have only two black bands on the wing-cases.

Admixture with (entire) exhausted beetles can be detected by the deficiency in cantharidin and in fat.

The drug is seldom adulterated, but accidental admixtures are occasionally met with.

[1] Maisch's figure is apparently erroneously named *M. cichorii*.

## COCHINEAL

### (Coccus)

**Source, &c.**—Commercial cochineal is the dried, fecundated, female insect, *Coccus cacti*, Linné (Phylum *Arthropoda*, Class *Insecta*, Order *Hemiptera*).

The cochineal insects are indigenous to Central America and Mexico, where they live upon the fleshy branches of various species of *Nopalea* (*Cactaceæ*). The production of aniline dyes has, however, largely diminished the trade in cochineal, and at the present time the insects bred in the Canary Islands form the bulk of the commercial drug.

The insects are of a bluish red colour and very minute, measuring about a millimetre in length, and the male alone is provided with wings. After fecundation the female insects rapidly increase in size and develop abundance of a red colouring matter. They are then brushed off the plants and killed either by the fumes of burning sulphur or charcoal, or by hot water or stove heat. They are then dried in the sun (during which process they shrivel to about one-third of their size) and sifted to remove foreign fragments, &c. The number of harvests varies with the climate; in the Canary Islands there are generally two.

**Description.**—The dried cochineal insects of commerce are about 5 mm. long, oval in outline, flattish or slightly concave on one side and arched on the other. They are transversely wrinkled and a purplish black or greyish white colour, according to the variety (see below). They scarcely show any resemblance to insects, but when macerated in water they swell considerably, and then the three pairs of legs can be discerned. They are brittle, and easily reduced to a dark red or puce-coloured powder.

**Constituents.**—Cochineal contains up to 10 per cent. of a red colouring matter, carminic acid, which is obtainable in small, red prismatic crystals; it is soluble in water, alcohol, and in alkaline solutions. The drug also contains fat (about 10 per cent. and wax (about 2 per cent.), together with albuminoids, inorganic matter, &c.

The exact methods by which commercial carmine is produced are trade secrets; the preparation contains about 15 per cent. of water, 50 per cent. of carminic acid, 7 per cent. of ash, and about 20 per cent. of nitrogenous substances. It appears to be produced by precipitating infusions of cochineal by alum, in the presence of lime salts and either albumen or gelatin.

**Varieties.**—Two chief varieties of cochineal are recognised, viz. black grain and silver grain; the former are of a uniform purplish black colour, the latter greyish white. The greyish white colour of the silver grain is usually attributed to the waxy matter that covers the insects and has not been melted by heat; it appears, however, to be largely due to inorganic matter.

**Adulterants.**—Cochineal is frequently adulterated by the addition of inorganic matter such as talc, barium carbonate and sulphate (for silver grain cochineal), or manganese dioxide, lead sulphide, magnetic iron sand, &c. (for black grain cochineal). These can readily be detected by means of the ash which should not exceed 6 per cent., genuine cochineal of good quality yielding generally about 2·5 per cent. Silver grain cochineal, heavily 'dressed,' may yield up to 50 per cent. of ash.

## ANIMAL GLANDS

## DUCTLESS GLANDS

The ductless glands include a group of organs of very varied functions. They have no special ducts to convey their secretions either into the digestive tract or elsewhere. They do, however, form substances which pass directly into the blood stream or into the lymph channels. These substances belong to the class of bodies known as hormones, which are also formed by organs such as the pancreas which discharge their secretions externally. The presence of these hormones in the body is in many cases essential to health and even to life, and the activity of the ductless glands is correlated with and regulates the functions of distant organs, the only link being the blood by which the hormone is carried from its place of origin to its place of action, each hormone exercising a specific function in exciting the activity of a particular organ or tissue. Those which have been studied are of comparatively simple nature, dialysable, readily soluble in water and not destroyed by boiling.

The chief ductless organs employed in modern medicine are the thyroid and parathyroid glands, the suprarenal glands, the pituitary body, the spleen and the thymus.

## SUPRARENAL GLAND

The suprarenal gland is a small, ductless, glandular organ situated above the kidney. It consists chiefly of an outer yellowish portion or cortex and an inner darker portion or medulla. This inner portion, and extracts made from it, possess remarkable hæmostatic properties which are due to a crystalline body, adrenalin or adrenin, which is present in very small proportion. This substance has been prepared synthetically in the form of a pale buff-coloured crystalline powder.

Adrenalin is the most powerful hæmostatic known. By subcutaneous injection the blood-vessels are locally constricted so as to render minor surgical operations almost bloodless. It is similarly used to arrest postpartum and other hæmorrhages.

Adrenalin is lævo-dihydroxyphenyl-ethanol-methylamine (lævo-methylamino-ethanol-catechol), $C_6H_3(OH)_2CHOHCH_2NHCH_3$, very slightly soluble in water, almost insoluble in alcohol, ether and chloroform ; a dilute acid solution gives with a trace of ferric chloride an emerald green colour which is changed to purple or carmine by the cautious addition of a dilute solution of sodium hydroxide.

## THYROID GLAND

### (Glandulæ Thyroidei)

**Source, &c.**—The official thyroid gland is obtained from the sheep, *Ovis aries*, Linné (N.O. *Ungulata*).

**Description.**—The thyroid gland is a highly vascular organ consisting of two almond-shaped lobes generally united towards their lower ends by a narrow bridge of tissue called the isthmus and forming a rounded projection on the trachea. It is distinguished from the surrounding tissue by its dark red colour and firm, coarsely granulated texture. Embedded in its substance are multitudes of closed vesicles containing a yellow glairy fluid termed colloid. This secretion contains the active constituent of the gland which probably passes into the system through the lymphatics. Each lobe of the fresh sheep's thyroid weighs about 2 grammes.

The lobes are usually removed separately, freed from surrounding fat and connective tissue, sliced or minced, and rapidly dried in a current of warm air at 30° to 40°. When sufficiently dry they are reduced to a coarse powder and the remaining fat removed by petroleum spirit. They are finally completely dried in a desiccator. Glands that are abnormal in size, or that exhibit cysts containing fatty matter when cut, should be rejected. The dry thyroid thus produced is a dull brown hygroscopic powder with a slight odour free from putresence.

**Constituents.**—The active constituent appears to be a proteid body, thyreoglobulin, which can be dissolved out by a weak solution of sodium chloride and precipitated by the addition of ammonium sulphate. This substance contains uniformly 1·6 per cent. of iodine but no phosphorus. Boiled with dilute sulphuric acid it yields iodothyrin (thyroiodin) which contains from 9 to 14 per cent. of iodine, the constituent containing the iodine being apparently resistant to heat. Iodothyrin has also been supposed to be the active constituent. Fresh thyroids yield about 30 per cent. of dry, containing an average of 0·34 per cent. of iodine.

**Uses.**—Thyroid gland is used in myxœdœma, goitre, obesity, and other cases in which a deficient production of the secretion is indicated.

**Note.**—*Parathyroid glands* are small, four in number and lie close to or embedded in the thyroid. Their removal induces tetany, and certain forms of tetany are benefited by the use of preparations of the parathyroids.

The *pituitary body*, a gland situated at the base of the brain, forms one or more secretions of great physiological importance. It consists of a posterior, an intermediate and an anterior portion. Preparations of the posterior and intermediate portions, of the anterior portion and of the whole gland are employed medicinally. Extracts of the posterior and intermediate portions injected hypodermically raise the blood pressure and promote contraction of the uterus; the anterior lobe, dried and powdered, influences growth and development. From neither portion has an active constituent been isolated.

The *spleen* is a solid organ enclosed in a capsule. Its functions are not fully known, but it appears to be concerned with the protection of the body against disease by removing micro-organisms from the blood.

The *thymus*, or throat-bread, is composed of lobules united by connective tissues. Its secretion has a marked influence on sexual development.

## GLAND WITH DUCT

### PANCREAS

(Pancreas)

**Source, &c.**—The pancreas used in medicine is obtained from the pig, *Sus scrofa*, Linné (N.O. *Ungulata*).

**Description.**—The pancreas is a gland, weighing (in man) from 60 to 100 grammes, situated near the stomach in a depression formed by the duodenum to which it is closely attached, and into which it discharges its secretion, the pancreatic juice, by means of a principal branching excretory duct called the pancreatic duct. Immediately after the introduction of food into the stomach, the pancreas begins to produce its secretion, which in the case of man averages about 150 grammes a day. The epithelial cells of the small intestine produce a substance termed prosecretin which, when acted upon by dilute hydrochloric acid, yields a hormone, secretin; this passes into the blood and causes the stimulation of the pancreas. The pancreatic juice contains xanthine, guanine, leucine, sodium carbonate and other substances, together with the zymogens of four enzymes, but not the enzymes themselves, these being liberated from the zymogens by means of the enzyme, enterokinase, which is present in the duodenum.

The pancreatic digestion of food is effected by the following three enzymes which act best in neutral or slightly alkaline solution:

(i) Trypsin, which digests proteids.

(ii) Amylase (amylopsin), which digests starch by converting it into maltose.

(iii) Lipase (steapsin), which digests fats by converting them into glycerin, and fat-acids, which unite with the alkalies to form soaps.

Commercial pancreatin is a mixture of these enzymes with other substances.

## ANIMAL SECRETIONS, &c.

### OX GALL

### (Ox Bile, Fel Bovinum)

**Source, &c.**—Ox gall is the liquid contained in the gall-bladder of the ox, *Bos taurus*, (Phylum *Chordata*, Class *Mammalia*, Order *Ungulata*).

The gall-bladder is a pyriform bag attached to the under surface of the liver. It receives the secretion from the liver and discharges it into the duodenum; this action is continuous, but is increased by the arrival of food in the duodenum. In man the quantity secreted varies from 500 to 1000 c.c. daily. Bile appears to exercise a favourable influence upon pancreatic digestion by increasing the rate of action of the pancreatic enzymes.

**Description.**—Fresh ox bile is a brownish yellow or brownish green, rather viscous liquid, with an unpleasant odour and disagreeable bitter taste. It is neutral or faintly alkaline in reaction, has a specific gravity of about 1·022, and is characterised by the following reaction: To 10 c.c. of a 5 per cent. aqueous solution of ox bile add a drop of solution of sucrose (1 in 4) and then gradually sulphuric acid until the precipitate at first formed is redissolved; the mixture will acquire a carmine colour changing to violet. This reaction is due to the action of furfuraldehyde (from the sulphuric acid and sucrose) upon cholalic acid (produced by the sulphuric acid acting upon taurocholic and glycocholic acids).

**Constituents.** Ox gall contains the sodium salts of taurocholic and glycocholic acids together with colouring substances (bilirubin, biliverdin), mucin, lecithin, cholesterol, fat, soaps, &c., in aqueous solution. For medicinal use the mucin is usually removed by precipitation with alcohol and the filtrate evaporated to a thick extract.

**Uses.**—Ox gall is given in cases of deficiency of bile, but it is of doubtful medicinal value.

## PEPSINUM
### (Pepsin)

**Source, &c.**—Commercial pepsin is a mixture of the enzyme pepsin with other substances, obtained from the mucous membrane lining the stomach of the pig, sheep, or calf.

The surface of this mucous membrane exhibits multitudes of minute pits, each of which is about 0·2 mm. in diameter, and is the common orifice of two or three minute, elongated, tubular ducts. The pit together with the ducts debouching into it is termed a gastric gland. The duct is lined with two kinds of cells, viz. central cells and parietal cells, the former being the more numerous. The central cells secrete substances termed zymogens which are the precursors of enzymes but are not the enzymes themselves, the zymogens comprising pepsinogens and rennin zymogen. The pepsinogen is converted into pepsin and the rennin zymogen into rennin. The parietal cells secrete hydrochloric acid. These substances, probably with others, in aqueous solution constitute the gastric juice which is discharged from the glands into the stomach when meat is introduced into the mouth and masticated. Their action on the food is to convert insoluble proteids into soluble peptones and thus permit of digestion. Peptones differ from (soluble) proteids in not being coagulated by heat or precipitated by nitric acid, &c.

Commercial pepsin is obtained by stripping the mucous membrane from the stomach, mincing it and digesting it with water acidified with hydrochloric acid by which insoluble proteids are dissolved. From the filtered liquid it is precipitated by saturation with sodium chloride or ammonium sulphate, the peptones being left in solution. The pepsin is collected, dissolved in water and freed from sodium chloride, &c., by dialysis ; it may then be precipitated by alcohol, or the aqueous solution may be evaporated to dryness in a vacuum and powdered.

**Description.** — Pepsin occurs in commerce as a pale yellowish powder, or in translucent scales or grains with a faint odour free from putrescence and a slightly saline bitterish taste. It is soluble in water, especially on the addition of a little hydrochloric acid. It contains the enzyme, pepsin, but does not consist of it. An acidified aqueous solution converts insoluble proteids into soluble parapeptone (or acid albumose), propeptone, and finally into peptone. Its action is inhibited by sodium chloride and by alcohol, and is completely destroyed at a temperature of 70°.

**Uses.**—It is employed in dyspepsia caused by deficient gastric secretion.

## GELATIN

### (Gelatinum)

**Source, &c.**—Gelatin is an albuminoid substance obtained by boiling skins, cartilages, &c., with water, straining, skimming, and otherwise purifying the solution, evaporating and finally drying by exposure to the air. During the boiling the collagen contained in the cartilages, &c., is converted into gelatin, which then passes into solution.

**Description.**—Gelatin occurs in thin sheets, or in shreds or powder, which may be nearly colourless or pale yellow and almost free from odour and taste. In cold water it swells and when heated dissolves; it is soluble also in acetic acid and glycerin, but not in alcohol, ether, &c. A two per cent. hot aqueous solution should gelatinise on cooling, but this property is destroyed by the prolonged action of heat. Boiling with diluted hydrochloric acid converts it into the hydrochloride of glutin-peptone. An aqueous solution is precipitated by solution of tannic acid but not by dilute solutions of alum, lead acetate, or ferric chloride. It is rendered insoluble when the aqueous solution is mixed with potassium bichromate and exposed to light; formic aldehyde produces a similar change.

**Constituents.**—Gelatin is composed chiefly of the nitrogenous substance glutin. It should not yield more than two per cent. of ash.

**Uses.**—Apart from its various technical uses, gelatin has been employed as a nutrient and as a styptic, but its value for these purposes has been over-rated. It is also used as a demulcent, and combined with glycerin as a basis for suppositories.

*Isinglass* (Ichthyocolla) is the dried prepared swimming bladder of the sturgeon, *Acipenser Huso*, Linné (N.O. *Sturiones*), and other species of *Acipenser* living in the Black Sea and Caspian Sea and the rivers which flow into them. The bladders are cut open, washed, soaked in water, spread out on a board and deprived of the outer, silvery membrane. Dried in sheets they form leaf isinglass, or several folded together, book isinglass, or rolled and folded, staple isinglass. It is prepared for use by cutting it into thin shreds.

Isinglass is whitish or pale yellow, semi-transparent, tasteless, but with a more or less perceptible odour. In cold water it softens and swells; with boiling water it forms a solution which (1 in 50), gelatinises on cooling. It consists chiefly of collagen, about 80 per cent., together with water, about 15 to 20 per cent.

## MUSK

### (Moschus)

**Source, &c.**—Musk is the dried secretion from the preputial follicles of the musk deer, *Moschus moschiferus*, Linné (Phylum *Chordata*, Class *Mammalia*, Order *Ungulata*).

The musk deer is a small, graceful animal about the size of the roebuck, and inhabits a large area in Central Asia, extending from the Caspian Sea to the eastern boundaries of the Chinese Empire. The male animal, which alone produces the musk, bears on its belly, a short distance behind the navel and just in front of the preputial orifice, a small sac produced by an infolding of the skin. This sac is the musk sac or musk pod, and it contains a treacly or soft, unctuous, brownish substance, musk, which is remarkable for its

FIG. 247.—Musk caddy.

intense, penetrating, and persistent odour. The musk is formed in alveoli produced on the inner lining of the sac and is discharged into the cavity. The outer surface of the pod is covered with hairs and is provided with a small canal, serving for the discharge of the secretion, debouching close to the preputial orifice. Very young animals do not secrete musk and old animals but little.

The animals are snared or shot, and the musk pods cut out trimmed and dried ; they are then wrapped singly in paper and packed in a small rectangular box covered with silk. This box is known as a 'caddy' and contains a 'catty' (21⅔ ounces) of musk pods (about 22). Of late musk pods have been packed in flat tins each containing two catties.

Most of the musk of European commerce is obtained from Tibet or from the Chinese province of Szechuen (Tonquin musk); it is conveyed down the Yangtse-Kiang river to Shanghai, whence it is exported. Smaller quantities are obtained from the southern Chinese province of Yunan (Yunan musk), and some finds its way *via* Nepaul or Assam to Calcutta (Nepaul musk, Assam musk).

The musk pods are examined in China and classified into three qualities, or 'piles' as they are termed. In London they are again examined, probed with a knife, &c., and again classified into piles, pile 1 consisting of genuine pods, whilst those of pile 3 are obviously sophisticated.

**Description.**—The best variety of the drug is that known as Tonquin. This is imported in pods packed in 'caddies' or latterly in larger tins. The pods are nearly circular or sometimes distinctly oval in outline and lenticular in shape; they resemble small dark flattened cakes about 5 to 7 cm. in diameter and 2 to 3 cm. thick. On one of the flattened surfaces (the lower surface of the pod as attached to the animal) is a circular or oval piece of brown skin about 4 cm. in diameter; this is part of the hide of the animal, and exhibits when closely examined a central or nearly central small orifice (the orifice of the pod), around which are arranged tangentially directed hairs. The latter vary from whitish to brown in colour; those near the orifice are naturally short, but those a little distance removed have been clipped and are stiff and bristly.

The remainder of the pod is covered with a very thin, soft, supple membrane, and appears dark brown in colour when the pod is filled with musk. It often exhibits a fine steel-blue iridescence, whence the term 'blue skin,' by which this variety of Tonquin pods is known. This thin blue skin is the inner skin of the pod, the outer skin, which is tough and fibrous, having been carefully stripped off with the exception of the circular piece surrounding the orifice. By this means the appearance of the pod is improved and its value enhanced. Good pods weigh about 30 to 40 grammes, and contain about half that weight of granular musk which fills them loosely.

The pods from which the outer skin has not been removed are also imported; these are known as 'natural skin' 'thick skin,' or 'old style' pods. They are usually convex on the lower surface, which is covered with the hide of the animal, but nearly flat on the upper surface, and there protected by a tough, fibrous but hairless skin.

The musk contained in the pods is always moist, and often has a strongly ammoniacal odour. It can be freed from moisture and ammonia by exposing it to the air, and then forms dark reddish brown unctuous grains ('grain musk') with which occasional short hairs are mixed; it possesses a strong characteristic odour and bitter taste.

**Constituents.**—Musk yields by distillation with steam and subsequent purification a small percentage of a viscid colourless oil with a very powerful and agreeable odour of musk ; this oil appears to be a ketone and has been termed muskone. The drug contains moisture, fatty matter, resin, proteids, and inorganic substances. Water dissolves from 50 to 75 per cent. of it, but alcohol only 10 to 12 per cent. It should not contain more than 15 per cent. of moisture, or yield on incineration more than 8 per cent. of ash.

**Adulterants.**—Musk is liable to gross adulterations, which is in some cases easy to detect, but in others exceedingly difficult. The

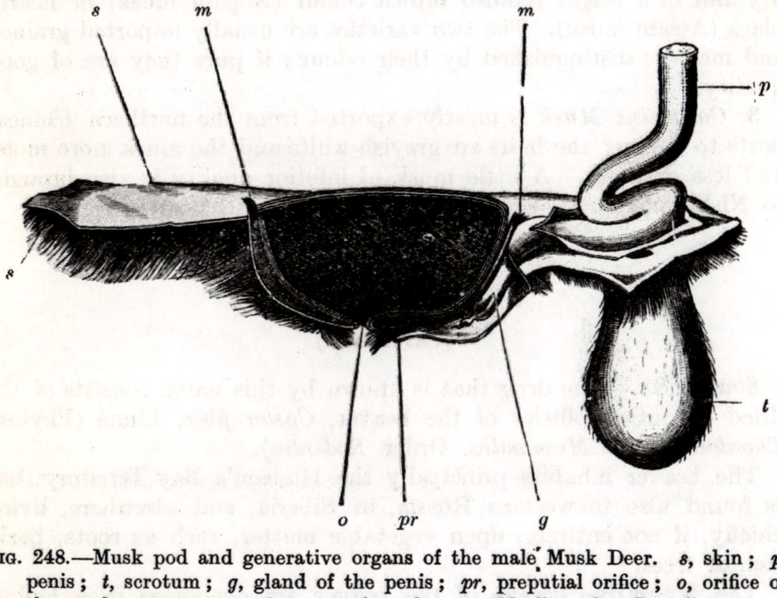


FIG. 248.—Musk pod and generative organs of the male Musk Deer. *s*, skin ; *p*, penis ; *t*, scrotum ; *g*, gland of the penis ; *pr*, preputial orifice ; *o*, orifice of the musk sac ; *m*, muscular coat of musk sac. (Moeller, after Brandt.)

sophistication of the drug is effected by the Chinese, and is practised to such an extent as to lead to the assertion that pure musk is scarcely procurable. The pods are skilfully opened, part of the musk is removed and replaced by some worthless substitute. Inorganic substances, such as small stones, leaden shot, &c., are comparatively easily detected, and so are such adulterations as scraps of leather or horn. Dried blood yields a red ash, whereas the ash of genuine musk is whitish. Resin and other substances soluble in spirit increase the alcoholic extract. Earthy matter raises the ash.

**Uses.**—Musk is occasionally used in medicine as a diffusible stimulant, but the bulk of the drug is employed in the manufacture of perfumes.

**Varieties.**—1. *Yunan Musk.*—This variety is imported in the pods, which are easily distinguished from Tonquin pods by their more nearly spherical or even pyriform shape. The skin rises towards the orifice, which is situated in a little depression ; near the margin of the pod are two small nipples which, with the depressed orifice, bear a fancied resemblance to a pig's eyes and snout, hence the common term ' pig-faced pods,' by which these are known. They are imported both in the natural pod and blue skin. The musk is nearly equal in value to Tonquin.

2. *Assam* and *Nepaul Musk.*—The pods are small, nearly spherical and about one-third of the weight of the Tonquin pods. The musk is dry and of a bright reddish brown colour (Nepaul musk) or nearly black (Assam musk). The two varieties are usually imported grained and may be distinguished by their odour ; if pure they are of good quality.

3. *Cabardine Musk* is mostly exported from the northern Chinese ports to Japan ; the hairs are greyish-white and the musk more moist and less granular. A little musk of inferior quality is also brought to Nishni-novgorod, whence it finds its way to London.

## CASTOR

### (Castoreum)

**Source, &c.**—The drug that is known by this name consists of the dried preputial follicles of the beaver, *Castor fiber*, Linné (Phylum *Chordata*, Class *Mammalia*, Order *Rodentia*).

The beaver inhabits principally the Hudson's Bay Territory, but is found also in western Russia, in Siberia, and elsewhere, living chiefly, if not entirely, upon vegetable matter, such as roots, bark, &c., of trees.

The generative organs of the beaver are concealed in a hollow or cloaca. Into the preputial (or vaginal) canal there open two large glands, one on each side. These glands, cut from both male and female animal and dried, form the commercial drug. In the fresh state they contain a whitish or yellowish creamy substance, but as they dry this becomes dark in colour. Our supplies are derived almost entirely from the Hudson's Bay Territory.

**Description.**—The dried glands are of a dark-brownish or greyish colour, pear-shaped, and about 8 to 10 cm. long, They are frequently connected in pairs by a portion of the preputial or vaginal canal which has been cut away with them. They are firm, heavy, and solid, and possess a characteristic, empyreumatic, and not altogether agreeable odour. They contain a brown or reddish brown resinous secretion, in which, under the microscope, spherical grains of crystalline calcium carbonate can be detected, often in considerable quantity. The secretion varies much both in quantity and

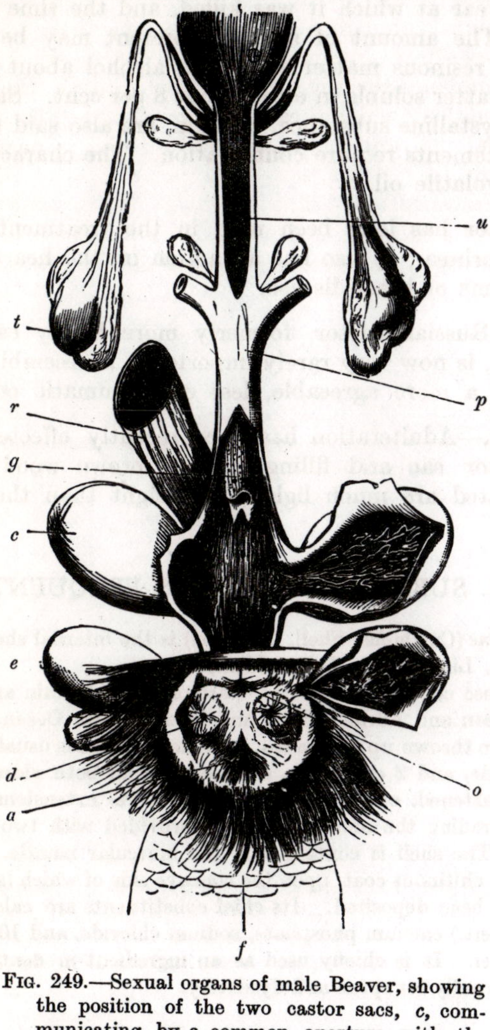


FIG. 249.—Sexual organs of male Beaver, showing the position of the two castor sacs, *c*, communicating by a common aperture with the preputial canal. The latter and one of the castor sacs have been opened. *t*, testicle; *u*, urethra; *c*, castor sacs; *g*, gland of penis; *p*, penis; *r*, rectum; *a*, anus; *e*, anal gland; *d*, aperture of anal gland; *o*, preputial orifice; *f*, undersurface of tail. (Moeller, after Brandt.)

appearance, being sometimes pale in colour and soft, sometimes hard and dark.

**Constituents.**—The composition of the resinous secretion appears to be subject to great variation, due probably to the age of the animal, the time of year at which it was killed, and the time the drug has been kept. The amount of moisture present may be as much as 40 per cent., resinous matter soluble in alcohol about 40 to 70 per cent., fatty matter soluble in ether about 8 per cent. Salicin, benzoic acid and a crystalline substance, castorin, are also said to be present, but these statements require confirmation. The characteristic odour is due to a volatile oil.

**Uses.**—Castor has long been used in the treatment of hysteria and dysmennorhœa ; it also has an action on the heart, and is used in certain forms of heart-disease.

**Varieties.**—Russian castor, formerly more highly esteemed than the American, is now very rarely imported. It resembles the American, but has a more agreeable, less empyreumatic odour.

**Adulterants.**—Adulteration has been recently effected by emptying the castor sac and filling it with brown wood fibre. Pods thus adulterated are much lighter in weight than the genuine.

## ANIMAL SUBSTANCES IN LESS FREQUENT USE

*Cuttle-fish Bone* (Cuttle-fish Shell, *Os Sepiæ*) is the internal shell of the cuttle, *Sepia officinalis*, Linné (N.O. *Dibranchiata*).

A large mollusc common round the coast of Great Britain and abundant in the Mediterranean and Adriatic Seas, and in the Indian Ocean. The shells of dead animals are thrown up by the sea. Cuttle-fish shell is usually about 20 cm. long, 7 cm. wide, and 2 cm. thick, oval, convex on both sides. The animal consists of a flattened, ovoid head with two lateral extensions, and ten long tentacles surrounding the mouth, which is provided with two large powerful horny beaks. The shell is contained in the muscular mantle. It consists of a hard, concave chitinous coat, upon the concave side of which layers of calcium carbonate have been deposited. Its chief constituents are calcium carbonate (80 to 85 per cent.) calcium phosphate, sodium chloride, and 10 to 15 per cent. of organic matter. It is chiefly used as an ingredient in dentifrices, but has been given internally for sprue and dysentery.

# INDEX

Abies balsamea, 489
„ canadensis, 489
„ Menziesii, 490
Abietic Acid, 456
Abrin, 191
Abrus precatorius, 191, 318
Absinthin, 222
Acacatechin, 445
Acacia arabica, 177, 249, 450, 471
„ Bark, 249
„ Catechu, 444
„ dealbata, 285, 450
„ decurrens, 249, 285
„ Gum, 448
„ gummifera, 450
„ horrida, 450
„ modesta, 450
„ pycnantha, 450
„ Senegal, 448
Acaciæ Gummi, 448
Accroides, Gum, 473
Achene, 84
Achillea millefolium, 69
Acid, Arabic, 449
„ Isogeddic, 449
„ Sinapine Sulphate, 149
„ Value, 457
Acipenser Huso, 538
Aconine, 297
Aconite Herb, 192
„ Leaves, 59
„ Root, 294
„ „ Indian, 298
„ „ Japanese, 298
Aconiti Herba, 192
„ Radix, 294
Aconitic Acid, 194
Aconitine, 297
Aconitum chasmanthum, 298
„ heterophyllum, 298
„ lycoctonum, 298
„ paniculatum, 298
„ Napellus, 192, 294
„ uncinatum, 298
Acori Calami Rhizoma, 392
Acorin, 394
Acorus Calamus, 392
Acrinyl Isothiocyanate, 149
Actææ Racemosæ Rhizoma, 298
Adenandra fragrans, 30
Adeps, 515
„ Lanæ, 519
Adhatoda, 60
Adrenalin, 533
Adrenin, 533
Æsculapius, 2
Æthusa Cynapium, 197
Agar-agar, 216
„ Ceylon, 217
„ Macassar, 217
Agathis australis, 473
Agathosma Cerefolium, 29
„ chortopila, 29
„ microphylla, 29
„ variabilis, 29
Agmatine, 221
Ailanthus glandulosa, 46
Ajacine, 145
Ajaconine, 145
Ajowan Fruits, 142
Alantolactone, 334
Alantolic Acid, 334
Albanans, 424
Albans, 424
Albaspidin, 397
Alder Buckthorn Bark, 34, 243
Aleurites laccifera, 471
Aleuritic Acid, 472
Alexander, 2
Algin, 216
Alginic Acid, 216
Alhagi Maurorum, 522
Alkanet Root, 347
Alkanna Root, 347
„ tinctoria, 347
Alkannic Acid, 348
Alkannin, 348
Allantoin, 349
Allspice, 103
„ Mexican, 104
Allyl Cyanide, 147
„ Isothiocyanate, 147
„ Thiocarbimide, 147
„ Thiocyanate, 147
Almond Oil, 509
Almonds, Bitter, 163
„ East Indian, 142, 162
„ Essential Oil of, 164
„ Sweet, 161
Alnus glutinosa, 246
Aloe, 431
„ chinensis, 435
„ ferox, 434
„ Perryi, 433
„ succotrina, 434
„ vera, 435
Aloe-emodin, 33, 365, 437
Aloes, 34, 431

Aloes, Barbados, 435
„ Cape, 434
„ Crown, 434
„ Curaçao, 435
„ East African, 433
„ glassy, 432
„ hepatic, 433
„ Jafferabad, 437
„ livery, 433
„ lucid, 432
„ Mocha, 437
„ Natal, 434
„ Reactions of, 436
„ Socotrine, 433
„ South African, 434
„ Uganda, 434
„ vitreous, 432
„ West Indian, 435
„ Zanzibar, 434
Alpinia Galanga, 381
„ officinarum, 39
Alpinin, 381
Alsophila lurida, 412
Alstonia, 268
„ constricta, 268
„ scholaris, 268
Alstonidine, 269
Alstonine, 269
Althæa officinalis, 314
Althææ Radix, 314
Amber, 459
Ambergris, 519
β-aminoethylglyoxaline, 21
Ammoniacum, 482
„ African, 484
Ammoresinotannol, 484
Amomum aromaticum, 141
„ Cardamomum, 141
„ Korarima, 141
„ Melegueta, 185
Amygdala dulcis, 161
Amygdalin, 163
Amylase, 536
Amylopsin, 536
Amylodextrin, 180
Amylum, 400
Amyris balsamifera, 231
„ Plumerii, 495
Amyrol, 232
Anabsinthin, 222
Anacardic Acid, 142
Anacardium occidentale, 142, 162
Anchusæ Radix, 347
Anchusic Acid, 348
Anacyclus officinalis, 336
„ Pyrethrum, 334
Andrographis paniculata, 205
Andira Araroba, 462
Anemone Pulsatilla, 222
Anemonin, 222
Anethi Fructus, 118
Anethol, 121
Angostura Bark, 238
„ Brazilian, 241
Angosturin, 240
Anhydrogitaligenin, 55
Anhydrogitalin, 55
Anhydrohapontigenin, 369
Animal Glands, 533
„ Secretions, 536
Animals, Classification, 526
Anise Fruits, 119
Aniseed, 119
Anisi Fructus, 119
„ Stellati Fructus, 85
Annatto, 141
Anogeissus latifolia, 450
Anthemene, 72
Anthemic Acid, 72
Anthemidis Flores, 70
Anthemis nobilis, 70
„ tinctoria, 75
Anthesterin, 72
Anthophylli, 68
Anthriscus sylvestris, 197
Aphis chinensis, 414
Apiol, 142
„ Yellow, 142
Apis mellifica, 516, 522
Apoatropine, 45, 359
Aporeine, 62
Apricot Kernels, 164
Arachis Oil, 509
Arachnoidiscus ornatus, 217
Araroba, 462
Ararobinol, 464
Arbutin, 42
Arctostaphylos Uva-ursi, 41
Areca Catechu, 190
„ Nuts, 190
Arecæ Semina, 190
Arecaidine, 191
Arecaine, 191
Arecoline, 191
Argel Leaves, 34
Aristolochia argentina, 372
„ reticulata, 370
„ Serpentaria, 370
Aristolochine, 372
Arnica Flowers, 73
„ montana, 336
„ Rhizome, 336
Arnicæ Flores, 73
„ Rhizoma, 336
Arnicin, 74, 338
Arnisterin, 74
Arrowroot, 402
„ Brazilian, 403
„ East Indian, 375, 403, 404
Artanthic Acid, 58
Artemisia Absinthium, 222
„ anthelmintica, 77
„ Cina, 75
„ gallica, 77
„ maritima, 75
Artemisin, 76
Asafetida, 487
Asaresinotannol, 488
Asclepias curassavica, 329
Aselline, 514

Asperula odorata, 157
Aspidin, 399
Aspidinol, 397
Aspidium Filix-fœmina, 398
Aspidosamine, 269
Aspidosperma, 269
„ quebracho-blanco, 269
„ quebracho-colorado, 285
Aspidospermatine, 269
Aspidospermine, 269
Astragalus cyllencus, 452
„ Glycyphyllos, 318
„ gummifer, 451
Atis Root, 298
Atisine, 298
Atropa Belladonna, 43, 356
„ lutescens, 360
Atropine, 45, 359
Atropurol, 243
Aurantiamaric Acid, 94
Aurantiamarin, 94
Aurantii Cortex Indicus, 92
Avicenna, 3
Avornic Acid, 246
Azadirach, Indian, 287
Azolitmin, 219

Babul Bark, 249
Bael Fruit, 91
Balata, 424
Ballota hirsuta, 207
„ nigra, 207
Balsam, Copaiba, 490
„ Gurjun, 493
„ Oregon, 490
„ Peru, 497
„ Tolu, 495
Balsamum Peruvianum, 497
„ Tolutanum, 495
Bambusa stricta, 522
Barbaloin, 436
β-barbaloin, 436
Barbaloresinotannol, 437
Barberry Bark, 235
Bark, Acacia, 249
„ Alder Buckthorn, 243
„ Alstonia, 268
„ Angostura, 238
„ Aspidosperma, 269
„ Australian Fever, 268
„ Barberry, 235
„ Bebeeru, 273
„ Bibiru, 273
„ Black Haw, 258
„ „ Willow, 287
„ Brazilian Angostura, 241
„ Canella, 236
„ Cascara Sagrada, 246
„ Cascarilla, 279
„ Cassia, 271
„ Cinchona, 259
„ Cinnamon, 269
„ Condurango, 288
„ Copalchi, 281
„ Coto, 275
Bark, Cotton Root, 238
„ Cuprea, 268
„ Cusparia, 238
„ Dita, 268
„ Elm, 288
„ Euonymus, 242
„ Greenheart, 273
„ Hemlock, 285
„ Larch, 288
„ Malambo, 237
„ Mangrove, 285
„ Margosa, 287
„ Mezereon, 277
„ Monesia, 318
„ Mudar, 288
„ Neem, 287
„ Oak, 282
„ Oliver, 273
„ Paracoto, 275
„ Piscidia, 288
„ Pomegranate, 256
„ Quebracho, 269, 285
„ Quillaja, 253
„ Sassy, 249
„ Simaruba, 241
„ Slippery Elm, 281
„ Wafer Ash, 243
„ Wattle, 285
„ Wild Cherry, 251
„ Wild Cinnamon, 236
„ Willow, 285
„ Winter's, 237
„ Witch-hazel, 255
Barks, 233
Barley, Pearl, 142
Barosma betulina, 27
„ crenulata, 29
„ Eckloniana, 29
„ ericifolia, 29
„ lanceolata, 29
„ pulchella, 29
„ serratifolia, 28
„ succulenta, 29
„ venusta, 29
Bassia latifolia, 508
„ longifolia, 508
„ Tallow, 508
Bassoric Acid, 452
Bassorin, 452
Bay Berries, 125
Bdellium, African, 480
„ Indian, 481
„ Opaque, 480
„ Perfumed, 480
„ Scented, 480
Bearberry Leaves, 41
Bebeeru Bark, 273
Beberine, 274
Beeswax, 516
Belæ Fructus, 91
Belladonna, Indian, 360
„ Japanese, 360
„ Leaves, 43
„ Root, 356
Belladonnæ Folia, 43

Belladonnæ Radix, 356
Belladonnine, 45, 359
Benne Oil, 511
Benzaldehyde-cyanhydrin, 164
Benzoin, 464
,, Palembang, 467
,, Penang, 466
,, Siam, 465
,, Sumatra, 464
Benzoinum, 464
Benzoresinol, 465
Benzoylaconine, 297, 298
Benzoyl-pseudotropine, 26
Benzyl Isothiocyanate, 148
Berbamine, 235
Berberidis Cortex, 235
Berberine, 235
Berberis, 224
,, aristata, 224
,, vulgaris, 235
Bergamot, 97
Berg and Schmidt, 4
Berries, Juniper, 135
Berry, 84
Betel, 58
,, Nuts, 190
Betel-phenol, 59
Betula alba, 505
Bhang, 208
Bibiru Bark, 273
Bilirubin, 536
Biliverdin, 536
Birch tar, 505
Bissabol, 480
Bissy Nuts, 151
Bitter Almonds, 163
,, Orange, 92
Bitter-sweet, 205
Bixa orellana, 141
Bixin, 141
Black-boy Gum, 473
Black Haw Bark, 258
,, Hellebore Rhizome, 292
,, Mustard Seeds, 145
,, Root, 399
,, Snakeroot, 298
Bladderwrack, 215
Bloodroot, 307
Blue Cohosh 399
Blumea balsamifera, 525
Boldine, 60
Boldo Leaves, 60
Boldoglucin, 60
Bornyl Isovalerianate, 331
,, Succino-abietate, 460
Boswellia Carterii, 481
,, Frereana, 495
Boswellic Acid, 482
Box Leaves, 43
Brachylana Hutchinsii, 232
Brasilein, 229
Brasilin, 229
Brassica alba, 148, 506
,, campestris, 507
,, juncea, 148
Brassica Napus, 507
,, nigra, 145
,, sinapioides, 145, 506
Brayera anthelmintica, 62
Brazil Wood, 229
Broom Tops, 194
Brucine, 171
Bryoidin, 494
Bryonia alba, 321
,, dioica, 320
Bryoniæ Radix, 320
Bryonol, 321
Bryony Root, 320
Buchu Folia, 27
Buchu Leaves, 27
,, ,, Long, 28
,, ,, Oval, 29
,, ,, Short, 27
Buckthorn Berries, 141
Bulb, Squill, 386
,, Urginea, 388
Bulbs, 289
Burgundy Pitch, 458
Bulnesia sp., 225
Butea frondosa, 159, 440, 471
,, Seeds, 159
Buteæ Gummi, 440
,, Semina, 159
Butyrospermum Parkii, 508
Butyrum Cacao, 508
Buxus, sempervirens, 43

CADE, HUILE DE, 504
,, Oil of, 504
Cæsalpinia brasiliensis, 229
,, corairia, 285
,, Echinata, 229
,, Sappan, 229
Caffea arabica, 22
Caffeine, 22
Cajanus indica, 471
Cake Saffron, 83
Calabar Beans, 157
Calabarine, 158
Calabarol, 158
Calamus Draco, 467
Calendula, 77
,, officinalis, 75, 77
Calendulæ Flores, 77
Calendulin, 77
Callitris quadrivalvis, 470
,, verrucosa, 471
Callitrolic Acid, 471
Calotropis, 288
,, procera, 288
Calumba, Root, 302
,, Ceylon, 304
,, Rhizome, 304
Calumbæ Radix, 302
Cambogia, 474
Camellia Thea, 21
Camphor, 524
,, Borneo, 525
,, Ngai, 525
,, Synthetic, 525

Canada Balsam, 489
,, Turpentine, 489
Canadine, 292
Canadinic Acid, 490
Canadinolic Acid, 490
Canadolic Acid, 490
Canado-resene, 490
Canaigre Root, 369
Canarium luzonicum, 494
Canella alba, 236
,, Bark, 236
Canellæ Cortex, 236
Canna edulis, 403
Cannabinol, 209
Cannabis Indica, 207
,, sativa, 207
Cannabinone, 209
Cantharides, 530
Cantharides, Chinese, 531
,, Mexican, 531
Cantharidin, 530
Cantharis, 530
,, quadrimaculatus, 531
,, vesicatoria, 530
Caoutchouc, 424
Caoutchouc-gutta, 426
Capaloin, 436
Capaloresinotannol, 437
Cape Saffron, 83
Capsacutin, 123
Capsaicin, 123
Capsici Fructus, 122
Capsicin, 123
Capsicum annuum, 122
,, frutescens, 122
,, grossum, 122
,, minimum, 122
,, tetragonum, 122
,, Fruits, 122
Capsule, 85
Capsules, Poppy 89
Caraway Fruits, 108
Cardamomi Fructus, 138
Cardamoms, 138
,, Bengal, 141
,, Ceylon, 139
,, Cluster, 141
,, Korarima, 141
,, Long Wild Natives, 140
,, Malabar, 140
,, Mangalore, 140
,, Mysore, 140
,, Round, 141
,, Wild, 139, 141
Cardol, 142, 162
Carminic Acid, 532
Carnaubic Acid, 520
Carotin, 82
Carpiline, 30
Carrageen, 213
Carrageenin, 214
Carthamus helenioides, 48
,, tinctorius, 82
Carui Fructus, 108
Carum Carvi, 108
Carum copticum, 142
,, Petroselinum, 142
Caryophylla, 66
Caroyphyllene, 67
Caryophyllin, 67
Caryopsis, 84
Cascara Sagrada, 34
Cascarilla, 279
,, Bark, 279
Cascarillæ Cortex, 279
Cascarillin, 281
Cascarilline, 281
Cascarin, 248
Cashew Nuts, 142
Cassia acutifolia, 32
,, angustifolia, 35
,, Bark, 271
,, Fistula, 34, 98
,, grandis, 99
,, holosericea, 35
,, lignea, 271
,, montana, 35
,, moschata, 99
,, obovata, 34
,, Pods, 98
,, Pulp, 34
,, vera, 272
,, Chinese, 271
,, Horse 99
Castilloa elastica, 427
Castor, 542
,, fiber, 542
,, Oil, 183, 512
,, Seeds, 182
Castoreum, 542
Catechin, 443
Catechu, 442, 444
,, Nigrum, 444
,, Black, 444
,, Pale, 442
,, red, 443
Catechutannic Acid, 443
Catha edulis, 23
Cathidine, 23
Cathine, 23
Cathinine, 23
Caulophylline, 399
Caulophyllosaponin, 399
Caulophyllum thalictroides, 399
Caulosaponin, 399
Cayenne Pepper, 122
Centaury, 222
Cephaëline, 325
Cephaëlis Ipecacuanha, 321
Cera Alba, 516
,, Flava, 516
Ceriops candolleana, 445
Cerolein, 517
Cerotic Acid, 517
Cetaceum, 518
Cetraria, 217
,, islandica, 217
Cetraric Acid, 218
Cetyl Palmitate, 519
Cevadilla Seeds, 189

Cevadillæ Semina, 189
Cevadilline, 190
Cevadine, 189
Cevine, 189
Chærophyllum temulum, 197
Chamomile Flowers, 70
,, German, 73
,, Roman, 71
Charas, 208
Chatinine, 331
Chaulmoogra Oil, 507
Chaulmoogric Acid, 507
Chavicin, 134
Chebulinic Acid, 103
Chelerythrine, 308
Chenopodium, 23, 142
,, ambrosioides, 142
,, hybridum, 48
Cherry-laurel Leaves, 36
Cherry Wood, 157
Chillies, 122
Chinese Vegetable Tallow, 508
Chirata, 203
Chiratin, 204
Chiratogenin, 204
Chiretta, 203
,, Japanese, 205
Chirettæ Herba, 203
Chittem Bark, 246
Chlora perfoliata, 346
Chlorogenic Acid, 172
Chlorogenine, 269
Chlorophora tinctoria, 229
Chondrodendron tomentosum, 300
Chondrus crispus, 213
Christembine, 122
Chrysanthemum cinerariæfolium, 78
,, coccineum, 79
,, Parthenium, 73
,, roseum, 79
Chrysarobin, 464
Chrysarone, 369
Chrysatropic Acid, 45, 343
Chrysophanic Acid, 34, 365, 464
Chrysophanol, 464
Chrysophanolanthranol, 464
Chrysopontin, 369
Chrysorhapontin, 369
Chrysotoxin, 221
Churrus, 208
Cibotium, 412
,, Barometz, 412
,, glaucum, 412
Cicuta virosa, 197
Cimicifuga, 298
,, racemosa, 298
,, Rhizoma, 298
Cinchona Bark, 259
,, ,, Cartagena, 265
,, ,, Columbian, 265
,, ,, Crown, 263
,, Calisaya, 261
,, lancifolia, 265
,, Ledgeriana, 263
,, officinalis, 263
Cinchona succirubra, 264
,, Red, 267
Cinchonæ Cortex, 259
Cinchonidine, 266
Cinchonine, 266
Cinchotannic Acid, 267
Cinnamon Bark, Java, 271
,, ,, Jungle, 271
,, ,, Saigon, 271
Cinnamomum Burmanni, 271
,, Cassia, 271
,, Oliveri, 273
,, zeylanicum, 269
Cinnamein, 498
Cinnamodendron corticosum, 237
Cinnamon Bark, 269
,, ,, Wild, 236
Cinnamyl-cocaine, 26
Cissampelos Pareiræ, 300
Citrullol, 108, 243
Citrullus Colocynthis, 106
Citrus Aurantium, 92
,, Bergamia, 97
,, Medica, 94, 97
Claviceps purpurea, 219
Clavicepsin, 221
Clitorea terneata, 177
Clove Stalks, 68
Cloves, 66
,, Blown, 68
,, Exhausted, 68
,, Mother, 68
Clusius, 4
Coal Tar, 503
Coca Leaves, Bolivian, 24
,, Huanuco, 24
,, Java, 26
,, Peruvian, 25
,, Truxillo, 25
Cocæ Folia, 23
Cocaine, 26
Cocamine, 26
Coccoloba uvifera, 439
Cocculi Fructus, 87
Cocculin, 88
Cocculus Indicus, 87
Coccus, 532
,, cacti, 532
,, cerifera, 518
,, Lacca, 471
Cochineal, 532
Cochlearia Armoracia, 309
,, officinalis, 148
Coco Nut Oil, 513
,, ,, Stearin, 513
Cocoa Butter, 508
,, Seeds, 152
Cocos butyracea, 513
,, nucifera, 513
Cod-liver Oil, 513
Codeine, 419
Coffee, 22
Cola acuminata, 153
,, alba, 153
,, astrophora, 153

Cola Ballayi, 152
„ Seeds, 22, 151
„ vera, 22, 151
Colchici Semina, 186
Colchiceine, 188
Colchicine, 188, 389
Colchici Radix, 389
Colchicoresin, 188
Colchicum autumnale, 186, 389
„ Corm, 389
„ Root, 389
„ Seeds, 186
Colocynth, 106
Colocynthidis Fructus, 106
Colocynthin, 108
Colophony, 454
Coltsfoot, 197
Colombo Root, 302
Columbamine, 303
Columbin, 304
Colutea arborescens, 35
Colza Oil, 507
Commiphora erythræum, 480
„ Myrrha, 478
Commiphoric Acid, 480
Commiphorinic Acid, 480
Comfrey Leaves, 56
„ Root, 349
Condurangin, 288
Condurango, 288
Cone, 85
Conhydrine, 113
Conii Folia, 195
„ Fructus, 111
„ Herba, 195
Coniine, 112
Conium maculatum, 111, 195
Consolidæ Radix, 349
Convallaria majalis, 83
Convallarin, 83
Convallamaretin, 83
Convallamarin, 83
Convallariæ Flores 83
Convolvulinolic Acid, 354
Convolvulus Scammonia, 349, 476
Copaiba, 490
„ African, 492
„ Maracaibo, 491
„ Pará, 492
„ resene, 491
Copaifera Lansdorfii, 490
Copaiva, 490
Copaivic Acid, 491
Copal, American, 473
„ Australian, 473
„ Manila, 473
„ Zanzibar, 473
Copalchi Bark, 281
Copernicia cerifera, 518
Coprah Oil, 513
Coriaria myrtifolia, 35
Coriander Fruits, 113
Coriandri Fructus, 113
Coriandrum sativum, 113
Corm, Colchicum, 389
Corms, 289
Cornutine, 221
Corrigiola littoralis, 336
Cortex Thymiamatis, 499
„ Winteranus, 237
Corydaline, 304
Corydalis tuberosa, 304
Corynocarpin, 38
Coscinium fenestratum, 304
Coto Bark, 275
Cotoin, 276
Cotton, 410
„ Root Bark, 238
„ Wool, 410
„ seed Oil, 507
Couch Grass Rhizome, 395
Coumaric Anhydride, 156
Coumarin, 156
Cousso, 62
Cowberry Leaves, 43
Cowhage, 414
Cremocarp, 109
Cress, 148
Crocetin, 82
Croci, Placentæ, 83
Crocin, 81
Crocose, 82
Crocus, 80
„ sativus, 80
„ in Placenta, 83
Crotin, 185
Crotolaria juncea, 177
Croton albumin, 185
„ Eleuteria, 279
„ Globulin, 185
„ lucidus, 281
„ Malambo, 237
„ niveus, 281
„ Oil, 512
„ Seeds, 184
„ Tiglium, 184, 512
Crotonyl Isothiocyanate, 148
Cryptocarya sp., 275
Cryptocoryne spiralis, 327
Cubeb Camphor, 131
Cubebæ Fructus, 128
Cubebic Acid, 131
Cubebin, 131
Cubebs, 128
„ African 132
„ Congo 132
Cucurbita maxima, 164
Cucurbitæ Semina Præparata, 164
Cudbear, 219
Culver's Root, 399
Cuminum Cyminum, 115
Cummin Fruit, 115
„ Persian, 116
Cuprea Bark, 268
Curare, 445
Curarine, 446
Curine, 446
Curcuma angustifolia, 404, 375
„ leucorhiza, 375, 404
„ Zedoaria, 375

Curcumæ Rhizoma, 373
Curcumin, 375
Cusparia Bark, 238
,, febrifuga, 238
Cuspariæ Cortex, 238
Cusparine, 240
Cusso, 62
Cutch, 444
,, Mangrove 445
Cuttle-Fish Bone, 544
Cyanin, 65
Cyanogenetic Drugs, 38
Cyclogallipharic Acid, 413
Cyclopia sp., 23
Cydonia Semina, 165
Cynips gallæ tinctoriæ, 412
Cynodon Dactylon, 396
Cynosbati Fructus, 102
Cypripedium parviflorum, 292
Cytisus Scoparius, 194

DÆMONOROPS PROPINQUUS, 467
,, ruber, 467
Damiana, 59
Dammar, 473
,, East Indian, 473
Dammara orientalis, 473
Dandelion Root, 338
Daphne Gnidium, 277
,, Laureola, 277
,, Mezereum, 277
Daphnin, 279
Datura fastuosa, 49
,, Leaves, 49
,, Metel, 49
,, Seeds, 176
,, Stramonium, 174
,, Tatula, 49
Daturæ Folia,
,, Semina, 176
Daturine, 48
Delphinine, 144
Delphinium Ajacis, 145
,, Consolida, 145
,, Staphisagria, 143
Delphinoidine, 144
Delphisine, 144
Dendrographa leucophæa, 218
Dextrin, 404
Dextrolichenin, 218
Dhurrin, 38
Dicotoin, 276
Digitalein, 55
Digitalic Acid, 55
Digitalin, 55
Digitalinum verum, 56
Digitalis Folia, 53
,, purpurea, 53
,, Thapsi, 56
Digitoflavone, 55
Digitonin, 55
Digitophyllin, 55
Digitoxigenin, 55
Digitoxin, 55
Digitoxose, 55
Digitsaponin, 55
Dill Apiol, 119
,, Fruits, 118
,, Indian, 119
Dimethylxanthine, 22
Dioscorides, 2
Diosmin, 28
Diosphenol, 28
Dioxymethylenephenylcumalin, 276
Dioxymethylenephloroglucinol, 276
Dita Bark, 268
Ditaine, 268
Ditamine, 268
Dipteryx odorata, 156
,, oppositifolia, 156
Divi-divi, 285
Dogwood, Jamaica, 288
Dolphin Oil, 514
Dorema Ammoniacum, 482
Doronicum pardalianches, 75
Dracæna Cinnabari, 468
Dracoalban, 468
Dracoresene, 468
Dracoresinotannol, 468
Dragon's Blood, 467
,, ,, Socotrine, 468
,, ,, Zanzibar Drop, 468
Drimys Winteri, 237
Drug Packages, 9
,, Routes, 8
Drugs, Drying and Preservation of, 17
,, Sale of, 12
Drupe, 84
Duboisia Leaves, 60
,, myoporoides, 46, 60
Duboisine, 46
Dulcamaræ Stipites, 205
Dulcamaretic Acid, 206
Dulcamaric Acid, 206
Dulcamarin, 206
Dulcitol, 243
Dryobalanops aromatica, 525
Dryopteris Filix-mas, 396
,, spinulosa, 398

EAST INDIAN ROOT, 327
Ebers, 2
Ecgonine, 26
Ecballium Elaterium, 441
Echicaoutchin, 268
Echicerin, 268
Echiretin, 269
Echitamine, 268
Echitein, 269
Echitenine, 268
Echitin, 269
Ecuelle à Piquer, 96
Elaterin, 441
α-Elaterin, 108
β-Elaterin, 108
Elaterium, 441
Elder Flowers, 69
Elecampane Leaves, 56
,, Root, 332
Elemi, 494

Elemi, Brazilian, 495
„ East African, 495
„ Manila, 494
„ Yucatan, 495
Elettaria Cardamomum, 138
Ellagic Acid, 284, 413
Elm Bark, 288
Embelia, 121
„ Ribes, 121
„ robusta, 121
Embelic Acid, 122
Emetine, 325
Emodin, 34, 365
„ -anthranol, 141
„ monomethyl ether, 365, 464
Emulsin, 162
Entada scandens, 159
Enterokinase, 535
Epicauta gorhami, 531
Epilobium angustifolium, 23
Ergochrysin, 221
Ergot, 219
Ergotamine, 221
Ergothioneine, 221
Ergotinic Acid, 221
Ergotinine, 221
Ergotodine, 221
Ergotoxine, 220
Ericolin, 43
Eriodictyol, 60
Eriodictyon, 60
„ californicum, 60
Erythrin, 219
Erythræa Centaurium, 222
Erythrocentaurin, 222
Erythrolein, 219
Erythrolitmin, 219
Erythrophlœi Cortex, 249
Erythrophlœine, 251
Erythrophlœum guineense, 249
Erythrolaccin, 472
Erythroresinotannol, 473
Erythroxylum Coca, 23
„ truxillense, 23
Esenbeckia febrifuga, 241
Eseramine, 158
Eseridine, 158
Eserine, 158
Ester Value, 457
Essence de Citron au Zeste, 96
Ethyl Piperidine, 113
Eucalypti Folia, 40
„ Gummi, 439
Eucalyptol, 41
Eucalyptus amygdalina, 439
„ calophylla, 439
„ citriodora, 41
„ Citron scented, 41
„ dumosa, 41
„ Globulus, 40
„ marginata, 439
„ rostrata, 439
„ Leaves, 40
Eucheuma spinosum, 217
Eugenia caryophyllata, 66
Eugenia Tabasco, 104
Euonymin, 243
Euonymol, 243
Euonymus atropurpureus, 242
„ Bark, 242
Euonysterol, 243
Eupatorin, 318
Eupatorium Rebaudianum, 318
Euphorbia pilulifera, 210
„ resinifera, 427
Euphorbic Acid, 429
Euphorbium, 427
Euphorbone, 429
Euphorbo-resene, 429
Eurotium oryzæ, 142
Exostemma sp., 268
Extracts, 442

Farfaræ Herba, 197
Fats, 506
Fel Bovinum, 536
Fenchone, 116
Fennel Fruits, 116
Fenugreek Seeds, 159
Feronia elephantum, 92
Ferula communis, 484
„ fœtida, 487
„ galbaniflua, 484
„ rubricaulis, 487
„ suaveolens, 321
„ Sumbul, 321
Ferulic Acid, 488
Festuca arundinacea, 221
„ elatior, 221
Ficus, 126
„ Carica, 126
„ elastica, 427
„ religiosa, 471
Figs, 126
Filicic Acid, 397
Filicin, 398
Filicis Maris Rhizoma, 396
Filicitannic Acid, 397
Filix Mas, 396
Filixnigrins, 398
Filmarone, 397
Fisetin, 229
Fish berries, 68
Fixed Oils, 506
Flavaspidic Acid, 397
Flavaspidinol, 397
Flax Seed, 149
Flea Seeds, 178
Flemingia congesta, 410
Flemingin, 410
Flowers, 61
„ Arnica, 73
„ Calendula, 77
„ Chamomile, 70
„ Cloves, 66
„ Dalmatian Insect, 78
„ Elder, 69
„ Feverfew, 73
„ Kousso, 62
„ Lavender, 79

Flowers, Lily of the Valley, 83
,, Marigold, 77
,, Pyrethrum, 78
,, Red Poppy, 61
,, Red-Rose, 65
,, Saffron, 89
,, Wormseed, 75
Fluavils, 424
Flückiger, 4
Fœniculi Fructus, 116
Fœni-græci Semina, 159
Follicle, 85
Fool's Parsley, 197
Foxglove Leaves, 53
Frangula-emodin, 34, 245
Frangulin, 34, 245
Frangulinic Acid, 246
Frankincense, 481
,, American, 455
,, Common, 455
Frasera caroliniensès, 304
Fraxin, 522
Fraxinus Ornus, 521
Fruits, Ajowan, 143
,, Anise, 119
,, Bael, 91
,, Bergamot, 97
,, Bitter Apple, 106
,, ,, Orange, 92
,, Black Pepper, 132
,, Capsicum, 122
,, Caraway, 108
,, Cardamoms, 138
,, Cashew, 142
,, Cassia Pods, 98
,, Chenopodium, 142
,, Classification of, 84
,, Cocculus Indicus, 87
,, Colocynth, 106
,, Coriander, 113
,, Cubebs, 128
,, Cummin, 115
,, Dill, 118
,, Divi-divi, 285
,, Embelia, 121
,, Fennel, 116
,, Figs, 126
,, Hemlock, 111
,, Hips, 102
,, Hops, 127
,, Juniper Berries, 135
,, Laurel, 125
,, Lemon, 94
,, Lime, 97
,, Long Pepper, 134
,, Mangosteen, 92
,, Myrobalans, 102, 285
,, Parsley, 142
,, Pimento, 103
,, Pomegranate, 105
,, Poppy Capsules, 89
,, Prunes, 101
,, Raisins, 97
,, Senna Pods, 100
,, Star Anise, 85
Fruits, Tamarinds, 99
,, Vanilla, 136
,, White Pepper, 134
,, Wood Apple, 92
Fucus nodusus, 216
,, serratus, 216
,, vesiculosus, 215
Funtumia elastica, 427
Fusanus acuminatus, 232
,, spicatus, 232
Fustic, 229
,, Hungarian, 229

GADUS MORRHUA, 513
Galangæ Rhizome, 379
Galangal Greater, 381
,, Lesser, 379
,, Rhizome, 379
Galangin, 381
Galangol, 380
Galbanum, 484
Galbaresinotannol, 486
Galbulus, 85
Galen, 3
Galipea officinalis, 238
Galipoidine, 240
Galipol, 240
Galipot, 457
Galla, 412
Gallic Acid, 284, 413
Gallotannic Acid, 413
Galls, 412
,, Aleppo, 412
,, Blue, 414
,, Chinese, 414
,, Japanese, 414
,, Tamarisk, 285
,, Turkey, 412
,, White, 414
Gambier, 442
Gambier-fluorescin, 444
Gamboge, 474
,, Indian, 476
,, Saigon, 476
Ganjah, 208
Garbee Beans, 159
Garcinia Hanburii, 474
,, indica, 508
,, Mangostana, 92
,, Morella, 476
,, parviflora, 467
Garcinolic Acid, 475
Gaultherase, 60
Gaultheria procumbens, 59
Gaultherin, 60
Galipine, 240
Geber, 3
Gelatin, 538
Gelatinum, 538
Gelidium Amansii, 217
,, elegans, 217
Gelsemic Acid, 343
Gelsemine, 343
Gelseminine, 343

Gelsemium nitidum, 342
,, Rhizome, 342
Gelsemoidine, 343
Geneserine, 158
Genisteine, 195
Gentiamarin, 346
Gentian Root, 344
Gentiana asclepiadea, 346
,, Burseri, 347
,, pannonica, 347
,, punctata 346
,, purpurea, 347
Gentianæ Radix, 344
Gentianic Acid, 346
Gentianose, 346
Gentiin, 346
Gentiobiose, 347
Gentiogenin, 346
Gentiopicrin, 346
Gentisic Acid, 346
Geoffroy, 4
Geum urbanum, 338
Gigartina mamillosa, 215
,, pistillata, 215
Gingelly Oil, 511
Ginger, 375
,, African, 379
,, Bengal, 379
,, Cochin, 379
,, Jamaica, 376
,, Japanese, 379
,, Ratoon, 379
Gingerol, 377
Gitalin, 55
Glandulæ Thyroidei, 534
Globularia Alypum, 35
Gloriosa superba, 188
Gluco-chrysarone, 369
Glucogallin, 365
Glucoresins, 454
Glutin, 538
Glycocholic Acid, 536
Glycuronic Acid, 318
Glycyrrhetinic Acid, 318
Glycyrrhiza glabra, 315
,, glandulifera, 315
Glycyrrhizæ Radix, 315
Glycyrrhizin, 317
Glycyrrhizinic Acid, 318
Goa Powder, 462
Golden Seal Rhizome, 290
Gonolobus Condurango, 288
Gooroo Nuts, 151
Gossypii Radicis Cortex, 238
Gossypium, 410
,, barbadense, 410
,, sp., 238
Gossypol, 508
Gracilaria lichenoides, 217
Grains of Paradise, 185
Graminis Radix, 395
Grana Paradisi, 185
Granati Cortex, 256
,, Fructus Cortex, 105
Grass-tree Gum, 473
Greenheart Bark, 273
Grindelia, 198
,, camporum, 198
,, cuneifolia, 199
,, robusta, 199
,, squarrosa, 199
Grindeliæ Herba, 198
Ground Nut Oil, 509
Guaiacic Acid, 225
Guaiac Blue, 225, 461
,, -β-resin, 461
,, Saponic Acid, 225
,, Saponin, 225
,, Yellow, 461
Guaiaci Lignum, 224
Guaiacic Acid, 461
Guaiaconic Acid, 225, 461
Guaiacum officinale, 224, 460
,, Peruvian, 462
,, Resin, 460
,, sanctum, 460
,, Wood, 224
Guaiaguttin, 225
Guaiaretic Acid, 225, 461
Guarana, 23, 154
Guaza, 208
Guibourt, 4
Guinea Grains, 185
Gulancha, 232
Gum, Accroides, 473
,, Aden, 450
,, Amrad, 450
,, Amritsar, 450
,, Animi, 473
,, Arabic, 448
,, Australian, 450
,, Butea, 440
,, Cape, 450
,, Caramania, 452
,, Gedaref, 450
,, Ghatti, 450
,, Ghezireh, 450
,, Hachab, 448
,, Hotai, 481
,, Indian, 450
,, Juniper, 470
,, Karayah, 452
,, Kauri, 473
,, Mogadore, 450
,, Red, 439
,, Sennaar, 450
,, Senegal, 450
,, Somali, 450
,, Sterculia, 452
,, Sweet, 501
,, Talka, 448, 450
,, Thus, 455
,, Tragacanth, 451
Gummi Indicum, 450
Gum-resins, 474
Gums, 447
Gurjun Balsam, 493
Gutta, 424
Guttapercha, 422
Guvacine, 191

Gynocardia odorata, 507
„ Oil, 507
Gynocardin, 38

HÆMATEIN, 228
Haematoxylin, 228
Haematoxylon Campechianum, 227
Hamamelidis Folia, 38
Hamamelin, 39
Hamamelis virginiana, 38, 255
Hamamelitannin, 256
Hanbury, 4
Hancornia speciosa, 427
Heerabo-myrrholol, 480
Helenin, 334
Hellebore Rhizome, Black, 292
„ „ Green, 294, 392
„ „ White, 390
Helleborein, 294
Helleboretin, 294
Hellebori nigri Rhizoma, 292
Helleborin, 294
Helleborus fœtidus, 294
„ niger, 292
„ viridis, 294
Hemidesmi Radix, 340
Hemidesmus indicus, 340
„ Root, 340
Hemlock Bark, 285
„ Fruits, 111
„ First Biennial, 50
„ Herb, 195
„ Leaves, 49
„ Second Biennial, 51
„ Seeds, 175
Hemp, Indian, 207
Henbane, Annual, 51
„ Biennial, 51
„ Egyptian, 53
Herb, Aconite, 192
„ Agar-agar, 216
„ Bladderwrack, 215
„ Broom Tops, 194
„ Carrageen, 213
„ Centaury, 222
„ Cetraria, 217
„ Chiretta, 203
„ Coltsfoot, 197
„ Cudbear, 219
„ Ergot, 219
„ Euphorbia, 210
„ Grindelia, 198
„ Hemlock, 195
„ Indian Hemp, 207
„ „ Pink, 202
„ Lettuce, 222
„ Lobelia, 200
„ Pilewort, 222
„ Pulsatilla, 222
„ Purging Flax, 222
„ Savin Tops, 212
„ White Horehound, 206
„ Wormwood, 222
Herbs, 192
Hesperetic Acid, 300
Hesperetin, 94
Hesperic Acid, 94
Hesperidin, 28, 94
Hevea brasiliensis, 424
Hippocrates, 2
Hips, 102
Hirudin, 529
Hirudo, 529
„ medicinalis, 529
„ quinquestriata, 529
Histamine, 221
Homochelidonine, 308
Homocinchonidine, 267
Homoeuonysterol, 243
Homoflemingin, 410
Homoparácopairic Acid, 492
Homopterocarpin, 230
Homotaraxasterol, 340
Honey, 522
Hordeum decorticatum, 142
Horehound Black, 207
„ White, 206
Horse-eye Beans, 159
Horseradish Root, 309
Hotai Gum, 481
Humulene, 128
Humulol, 128
Humulus, 127
„ Lupulus, 127, 407
Hydrastine, 292
Hydrastis canadensis, 290
„ Rhizoma, 290
„ Rhizome, 290
Hydrocinchonidine, 267
Hydrocotoin, 276
Hydroergotinine, 220
Hydroipecamine, 325
Hydroquinine, 266
Hydroxybrasilein, 228
Hydroxybrasilin, 228
Hymenæa Courbaril, 473
Hyoscyamine, 45, 359
Hyoscami Folia, 49
„ Semina, 175
Hyoscyamus muticus, 53
„ niger, 49, 175
Hyperoodon rostratus, 518
Hypoquebrachine, 269

IBN BAITAR, 3
Iceland Moss, 217
Ichthyocolla, 538
Igasuric Acid, 171
Ignatius Beans, 172
Ilex paraguayensis, 22, 23, 59
Illicium religiosum, 86
„ verum, 85
Illipe Butter, 508
Illurinic Acid, 492
β-iminazolylethylamine, 221
Indaconitine, 298
Indian Aconite Root, 298
„ Bael, 91
„ Hemp, 207

Indian Mustard Seeds, 148
„ Pink, 202
„ Senna Leaves, 35
„ Tobacco, 200
Inflatin, 202
Inga flowers, 25
Insect Flowers, Dalmatian, 78
„ „ Persian, 79
Inula britannica, 75
„ Conyza, 56
„ Helenium, 56, 332
Inulæ Radix, 332
Inulin, 334
Iodothyrin, 534
Ionidium Ipecacuanha, 328
Ipecacuanha, Cartagena, 326
„ Greater Striated, 328
„ Lesser Striated, 328
„ Indian, 326
„ Matto Grosso, 326
„ Minas, 326
„ Root, 321
„ Undulated, 328
„ White, 328, 329
Ipecacuanhic Acid, 326
Ipecacuanhin, 326
Ipecamine, 325
Ipomœa Batatas, 403
„ hederacea, 176
„ muricata, 177
„ orizabensis, 355
„ Purga, 351
„ purpurea, 354, 356
„ simulans, 354
„ Turpethum, 356
Ipuranol, 251, 351, 355
Ipurganol, 353
Ipurolic Acid, 354
Iridin, 382
Iridis Rhizoma, 381
Iris florentina, 381
„ germanica, 381
„ pallida, 381
„ versicolor, 382
Irish Moss, 213
Irone, 382
Isatropylcocaine, 26
Isinglass, 538
„ Japanese, 216
Isoalantolactone, 334
Isoamylamine, 221
Isoanemonic Acid, 222
Isobarbaloin, 436
Isobutyl Isothiocyanate, 148
Isocodamine, 26
Isoferulic Acid, 300
Isohesperidin, 94
Isolichenin, 218
Isopelletierine, 258
Isopilocarpine, 30
Isorhamnetin, 33
Isorottlerin, 410
Isotrachylolic Acid, 473
Ispaghula, 177
Jaborandi Folia, 30
„ Leaves, 30
„ „ Aracati, 31
„ „ Guadeloupe, 31
„ „ Maranham, 30
„ „ Paraguay, 31
„ „ Rio Janeiro, 31
Jaborine, 31
Jalap, 351
„ Brazilian, 356
„ Mexican, 351
„ Orizaba, 355
„ Tampico, 354
„ Vera Cruz, 351
Jalapæ Radix, 351
Jalapinolic acid, 351, 355
Jamaica Dogwood, 288
„ Pepper, 103
Japaconitine, 298
Japonic Acid, 443
Jasmine Root, Yellow, 342
Jasminum nudiflorum, 342
Jateorhiza Columba, 302
Jateorhizine, 303
Jatropha Curcas, 184
Jecolein, 514
Jequirity Seeds, 191
Jervine, 391
Juices, Dried, 431
Juniperi Fructus, 135
„ Roob, 136
„ Succus, 136
Juniper-tar Oil, 504
Juniperus communis, 135
„ Oxycedrus, 504
„ phœnicea, 213
„ Sabina, 212
Juro-resene, 458
Kæmpferide, 381
Kæmpferin, 33
Kæmpferol, 33
Kaladana, 176
Kamala, 408
Kano, 438
Kauri Gum, 473
Kava Rhizome, 369
„ Resin, 370
Kavæ Rhizoma, 369
Kavaine, 370
Kava-kava, 369
Kesso Root, 332
Kickxia elastica, 427
Kinic Acid, 267
Kino, 436
„ African, 439
„ Bengal, 440
„ Botany Bay, 439
„ Cochin, 438
„ East Indian, 438
„ Eucalyptus, 439
„ Jamaica, 439
„ Madras, 438
„ Malabar, 438
„ -red, 439

Kinotannic Acid, 438
Kokum Butter, 508
Kola, *see* Cola
Kolanin, 152
Kolatein, 152
Kolatin, 152
Kola-red, 152
Kombic Acid, 168
Kousso, 62
Kosidin, 63
Kosin, 63
Kosotoxin, 63
Krameria argentea, 312
„ triandra, 312
Krameria-red, 314
„ -tannic Acid, 314

Lac, 471
„ Button, 472
„ Dye, 472
„ Garnet, 472
„ Seed, 472
„ Stick, 472
Lacca, 471
Laccaic Acid, 472
Lactuca virosa, 222, 421
Lactucarium, 421
Lactucerin, 422
Lactucerol, 422
Lactucic Acid, 422
Lactucin, 422
Lactucol, 422
Lactucone, 422
Lactucopicrin, 422
Lævo-mandelonitrile glucoside, 252
Landolphia florida, 427
„ Kirkii, 427
Lanoceric Acid, 520
Lanolin, 519
Lanopalmitic Acid, 520
Larch Bark, 288
Lard, 515
Larinolic Acid, 458
Larix europœa, 288, 457, 522
Larixin, 288
Laserpitium latifolium, 347
Latex, dried, 415
Laurel Berries, 125
„ Leaves, 56
Lauri Folia, 56
„ Fructus, 125
„ Oleum Expressum, 126
Laurocerasi Folia, 36
Laurocerasin, 37
Laurus nobilis, 56, 125
Lavandula spica, 80
„ vera, 79
Lavandulæ Flores, 79
Lavender Flowers, 79
Leaves, 20
„ Aconite, 59
„ Adhatoda, 60
„ Ailanthus, 46
„ Alexandrian Senna, 32
„ Argel, 34
Leaves, Bearberry, 41
„ Belladonna, 43
„ Betel, 58
„ Boldo, 60
„ Box, 43
„ Buchu, 27
„ Carthamus, 48
„ Chenopodium, 48
„ Cherry-laurel, 36
„ Coca, 23
„ Comfrey, 56
„ Cowberry, 43
„ Damiana, 59
„ Datura, 49
„ Duboisia, 46, 60
„ Elecampane, 56
„ Eriodictyon, 60
„ Eucalyptus, 40
„ Foxglove, 53
„ Henbane, 49
„ Indian Senna, 33
„ Jaborandi, 30
„ Laurel, 56
„ Matico, 57
„ Mullein, 56
„ Phytolacca, 46
„ Primrose, 56
„ Scopola, 46
„ Stramonium, 46
„ Sumac, 285
„ Tea, 21
„ Tinnevelly Senna, 35
„ Tylophora, 60
„ Wintergreen, 59
„ Witch-hazel, 38
„ Xanthium, 48
Lecanoric Acid, 219
Ledum palustre, 23
Leech, 529
Legume, 85
Lémery, 4
Lemon, 94
„ Juice, 95
„ Oil of, 95
„ Peel, 95
Lepidium sativum, 148
Leptandra, 399
„ virginica, 399
Leptandrin, 399
Lesser Celandine, 222
Lettuce, 222
Leucotin, 276
Levant Berries, 87
Lichenin, 218
Lichenostearic Acid, 218
Lignum Vitæ, 224
Ligustrum lucidum, 518
Lily of the Valley Flowers, 83
Lime, 97
Limonis Fructus, 94
Linalol, 80
Linalyl Acetate, 80
Linamarin, 38, 150
Lini, Oleum, 150
„ Semina, 149

Lini, Semina contusa, 151
Linin, 222
Linseed, 149
,, Crushed, 151
,, Oil, 508
Linum catharticum, 222
,, usitatissimum, 149
Liquidambar orientalis, 499
,, styraciflua, 501
Liquorice Root, 315
,, ,, Anatolian, 319
,, ,, Indian, 318
,, ,, Mountain, 318
,, ,, Persian, 319
,, ,, Russian, 318
,, ,, Spanish, 315
,, ,, Syrian, 319
,, ,, Stick, 320
Litmus, 218
Lobelacrin, 202
Lobelia, 200
,, inflata, 200
Lobeliæ Herba, 200
Lobeline, 202
Loganin, 171
Logwood, 227
Lodium multiflorum, 221
,, perenne, 221
Long Pepper, 134
Lotusin, 38
Loxopterygium Lourenzii, 269
Lubanol, 465, 466
Lupuli Strobili, 127
Lupulin, 407
Lupulinum, 407
Lupulus, 127
Luteolin, 251
Lycaconitine, 298
Lycopodium, 405
,, clavatum, 405
,, hungaricum, 407
,, -oleic Acid, 405
Lyperia atropurpurea, 83
Lytta vesicatoria, 530

MACE, 181
Maclurin, 229
Macrotomia cephalotes, 348
Madaralbin, 288
Madarfluavil, 288
Mahua Butter, 508
Malambo Bark, 237
Male Fern Rhizome, 396
Mallotus philipinensis, 408
Manamyrin, 494
Mandelic Acid 38
Manelemic Acid, 494
Maneleresene, 494
Mangosteen Fruits, 92
Mangrove Bark, 285
Manihot Glaziovii, 427
,, utilissima, 403, 405
Manihotixin, 38
Manna, 521
,, Alhage, 522
Manna, Australian, 522
,, Bamboo, 522
,, Briançon, 522
,, Oak, 522
,, Tamarisk, 522
Manneotetrose, 522
Manninotriose, 522
Mannite, 522
Mannitol, 522
Margosa, Bark, 287
,, Oil, 288
Margosic Acid, 288
Margosine, 288
Margiold Flowers, 77
Marking Nuts, 142
Marrubiin, 207
Marrubium candidissimum, 207
,, peregrinum, 207
,, vulgare, 206
Marshmallow Root, 314
Mastich, 468
,, Bombay, 470
,, East Indian, 470
Masticinic Acid, 469
Masticolic Acid, 469
Masticonic Acid, 469
Masticoresene, 469
Maté, 20, 22
Matico Camphor, 58
,, Leaves, 57
Matricaria Chamomilla, 73
Matthiolus, 4
Mayapple Root, 304
Meconic Acid, 419
Meconin, 419
Meconoiosin, 419
Mekocyanin, 62
Mel, 522
Melilot, 157
Melilotus officinalis, 157
Melissyl Palmitate, 517
Melon Pumpkin Seeds, 164
Menispermine, 88
Mericarp, 109
Metacopaivic Acid, 491
β-Methylæsculetin, 45
Methylarbutin, 42
Methylarecaidine, 191
Methylconiine, 113
Methylcytisine, 399
*n*-Methylgranatonine, 258
Methylhydrocotoin, 276
Methylpelletierine, 258
Methylprotocotoin, 276
Methysticin, 370
Metroxylon Rumphii, 405
,, Sagu, 405
Mezereic Acid, 278
Mezerein, 278
Mezereon Bark, 277
Mimusops globosa, 424
Monardes, 4
Monesia Bark, 318
Moric Acid, 229
Morin, 229

Morphine, 419
Morrhuine, 514
Moschus, 538
„ moschiferus, 538
Mucuna, 414
„ pruriens, 414
„ urens, 159
Mudar Bark, 299
Mullein Leaves, 56
Musk, 538
„ Assam, 542
„ Cabardine, 542
„ Nepaul, 542
„ Root, 321
„ Yunan, 542
Muskone, 541
Mustard, Expressed Oil of, 145, 506
„ Volatile Oil of, 147
Mylabris bifasciata, 531
„ cichorii, 531
„ lunata, 531
„ phalerata, 531
„ sidæ, 531
Myoctonine, 298
Myoporum platycarpum, 522
Myricyl Alcohol, 33
Myricin, 517
Myristica argentea, 181
„ fragrans, 178
„ malabarica, 181
Myristicæ Semina, 178
Myristicin, 180
Myrobalanin, 103
Myrobalans, 102, 285
„ Black Chebulic, 102
Myrobalanum, 102
Myrosin, 146
Myroxylon Pereiræ, 497
„ Toluifera, 495
Myrrh, 478
„ Arabian, 480
„ Fadhli, 480
„ Yemen, 480
Myrrha, 478
Myrrholic, Acid, 480
Myrrhis odorata, 318

NARCEINE, 419
Narcotine, 419
Nardostachys Jatamansi, 332
Nardus Root, 332
Nasturtium officinale, 148
„ Seed, 148
Nectandra Rodiæi, 273
Nectandræ Cortex 273
Neem Bark, 287
Neroli, Oil of, 94
Nut, 84
„ Oil, 509
Nutmeg, Expressed Oil of, 180
Nutmegs, 178
„ Bombay, 181
„ Long, 181
„ Macassar, 181
„ Papua, 181
Nutmegs, Wild, 181
Nux moschata, 178
„ Vomica, 170

OAK BARK, 282
Oenanthe crocata, 197
Oil, Almond, 509
„ Arachis, 509
„ Benne, 511
„ Castor, 512
„ Chaulmoogra, 507
„ Coco Nut, 513
„ Cod Liver, 513
„ Colza, 507
„ Coprah, 513
„ Cotton-seed, 507
„ Croton, 512
„ Dolphin, 514
„ Ground Nut, 509
„ Gynocardia, 507
„ Juniper-tar, 504
„ Linseed, 508
„ Margosa, 288
„ Mustard, 506
„ Nut, 509
„ Olive, 510
„ Pea Nut, 509
„ Peach Kernel, 164
„ Poppy, 506
„ Rape, 507
„ Seal, 514
„ Sesamé, 511
„ Shark, 514
„ Teel, 511
„ Theobroma, 508
„ Turkey Red, 512
„ Whale, 514
Olea europœa, 510
Oleo-resins, 489
Oleum Amygdalæ, 509
„ Amygdalæ Persicum, 164
„ Arachidis, 509
„ Betulæ Empyreumaticum, 505
„ Cadinum, 504
„ Chaulmoogræ, 507
„ Cocois, 513
„ „ Nuciferæ, 513
„ Crotonis, 512
„ Gossypii, 507
„ Lini, 508
„ Morrhuæ, 513
„ Olivæ, 510
„ Papaveris, 506
„ Rapae, 507
„ Ricini, 512
„ Rusci, 505
„ Sesami, 511
„ Sinapis Expressum, 506
„ Theobromatis, 508
Olearia Traversii, 232
Olibanoresene, 482
Olibanum, 481
Olive Oil, 510
Oliver Bark, 273
Oliveri Cortex, 273

Ononis spinosa, 318
Onosma echioides, 348
Ophelic Acid, 204
Opium, 415
„ Indian, 418
„ Persian, 418
„ Soft Shipping, 418
„ Turkey, 417
Orange, Bitter, 92
„ Flower Water, 94
„ Oil of, 94
„ Seville, 92
Orcin, 219
Ordeal Beans, 157
Oregon Balsam, 490
Orizaba Jalap Root, 355
Orris, Butter of 382
„ Rhizome, 381
„ Root, Florentine, 383
„ „ Mogadore, 393
„ „ Veronese, 383
Orsellinic Acid, 219
Orsin, 219
Ortho-oxycinnamic Aldehyde, 156
Oryza sativa, 400
Os Sepiæ, 544
Osyris tenuifolia, 232
Ovis Aries, 515, 519, 534
Ox Bile, 536
„ Gall, 536
Oxyacanthine, 235
Oxyleucotin, 276
Oxymethylanthraquinone Drugs, 34

PALAQUIUM BORNEENSE, 422
„ oblongifolium, 422
„ Treubii, 422
Palmatine, 303
Panama Wood, 253
Pancreas, 535
Papaver dubium, 61
„ Rhœas, 61
Papaver somniferum, 89, 415, 506
Papaverine, 419
Papaveris Fructus, 89
Paprika, 125
Pará Cress, 336
Parácopaivic Acid, 492
Paracoto Bark, 275
Paracotoin, 276
Paradol, 186
Paraguay Tea, 59
Parahydroxyphenylethylamine, 221
Paramenispermine, 88
Parathyroid Glands, 535
Pareira Brava, 300
Pareiræ Radix, 300
Parillin, 385
Parsley Camphor, 142
„ Fruits, 142
Pasque Flower, 222
Pasta Guarana, 154
Paullinia Cupana, 23, 154
Payena sp., 422
Pea Nut Oil, 509
Peachwood, 229
Pelletierine, 258
Pellitorine, 336
Pellitory Root, 334
„ „ German, 336
Pentaclethra macrophylla, 159
Penghwar Djambi, 412
Pepper, Bird, 125
„ Cayenne, 122
„ Jamaica, 103
„ Long, 134
„ White, 134
Pepsin, 537
Pepsinum, 537
Pereira, 4
Periandra dulcis, 318
Persio, 219
Peru, Balsam of, 497
Perugen, 499
Peruresinotannol, 498
Peruscabin, 499
Peruviol, 498
Petit Grain, Oil of, 94
Peucedanum graveolens, 118
„ Sowa, 119
Peumus Boldus, 60
Pharbitis Seeds, 176
Pharmacognosy, Commerce, 4
„ Definition of, 1
„ History of, 2
Phaseolunatin, 38
Phenylcumalin, 276
Phenylethyl Isothiocyanate, 148
Phenylglycollic Acid, 38
Phenylpropyl Cinnamate, 465
Philodendron sp., 386
Phloroglucinol, 284
Photosantonin, 76
Physeter macrocephalus, 518
Physic Nuts, 184
Physostigma cylindrospermum, 158
Physostigmatis Semina, 157
Physostigmine, 158
Physovenine, 158
Phytolacca decandra, 46, 360
„ Root, 360
Picea excelsa, 458
Picea-pimaric Acid, 458
Picea-pimarinic Acid, 458
Picea-pimarolic Acid, 458
Piraconitine, 297
Picræna excelsa, 226
Picrasmin, 226
Picrocrocin, 81
Picropodophyllin, 306
Picrorhiza, 343
„ Kurroa, 343
Picrotin, 88
Picrotoxin, 88
Picrotoxinin, 88
Pignons d'Inde, 184
Pilewort, 222
Pili Hæmostatici, 412
Pitocarpidine, 31
Pilocarpine, 30

Pilocarpus Jaborandi, 31
„ pennatifolius, 31
„ racemosus, 31
„ selloanus, 31
„ spicatus, 31
„ trachylophus, 31
Pilosine, 30
Pimaric Acid, 470, 457
Pimarinic Acid, 457
Pimarolic Acid, 457
Pimenta acris, 57, 104
„ officinalis, 103
Pimentæ Fructus, 163
Pimento, 103
Pimpinella Anisum, 119
Pinus cubensis, 455
„ echinata, 455
„ maritima, 457
„ palustris, 455
„ succinifera, 459
„ sylvestris, 502
„ Tæda, 455
Piper acutifolium, 58
„ angustifolium, 57
„ Betle, 58, 442
„ camphoriferum, 58
„ Clusii, 132
„ crassipes, 132
„ Cubeba, 128
„ Jaborandi, 31
„ lineatum, 58
„ Lowong, 132
„ methysticum, 369
„ mollissimum, 132
„ nigrum, 132
„ officinarum, 134
„ ribesioides, 132
Piperic Acid, 132
Piperidine, 134
Piperine, 134
Piptostegia Pisonis, 356
Piscidia, 288
„ Erythrina, 288
Piscidin, 288
Pistacia Khinjuk, 470
„ Lentiscus, 468
Pitch, Burgundy, 458
Pituitary Body, 535
Pix Burgundica, 458
„ Carbonis, 503
„ Liquida, 502
Plantago ovata, 177
„ Psyllium, 178
Pliny, 3
Ploughman's Spikenard, 56
Poa annua, 221
Podophylli Rhizoma, 304
„ Indici Rhizoma, 306
Podophyllin, 306
Podophylloresin, 306
Podophyllotoxin, 306
Podophyllum Emodi, 306
„ peltatum, 304
„ Rhizome, 304
Rhizome, Indian, 306
Polychroite, 81
Polygala alba, 312
„ Senega, 309
Polygalic Acid, 311
Polystichalbin, 399
Polystichin, 399
Polystichinin, 399
Polypodium vulgare, 318
Pome, 85
Pomegranate, Bark, 256
„ Rind, 105
Pomet, 4
Poppy Capsules, 89
„ Heads, 89
„ Oil, 506
Porphyrine, 269
Porphyrosine, 269
Porlieria sp., 225
Portugal, oil of, 94
Potassium myronate, 147
Pradosia latescens, 318
Prayer Beads, 191
Primrose Leaves, 56
Primula vulgaris, 56
Protium heptaphyllum, 495
Prosecretin, 535
Protocatechuic Acid, 39
Protocetraric Acid, 218
Protocotoin, 276
Protocurarine, 446
Protocuridine, 446
Protocurine, 446
Protokosin, 63
Protopine, 308
Protoveratridine, 392
Protoveratrine, 391
Prulaurasin, 37
Prunase, 37
Prunes, 101
Prunum, 101
Prunus Amygdalus, 163
„ domestica, 101
„ Laurocerasus, 36
„ Mahaleb, 157
„ serotina, 251
Pruni Virginianæ Cortex, 251
Pseudaconitine, 298
Pseudoconhydrine, 113
Pseudojervine, 392
Pseudomethysticin, 370
Pseudopelletierine, 258
Pseudotropine, 26
Psoralea obliqua, 29
Psychotria emetica, 328
„ Ipecacuanha, 321, 326
Psychotrine, 325
Psyllium Seeds, 178
Ptelea trifoliata, 243
Pteris sp., 386
Pterocarpin, 230
Pterocarpus, erinaceus 438, 439
„ Marsupium, 438
„ santalinus, 229
Pulsatilla, 222
Punica Granatum, 105, 256

Punicine, 258
Purging Cassia, 98
„ Flax, 222
„ Nuts, 184
Purin, 22
Purshianin, 248
Pyrethri Flores, 78
„ Radix, 334
Pyrethrine, 336
Pyrethrol, 79
Pyrethrone, 79
Pyrethrotoxic Acid, 79
Pyrethrum Flowers, 78
Pyretol, 79
Pyrus Cydonia, 165

Quassia amara, 226
„ Jamaica, 226
„ Surinam, 227
„ Wood, 226
Quassiæ Lignum, 226
Quassiin, 227
Quebracha, 269
Quebrachamine, 269
Quebrachine, 269
Quebrachite, 269
Quebracho, 269
„ Bark, 285
„ Colorado, 269
„ Extract, 285
„ White, 269
Quercetin, 42, 141
Quercite, 284
Quercitannic Acid, 284
Quercus Aegilops, 285
„ Cortex, 282
„ infectoria, 412
„ robur, 282
„ Suber, 284
„ Vallonea, 522
Quillaja Bark, 253
„ Poeppigii, 255
„ saponaria, 253
„ smegmadermos, 255
Quillajæ Cortex, 253
Quillajasapotoxin, 254
Quillajic Acid, 254
Quinamine, 267
Quince Seeds, 165
Quinic Acid, 267
Quinidine, 266
Quinine, 266
Quinovin, 267

Racemosin, 300
Radish, 148
Raisins, 97
Ranunculus Ficaria, 222
Rape Oil, 507
Raphanus sativus, 148
Rebaudin, 318
Red Gum, 439
„ Poppy Petals, 61
„ -rose Petals, 65
Red Sanders Wood, 229
„ Water Bark, 249
Remijia pedunculata, 268
„ purdicana, 268
Resenes, 454
Resin, 454
„ Amber, 454
Resin-acids, 454
Resina, 454
„ Elemi, 494
„ Guaiaci, 460
Resinols, 454
Resinotannols, 454
Resins, 453
Rhabarberone, 369
Rhamni Frangulae Cortex, 243
„ Purshiani Cortex, 246
„ Succus, 141
„ Syrupus, 141
Rhamnocarthartin, 141
Rhamnochrysin, 141
Rhamnocitrin, 141
Rhamno-emodin, 141
Rhamnofluorin, 248
Rhamnolutin, 141
Rhamnosterin, 248
Rhamnoxanthin, 141, 246
Rhamnus californicus, 248
„ carniolicus, 246
„ catharticus, 141, 246, 248
„ Frangula, 34, 243
„ Purshianus, 34, 246
Rhapontic Acid, 369
Rhaponticin, 338
Rhapontin, 338
Rhatany, Para, 312
„ Peruvian, 312
„ Root, 312
Rhazes, 3
Rhein, 33, 34, 365
Rhei Radix, 360
„ Rhizoma, 360
Rheopurgarin, 365
Rheum officinale, 34, 360
„ palmatum, 360
„ rhaponticum, 367
Rhizome, Arnica, 336
„ Black Hellebore, 292
„ Blue Cohosh, 399
„ Cimicifuga, 298
„ Couch Grass, 395
„ Galangal, 379
„ Greater, 381
„ Gelsemium, 342
„ Ginger, 375
„ Golden Seal, 290
„ Green Hellebore, 392
„ Hydrastis, 290
„ Indian Podophyllum, 306
„ Kava, 369
„ Leptandra, 399
„ Male Fern, 396
„ Orris, 381
„ Picrorhiza, 343
„ Podophyllum, 304

Rhizome, Rhubarb, 360
,, Scopola, 359
,, Serpentary, 370
,, Sumbul, 321
,, Sweet Flag, 392
,, Turmeric, 373
,, Valerian, 329
,, White Hellebore, 590
Rhizomes, 289
Rhizophora mangle, 285
Rhœadine, 62
Rhœados Flores, 61
Rhubarb, 34
,, Canton, 361, 366
,, Chinese 361
,, Chinese, Rhapontic, 369
,, East Indian 361
,, English, 367
,, High dried 361, 367
,, Rhizome, 360
,, Shensi, 366
,, Turkey, 361
Rhus Cotinus, 229, 285
,, coriaria, 285
,, semialata, 414
,, succedanea, 518
Richardsonia scabra, 327
Ricin, 183
Ricini Semina, 182
,, ,, Majores, 184
Ricinine, 183
Ricinone, 184
Ricinus communis, 182, 512
Rinoe Badak, 132
Roccella Montagnei, 218
,, tinctoria, 218
Rood Scorce, 287
Root, Aconite, 294
,, Alkanet, 247
,, Alkanna, 347
,, Arnica, 336
,, Atis, 298
,, Belladonna, 356
,, Bryony, 320
,, Calumba, 302
,, Canaigre, 369
,, Colchicum, 389
,, Colombo, 302
,, Comfrey, 349
,, Dandelion, 338
,, East Indian, 327
,, Elecampane, 332
,, Gentian, 344
,, Hemidesmus, 340
,, Horseradish, 309
,, Indian Aconite, 298
,, Ipecacuanha, 321
,, Japanese Aconite, 298
,, Kesso, 332
,, Liquorice, 315
,, Marshmallow, 314
,, Mayapple, 304
,, Musk, 321
,, Nardus, 332
,, Orizaba Jalap, 355
Root, Pareira, 300
,, Pellitory, 334
,, Phytolacca, 360
,, Rhatany, 312
,, Sarsaparilla, 383
,, Sassafras, 372
,, Scammony, 349
,, Senega, 309
,, Serpentary, 370
,, Turpeth, 356
,, Yellow Jasmine, 342
,, Zedoary, 375
Roots, 289
Rosa arvensis, 102
,, canina, 102
,, centifolia, 66
,, damascena, 66
,, gallica, 65
Rosæ caninæ Fructus, 102
,, gallicæ Petale, 65
Rose, Oil of, 66
Rosin Oil, 458
,, Spirit, 458
Rottlerin, 409
Rough Chervil, 197
Rubber, Assam, 427
,, Ceara, 427
,, Central American, 427
,, Ceylon, 426
,, Mangabeira, 427
,, Mozambique, 427
,, Pará, 424
,, Plantation, 426
,, West African, 427
Rubia Cordifolia, 205
Rubijervine, 392
Rubinic Acid, 443
Ruellia ciliosa, 202
Rumex alpinus, 347, 369
,, hymenosepalus, 369
,, obtusifolius, 369
Rusci, Oleum, 505

SABADILLA SEEDS, 189
Sabadilline, 190
Sabadine, 190
Sabadinine, 190
Sabatrinue, 190
Sabinæ Cacamina, 212
Saccharine Substances, 521
Sacred Bark, 246
Safflower, 82
Saffron, 80
,, Cake, 83
,, Cape, 83
Sago, 405
Salicin, 286
Salicis Cortex, 285
Saligenin, 286
Saliretin, 287
Salix alba, 285
,, discolor, 287
,, fragilis, 287
,, nigra, 287
,, purpurea, 287

Salix triandra, 287
,, viminalis, 287
,, vitellina, 287
Samaderin, 242
Sambuci Flores, 69
Sambucine, 70
Sambucus Ebulus, 69
,, nigra, 69
Sambunigrin, 38
,, 70
Sandal Wood, 230
,, ,, Fiji, 232
,, ,, West Australian, 232
,, ,, West Indian, 231
,, ,, South Australian, 232
Sandarac, 470
,, Australian, 471
Sandaracinic Acid, 470
Sandaraco-pimaric Acid, 470
Sanguinariæ Rhizoma, 307
Sanguinarine, 308
Sanguis Draconis, 467
Santal, 230
Santalic Acid, 230
Santalin, 230
Santalol, 231
Santalum album, 230
,, Freycinetianum, 232
,, Preissianum, 232
,, Yasi, 232
Santonic Acid, 76
Santonica, 75
Santonin, 76
Saponification Value, 457
Saponins, 254
Sapotoxins, 255
Sappan, 229
Sarothamnine, 195
Sarsæ Radix, 383
Sarsaparilla, 383
,, Costa Rica, 383
,, Guayaquil, 386
,, Honduras, 385
,, Indian, 340
,, Jamaica, 383
,, Lima, 385
,, Mexican, 386
,, Native Jamaica, 386
,, Vera Cruz, 386
Sarsaparillæ Radix, 383
Sarsapic Acid, 385
Sarsasapogenin, 385
Sarsasaponin, 384
Sassafras officinale, 372
,, Radix, 372
,, Root, 372
Sassy Bark, 249
Savin Tops, 212
Scammoniæ Radix, 349
Scammonium, 476
Scammonolic Acid, 351
Scammony, 476
,, Root, 349
Schikimic Acid, 87
Schizocarp, 84, 109
Schleichera trijuga, 471
Schleiden, 4
Schœnocaulon officinale, 189
Scilla, 386
Scillæ Radix, 386
Scillain, 388
Scillamarin, 388
Scillidiuretin, 388
Scillin, 388
Scillinin, 388
Scillipicrin, 388
Scillitin, 388
Scillitoxin, 388
Sclererythrin, 221
Sclerocrystalline, 221
Scleroxanthin, 221
Scoparii Cacumina, 194
Scoparin, 195
Scopola carniolica, 46, 359
,, japonica, 360
,, Rhizome, 359
Scopolamine, 359
Scopoletin, 45, 343, 359
Scorzetta, 96
Scrape, 455
Scurvy Grass, 148
Seal Oil, 514
Secale cereale, 219
Secalintoxin, 221
Secretin, 535
Seeds, 143
,, Areca Nuts, 190
,, Bitter Almonds, 163
,, Black Mustard, 145
,, Butea, 159
,, Calabar Beans, 157
,, Castor, 182
,, Cevadilla, 189
,, Cocoa, 23, 153
,, Cola, 22, 151
,, Colchicum, 186
,, Croton, 184
,, Datura, 176
,, Fenugreek, 159
,, Flax, 149
,, Grains of Paradise, 185
,, Henbane, 175
,, Ignatius Beans, 172
,, Ispaghula, 177
,, Jequirity, 191
,, Kaladana, 176
,, Linseed, 149
,, Melon Pumpkin, 164
,, Nutmegs, 178
,, Nux Vomica, 170
,, Quince, 165
,, Sabadilla, 189
,, Strophanthus, 167
,, Sweet Almonds, 161
,, Stavesacre, 143
,, Stramonium 174
,, White Mustard, 148
Semecarpus Anacardium, 142
Semen Cinæ, 75
,, Contra, 75

Senega Root, 309
,, ,, White, 312
,, ,, Northern, 311
Senegæ Radix, 309
Senegin, 311
Senna, Arabian, 35
,, Bombay, 35
,, Dog, 34
,, Leaves, Alexandrian, 32
,, ,, Indian, 35
,, ,, Tinnevelly, 35
,, Legumes, 100
,, Mecca, 35
,, Pods, 34, 100
Sennæ Fructus, 100
Sepia officinalis, 544
Serpentariæ Rhizoma, 370
Serpentary Rhizome, 370
Sesamé Oil, 511
Sesamum indicum, 511
Sevum, 515
Shark Oil, 514
Shea Butter, 508
Shellac, 417
,, Bleached, 473
Siaresinol Benzoate, 466
Sikimic Acid, 87
Sikimin, 87
Sikimipicrin, 87
Simaba suffruticosa, 241
Simaruba amara, 241
,, glauca, 241
,, officinalis, 241
,, Bark, 241
Simarubæ Cortex, 241
Sinalbin, 149
Sinapine, 149
,, Acid Sulphate, 146
Sinapis albæ Semina, 148
,, nigræ Semina, 145
Sinigrin, 146
Sinistrin, 388
Siperine, 274
Sitosterol, 158
Slippery Elm Bark, 281
Smilacin, 385
Smilax ornata, 383
Snakeroot, Black, 298
,, Red River, 370
,, Texan, 370
,, Virginian, 370
Soap Bark, 253
Socaloin, 436
Solanine, 206
Solanum Dulcamara, 205
,, nigrum, 49
Solenostemma Argel, 34
Spaniolitmin, 219
Spanish Flies, 530
Sparteine, 195
Spermaceti, 518
Sphacelinic Acid, 221
Sphacelotoxin, 221
Spigelia marilandica, 202
Spigeliæ Herba, 202
Spigeline, 202
Spilanthes oleracea, 336
Spleen, 535
Spogel Seeds, 177
Squill, 386
Indian, 388
Staphisagriæ Semina, 143
Staphisagrine, 144
Staphisagroine, 144
Star Anise Fruit, 85
,, ,, Japanese, 86
Starch, 400
,, Carcuma, 404
,, Maize, 402
,, Maranta, 402
,, Potato, 402
,, Rice, 402
,, Wheat, 401
Stavesacre, Expressed Oil of, 144
Stenostemma sp., 268
Sterculia urens, 452
Stigmasterol, 158
Stillingia sebifera, 508
Stockholm Tar, 502
Storax, 499
,, American, 501
Storesinol, 500
Stramonii Folia, 46
,, Semina, 174
Stramonium Leaves, 46
,, Seeds, 174
Strobile, 85
Strophanthi Semina, 167
Strophanthic Acid, 168
Strophanthidin, 168
Stropanthin, 168
Strophanthus Seeds, 167
,, Courmontii, 169
,, Emini, 169
,, gratus, 169
,, hispidus, 169
,, Kombé, 167
,, Nicholsoni, 169
Struxine, 171
Strychni Semina, 170
Strychnicine, 171
Strychnine, 171
Strychnos Castelnæi, 446
,, Crevauxii, 446
,, Gubleri, 446
,, Ignatii, 172
,, Nux-blanda, 172
,, Nux-vomica, 170
,, potatorum, 172
,, toxifera, 446
Stylophorum diphyllum, 292
Styracin, 465, 500
Styrax, 499
,, Benzoin, 464
,, calamitus, 499
Styrol, 465, 500
Succinin, 460
Succinite, 459
Succino-abietic Acid, 459
Succino-abietinolic Acid, 460

Succino-abietol, 460
Suceino-resinol, 460
Succino-sylvic Acid, 460
Succinum, 459
Succoxyabietic Acid, 460
Suet, 515
Sumac, 285
Sumaresinotannol, 465
Sumbul Radix, 321
,, Rhizome, 321
Suprarenal Gland, 533
Sus scrofa, 515, 535
Swartzia sp., 31
Sweet Almonds, 161
,, Flag Rhizome, 392
,, Gum, 501
Swertia alata, 205
,, angustifolia, 205
,, chinensis, 205
,, Chiretta, 203
,, trichotoma, 205
Syconus, 85
Symphyti Radix, 349
Symphytum officinale, 56

TABASHIR, 522
Tachardia Lacea, 471
Taka-diastase, 142
Tamarindi Fructus, 99
Tamarinds, East Indian, 100
,, West Indian, 99
Tamarindus indica, 99
Tamarisk Galls, 285
Tamarix gallica, 285, 522
Tampicin, 355
Tapioca, 404
Tamus communis, 320
Tannic Acid, 413
Tar, 502
,, Birch, 505
,, Coal, 503
,, Stockholm, 502
,, Wood, 502
Taraktogenos Kurzii, 507
Taraxaci Radix, 338
Taraxacin, 340
Taraxasterol, 340
Tars, 502
Taurocholic Acid, 536
Tea, 21
,, Abyssinian, 23
,, African, 23
,, Arabian, 23
,, Bush, 23
,, Kaporie, 23
,, Marsh, 23
,, Paraguay, 59
Teel Oil, 511
Tephrosia apollinea, 35
Terebinthina canadensis, 489
Terminalia Chebula, 102, 285
Terra Japonica, 442
Tetrarin, 365
Theæ Folia, 21
Thease, 21
Theobroma Cacao, 23, 153
Theobromatis Oleum, 154
,, Semina, 153
Theobromine, 22, 154
Theophrastus, 2
Theophylline, 22,
Therapin, 514
Thornapple, 46
,, Seeds, 174
Thymiamatis Cortex, 499
Thymus, 535
Thyreoglobulin, 534
Thyroid Gland, 534
Thyroiodin, 534
Tinnevelly Senna Leaves, 35
Tinospora, 232
,, cordifolia, 232
Tonco Beans, 156
,, Semina, 156
Tonka Beans, 156
Tonquin Beans, 156
Tolu, Balsam of, 495
Toluresinotannol, 496
Trachylobium Hornemannianum, 473
Trachylolic Acid, 473
Tragacanth, Indian, 452
,, Persian, 451
,, Symrna, 451
,, Syrian, 451
Trifolianol, 158
Trifolium alpnum, 318
Trigonella Fœnum-græcum, 159
Trigonelline, 160
Trimethylxanthine, 22
Tritici Rhizoma, 395
Triticin, 395
Triticum sativum, 400
Tropacocaine, 26
Tropæolum majus, 148
Truxilline, 26
Trypsin, 536
Tsuga canadensis, 285
Tubocurarine, 446
Turnera diffusa, 59
Turpeth, 356
Turpethein, 356
Turpethi Radix, 356
Turpethin, 356
Turmeric, 373
Turpentine, American, 455
,, Bordeaux, 457
,, Canada, 489
,, Oil of, 458
Turkey Red Oil, 512
Turpentine, Venice, 457
Turpethum, 356
Tussilago Farfara, 197
Tylophora asthmatica, 60
,, Leaves, 60
Tylophorine, 60
Tyrosamine, 221

ULMI FULVÆ CORTEX, 281
Ulmus fulva, 281
,, campestris, 288

Umbellic Acid, 486
Umbelliferone, 486
Uncaria Gambier, 442
Urginea, 388
,, indica, 388
,, Scilla, 386
Ursone, 42
Uvæ Passæ, 97
,, Ursi Folia, 41

Vaccinium Vitis-idæa, 43
Valonia, 285
Valerian, Indian, 332
,, Japanese, 332
,, Mexican, 332
,, Rhizome, 329
Valeriana Wallichii, 332
,, Phu, 332
,, officinalis, 329
Valerianæ Indicæ Rhizoma, 332
,, Majoris Radix, 332
,, Rhizoma, 329
Valerianine, 331
Vanilla, 136
Vanillæ Fructus, 136
Vanillin, 137
Vasicine, 60
Veratri Albi Rhizoma, 390
,, Viridis Rhizoma, 392
Veratridine, 189
Veratrine, 189
Veratrum viride, 392
,, album, 390
,, American, 392
Verbascum Thapsus, 56
Veronica virginica, 399
Viburni prunifolii Cortex, 258
Viburnin, 259
Virburnum prunifolium, 258
Virginian Prune Bark, 251
Viridic Acid, 172
Vitis vinifera, 97
Vogl, 4

Wafer Ash Bark, 243
Wahoo Bark, 242
Wars, 408, 410
Watercress, 148
Wattle Bark, 285
Wax, Carnauba, 518
,, Chinese Insect, 518
,, Japan, 518
Waxes, 506
Whale Oil, 514
White Horehound, 206
,, Mustard Seeds, 148
White Pepper, 134
Wild Cherry Bark, 251
,, Chervil, 197
Willow Bark, 285
,, ,, Black, 287
Winter's Bark, 237
Wintergreen Leaves, 59
Witch-hazel Bark, 255
,, Leaves, 38
Wood Apple, 92
Wood, Berberis, 224
,, Brazil, 229
,, Fiji Sandal, 232
,, Fustic, 229
,, Guaiacum, 224
,, Hungarian Fustic, 229
,, Logwood, 227
,, Peachwood, 229
,, Quassia, 226
,, Red Sanders, 229
,, ,, Sandal, 229
,, Sandal, 230
,, Sappan, 229
,, South Australian Sandal, 232
,, Surinam Quassia, 227
,, Tinospora, 232
,, West Australian Sandal, 232
,, ,, Indian Sandal, 231
,, Yellow Sandal, 230
,, Tar, 502
Woodruff, 157
Woods, 223
Wool, Absorbent, 411
,, Fat, 519
Wormwood, 222
Wormseed, 75
,, American, 142
Wurus, 410

Xanthium macrocarpum, 48
,, Strumarium, 48
Xanthohumol, 128
Xanthoresinotannol, 473
Xanthorhamnin, 141
Xanthopuccine, 292
Xanthorrhœa australis, 473
,, hastilis, 473

Yangonin, 370

Zanaloin, 436
Zea Mays, 400
Zedoary, 375
Zingerone, 378
Zingiber Mioga, 379
Zingiberis Rhizoma, 375


Printed by SPOTTISWOODE, BALLANTYNE & CO. LTD.
Colchester, London & Eton, England